Current Methods in Muscle Physiology

Advantages, Problems, and Limitations

Prepared under the auspices of the
International Union of Physiological Sciences

Edited by

Haruo Sugi

Chairman of the IUPS Commission on Muscle Physiology

OXFORD • NEW YORK • TOKYO

OXFORD UNIVERSITY PRESS

1998

Oxford University Press, Great Clarendon Street, Oxford OX2 6DP

Oxford New York
Athens Auckland Bangkok Bogota Bombay
Buenos Aires Calcutta Cape Town Dar es Salaam
Delhi Florence Hong Kong Istanbul Karachi
Kuala Lumpur Madras Madrid Melbourne
Mexico City Nairobi Paris Singapore
Taipei Tokyo Toronto Warsaw
and associated companies in
Berlin Ibadan

Oxford is a trade mark of Oxford University Press

Published in the United States
by Oxford University Press, Inc., New York

A catalogue record for this book is available from the British Library

Library of Congress Cataloging in Publication Data
Current methods in muscle physiology : advantages, problems, limitations / edited by Haruo Sugi.
"Prepared under the auspices of the International Union of Physiological Sciences."
Includes index.
1. Muscles – Research – Methodology. I. Haruo, Sugi, 1933–
II. International Union of Physiological Sciences.
QP321.C87 1998 573.7'5–dc21 97-41612

ISBN 0 19 852397 1

Typeset by Newgen Imaging Systems (P) Ltd, Chennai, India
Printed in Great Britain by
Biddles Ltd, Guildford & King's Lynn

Preface

Muscle can be considered to be an engine, which generates force and motion using ATP as its fuel. This engine operates at room temperature without noise and vibration. Its efficiency is very high (up to 60% under optimum conditions), and its final products are CO_2 and H_2O that are harmless to living organisms. Despite extensive physiological, biochemical and structural studies for many decades, the mechanism of muscle contraction still remains mysterious, so it seems unlikely that anyone will create such a remarkable engine in the near future. The history of research in any field of natural science tells us that Nature is much wiser and much cleverer than is generally thought by investigators, and that one should be very modest and careful in making inferences about underlying mechanisms. In every research field, a dogma can be effective and stimulating to investigators, but it can also retard progress if it is held with excessive conviction.

Typically, authors of scientific papers advertise and emphasize the advantages of the methods they develop, but tend not to mention their disadvantages and limitations, perhaps reflecting the intense competition among investigators. As readers cannot properly evaluate all current experimental methods, they tend to accept a writer's conclusions as correct, thus overlooking the problems and mysteries behind the reports. After the famous 'Symposium on the Mechanism of Muscle Contraction' held in 1972 at Cold Spring Harbor, many participants, as well as readers of the proceedings, felt that the contraction problem had been solved in principle. At that time, I applied for a grant to support my research, but it was rejected because, as one reviewer put it, 'Research work in this field is coming to an end'. Also at that time, however, Sir Andrew Huxley was remarking that, 'Even when a set of ideas becomes well-established, there is a distinct possibility that it will be overthrown by an unexpected discovery' (Review Lecture on Molecular Contraction).

For the past twenty years, I have had the ambition to edit a book dealing with the advantages, problems and limitations of currently used experimental methods, so that investigators can more correctly evaluate papers reporting results with these methods. Fortunately, I had the chance to implement my idea as a member of the Commission on Muscle Physiology in the International Union of Physiological Sciences (IUPS). This book is published as an activity of the IUPS Muscle Commission, and is prepared with the viewpoint of 'traditional' muscle physiology, whose ultimate goal is to understand the mechanism operating in living muscle fibres.

This book aims at describing critically various experimental methods currently used in the field of muscle research, along with central results obtained, but not at providing a comprehensive guidebook or an exhaustive survey of the literature. From the standpoint of muscle physiology, most methods described in this book are used for experiments with intact or demembranated muscle fibres containing three-dimensional myofilament lattice structures, as these structures are essential for converting chemical energy derived from ATP hydrolysis into mechanical work.

Chapters in Part I are devoted to various methods, developed to study biochemical (actomyosin ATPase) reactions producing muscle contraction, starting with introductory Chapters 1 and 2 on the problem of muscle mechanics. In Chapters 3 and 4, the use of caged compounds and fluorescent nucleotides, being effectively used to obtain information about actomyosin ATPase reactions and their rate constants under various mechanical conditions, are described. The use of peptide mimetics, described in Chapter 5, is useful for determining functionally essential parts of actin, myosin and other muscle proteins. Reflecting the recent explosive development of genetic engineering, various recombinant actin and myosin molecules are being prepared as explained in Chapter 6, and these recombinant molecules are powerful tools to determine essential amino acid residues for the function of actin and myosin used in the *in vitro* motility assay systems described in Chapter 7.

Chapters in Part II deals with the methods developed to detect structural changes associated with production of force and motion in muscle fibres. Chapters 8 and 9 are concerned with the use of probes attached to myosin heads to obtain information about their changes in orientation during contraction. Chapter 10 deals with techniques for time-resolved X-ray diffraction, the only means to obtain information about structural changes taking place in living and contracting muscle fibres. In Chapter 11, electron microscopic techniques to study structural changes of myofilament during contraction are discussed.

While chapters in Parts I and II are related to research work on the molecular mechanism of muscle contraction, the chapters in Part III are devoted to methods for studying other aspects of muscle functions. In Chapter 12, the uses of various Ca^{2+} indicators for studying movements of Ca^{2+} regulating the contraction–relaxation cycle in muscle are described. Chapter 13 deals with the patch clamp method, which has been widely used by investigators studying membrane ionic channels. Finally, various techniques for muscle energetics are discussed in Chapter 14.

The authors of these chapters are all experts in their respective fields, and I thank them sincerely for their contributions, especially for their comprehensive description on the advantages, problems and limitations of the methods, responding to my ambition. I hope that the book will be widely read by both young and senior investigators in the field of muscle research, and that it will help them to further develop new experimental methods leading to new discoveries.

Tokyo H. S.
January 1997

Contents

Contributors

Numbers in parentheses indicate pages on which the authors' contributions begin.

Toshiaki Arata (223), Department of Biology, Faculty of Science, Osaka University, Toyonaka, Osaka 560, Japan.

Clive R. Bagshaw (91), Department of Biochemistry, University of Leicester, Leicester LE1 7RH, UK.

Bernhard Brenner (33), Abteilung für Klinische Physiologie, Medizinische Hochschule Hannover, D-30625 Hannover, Germany.

Earl Homsher (71), Department of Physiology, School of Medicine, UCLA Center for Health Sciences, Los Angeles, California 90024, USA.

Shin'ichi Ishiwata (199), Department of Physics, School of Science and Engineering, Waseda University, Shinjuku-ku, Tokyo 160, Japan.

Eisaku Katayama (287), Department of Fine Morphology, Institute of Medical Science, The University of Tokyo, Minato-ku, Tokyo 108, Japan.

Harald Kögler (133), II. Physiologisches Institut, Universität Heidelberg, D-69120 Heidelberg, Germany.

Masato Konishi (305), Department of Physiology, The Jikei Medical School, Minato-ku, Tokyo 105, Japan.

Satoshi Kurihara (305), Department of Physiology, The Jikei Medical School, Minato-ku, Tokyo 105, Japan.

Jean Mironneau (329), Laboratoire de Physiologie Cellulaire et Pharmacologie Moléculaire, CNRS ESA 5017, Université de Bordeaux II, 146 rue Léo Saignat, 33076 Bordeaux, France.

Carl Morris (71), Department of Physiology, School of Medicine, UCLA Center for Health Sciences, Los Angeles, California 90024, USA.

J. Caspar Rüegg (133), II. Physiologisches Institut, Universität Heidelberg, D-69120 Heidelberg, Germany.

John M. Squire (241), Biophysics Section, The Blackett Laboratory, Imperial College of Science, Technology and Medicine, London SW7 2BZ, UK.

Haruo Sugi (3, 173), Department of Physiology, School of Medicine, Teikyo University, Itabashi-ku, Tokyo 173, Japan.

Kazuo Sutoh (163), Department of Pure and Applied Sciences, College of Arts and Sciences, University of Tokyo, Meguro-ku, Tokyo 153, Japan.

Teizo Tsuchiya (3), Department of Biology, Faculty of Science, Kobe University, Naka-ku, Kobe 657, Japan.

Roger C. Woledge (343), UCL Institute of Human Performance, Royal National Orthopaedic Hospital Trust, Middlesex HA7 4LP, UK.

PART I

Combined biochemical and mechanical experiments

1 Muscle mechanics I: intact single muscle fibres

Haruo Sugi and Teizo Tsuchiya

1.1 Historical background

Since the nineteenth century, intact whole skeletal muscles isolated from amphibian animals (frog and toad) have been widely used by physiologists, because they can be kept in a living state in physiological saline solution, and respond to electrical stimulus with contraction. Contraction of skeletal muscle was first characterized systematically by A.V. Hill and his colleagues, mainly with frog sartorius muscles, which are composed of muscle fibres running in parallel from one end to the other, thus providing a material suitable for analysing mechanical responses. A muscle at rest is highly extensible, while it becomes much stiffer during contraction in response to stimulation. This gave investigators the idea that an active muscle is in a state analogous to a stretched spring, so that it exerts force when kept isometric or lifts a load when allowed to shorten. To explain the fact that an active muscle shortens against a constant load with a steady velocity, Hill constructed a visco-elastic model in which the stretched spring is surrounded by a viscous medium generating a force opposing muscle shortening. The visco-elastic model predicts that the total elastic energy (heat plus work) released from an actively shortening muscle is only determined by the distance shortened but is independent of the amount of load on it.

In 1924, Fenn measured both the amount of work done and the amount of heat produced during a twitch or a short tetanus, and found that the energy released for a given distance of shortening increased when the load was increased from zero to a certain value. This finding, known as the 'Fenn effect', clearly demonstrated that a contracting muscle can vary its energy output (heat plus work) depending on the amount of external load. The Fenn effect implies that a contracting muscle is not analogous to the stretched spring, but behaves like a motor or an engine, which continuously converts energy into mechanical work and heat at a rate determined by the external load (Fenn 1924). In fact, it is now well established that muscle contraction is continuously driven by cyclic chemical reactions coupled with ATP hydrolysis.

In 1954, a monumental discovery was made (Huxley and Hanson 1954; Huxley and Niedergerke 1954): that muscle contraction results from relative sliding between the thick (myosin) and thin (actin) filaments, indicating that the sarcomere is not only the structural unit of striated muscle, but also its functional unit. Since the discovery of the sliding-filament mechanism, the attention of investigators has been focused on the mechanism by which the thick and thin filaments slide past each other, utilizing chemical energy derived from ATP hydrolysis.

In the field of muscle mechanics, muscle physiologists attempt to obtain information about molecular events responsible for muscle contraction, i.e. the myofilament sliding, mainly by applying mechanical perturbations to contracting muscles or muscle fibres and analysing the results obtained on the basis of specific contraction models. In this chapter, we shall describe the main methods developed in the field of muscle mechanics to solve some basic problems of muscle contraction, using mostly intact single skeletal muscle fibres (or small fibre bundles). Due to limited space, this article deals only with selected papers out of many publications. Experiments in the field of muscle energetics, including measurement of heat production and estimation of the efficiency of chemo-mechanical energy conversion are described in Chapter 14 of this volume.

Basic contraction characteristics: A.F. Huxley's contraction model

The most basic contraction characteristics, obtained with intact whole muscles or single muscle fibres are:

(1) the isometric force versus muscle length relation is bell-shaped, with a peak at slack length where the resting force is just barely detectable (isometric force–length relation; Ramsey and Street 1940);

(2) the relation between the steady velocity of shortening V and the external load (= force) P can be fitted to a hyperbola, and can be expressed by the Hill equation,

$$(P+a)(V+b)=(P_0+a)b,$$

where P_0 is the maximum isometric force and a and b are constants (force–velocity relation; Hill 1938);

(3) the contractile component in muscle is in series with a spring (series elastic component), so that the force in an isometrically contracting muscle drops quickly in response to a quick decrease in muscle length and then redevelops again after the completion of the length change (Gasser and Hill 1924);

(4) the rate of heat production in a contracting muscle is greater when it is allowed to shorten than when it is kept isometric (extra heat production during shortening; Hill 1938).

Among a number of theories put forward to explain the foregoing characteristics, A.F. Huxley's contraction model (Huxley 1957) has been central in the field of muscle mechanics. All the papers to be described in this chapter are based on this contraction model. The main essence of the Huxley contraction model is as follows (Fig. 1.1):

1. Muscle contraction results from relative sliding between the thick and thin filaments. This sliding is driven by cyclic attachment and detachment between the myosin heads (cross-bridges) on the thick filaments and the active sites on the thin filaments. The cross-bridges act independently of each other to produce force and motion.
2. A cross-bridge M fluctuates around its equilibrium position, 0 (Fig. 1.1A). When it attaches to an active site A on the thin filament, it exerts a positive or negative force proportional to its attached distance x from 0, as determined by its stiffness k (force function).
3. The rate constants for making and breaking the A–M links (f and g, respectively) are also functions of x (rate functions). A–M links are formed in the range $0 \leqslant x \leqslant h$, where f has a finite value, to exert a positive force for filament sliding, while they break fairly rapidly at $x < 0$, where f is zero and g has a large value (Fig. 1.1B).

With appropriately chosen values of rate functions to explain most of the basic contraction characteristics, steady-state distributions of A–M links, expressed by fractional numbers $0 \leqslant n \leqslant 1$ along the x axis (distribution function), at three different steady velocities of filament sliding ($V = 0$, $0.25\ V_{max}$ and V_{max}) are also shown in Fig. 1.1C_1–C_3. The value of f is assumed to be moderate, so that the proportion of A sliding past M without A–M link formation increases with increasing myofilament sliding velocity. At $V = 0$ (isometric condition), A–M links exist only at $0 \leqslant x \leqslant h$ (Fig. 1.1C_1). With increasing V, A–M links exerting a positive force decrease at $0 \leqslant x \leqslant h$, while those brought into $x < 0$ and exerting a negative force tend to increase (Fig. 1.1C_2). Finally, V reaches a maximum value V_{max} (the maximum velocity of shortening) when the positive and negative forces exerted by A–M links are equal (Fig. 1.1C_3).

In the Huxley contraction model, it is thermal agitation that moves A–M links from positive to negative values of x. Huxley and Simmons (1971) made the additional assumption that the sequential movement of A–M links occurs along the potential energy profile of the x axis, though the main framework of the Huxley contraction model remains unchanged. In this model, the filaments are assumed to be much stiffer than the cross-bridges, wherein the series elastic component is assumed to reside largely (Huxley and Simmons 1973). Based on the result that the amount of quick decrease in muscle length required to just drop the isometric force to zero is about 1% of the muscle length (Jewell and Wilkie 1958), the value of h is estimated to be ~ 20 nm ($\sim 2\%$ of the half sarcomere length at slack muscle length), because a rapid shift of A–M link distribution at $V = 0$ by $h/2$ to the left results in a balance between positive and negative forces, as illustrated in Fig. 1.1C_1.

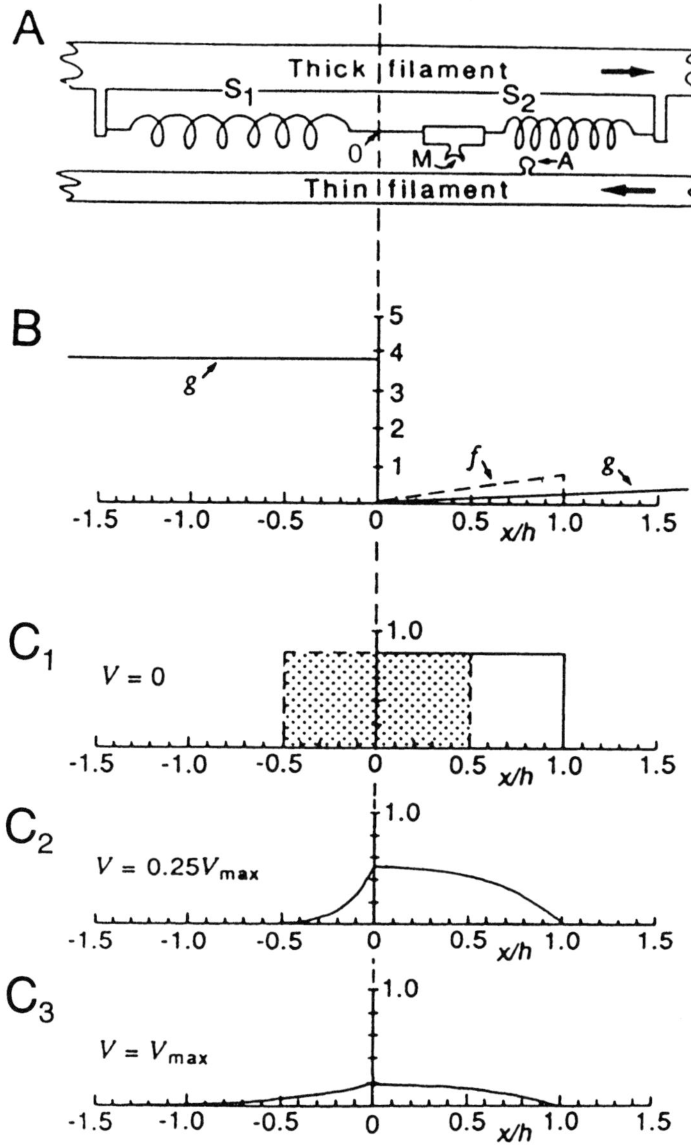

Fig. 1.1 A.F. Huxley's contraction model (Huxley 1957). (A) Diagram showing the myosin head M connected to the thick filament with springs S_1 and S_2, and the active site A on the thin filament. Direction of relative myofilament sliding is indicated by arrows. (B) Dependence of f and g on the distance x of M from the equilibrium position 0. M can only attach to A to form an A–M link at $0 \leqslant x \leqslant h$, where f has a finite value. (C_1–C_3) Diagram showing distribution of A–M links with respect to x at three different steady-shortening velocities: in C_1, the broken line and dotted area show a rapid shift of A–M link distribution in response to a quick length decrease of an isometrically tetanized muscle fibre to drop the force to zero; C_2 and C_3 show changes in A–M link distribution with increasing steady velocity of shortening.

Advantages of using single muscle fibres

Although the basic contraction characteristics were first obtained mostly with frog sartorius muscles, all current experiments on the mechanism of muscle contraction are carried out using intact single muscle fibres (or small bundles consisting of a few muscle fibres). The problems arising from the use of whole muscle are:

(1) whole muscles are composed not only of muscle fibres, but also of nerve fibres, connective tissues, blood vessels, etc., and this complicates the analysis of experimental results;

(2) whole muscles contain a large number of muscle fibres whose characteristics and mechanical responses are not necessarily uniform;

(3) whole muscles tend to deteriorate fairly rapidly with repeated stimulation, partly due to overstimulation of surface muscle fibres close to, or in contact with, stimulating electrodes, and partly due to an insufficient oxygen supply to the fibres located distant from muscle surface;

(4) it is difficult to measure and control sarcomere spacings in whole muscles.

The advantages of single muscle fibres (or small bundles consisting of a few muscle fibres) over whole muscles are:

(1) a single muscle fibre is a structural and functional unit of muscle at the cellular level, and is free from the problems stated above;

(2) in single muscle fibres isolated from frog skeletal muscle, perfectly reproducible mechanical responses to repeated stimulation can be obtained for many hours;

(3) striation patterns are clearly visible under a light microscope, and their changes can be recorded in various ways with high spatial and time resolution.

As it is not possible to prepare intact single fibre preparations from mammalian skeletal muscles, all the results discussed in this chapter have been obtained with single muscle fibres isolated from frog skeletal muscles.

1.2 Factors affecting the isometric force

On the basis of the Huxley contraction model, in which the cross-bridges are regarded as independent force generators, the magnitude of the isometric force in a tetanized muscle fibre should be proportional to the amount of overlap between the thick and thin filaments. Dependence of the isometric force on the sarcomere length of the fibre is therefore crucial for proving the validity of the contraction model. On the other hand, it has long been known that the isometric force at a given fibre length is not constant but varies depending on preceding length changes during tetanic stimulation. Although it is easy to measure the isometric force generated by a muscle fibre with both ends fixed in position (fixed-end contraction) using various force transducers, the results are complicated by inhomogeneities in sarcomere spacings, which are inherent in single-fibre experiments.

Sarcomere length–force relationship

By measuring sarcomere spacings along the entire fibre length, Huxley and Peachey (1961) showed that sarcomeres close to the fibre ends are much less extensible than those in the rest of the fibre. Thus, when a muscle fibre at rest is sufficiently stretched, the sarcomere lengths at the middle can be increased to $\sim 3.65\,\mu m$, i.e. the length at which the thick and thin filaments no longer overlap, sarcomere lengths near the fibre ends are still below $3.65\,\mu m$. As a result, the fibre-end sarcomeres shorten on stimulation by stretching others, producing a slowly rising isometric force (force creep). In order to measure the isometric force at controlled sarcomere lengths, Gordon *et al.* (1966*a*) developed a method, in which the length of a middle-fibre segment with reasonably uniform sarcomere spacings was clamped by a feedback device during the course of tetanic stimulation (segmental length-clamp method). Since the force is the same in every fibre cross-section during steady isometric force generation, the sarcomere length–force relationship can be obtained if the sarcomere length within a fibre segment is controlled. The feedback device consisted of an optical system detecting the segment length (distance between two position markers on the fibre surface) and a mechanical system moving one fibre end. With this method, Gordon *et al.* (1966*b*) determined the relationship between the isometric force developed in a tetanized fibre and the sarcomere length within the length-clamped fibre segment (length, 6–8 mm), and found (Fig. 1.2A, solid line):

(1) the isometric force was maximum at a sarcomere length of 2.0–2.2 μm and decreased linearly to zero with increasing sarcomere length from 2.2 to 3.65 μm (descending limb), a result consistent with the contraction model;

(2) the isometric force fell abruptly towards zero with a decreasing sarcomere length below 2.0 μm (ascending limb). In the descending limb, the isometric force was estimated from the initial rapid rise in force in response to tetanic stimulation, while the late force creep was not taken into consideration, since it was thought to arise from sarcomere length inhomogeneities occurring within the length-clamped segment.

Sarcomere length of the fibre can be measured either directly under a light microscope or indirectly from diffraction of the He–Ne laser light. If a muscle fibre is irradiated with a He–Ne laser light at right angles to its long axis, the light diffraction is given by the Bragg equation,

$$d \sin \theta = n\lambda,$$

where d is the sarcomere length, θ is the angle of diffraction, n is the order (integral number) of diffraction, and λ is the wavelength of the He–Ne laser light (632.8 nm). Sarcomere length d is obtained from the distance x between zero and the first diffraction lines on a plane at distance ℓ from the fibre by

$$x = n\lambda\ell / \sqrt{d^2 - n^2\lambda^2} \approx k/dx.$$

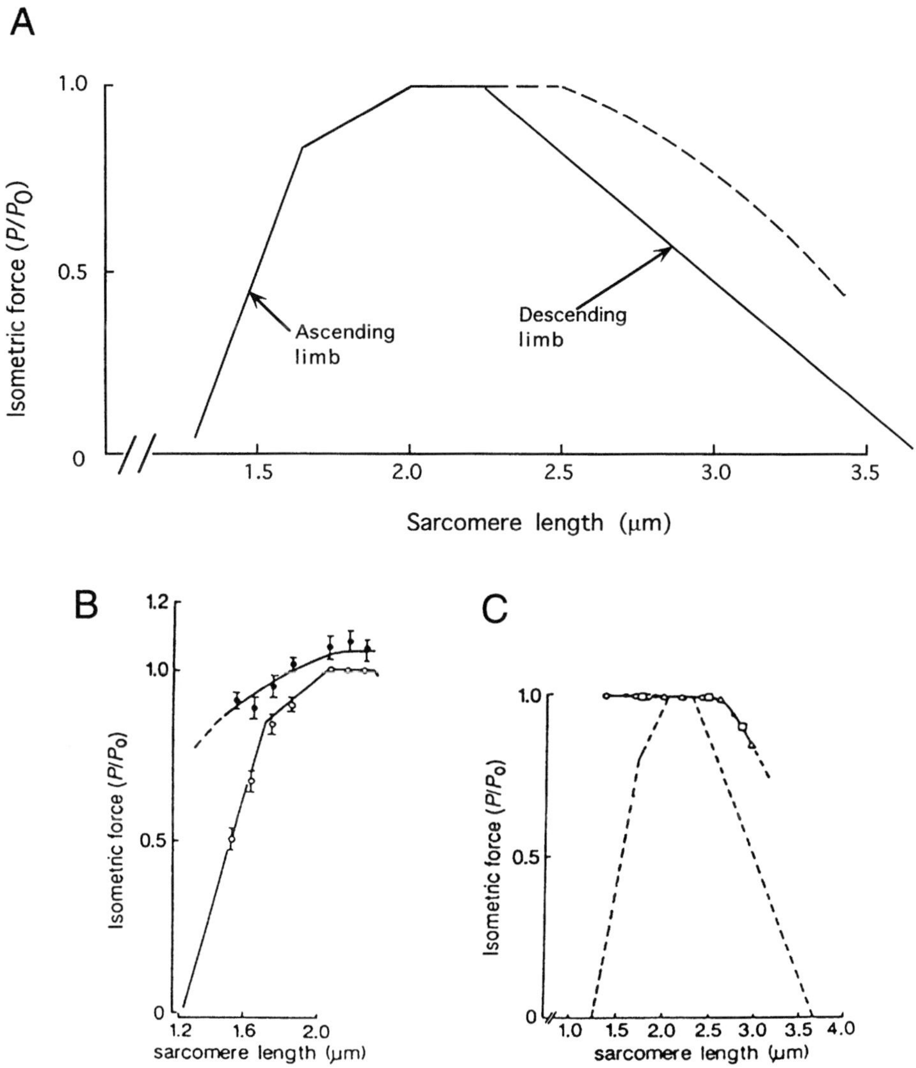

Fig. 1.2 Relationship between sarcomere length and isometric force. (A) Sarcomere length–force relationship of single tetanized muscle fibres: (*solid line*) whole range of sarcomere length–force relationship consisting of ascending and descending limbs obtained by the spot-follower method (Gordon *et al.* 1966*b*); (*broken line*) descending limb of sarcomere length–force relationship obtained from fixed-end tetani of single muscle fibres with uniform sarcomere length along the fibre length (ter Keurs *et al.* 1978).
(B) Zn^{2+}-induced changes of ascending limb of sarcomere length–force relationship in single tetanized fibres (Lopez *et al.* 1981). (C) Absence of ascending limb in sarcomere length–force relationship of maximally Ca^{2+}-activated myofibril bundles (Sugi *et al.* 1983).

Muscle fibres with fairly uniform sarcomere spacings (including fibre ends) were selected by ter Keurs *et al.* (1978), who showed that, in such fibres, the sarcomere spacings were kept fairly uniform during the course of a fixed-end tetanus (a condition in which a muscle fibre is simply tetanized with both ends fixed in position); they claimed that the force creep was not due to sarcomere length inhomogeneities, and estimated the isometric force from the plateau level attained after completion of the force creep. In the sarcomere length–force relationship thus obtained, the force was maximum up to 2.6 μm sarcomere length, suggesting that the rate of cross-bridge attachment to the thin filament, or the force generated by each cross-bridge, might depend on the amount of filament overlap (Fig. 1.2A, broken line). Sarcomere length measurement by laser light diffraction, however, has the problem that the diffraction pattern only represents the sarcomere length of myofibrillar planes that fulfil the Bragg condition (Rüdel and Zite-Ferenczy 1979), and can not detect sarcomere length inhomogeneities.

The linear descending limb of the sarcomere length–isometric force relationship has been obtained by many authors using the segmental length-clamp method with variable lengths of the clamped fibre segment (1.5–3.0 mm, Julian *et al.* 1978; 0.5–0.6 mm, Edman and Reggiani 1984, 1987; 100–200 μm, Altringham and Bottinelli 1985; 800 μm, Granzier and Pollack 1990). The linear descending limb has also been obtained by Bagni *et al.* (1988*a*) with a sophisticated method, in which the sarcomere length in a very short fibre segment (length, 70–80 μm) is clamped with a feedback device to keep the distance between the zero and the first order diffraction lines constant (sarcomere length-clamp method). It should be noted that, in these studies, the force creep becomes less pronounced as the clamped segment length is decreased, virtually disappearing in the experiment of Bagni *et al.*; it may therefore be safe to conclude that the descending limb of the sarcomere length–force relationship is linear, indicating that the isometric force is proportional to the amount of overlap between the thick and thin filaments. Thus, for the determination of the steady isometric force corresponding to a given sarcomere length without complications arising from sarcomere length inhomogeneities, the segmental length-clamp and the sarcomere length-clamp methods are useful, since the force is the same in every fibre cross-section (at force equilibrium). However, these methods have problems when applied to rapid, transient mechanical responses. These problems will be discussed latter.

Gordon *et al.* (1966*b*) explained the ascending limb as being due to double overlap of the thin filaments and collision of the thick filaments with the Z-line at short sarcomere lengths. To tetanize the fibre isometrically at a sarcomere length below 2.0 μm, the fibres have to shorten for a distance (freely in the experiments of Gordon *et al.*) before generating isometric force under the length-clamped condition. Although it is not clear how the preceding shortening affects the subsequent isometric force generation, it has been shown that fibres tetanized at short length are not fully activated (Rüdel and Taylor 1971). As a matter of fact, the ascending limb of a tetanized fibre is made less steep in the presence of twitch potentiators, such as Zn^{2+} (Lopez *et al.* 1981; Fig. 1.2B). The ascending limb is virtually absent up to sarcomere of length of 1.4 μm if a small myofibril bundle (sarcomere spacings can be measured directly under a light microscope) is fully Ca^{2+}-activated (Sugi *et al.* 1983; Fig. 1.2C). Therefore, the ascending limb observed in the tetanized fibre seems to result mainly from incomplete activation at short sarcomere lengths.

Effect of preceding length changes

If an isometrically tetanized muscle is slowly stretched, the force rises to a peak at the completion of stretch and then decays to a steady level appreciably higher than normal isometric force at the same muscle length. Conversely, if a tetanized muscle is slowly released, the isometric force attained after completion of stretch is below normal isometric force at the same muscle length (Abbot and Aubert 1952). These phenomena are called 'force enhancement by stretch' and 'force deficit by release', respectively, and have been studied by applying controlled length changes to one end of a single muscle fibre with a servo-motor system and recording the resulting force changes at the other (fixed) end. Though whole-fibre length changes are recorded with a light source–photodiode system or a differential transformer, it is necessary to ascertain whether each elementary fibre segment changes its length uniformly or not. Edman *et al.* (1978) recorded sarcomere length changes at the centre of the fibre during a slow constant-velocity stretch (duration, >1 s), and showed that the fibre length and the central sarcomere length changed reasonably in parallel with each other. Of course, this does not exclude the possibility of inhomogeneous segmental length changes along the whole fibre length. Sugi and Tsuchiya (1988), on the other hand, recorded the whole image of the fibre with position markers on its surface during the course of a slow stretch with a high-speed video system (at 200 frames/s), and found that the length changes were reasonably uniform at each elementary fibre segment divided by the markers, indicating the validity of interpreting the results in terms of cross-bridges forcibly displaced in the direction opposite to that for muscle shortening.

The effects of slow stretches obtained by the foregoing authors are as follows:

(1) the force enhancement, as measured by the increment of isometric force after completion of stretch above the normal isometric force at the same fibre length, is more pronounced as the initial sarcomere length before application of stretch is increased, indicating that it is not an increase in the number of cross-bridges that causes the stretch-induced enhancement of isometric force (Edman *et al.* 1978);

(2) muscle fibre stiffness attained after completion of stretch is the same as that during normal isometric force generation as the same fibre length (Sugi and Tsuchiya 1988, see Fig. 1.4C; for techniques of stiffness measurement, see Section 2.1.4);

(3) the stretch-induced force enhancement is associated with an increase of shortening velocity for a given amount of external load (as well as by an increase of the maximum isometric force), but the maximum shortening velocity remains unchanged (Edman *et al.* 1978).

Edman *et al.* (1978) have suggested that stretch-induced force enhancement results from recruitment of some (passive) elements generating additional force. This idea seems, however, to be inconsistent with the stiffness data of Sugi and Tsuchiya, since the additional force-generating elements should be detected as an increase in the fibre stiffness. Based on the results of time-resolved X-ray diffraction experiments, namely that the intensity of 1, 1 equatorial reflection decreases with stretch, Sugi and his co-workers have suggested that the force enhancement was due to distortion of the hexagonal myofilament-lattice structure (Amemiya *et al.* 1988). Accordingly, if the stretch-induced 1, 1 intensity decrease is associated with movement of the thin filaments from their trigonal position in the hexagonal myofilament-lattice, this would cause an increase of overall electrostatic repulsion force within the whole filament-lattice, which would show up as an enhanced isometric force but would not be detected as a stiffness change using the vibration method.

To obtain information about the mechanism underlying the stretch-induced enhancement of muscle fibre performance, Cavagna *et al.* (1994) focused attention on to the shortening of a tetanized fibre against a load equal to the normal maximum isometric force P_0 as a result of force enhancement after a slow stretch, and showed that the shortening was more pronounced as the amount of stretch was increased (up to 8 nm per half sarcomere). To try to explain the results in terms of cross-bridges interacting with the thin filament, they confirmed that the length change of the fibre segments closely followed the time course of a slow stretch applied to the fibre, by the novel method of 'striation follower' developed by Huxley *et al.* (1981). In this method, the images of ~ 10 consecutive sarcomeres were optically averaged to form a single cycle sinusoidal light intensity distribution, which was then projected on an array of five photodiodes; the photocurrents of these photodiodes were then combined to give the fibre displacement in terms of sarcomere length.

Meanwhile, Julian and Morgan (1979) examined development of sarcomere length non-uniformities during a fairly rapid stretch (duration, ~ 0.1 s) by recording length changes of the central fibre segment with the spot-follower method and those near the fibre end by taking photographs. They observed that the middle segment was stretched more than the ends, and that the level of isometric force attained after stretch was nearly the same as the isometric force before stretch, despite the decrease in the amount of myofilament overlap by stretch. In the descending limb of the sarcomere length–force relationship, stretched sarcomeres generated less and less force as they were being stretched by other (stronger) sarcomeres. As sarcomere length inhomogeneities in isometrically tetanized fibres were not as marked as might be expected from the unstable nature of the sarcomere length–force relationship, it seems clear that stretched sarcomeres generate forces larger than the normal isometric force at a given sarcomere length, so that a balance in force is established among all sarcomeres along the fibre length. This stabilizing effect, limiting the sarcomere length inhomogeneities, is the mechanism of 'longitudinal stability', an important mechanism to be further investigated.

The deficit in isometric force after a slow release, was studied by Julian and Morgan (1979) by the spot-follower method combined with photographic recording, and by Sugi and Tsuchiya (1988) by high-speed video recording of a whole fibre with position markers. It has been shown that the applied release is mostly taken up by segments near the fibre end, while the central segment length stays unchanged, suggesting that the deficit in isometric force is associated with sarcomere length inhomogeneities. The stiffness of the whole fibre measured by the sinusoidal vibration method always changed in parallel with the force during the course of slow releases (Sugi and Tsuchiya 1988).

1.3 Factors affecting the velocity of shortening

A characteristic feature of contraction in skeletal muscle is that, when the force in a tetanized muscle fibre rises to a value equal to a constant external load (isotonic condition), the fibre shortens with a constant velocity over a considerable distance, while the force it generates also stays constant. During such steady isotonic shortening, the length of the series elastic component remains unchanged, so that the velocity of isotonic shortening directly reflects the velocity of myofilament sliding in each sarcomere resulting from cyclic actin–myosin interaction (assuming the myofilament shortening is uniform in every sarcomere). As the Huxley contraction model is mainly concerned with the kinetics of actin–myosin interaction, depending on the velocity of myofilament sliding (and therefore also on the amount of isotonic load), a number of experiments have been made on the relationship between the shortening velocity and the external load to obtain information about factors affecting the kinetics of actin–myosin sliding.

Force–velocity relationship

To obtain the force–velocity relationship in the isotonic condition, an isometrically tetanized muscle fibre is first subjected to a quick decrease in load (= force) from the maximum isometric force P_0 to $P < P_0$. This procedure is now made most readily with a servo-motor (or a vibrator) system, which first operates in an 'isometric mode' until the force generated by the fibre reaches P_0, and then operates in an 'isotonic mode' to impose a constant load on the fibre during the course of fibre shortening. With this load-clamp technique, Edman (1988) showed that the force–velocity relationship in both whole tetanized muscle fibres and fibre segments could not be fitted by a single hyperbola, as the data points deviated from the hyperbola with external loads above $\sim 0.8\,P_0$ (Fig. 1.3A). He also measured muscle fibre stiffness during shortening by applying small sinusoidal vibrations (see Section 2.1.4, Edman 1993), and found that the rate of stiffness increase with increasing load became suddenly much larger above $0.8\,P_0$. Based on these results, Edman (1993) suggested that, when the velocity of myofilament sliding is reduced below a critical value (with loads $\sim 0.8\,P_0$), the number of cross-bridges attached to the thin filament increase abruptly but produce less force-driving myofilament sliding than they do for loads $< 0.8\,P_0$.

Both the maximum shortening velocity $V_{\max}$ and the maximum isometric force P_0 in a tetanized fibre increase in hypotonic solutions and decrease in hypertonic solutions, while the value of a/P_0, representing curvature of the force–velocity curve, does not change appreciably (Edman and Hwang 1977). These changes are not due to the change in myofilament spacing resulting from volume changes of the fibre, and are therefore associated with changes in ionic strength of the myoplasm.

If an isometrically contracting fibre is released at various constant velocities (controlled-velocity release), the force in the fibre reaches steady values during the

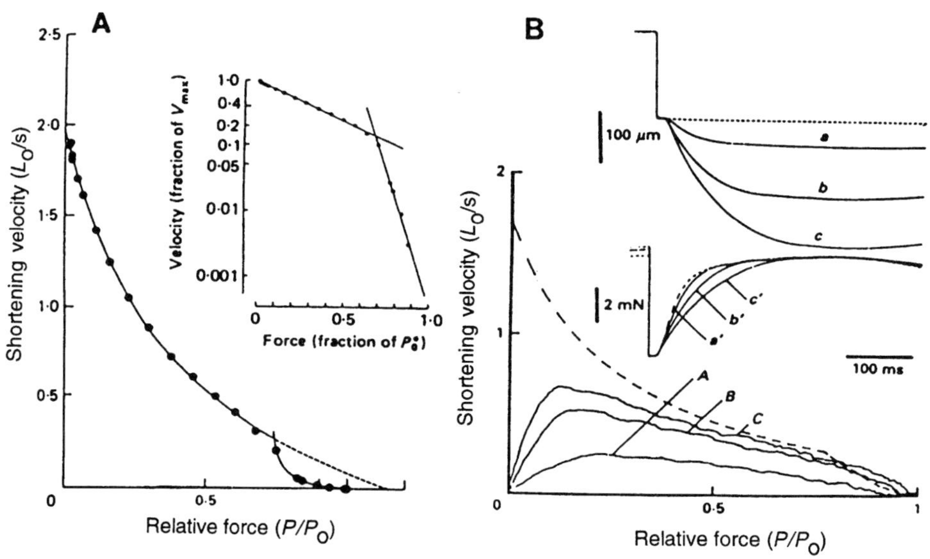

Fig. 1.3 Force–velocity relationship of single tetanized muscle fibres.
(A) Double-hyperbolic force–velocity relationship of single tetanized muscle fibres obtained in the isotonic condition. Inset shows semi-logarthmic plot of the same force–velocity data (Edman 1988). (B) Force–velocity relationship of single tetanized muscle fibres obtained in the auxotonic condition. Curves A, B and C were obtained from the length (a, b and c) and force (a′, b′ and c′) records shown in the inset, respectively. The fibre was first tetanized isometrically, and then slackened to let it shorten against three different auxotonic load (compliance, 1.2 (a, a′), 3.5 (b, b′) and 5.9% fibre length/P_0 (c, c′), respectively). Broken line indicates isotonic force–velocity curve (Iwamoto *et al.* 1990).

applied releases, thus giving a force–velocity relationship similar to that obtained in the isotonic condition. With this controlled-release method, Cecchi *et al.* (1978) determined the force–velocity relationships at various times during the rise of isometric force, and found that the value of V_{max} was the same irrespective of the time when the fibre was released during its isometric force development, a result consistent with the Huxley contraction model, in which the V_{max} reached at zero external load results from a balance between positive and negative forces generated by the cross-bridges during myofilament sliding, and is therefore independent of the level of activation, i.e. the number of cross-bridges involved (see Fig. 1.2C_3).

A.V. Hill (1938) thought that the force–velocity relationship in contracting muscle was instantaneous, so that the relationship was valid not only for isotonic and isovelocity shortening, but also under conditions in which the load was continuously changing with time. Iwamoto *et al.* (1990) studied the force–velocity relationship when a tetanized fibre was made to shorten against linear springs; in this condition, the amount of load increases linearly with increasing distance of shortening (auxotonic condition). The auxotonic condition was imposed by feeding a force development signal from the force transducer to an amplifier driving the servo-motor in such a way that resistance of the motor arm against fibre shortening increased in proportion to the isometric force developed. The authors demonstrated that, contrary to Hill's expectation, the force–velocity relationships in the auxotonic condition were convex upwards and varied depending on the compliance of auxotonic load (Fig. 1.3B). They further showed that these results can be accounted for in the framework of the Huxley contraction model (Fig. 1.2) indicating that the kinetics of actin–myosin interaction producing myofilament sliding are the same in both steady and non-steady shortening.

Maximum velocity of shortening

When the amount of load on a contracting muscle is reduced to zero, the muscle shortens at its maximum velocity V_{max}. While P_0 simply reflects the number of cross-bridges involved in isometric force generation, the value of V_{max} reflects the maximum rate of cross-bridge cycling and therefore the rate constants in the Huxley contraction model (Fig. 1.2). In practice, there are difficulties in determining V_{max} because the initial time course of fibre shortening under zero external load is masked by oscillation of the experimental apparatus following a quick decrease in load from the maximum isometric force P_0 to zero. Consequently, V_{max} is normally determined by extrapolation of the force–velocity data back to the velocity axis (assuming a hyperbolic force–velocity curve). Edman (1979) developed a convenient method of obtaining V_{max}; an isometrically tetanized fibre is suddenly released to various extents to make the fibre slack, the slackened fibre then shortens in the unloaded condition to become taut and redevelops isometric force. V_{max} is calculated from the relationship between the distance of release and the time of onset of force redevelopment. This method gives the same V_{max} values as those obtained by extrapolation of the force–velocity data points.

Using his 'slack test' method, Edman (1979) showed that V_{max} was constant in the range of sarcomere lengths from 1.65 to 2.7 µm. At sarcomere lengths >2.7 µm, the measured value of V_{max} increased steeply due to development of resting force. These results are also consistent with the Huxley contraction model, if they are simply taken to indicate that V_{max} is independent of the number of cross-bridges involved.

Julian *et al.* (1986) compared the values of V_{max} obtained by four different methods:

(1) extrapolation of force–velocity data points in the isotonic condition;

(2) extrapolation of force–velocity data points obtained by constant–velocity releases;

(3) the slack test with photographic recording of the whole fibre to ascertain whether the fibre becomes taut or not;

(4) photographic recording of the whole fibre subjected to constant–velocity releases.

They pointed out that:

(1) the V_{max} value was better obtained by linear extrapolation of the data points under very small loads;

(2) the accuracy of the slack test depends on the sensitivity of force measurement;

(3) the four methods give the same value of V_{max} if care is taken to check the fibre shape. They found a good agreement among the V_{max} values if great care is taken with each method.

Since a muscle fibre consists of thousands of sarcomeres connected in series, its elementary segments may shorten with variable velocities even under zero external load. Edman *et al.* (1988) measured V_{max} in a fibre segment between the two position markers (0.5–0.7 mm long), and found that V_{max} actually differed from segment to segment. Based on the good correlation between the segmental V_{max} and ATPase activity of the same segment, they considered the variation of V_{max} along the fibre length to be due to different myosin isozyme compositions along the fibre.

1.4 Muscle fibre stiffness

When an active muscle generating full isometric force is suddenly released by an appropriate amount, the force in it quickly drops to zero and then develops again after completion of release (Gasser and Hill 1924), indicating the presence of an elastic component in series with the contractile component. The series elastic component (SEC) originates primarily from tendons and recording apparatus, but also distributes along muscle length (Jewell and Wilkie 1958). In the Huxley contraction model, myofilaments are assumed to be rigid, and the series elasticity in each sarcomere is regarded as largely residing in each cross-bridge (more specifically, in the link connecting myosin heads to the thick filament backbone). On this basis, small length changes applied to a tetanized muscle fibre are taken up largely by the elastic part of each cross-bridge attached to the thin filament, provided the compliance of tendons and of the recording system are reasonably reduced or eliminated by the segmental length-clamp or sarcomere length-clamp methods.

As the attached cross-bridges may not break with small applied displacements, muscle fibre stiffness (relative value) is generally measured by applying small sinusoidal vibrations (peak-to-peak amplitude, $\sim 0.1\%$ of fibre length, frequency, kHz range) to one fibre end with a vibrator or a servo-motor, and measuring the resulting amplitude of force oscillations with a force transducer at the other fibre end (vibration method). With the vibration method, the stiffness can be measured continuously during the course of experiments. The time course of the stiffness change is obtained by digital subtraction of the force record without vibration from that with vibration. The stiffness can also be estimated by applying small quick changes in fibre length (amplitude $\leqslant 1\%$ of fibre length, complete in $\leqslant 1$ ms) and measuring the coincident force changes (step length change method).

Factors affecting muscle fibre stiffness

Julian and Sollins (1975) measured muscle fibre stiffness during steady isotonic shortening, and showed that the stiffness decreased with decreasing isotonic load (= force), approaching a minimum value (about 20% of the maximum value attained at the plateau of isometric tetanus) as the load tended to zero (Fig. 1.4B). This result is in fairly good agreement with the Huxley contraction model with respect to the number of attached cross-bridges at various steady shortening velocities (Fig. 1.2). On the other hand, the fibre stiffness rises from zero to the maximum value during isometric force development, reflecting an increase in the number of attached cross-bridges involved in isometric force generation. Cecchi *et al.* (1982, 1986) have shown that muscle fibre stiffness rises ahead of force by more than 15 ms (1–5 °C) during isometric force development. This suggests that cross-bridge attachment to the thin filament, as detected by stiffness, is followed by subsequent force generation after a long delay (Fig. 1.4A). According to Bagni *et al.* (1988*b*), the delay is too long to the explained by the Huxley–Simmons contraction model to be described later. They also point out that the stiffness versus force relationship is approximately the same during both force rise and relaxation in an isometric tetanus, and also during force redevelopment preceded by fibre shortening.

Cecchi *et al.* (1987) measured the stiffness with the step-length change method during the rise, plateau, and relaxation of isometric tetani in the ordinary fixed-end condition, and also in the segmental length-clamp condition, in which the length and the displacement of a fibre segment were recorded with the striation-follower method. In the former condition, the applied step-length change is partly taken up by tendon connections, while in the latter method compliance of tendon connections can be excluded. As expected, the stiffness values were larger in the

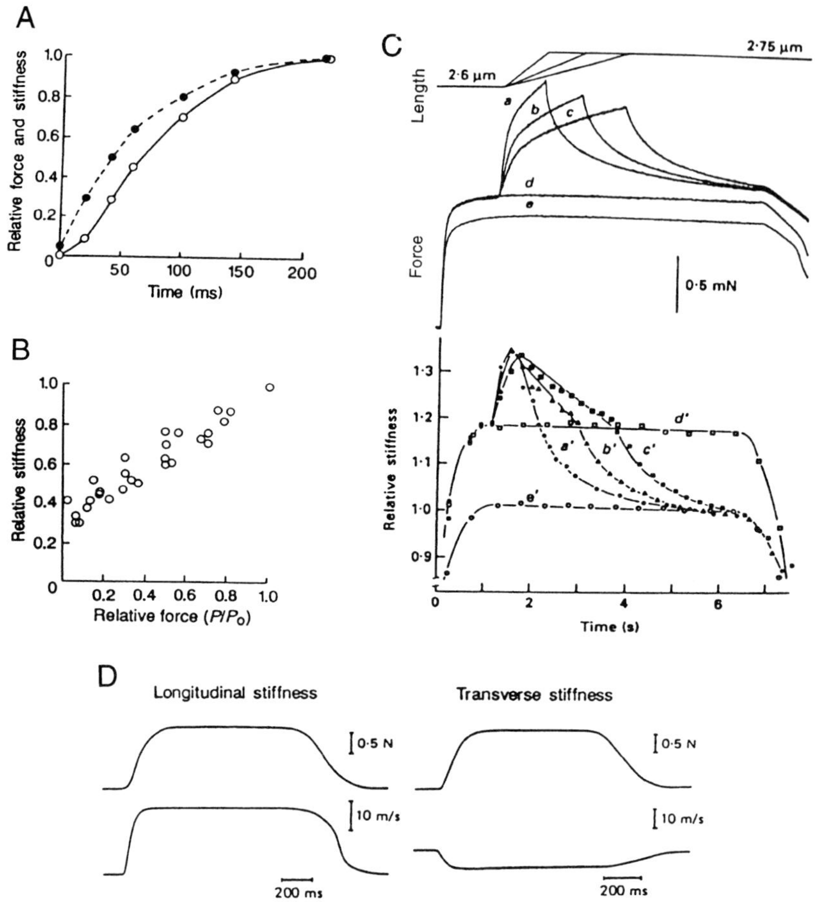

Fig. 1.4 Muscle and muscle fibre stiffness changes in various experimental conditions. (A) Increase in muscle fibre stiffness (*filled circles*) and force (*open circles*) during rise of isometric tetanic force (Cecchi *et al.* 1984). (B) Dependence of muscle fibre stiffness on the amount of load (= force) during steady isotonic shortening (Julian and Sollins 1975). (C) Force and stiffness changes of a tetanized fibre in response to slow stretches. The fibre was first tetanized isometrically at 2.6 µm sarcomere length, and then slowly stretched to 2.75 µm sarcomere length with three different velocities. *Upper, middle* and *lower traces* show time course of changes in the fibre length, the force and the stiffness, respectively, Trances a, b and c in the force records correspond to traces a′, b′ and c′ in the stiffness traces, respectively. Traces d and e are records of ordinary isometric forces at sarcomere lengths of 2.6 µm and 2.75 µm, respectively. Note that the stiffness increases abruptly on application of stretch, then decreases linearly during stretch, and eventually decreases to a steady level equal to that of ordinary isometric tetanus at a sarcomere length of 2.75 µm after completion of stretch, whereas the isometric force attained after completion of stretch is appreciably higher than the ordinary isometric force at a sarcomere length of 2.6 µm despite the decrease in the amount of myofilament overlap (Sugi and Tsuchiya 1988). (D) Changes in the force (*upper traces*) and the stiffness (*lower traces*) of a muscle during isometric tetanus. The stiffness was measured as propagation velocity of ultrasonic waves (3–7 MHz) between the two piezoelectric transducers. Upward and downward deflections of the stiffness records represent increase and decrease of stiffness, respectively. Note that the transverse stiffness decrease during contraction (Hatta *et al.* 1988).

segmental length-clamp condition than in the fixed-end condition, an exponential stress–strain relationship of tendon connections being obtained from the difference of data points in the two conditions. On the other hand, the stiffness versus force relationship obtained during isometric tetani was the same in both conditions, probably because the stiffness of tendon connections (corresponding to the slope of the stress–strain relationship) rises with increasing isometric force in nearly the same manner as does the stiffness of the tendon-free fibre with increasing force.

In contrast with the above similarity of stiffness versus force relationship during the whole period of isometric tetani, Sugi and Tsuchiya (1988) have demonstrated a marked dissociation between the stiffness and the isometric force when an isometrically tetanized fibre is subjected to a slow stretch. As illustrated in Fig. 1.4C, the stiffness first rises to a steady level as the fibre develops a steady isometric force on tetanic stimulation. On application of a slow stretch at the plateau of the preceding isometric tetanus, the stiffness shows an abrupt rise (probably due to synchronized extension of attached cross-bridges), which is then followed by a gradual decrease to a new steady level appreciably below the level of the preceding tetanus, reflecting a decrease in the amount of myofilament overlap caused by the stretch. The marked decrease of the stiffness starts when the stretch is still going on, and the final stiffness level below the initial stiffness level is attained while the isometric force attained after stretch is appreciably higher than the normal isometric force at the stretched fibre length. These results indicate that the parallel relationship between the stiffness and the force, observed during the whole period of isometric tetani, does not hold when a slow stretch is applied on isometrically tetanized fibres.

The stiffness measurements described so far are concerned with relative values expressed as the magnitude of force changes caused by small constant length changes in the kHz region. Sugi and his co-workers (Tamura *et al.* 1982; Hatta *et al.* 1988) developed a method, whereby muscle stiffness was measured with ultrasonic waves in the MHz region, with negligibly small perturbations to the muscle. The stiffness measurement is based on the equation,

$$V_f = (C_f/\rho)^{1/2},$$

where V_f is the phase velocity of ultrasonic waves, C_f is the muscle stiffness, and ρ is the density of muscle; this is essentially the measurement of propagation velocity of ultrasonic waves in the muscle. The advantage of this method is that the stiffness can be measured not only along the fibre length as with other stiffness measurements (longitudinal stiffness), but also across the muscle (transverse stiffness). On tetanic stimulation, the longitudinal stiffness rises ahead of force by more than 20 ms (20 °C) during isometric force development, while it lags behind the force during relaxation of isometric force (Fig. 1.4D, left records). On the other hand, the transverse stiffness decreases during isometric tetanus (Fig. 1.4D, right records). This interesting phenomenon, possibly reflecting change in the state of water between the myofibrils as ATP hydrolysis goes on, should be further studied using single fibres.

The origin of muscle fibre stiffness

In the segmental length-clamp condition monitored by the spot-follower method, Huxley and Simmons (1973) applied step decreases in fibre length during isometric tetani, and showed that the curve relating the magnitude of length change and that of the coincident force change (T_1 curve) was scaled in direct proportion to the isometric force, i.e. to the amount of overlap between the myofilaments. This result has been taken as evidence that the stiffness (or instantaneous elasticity) largely resides in the cross-bridges involved in isometric force generation. A change in the amount of myofilament overlap results, not only in a change in the number of cross-bridges interacting with the thin filament, but also in a change in the length of the thin filament in the I-band, i.e. in the myofilament non-overlap region. If the thin filament is not completely rigid, its stiffness in the I-band may also be expected to scale according to the I-band length. This implies that the scaling of the T_1 curves described above does not necessarily exclude the contribution of the thin filament stiffness to the instantaneous elasticity.

Suzuki and Sugi (1983) explored the extensibility of sarcomere structures by stretching rigor muscle fibres and examining the resulting change in sarcomere structures electron microscopically, based on the idea that, in rigor linkages, an externally applied stretch is taken up by all sarcomere structures, thus reflecting their extensibilities. As shown in Fig. 1.5A, the applied stretch was found to be taken up primarily by extension of the thin filaments in the I-band (and also by extension of the H-zone), while the length of myofilament overlap and the width of the Z-line did not change appreciably, indicating that the thin filaments are actually extensible. Based on their electron microscopic data and the force–extension curve of rigor fibres, Suzuki and Sugi calculated the thin filament stiffness per unit length to be $\sim 2.4 \times 10^4$ pN, suggesting that extension of the thin filament at P_0 amounts to $\sim 0.4\%$ of the fibre length. Suzuki *et al.* (1993) also examined the elastic extension of the thin filament in stretched rigor fibres electron microscopically, using rapid-freezing and freeze-fracture techniques, and could demonstrate a force-dependent increase in the actin monomer spacing (Fig. 1.5B).

Bagni *et al.* (1990) measured the stiffness of a tetanized fibre by the sinusoidal vibration method in the sarcomere length-clamp condition. They focused attention on the stiffness at the plateau region of the sarcomere length–force relationship, i.e. in the range of sarcomere lengths from 1.96 to 2.16 μm. They found that the stiffness was not constant but increased with decreasing sarcomere length within the plateau region, also suggesting that the stiffness includes the thin filament stiffness. As a matter of fact, recent time-resolved X-ray diffraction studies using intense X-ray sources from synchrotron radiation (Huxley *et al.* 1994; Wakabayashi *et al.* 1994) have clearly demonstrated that the thin filaments are actually elastic with the stiffness value not much different from that estimated by Suzuki and Sugi (1983). They further showed that the thick filaments are also elastic. The fact that the thick and thin filament are not completely rigid but have

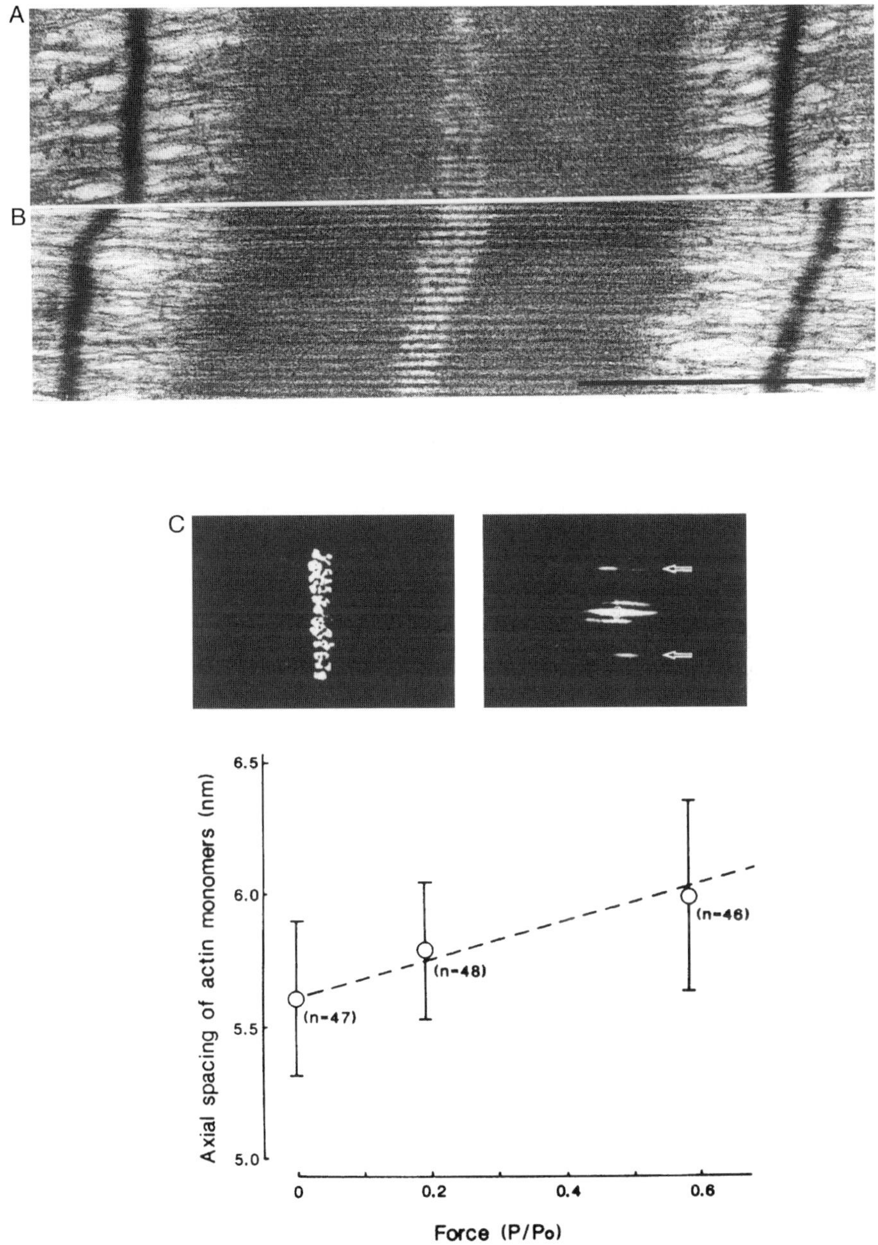

Fig. 1.5 Extensibility of the thin filaments. (A, B) Electron micrographs of sarcomeres in rigor skeletal muscle fibres before (A) and after (B) an applied stretch. Note marked extension of the thin filaments in the I-band and the thick filaments in the H-zone (Suzuki and Sugi 1983). (C) Relationship between the force in a rigor muscle fibre immediately before quick freezing and the axial spacing of actin monomers in the thin filament in the I-band (mean $\pm$ S D). Inset shows a replica image of thin filament segment (*left*) and its optical diffraction pattern (*right*). *Arrows* indicate layer lines based on axial spacing of actin monomers (Suzuki *et al.* 1993).

finite stiffness values indicates that their overall extension with the maximum isometric force P_0 might amount to 2–3 nm per half sarcomere (Huxley *et al.* 1994) or 70% of the muscle stiffness (Wakabayashi *et al.* 1994). This will more or less affect the interpretation of many muscle mechanics data, especially in the case of 'isometric force transients' to be discussed later.

1.5 Transient mechanical events

Although it is in principle difficult to obtain direct information about the kinetic properties of the ATP-dependent actin–myosin interaction producing force and motion in muscle from experiments with intact muscles or muscle fibres, muscle physiologists have attempted to give insight into the actin–myosin reactions responsible for muscle contraction by applying step (quick) changes in load (= force) or step changes in fibre length to isometrically tetanized fibres, and recording the resulting transient mechanical events. The former method gives the 'isotonic velocity transients', while the latter method provides the 'isometric force (tension) transients'. The isotonic velocity transients are obtained with the technique of 'load-clamp' (or 'force-clamp'), which corresponds to the technique of 'voltage-clamp' in the field of electrophysiology, and are free from complications arising from the myofilament stiffness, since the length of any elastic sarcomere structures stays constant under the load-clamp condition. The isometric force transient experiments, on the other hand, involve length changes of all elastic structures in each sarcomere other than the cross-bridges. In both types of experiments, the results obtained have been interpreted in terms of specific models envisioned by the authors.

Isotonic velocity transients

When the load on an isometrically contracting muscle is suddenly changed from P_0 to $P < P_0$, the muscle starts shortening with a constant velocity characteristic of the isotonic load P. The initial time course of isotonic shortening is, however, largely masked by inertial oscillations. Civan and Podolsky (1966) constructed a mechanical system, consisting of a light magnesium lever (with a very small equivalent mass) and a spring (much longer than the fibre length) to impose a constant isotonic load on a tetanized fibre. With this system, they could record the initial isotonic shortening following a quick decrease in load from P_0 to P without marked inertial oscillations. Contrary to the concept (Hill 1938) that the shortening velocity is instantaneously determined by the load, they found oscillatory changes in the shortening velocity while the load stayed constant (see Fig. 1.6A); the velocity was initially fast, then slowed down and eventually increased to reach constant. Similar oscillatory velocity transients following step changes in load have been recorded by Huxley (1971) with the spot-follower method, indicating that these phenomena are observable not only at the moving end of a tetanized fibre but also in every fibre segment.

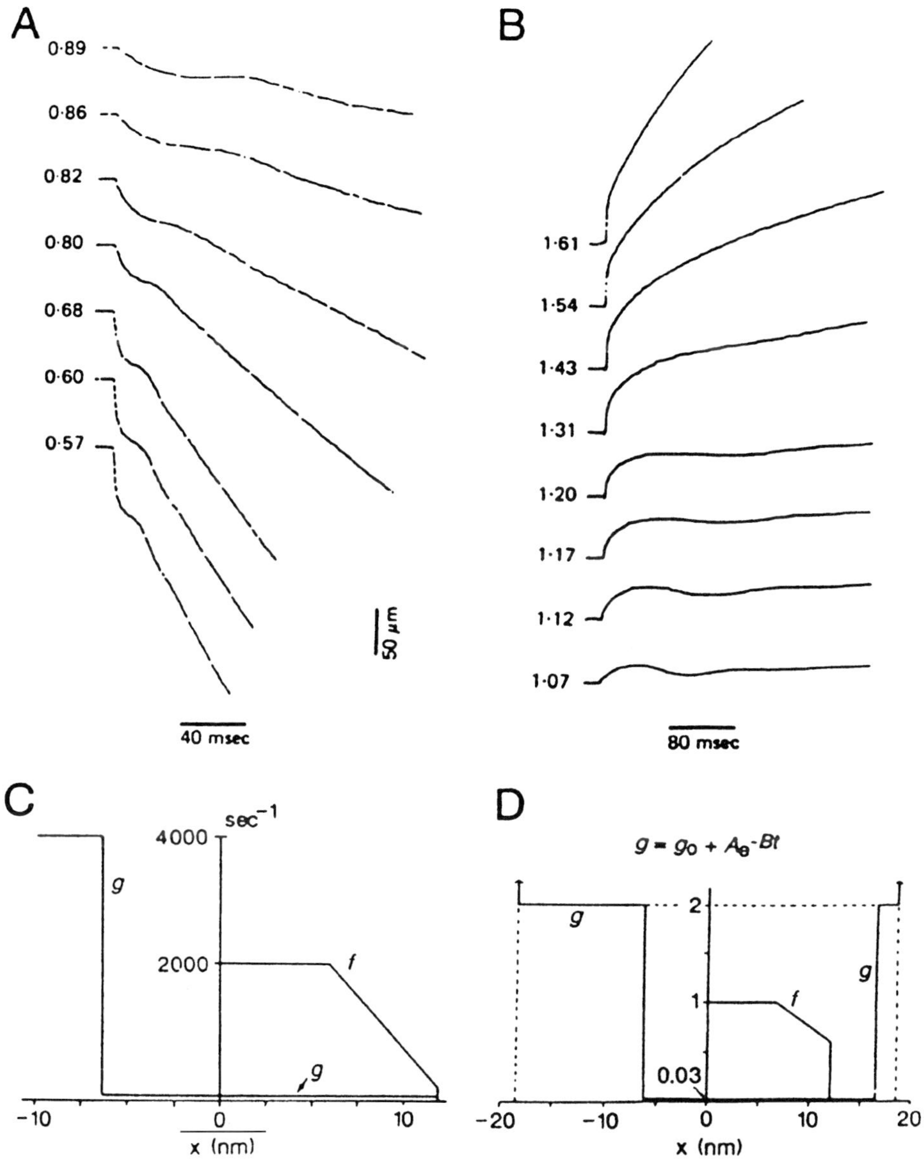

Fig. 1.6 Isotonic velocity transients and the contraction models to simulate the transients. (A) Isotonic velocity transients following step decreases in load (= force) from P_0 to $P < P_0$. Note oscillatory changes in shortening velocity until steady velocities and reached. (B) Isotonic velocity transients following step increases in load from P_0 to $P > P_0$. Note alternate fibre lengthening and shortening following small step increases in load. In both A and B, the value of P relative to P_0 is indicated alongside each record (Sugi and Tsuchiya 1981*a*). (C) Podolsky–Nolan contraction model to simulate the transients shown in A (Podolsky and Nolan 1973). (D) Sugi–Tsuchiya contraction model to simulate the transients shown in B as well as those shown in A (Sugi and Tsuchiya 1981*a*).

Podolsky and Nolan (1973) sought to explain these isotonic velocity transients in terms of the force and rate functions of the Huxley contraction model (Fig. 1.2), and presented a contraction model (Fig. 1.6C). The main features of their model are:

(1) f has large values, so that all A slide past M to form A–M links;

(2) g has a very small value in the positive x region and also in a part of the negative x region ($-6.3\,\text{nm} < x < 0$), but has a large value in the rest of the negative x region (this gap in the g function is essential for simulating the oscillatory velocity transient);

(3) breaking of A–M links is coupled with ATP hydrolysis only in the x region with a small value of g, while it occurs without ATP splitting in the other x regions.

Using a mechanical system analogous to that of Civan and Podolsky (1966), Sugi and Tsuchiya (1981*a*) studied the isotonic velocity transients following step changes in load. While the velocity transients following step decreases in load were very similar to those reported by Civan and Podolsky (1966) (Fig. 1.6A), the velocity transients following small step increase in load were markedly oscillatory, consisting of alternate lengthening and shortening of the fibre, while those following larger load steps were simple lengthening of the fibre (Fig. 1.6B). Sugi and Tsuchiya (1981*b*) further studied the velocity transients when the load of a tetanized fibre was changed in two steps. An interesting feature found in the two-load step experiments was that, when the load on an isotonically lengthening fibre was reduced to $P \leqslant P_0$, the fibre could shorten even against a load equal to P_0, indicating enhancement of mechanical performance of the fibre after a period of lengthening, a phenomenon analogous to the stretch-induced enhancement of mechanical performance described previously. To simulate the marked oscillatory velocity transients following small step increases in load, Sugi and Tsuchiya further modified the Podolsky–Nolan model, by adding a gap of g function in the positive x region together with an assumption that, after a step increase in load, g increases abruptly and then decays to the initial value with an exponential time course (Fig. 1.6D). The construction of the contraction models with different rate functions may seem somewhat arbitrary, since the models are based only on the transient mechanical events of intact single muscle fibres. Indeed, such models are only useful when they provide suggestions and ideas for designing and performing new experiments in demembranated fibres, in which behaviour of the cross-bridges can be studied more directly.

Isometric force transients

Using the spot-follower method, Huxley and Simmons (1971) applied small step length changes to an isometrically tetanized fibre, and analysed the resulting transient force changes, i.e. the isometric force transients. The transient force changes following a step decrease in length of a middle fibre segment consisted of three phases:

(1) the initial quick force reduction (from T_0 to T_1) coincident with the length step;
(2) a quick force recovery (from T_1 to T_2) level.

They thought phase 1 to be due to elastic recoil of the link between the myosin head attached to the thin filament and the thick filament backbone, assuming that other sarcomere structures were much more rigid than the elastic link. Phase 2 was thought to be the quick tilting of the myosin head to restretch the slackened elastic link. Phase 3 was thought to result from several cycles of attachment–detachment between the myosin head and the thin filament to restore the original isometric force.

The Huxley–Simmons model has been central in muscle physiology, and many attempts have been made to prove the myosin head tilting motion, responsible for producing force and motion in muscle. Since such experiments are described and discussed in almost every chapter in this volume, we shall discuss here possible problems arising from experiments to study the isometric force transients. First, it is an oversimplified assumption that, among various sarcomere structures, elasticity largely resides in the link connecting the myosin head to the thick filament. As discussed in the section on muscle fibre stiffness, myofilament elasticity is not negligible compared with cross-bridge elasticity, and the parameters concerning the magnitude of myosin head tilting movement should be altered on this basis (Goldman and Huxley 1994).

Second, the length changes of a middle fibre segment are not necessarily directly correlated with the force changes recorded with the force transducer at the fixed fibre end, if a length change is applied too rapidly at the moving fibre end with the servo-motor. The length change applied to one fibre end produces a mechanical impulse to be transmitted along the whole fibre length, and the force transducer detects force changes only after arrival of the mechanical impulse at the force transducer. Since it is impossible to record 'segmental' force changes from the same length-controlled segment, direct comparison of the segmental length changes with the force changes, recorded at the fixed fibre end, may not always provide direct information about the behaviour of the cross-bridges.

Haugen and Sten-Knudsen (1976) developed a method, with which very small sarcomere length changes (theoretically ~ 0.1 nm) could be detected with a high time resolution. They accomplished this by splitting the first order diffraction line of He–Ne laser light passed through a muscle fibre into two parts, and focused on two separate photodiodes; the diode voltage difference signal was a linear function of the shift of the first order diffraction line, reflecting sarcomere length changes (up to ~ 30 nm). As already mentioned, the laser light diffraction method for sarcomere length measurement detects only the sarcomere spacings of myofibrils satisfying the Bragg condition, and such myofibril groups may change with time during the applied fibre length changes. As a matter of fact, stepwise sarcomere length-change signals are occasionally recorded from a muscle fibre during constant-velocity fibre length changes. A way of effectively avoiding this problem is to slightly change the incident angle of the laser light with a sinusoidally vibrating mirror (frequency, kH range) and taking the centroid position of the first order diffraction as the sarcomere length change signal (Goldman and Simmons 1984).

Sugi and Kobayashi (1983), on the other hand, found that they could sometimes prepare 'good' fibres, in which sarcomere length changes recorded by the method of Haugen and Sten-Knudsen closely followed the fibre length changes along the entire fibre length, indicating that such fibres consisted of myofibrils with uniform sarcomere length and orientation. Using only such good fibres, they examined sarcomere length changes along the fibre length in response to a quick decrease of fibre length (Fig. 1.7A). The main results obtained were:

(1) after the onset of a fibre length change (applied to the moving fibre end), the force change (recorded at the fixed fibre end) started with a delay resulting from the time of propagation of the mechanical impulse along the fibre (at ~ 180 m/s);

(2) the delay of the onset of the sarcomere length change was longer than the delay of the onset of the force change, indicating that the mechanical impulse propagates faster than the sarcomere length change;

(3) the sarcomere shortening recorded close to the fixed fibre end still continued while the quick force recovery was already in progress.

Based on these results, Sugi and Kobayashi emphasized the importance of analysing the transient mechanical events, taking the multi-segmental nature of muscle fibre into consideration, because a muscle fibre consists of thousands of sarcomeres connected in series (Fig. 1.7B). In other words, they think that it is important to use distributed models, in which elements of sarcomere structures are distributed along the fibre length, rather than lumped models, in which all sarcomere elements are united into a single lump. Such a multi-segmental analysis can actually compute non-uniform sarcomere length changes along the fibre (Nishiyama 1984) (Fig. 1.7C).

1.6 Future prospects

Experiments with intact single muscle fibres will remain central in the field of muscle physiology, since the behaviour of such fibres is no doubt a manifestation of the properties of actin–myosin sliding coupled with ATP hydrolysis. The limitation of the intact single-fibre experiments is that it is not possible to obtain from them direct information about the biochemical and structural changes taking place in muscle fibres. Although Hill and his colleagues intensively studied muscle heat production under a variety of experimental conditions, from the 1930s to the 1960s, to obtain insight into the mechanism of chemo-mechanical energy conversion, progress was limited mainly by difficulties in specifying the sources of heat production unambiguously.

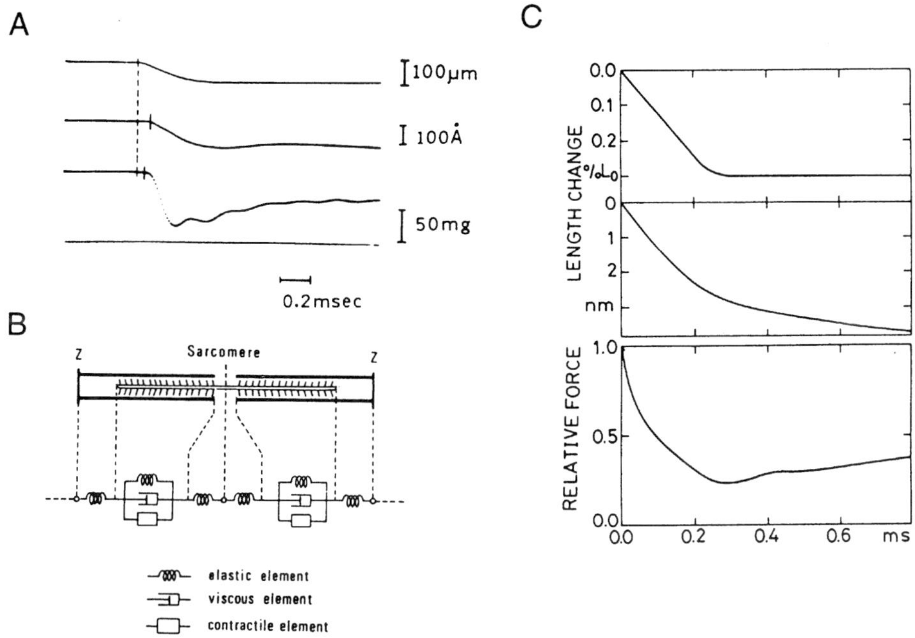

Fig. 1.7 Analysis of the mechanical transients based on the multi-segmental model. (A) Changes in fibre length (*upper trace*), sarcomere length (*middle trace*) recorded close to the fixed fibre end, and force (*lower trace*) recorded at the fixed fibre end. Note that the delay of onset of force change in response to the applied change in fibre length is shorter than the delay of onset of sarcomere length change close to the fixed fibre end (Sugi and Kobayashi 1983). (B) Elements of the multi-segmental model constructed on the basis of electron microscopic observations in Fig. 1.5. Elastic elements in series with contractile elements represent elasticity of myofilaments in the I-band and in the H-zone (Fig. 1.5A, B). (C) Calculated force change at the fixed fibre end (*lower trace*) and sarcomere length change close to the fixed fibre end (*middle trace*) in response to fibre length change applied to the moving fibre end (*upper trace*). Calculations were made with a model consisting of 30 sarcomeres with appropriately chosen values for the component elements (Nishiyama 1984).

Unlike their successful application to other fields of biology and medicine, the methods of biochemistry (and molecular biology) have not been so effective in clarifying the molecular mechanism of muscle contraction because:

(1) the function of chemo-mechanical energy conversion is lost in muscle protein samples in solution due to the loss of the myofilament-lattice structures;

(2) no theory has been developed to deal effectively with the rates of actomyosin reactions in the myofilament-lattice system;

(3) the protein concentrations and the ionic strength within a muscle fibre are much higher than those in which enzyme kinetics of actomyosin ATPase can be studied.

In order to connect muscle biochemistry and muscle physiology, demembranated muscle fibres have been used together with a number of methods to be described in the many chapters of this volume.

As has been described in the previous sections, experimental methods have now almost reached their limit of accuracy with respect to time and spatial resolution; this reflects recent development of electronic apparatus. A serious difficulty inherent in intact single fibre (and also demembranated fibre) experiments is sarcomere length inhomogeneities; this reduces the accuracy of mechanical measurements and more or less prevents interpretation of experimental data from going into elementary molecular events. Another factor, which makes interpretation of dynamic experiments complicated, is that the segmental force changes cannot be recorded while the segmental and sarcomere length changes are well controlled. Nevertheless, we believe that many well-designed experiments are still possible and that these will lead to 'unexpected discoveries' or to 'new mysteries'.

Acknowledgements

We wish to thank our colleagues who carried out research work with the various techniques described in this article. Our thanks are also due to Professor Manuel F. Morales for his valuable comments on the manuscript.

References

Abbott, B.C. and Aubert, X.M. (1952) The force exerted by active striated muscle during and after change of length. *J. Physiol.* **117**, 77–86.

Altringham, J.D. and Bottinelli, R. (1985) The descending limb of the sarcomere length–force relation in single muscle fibres of the frog. *J Muscle Res. Cell Motil.* **6**, 585–600.

Amemiya, Y., Iwamoto, H., Kobayashi, T., Sugi, H., Tanaka, H. and Wakabayashi, K. (1988) Time-resolved X-ray diffraction studies on the effect of slow length changes on tetanized frog skeletal muscle. *J. Physiol.* **407**, 231–41.

Bagni, M.A., Cecchi, G., Colomo, F. and Poggesi, C. (1990) Tension and stiffness of frog muscle fibres at full filament overlap. *J. Muscle Res. Cell Motil.* **11**, 371–7.

Bagni, M.A., Cecchi, G., Colomo, F. and Tesi, C. (1988*a*) Plateau and descending limb of the sarcomere length–tension relation in short length-clamped segments of frog muscle fibres. *J. Physiol.* **401**, 581–95.

Bagni, M.A., Cecchi, G. and Schoenberg, M. (1988*b*) A model of force production that explains the lag between cross-bridge attachment and force after electrical stimulation of striated muscle fibers. *Biophys. J.* **54**, 1105–14.

Cavagna, G.A., Heglund, N.C., Harry, J.D. and Mantovani, M. (1994) Storage and release of mechanical energy by contracting frog muscle fibres. *J. Physiol.* **481**, 689–708.

Cecchi, G., Colomo, F. and Lombardi, V. (1978) Force–velocity relation in normal and nitrate-treated frog single muscle fibres during rise of tension in an isometric tetanus. *J. Physiol.* **285**, 257–73.

Cecchi, G., Griffiths, P.J. and Taylor, S. (1982) Muscular contraction: kinetics of crossbridge attachment studied by high-frequency stiffness measurements. *Science* **217**, 70–2.

Cecchi, G., Griffiths, P.J. and Taylor, S. (1986) Stiffness and force in activated frog skeletal muscle fibers. *Biophys. J.* **49**, 437–51.

Cecchi, G., Colomo, F., Lombardi, V. and Piazesi, G. (1987) Stiffness of frog muscle fibers during rise of tension and relaxation in fixed-end or length-clamped tetani. *Pflüg. Arch.* **409**, 39–46.

Civan, M.M. and Podolsky, R.J. (1966) Contraction kinetics of striated muscle fibres following quick changes in load. *J. Physiol.* **184**, 511–34.

Edman, K.A.P. (1979) The velocity of unloaded shortening and its relation to sarcomere length and isometric force in vertebrate muscle fibres. *J. Physiol.* **291**, 143–59.

Edman, K.A.P. (1988) Double-hyperbolic force–velocity relation in frog muscle fibres. *J. Physiol.* **404**, 301–21.

Edman, K.A.P. (1993) Mechanism underlying double-hyperbolic force–velocity relation in vertebrate skeletal muscle. In *Mechanism of myofilament sliding in muscle contraction* (ed. H. Sugi and G.H. Pollack), pp. 667–78. Plenum, New York.

Edman, K.A.P., Elzinga, G. and Noble, M.I.M. (1978) Enhancement of mechanical performance by stretch during tetanic contractions of vertebrate skeletal muscle fibres. *J. Physiol.* **281**, 139–55.

Edman, K.A.P. and Hwang, J.C. (1977) The force–velocity relationship in vertebrate muscle fibres at varied tonicity of the extracellular medium. *J. Physiol.* **269**, 255–72.

Edman, K.A.P. and Reggiani, C. (1984) Redistribution of sarcomere length during isometric contraction of frog muscle fibres and its relation to tension creep. *J. Physiol.* **351**, 169–98.

Edman, K.A.P. and Reggiani, C. (1987) The sarcomere length–tension relation determined in short segments of intact muscle fibres of the frog. *J. Physiol.* **385**, 709–32.

Edman, K.A.P., Reggiani, C., Schiaffino, S. and te Kronnie, G. (1988) Maximum velocity of shortening related to myosin isoform composition in frog skeletal muscle fibres. *J. Physiol.* **395**, 679–94.

Fenn, W.O. (1924) The relation between the work performed and the energy liberated in muscular contraction. *J. Physiol.* **58**, 373–95.

Gasser, H.S. and Hill, A.V. (1924) The dynamics of muscular contraction. *Proc. Roy. Soc.* **B96**, 398–437.

Goldman, Y.E. and Huxley, A.F. (1994) Actin compliance: are you pulling my chain? *Biophys. J.* **67**, 2131–6.

Goldman, Y.E. and Simmons R.M. (1984) Control of sarcomere length in skinned muscle fibres of *Rana temporaria* during mechanical transients. *J. Physiol.* **350**, 497–518.

Gordon A.M., Huxley, A.F. and Julian, F.J. (1966*a*) Tension development in highly stretched vertebrate muscle fibres. *J. Physiol.* **184**, 143–69.

Gordon A.M., Huxley, A.F. and Julian, F.J. (1966*b*) The variation in isometric tension with sarcomere length in vertebrate muscle fibres. *J. Physiol.* **184**, 170–92.

Granzier, H.L.M. and Pollack, G.H. (1990) The descending limb of the force–sarcomere length relation revisited. *J. Physiol.* **421**, 595–615.

Hatta, I., Sugi, H. and Tamura, Y. (1988) Stiffness changes in frog skeletal muscle during contraction recorded using ultrasonic waves. *J. Physiol.* **403**, 193–209.

Haugen, P. and Sten-Knudsen, O. (1976) Sarcomere lengthening and tension drop in the latent period of isolated frog skeletal muscle fibres. *J. Gen. Physiol.* **68**, 247–65.

Hill, A.V. (1938) The heat of shortening and the dynamic constants of muscle. *Proc. Roy. Soc.* **B126**, 136–95.

Huxley, A.F. (1957) Muscle structure and theories of contraction. *Prog. Biophys. Biophys. Chem.* **7**, 255–318.

Huxley, A.F. (1971) The activation of muscle and its mechanical response. *Proc. Roy. Soc.* **B178**, 1–27.

Huxley, A.F., Lombardi, V. and Peachey, L.D. (1981) A system for fast recording of longitudinal displacement of a striated muscle fibre. *J. Physiol.* **317**, 12–13P.

Huxley, A.F. and Niedergerke, R. (1954) Interference microscopy of living muscle fibres. *Nature* **173**, 971–3.

Huxley, A.F. and Peachey, L.D. (1961) The maximum length for contraction in vertebrate striated muscle. *J. Physiol.* **156**, 150–65.

Huxley, A.F. and Simmons, R.M. (1971) Proposed mechanism of force generation in striated muscle. *Nature* **233**, 533–8.

Huxley, A.F. and Simmons, R.M. (1973) Mechanical transients and the origin of muscular force. *Cold Spring Harb. Symp. Quant. Biol.* **37**, 669–80.

Huxley, H.E. and Hanson, J. (1954) Changes in the cross-striation of muscle during contraction and stretch and their structural interpretation. *Nature* **173**, 973–6.

Huxley, H.E., Stewart, A., Sosa, H. and Irving, T. (1994) X-ray diffraction measurements of the extensibility of actin and myosin filaments in contracting muscle. *Biophys. J.* **67**, 2411–21.

Iwamoto, H., Sugaya, R. and Sugi, H. (1990) Force–velocity relation of frog skeletal muscle fibres shortening under continuously changing load. *J. Physiol.* **422**, 185–202.

Jewell, B.R. and Wilkie, D.R. (1958) An analysis of the mechanical components in frog's striated muscle. *J. Physiol.* **143**, 515–40.

Julian, F.J. and Morgan, D.L. (1979) The effect on tension of non-uniform distribution of length changes applied to frog muscle fibres. *J. Physiol.* **293**, 379–92.

Julian, F.J., Rome, L.C., Stephenson, D.G. and Stritz, S. (1986) The maximum speed of shortening in living and skinned frog muscle fibres. *J. Physiol.* **370**, 181–99.

Julian F.J. and Sollins, M.R. (1975) Variation of muscle stiffness with force at increasing speeds of shortening. *J. Gen. Physiol.* **66**, 287–302.

Julian, F.J., Sollins, M.R. and Moss, R.L. (1978) Sarcomere length non-uniformity in relation to tetanic responses of stretched skeletal muscle fibres. *Proc. Roy. Soc.* **B200**, 109–16.

Lopez, J.R., Wank, L.A. and Taylor, S.R. (1981) Skeletal muscle: length-dependent effects of potentiating agents. *Science* **214**, 79–82.

Nishiyama, K. (1984) Analysis of mechanical behaviour of muscle by a multi-sarcomere model. In *Contractile mechanism in muscle* (ed. G.H. Pollack and H. Sugi), pp. 637–9. Plenum, New York.

Podolsky, R.J. and Nolan, A.C. (1973) Muscle contraction, cross-bridge kinetics, and the Fenn effect. *Cold Spring Harb. Symp. Quant. Biol.* **37**, 661–8.

Ramsey, R.W. and Street, S.F. (1940) The isometric length–tension diagram of isolated skeletal muscle fibres of the frog. *J. Cell. Comp. Physiol.* **15**, 11–34.

Rüdel, R. and Taylor, S.R. (1971) Striated muscle fibres: facilitation of contraction at short lengths by caffeine. *Science* **172**, 387–8.

Rüdel, R. and Zite-Ferenczy, F. (1979) Interpretation of light diffraction by cross-striated muscle as Bragg reflexion of light by the lattice of contractile proteins. *J. Physiol.* **290**, 317–30.

Sugi, H. and Kobayashi, T. (1983) Sarcomere length and tension changes in tetanized frog muscle fibers after quick stretches and releases. *Proc. Natl. Acad. Sci. USA* **80**, 6422–5.

Sugi, H., Ohta, T. and Tameyasu, T. (1983) Development of the maximum isometric force at short sarcomere length in calcium-activated muscle myofibrils. *Experientia* **39**, 147–8.

Sugi, H. and Tsuchiya, T. (1981*a*) Isotonic velocity transients in frog muscle fibres following quick changes in load. *J. Physiol.* **319**, 219–38.

Sugi, H. and Tsuchiya, T. (1981*b*) Enhancement of mechanical performance in frog muscle fibres after quick increases in load. *J. Physiol.* **319**, 239–52.

Sugi, H. and Tsuchiya, T. (1988) Stiffness changes during enhancement and deficit of isometric force by slow length changes in frog skeletal muscle fibres. *J. Physiol.* **407**, 215–29.

Suzuki, S., Oshimi, Y. and Sugi, H. (1993) Freeze-fracture studies on the cross-bridge angle distribution at various states and the thin filament stiffness in single skinned frog muscle fibers. *J. Electron Microsc.* **42**, 107–16.

Suzuki, S. and Sugi, H. (1983) Extensibility of the myofilaments in vertebrate skeletal muscle as revealed by stretching rigor muscle fibres. *J. Gen. Physiol.* **81**, 531–46.

Tamura, Y., Hatta, I., Matsuda, T., Sugi, H. and Tsuchiya, T. (1982) Changes in muscle stiffness during contraction recorded using ultrasonic waves. *Nature* **299**, 631–3.

ter Keurs, H.E.D.J., Iwazumi, T. and Pollack, G.H. (1978) The sarcomere length–tension relation in skeletal muscle. *J. Gen. Physiol.* **72**, 565–92.

Wakabayashi, K., Sugimoto, Y., Tanaka, H., Ueno, Y., Takezawa, Y. and Amemiya, Y. (1994) X-ray diffraction evidence for the extensibility of actin and myosin filaments during muscle contraction. *Biophys. J.* **67**, 2422–35.

2 Muscle mechanics II: skinned muscle fibres

Bernhard Brenner

2.1 Introduction

The goal of this chapter is not to provide a comprehensive review of mechanical studies that have been performed on skinned fibres; instead, the idea is to provide the reader with a critical discussion of the various approaches used in this field, their technical limitations, or limitations in evaluation/interpretation of the observed data, such that readers can form their own critical opinion of the literature. Some controversial issues will be presented, not to resolve the controversies, but as representative examples to illustrate how technical shortcomings or lack of in-depth analysis can result in apparent inconsistencies that might be resolved by critical awareness of the limitations of a technique.

2.2 Historical background

To develop a detailed understanding of the molecular processes underlying muscle contraction and its regulation, characterization of mechanical properties, chemical identity, and structural features of all relevant cross-bridge states is needed, as well as evaluation of kinetics (and their strain-dependence) of all relevant steps in the cross-bridge cycle. Only on this basis, i.e. by integration of mechanical, biochemical and structural data, quantitative models of cross-bridge action in muscle and their regulation can be developed.

Pursuing this goal, one major limitation of studies with intact fibres is the difficulty of modifying and precisely controlling the chemical and ionic conditions to which the contractile proteins are exposed. Studies with actomyosin in solution cannot serve this purpose since the three-dimensional filament-lattice structures are destroyed, so that ATP-hydrolysis is no longer coupled to the generation of mechanical work which, however, is the crucial product of ATP-hydrolysis during muscle contraction.

To overcome the limitations of intact fibres, while keeping the three-dimensional filament-lattice structures intact, various 'skinned fibre' preparations were developed. These are fibre preparations in which membranes were removed or made highly permeable, either mechanically or by chemical means, while the contractile system was left intact.

The first method for generating skinned fibres was chemical skinning with aqueous glycerol solutions, as developed by A. Szent-Györgyi (1949), which also served as a method of preserving skinned fibres at $-20\,^{\circ}C$ for at least some time. This system was used to demonstrate the relevance of ATP for muscle contraction (e.g. Weber 1951; Weber and Weber 1951; Weber and Portzehl 1954). The method of generating mechanically skinned fibres was first introduced by Natori who mechanically removed the sarcolemma of frog fibres. Such mechanically skinned fibres were used to study the effects of Ca^{++} on isometric tension (Hellam and Podolsky 1969) and on maximum unloaded shortening velocity (Podolsky and Teichholz 1970). Detergent treatment was used by Julian (1971) when studying the effect of Ca^{++} on the force–velocity relationship. The most recent developments for generating highly permeable fibres are incorporation of suitable proteins into fibre membranes so that pore-like aggregates are formed (Nishimura *et al.* 1988; Kobayashi *et al.* 1989), e.g. by using α-toxin (Ahnert-Hilger *et al.* 1989) or β-escine (Iizuka *et al.* 1994). This method became widely used for permeabilizing smooth muscle tissue, since membrane associated channels and receptor systems, as well as intracellular signal pathways, remain intact except for small diffusible components that are lost through the well-defined pore-like aggregates. Such systems are suitable for controlled studies into signal pathways associated with regulation of smooth muscle contraction. This approach has been the subject of a recent, detailed review (Pfitzer 1996) and thus will not be covered in this chapter.

2.3 The chemically skinned fibre

Several methods for chemical skinning have been developed, e.g. glycerol treatment (Szent-Györgyi 1949), treatment with EDTA (Winegrad 1971) and EGTA (Eastwood *et al.* 1975). For some of these methods the question has been raised as to whether or not they are sufficient to rapidly and permanently permeabilize muscle fibre membranes. Consequently, after only short treatment, diffusion of large molecules, like creatine kinase, might still be impaired. To rapidly and permanently permeabilize the fibres, glycerol or detergents were added to EGTA-containing solutions (e.g. Magid and Reedy 1980; Brenner 1983). In the following discussion we will focus on the treatment with Triton X-100 that is now widely used. Incubation in glycerol-containing solutions may introduce irreversible structural damage (detectible in X-ray diffraction studies) due to rapid compression or swelling of the myofilament lattice induced by osmotic gradients when increasing or decreasing the concentration of glycerol, respectively (own unpublished results).

Permeabilizing fibres

For chemical skinning, muscle fibre bundles of 0.5–1 mm in diameter are usually incubated for 30 min in skinning solution with 0.5% Triton X-100. After incubation with the Triton X-100-containing skinning solution, the bundles are transferred to skinning solution without the detergent. Before addition to the skinning solution, however, the Triton X-100 has to be pre-treated to remove degraded material, e.g. by equilibration with an excess of the mixed-bed resin AG 501-X8 (Bio-Rad Laboratories, Richmond, CA; cf. Kraft *et al.* 1995*a*). Without such pre-treatment, incubation with Triton X-100 can cause both substantial reduction of active force and impaired regulation, as a result of which the fibres do not fully relax at low Ca^{++}-concentrations (own unpublished results). Some of these effects can be partially reversed by incubation in dithiothreitol (DTT)-containing solutions.

To optimize structural stability of fibres, inhibition of proteolysis by protease inhibitors was introduced (Chase and Kushmerick 1988). In our experience, addition of a cocktail of 4-(2-aminoethyl)-benzolsulfonylfluorid (AEBSF), aprotinin, leupeptin, antipain, trans-epoxysuccinyl-***L***-leucylamido-(4-guanidino)butane (E64), and pepstatin yields the best results (Kraft *et al.* 1995*a*). With these protease inhibitors, the structural stability of fibres, especially for activation at high temperature, is much improved.

Using X-ray diffraction patterns it was found that exchange of propionate for creatine phosphate (CrP), i.e. exchange for the major physiological anion, in the skinning solution very much improved the structural order of the skinned fibres (Brenner and Yu 1993*a*,*b*), as indicated by much increased (about two-fold) intensities in the higher order equatorial reflections. The resulting skinning solution used in our laboratory has the following composition (cf. Kraft *et al.* 1995*a*): 5 mM KH_2PO_4, 3 mM magnesium acetate, 5 mM EGTA, 1 mM sodium ATP (Merck, Darmstadt, Germany), 50 mM sodium creatine phosphate, 5 mM NaN_3, 10 mM glutathione, 2 mM DTT, 100 μM AEBSF (Calbiochem, La Jolla, CA), 1 μg/ml aprotinin, and 10 μM of each of leupeptin, antipain, E64, and pepstatin.

Upon mechanical skinning or chemical permeabilization of membranes, the myofilament lattice spacing expands. To restore the physiological lattice spacing, Magid and Reedy (1980) proposed adding appropriate amounts of macromolecules (e.g. polyvinylpyrrolidone av. mol. wt. 40 000, PVP 40 or L dextran av. mol. wt. 500 000, Dextran T500). In doing this, however, it is important to be aware that upon addition of more than about 4% of Dextran T500, the sarcomeric order is irreversibly lost rather rapidly (within a few minutes) when such fibres are maximally activated by saturating Ca^{++}-concentrations (Brenner and Yu 1993*b*).

Preservation of skinned fibre material

Overall, three approaches are used for preserving skinned fibres over longer periods of time.

One is the incubation in a 1 : 1 (vol : vol) mixture of glycerol with relaxing, rigor, or skinning solution. Fibre bundles are kept in such solutions at −20 °C for up to several months. This method was originally introduced by Szent-Györgyi (1949) and is still used by many groups. One disadvantage is the slow decay in the amount of actively developed tension and the slow impairment of full relaxation that can result in some partial activation even under relaxing conditions, specifically at lower ionic strengths (own unpublished results).

Rapid freezing in liquid nitrogen and subsequent freeze-drying was introduced by Stienen *et al.* (1983). One limitation of this method appears to be an impaired structural order. This becomes evident in X-ray diffraction studies which show equatorial intensities with no clearly resolved higher order reflections, even at lower ionic strength (Jung *et al.* 1988); usually higher order reflections are quite prominent under such conditions.

More recently we have developed a method to preserve skinned muscle material for many months with fully intact structural and functional properties (cf. Kraft *et al.* 1995*c*). After skinning in Triton X-100, the fibre bundles were successively incubated for at least 30 min each in skinning solution with 0.5 M, 1.0 M, 1.5 M, and 2 M sucrose. After equilibration in 2 M sucrose, the bundles were rapidly frozen in liquid propane and transferred to liquid nitrogen for long-term storage. For thawing, the bundles were directly transferred from liquid nitrogen into skinning solution with 2 M sucrose. Subsequently, the bundles were equilibrated successively for 30 min each in skinning solution with 1.5 M, 1.0 M, and 0.5 M sucrose, and finally transferred to normal skinning solution. The sucrose serves as a cryoprotectant allowing bundles of 0.5–1 mm diameter to be vitrified in liquid propane. The stepwise equilibration of increasing or decreasing sucrose concentration proved essential to fully preserve structural properties as revealed by 2-D X-ray diffraction patterns (Kraft *et al.* 1996). Otherwise, large osmotic gradients result in irreversibly increased myofilament lattice spacing and loss of higher order reflections in equatorial X-ray diffraction patterns. With this protocol, none of the parameters studied (e.g. 2-D diffraction patterns, isometric force, shortening velocity, fibre ATPase, fibre stiffness under various conditions, equilibration of fluorescently labelled proteins) were detectably affected for a storage period of up to some 3 years (the time over which this method has been tested so far).

2.4 Problems and limitations when using skinned fibres

Structural instabilities

One major difficulty in using skinned muscle fibres is the tendency for progressively increasing disorder during Ca^{++}-activation, both at the microscopic and submicroscopic level. This, however, is not a specific feature of skinned fibres. For instance, in native single frog fibres, a rather comparable increase in sarcomeric disorder during tetanus rise and the plateau phase has been reported (Julian *et al.* 1978).

Since many parameters (e.g. shortening velocity, isometric and isotonic transients, fibre stiffness, T-plots, force re-development) can be seriously affected by structural disorder, structural stability (e.g. stable striation pattern) and its continuous control during experiments are essential for the recording of reliable and reproducible data on skinned fibres.

To cope with the problem of structural instabilities, two main sources for such instabilities need to be considered:

1. Instability due to structural damage caused by the various procedures to attach the skinned fibre segments to the experimental equipment, whether by various types of glues (e.g. cyanoacrylate-based glues) or by T-clip mounting (Ford *et al.* 1977).
2. Instability inherent in the organization of the sarcomeres, i.e. beyond the plateau area in the length–tension diagram an increase in sarcomere length results in reduced generation of active force. Any prolonged (non-physiological) isometric steady-state contraction is thus expected to result in an increasing disorder in the striation pattern as is seen, not only in skinned fibres, but also in native isolated fibres, i.e. fibres with intact membrane systems (Julian *et al.* 1978).

To reduce the first type of instability, Chase and Kushmerick (1988) developed a procedure to cross-link the fibre ends with glutaraldehyde before mounting the fibres on to the experimental equipment. One problem of this technique can be the introduction of a small zone between the fully fixed and the unfixed areas. This zone appears either permanently relaxed, if glutaraldehyde fixation is done under relaxing conditions, or this zone is permanently activated, if fixation occurred in rigor. Permanent relaxation, however, results in overstretch and possible rupture when fibres are Ca^{++}-activated. Therefore, in our lab, fixation of fibre ends is always done under rigor conditions (cf. Kraft *et al.* 1995*b*).

The second cause of instability can be coped with by interrupting isometric steady-state contraction with periods of unloaded or lightly loaded shortening. During each period of reduced load, the sarcomeres can re-register, i.e. re-order (Brenner 1983). Re-stretching fibres from the low-load condition allows re-establishment of the initial isometric sarcomere length, thereby allowing this manoeuver to be repeated many times. As long as sarcomeric order has not deteriorated, the re-development of force after the re-stretch starts from low force levels (comparable to the load imposed during the period of isotonic shortening). In contrast, progressive structural disorder results in the re-development of force starting from much higher force levels. For technical simplification, a large stepwise length release can be imposed to slacken the fibres. This has the same ordering effect provided that the re-stretch manoeuver occurs before the fibres begin to re-develop tension, when the imposed slack has been taken up by active unloaded shortening of the fibres.

End compliance

Another difficulty for mechanical measurements with muscle fibres, whether skinned or native, originates from the different behaviour of the fibre ends attached to the setup as opposed to the central part of the preparation. As a consequence, the overall length of the fibre can behave quite differently compared to the sarcomere length. For example, even if the overall length is held constant ('isometric contraction'), there can still be substantial changes in sarcomere length, especially when the level of active force is changing. The associated filament sliding can become a serious limitation when evaluating kinetic parameters for defined conditions (e.g. isometric conditions or specified speed of filament sliding).

The problem is reduced by the fixation procedure of Chase and Kushmerick (1988), specifically when fixation is done under rigor conditions (Kraft *et al.* 1995*b*).

To cope with the remaining end compliance, control of sarcomere length within the central part of the fibre (or length of the central fibre segment) is necessary. This can be achieved by different techniques to be described in the next section.

2.5 Techniques to measure and control sarcomere length or fibre segment length

Laser light diffraction

One widely used technique to monitor the sarcomere length of a central fibre segment is laser light diffraction (Cleworth and Edman 1972). When deriving sarcomere length from laser light diffraction, however, the three-dimensional arrangement of diffracting planes has to be considered. It has been demonstrated that fibres are organized in domains, i.e. areas of limited dimensions over which orientation of z-discs (tilt) and sarcomere length are identical (Rüdel and Zite-Ferenczy 1979). As a result, the Bragg condition is only fulfilled for a few domains, particularly at normal beam incidence. At a fixed incidence angle, only a few domains contribute to the recorded signal, thus raising the question of whether or not these few domains generate a signal representative for all domains within the fibre cross-section. Non-linearity in the recorded signal (e.g. steps) might arise when changes in sarcomere length and/or tilt of the striation pattern occur during the length changes, thus resulting in changes in the relative contribution of different domains to the recorded signal (Fig. 2.1A). The correct way to cope with these two problems is to scan the incidence angle of the laser beam over the complete relevant range, i.e. the complete range over which a significant intensity is recorded in the diffraction maximum used to derive sarcomere length (at least some 10°; cf. Rüdel and Zite-Ferenczy 1979; Brenner 1985*a*). In principle, this can be achieved by calculating a weighted average of all signals recorded at different incidence angles. In practice, sufficient precision is achieved when averaging the signals recorded at different incidence angles, about 0.5° apart, over the range of significant intensity in the observed diffraction maximum (Fig. 2.1B; Brenner 1991). If sarcomere length is to be used in a feedback mode, scanning of the beam by a 14 kHz torsion scanner can be used to obtain an ω-averaged signal (own unpublished results).

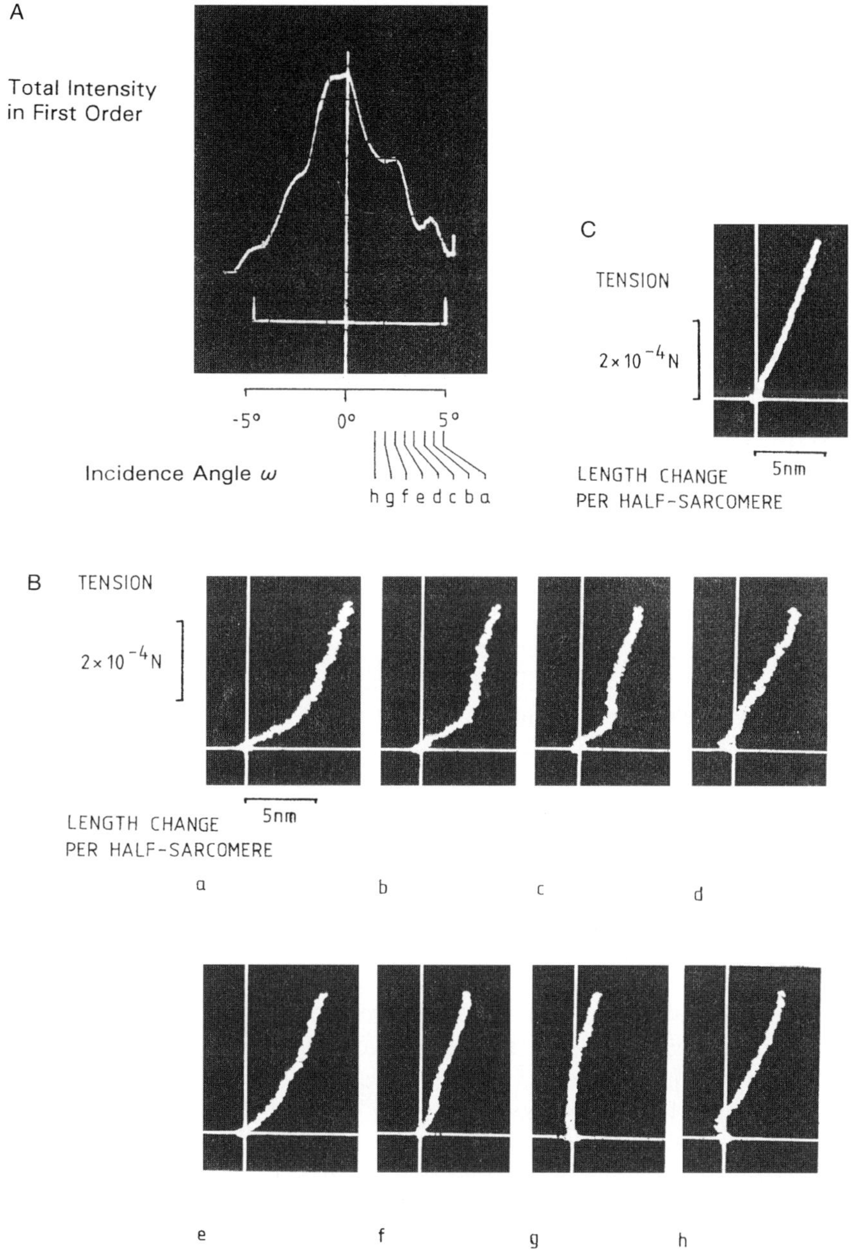

Fig. 2.1 (A) ω-scan of a rigor fibre. Intensity in one first-order diffraction maximum versus incidence angle (ω) between laser beam and normal on fibre: (a)–(h) illustrate the angles at which the sarcomere length signal (via centroid of the first-order diffraction maximum) was recorded. (B) (a)–(h). Force versus sarcomere length, recorded at the different incidence angles. (C) Force versus sarcomere length, when averaging all individual sarcomere length records collected between $-5°$ and $5°$ at $0.5°$ intervals.

As an alternative, white light diffraction has been proposed for recording of a sarcomere length signal originating from different domains of the illuminated fibre section (Goldman 1987). A limitation of white light diffraction, however, is the systematic coupling of incidence angle and wave length, such that the effective scan range is narrower than the range of incidence angles of the probing beam (some 5° in the system of Goldman 1987). The minimum necessary scan range, however, varies from fibre to fibre but is in the order of 7–10°. Thus, white light diffraction can certainly reduce limitations originating from the domain structure of the striation pattern, but may not fully exclude limitations originating from the domain structure.

The striation follower

Another system that has been developed for precise recording of sarcomere length is the striation-follower device (Huxley *et al.* 1981). Two narrow laser beams are used to project an average 'image' (formed by using the zeroth and one first-order diffraction maximum) of some 10–20 sarcomeres onto a device that counts the number of sarcomeres passing by the two probing beams. With interpolation for fraction of sarcomeres, the difference in the number of sarcomeres passing by the two illuminated spots, multiplied by the sarcomere length, allows the length change of the fibre segment, defined by the separation between the two illuminated spots, to be derived. Again, laser beams with fixed incidence angles are used, in addition to looking at only a small spot on the fibre. Thus, the averaged transforms are expected to be again dominated by only a few domains of the striation pattern, i.e. some limitations might arise from recording the number of bypassing sarcomeres from possibly a non-representative domain. Secondly, calculating the movement at each spot as the product of the number of bypassing sarcomeres, multiplied by the sarcomere length, requires a precise knowledge of the sarcomere length at each spot. However, only an average sarcomere length is assumed, at least in the original design. Variations in sarcomere length along the length of a fibre are thus expected to limit the precision of this device.

In summary it should be noted that all diffraction-based systems that have been introduced so far, have some limitations that should be taken into account when interpreting sarcomere length signals recorded by these devices.

Some attempts have been made to use the substructure of diffraction maxima to determine and follow the sarcomere length of individual domains and the variance of sarcomere length among different domains (Tameyasu *et al.* 1982; Leung 1983; Sundell *et al.* 1984; Zite-Ferenczy *et al.* 1984). These trials were based on the assumption that each substructural line represents the diffraction maximum of one individual sarcomeric domain. It was, however, demonstrated that the substructural lines within the diffraction maxima of a fibre represent interference fringes arising from simultaneous illumination of several domains by laser light (Brenner 1985*a*), i.e. light with high spatial coherence. Thus, the individual substructural lines do not represent diffraction maxima of individual domains.

Segment-length control techniques

A principally different approach in following the length of a central segment of both intact and skinned fibres involves the attachment of markers. Imaging of the markers on suitable detecting devices enables monitoring of the separation between two markers and therefore the behaviour of the fibre segment delineated by these markers (e.g. Gordon *et al.* 1966; Edman and Reggiani 1984). Attaching markers to the surface of skinned fibres, however, might introduce an additional diffusion barrier, resulting in uneven diffusion of Ca^{++}, substrate and products. This could cause local instabilities in activated skinned fibres. Nevertheless, this method is sometimes employed instead of diffraction techniques when deterioration of the striation pattern at high activation levels makes the use of diffraction techniques difficult or impossible. In such disordered fibres, the segment length signal only represents an average behaviour of a population of sarcomeres with large dispersion of sarcomere lengths, a problem by which mechanical parameters might be significantly altered when compared with behaviour in well-ordered systems (e.g. amplitude of instantaneous length change versus amplitude of active shortening in isotonic releases). Use of segment length control devices when sarcomeric order is lost, may result in ill-controlled experimental conditions and observations at variance with studies on well-ordered systems.

2.6 Diffusion problems

An important factor in skinned fibre experiments is the concentration gradient between bath and myofilament space for substrates and products, i.e. Mg adenosine 5′-triphosphate (MgATP), Mg adenosine-5′-diphosphate (MgADP) and inorganic phosphate (Yu and Podolsky 1990). To cope with MgATP-gradients in activated fibres, i.e. when ATPase activity is high, ATP-regenerating systems, like creatine kinase and creatine phosphate, are widely used. However, it is important to be aware that, assuming sufficient activity of the creatine kinase, the limiting factor is now diffusion of creatine phosphate. Thus, a substantial gain upon addition of an ATP-regenerating system is only obtained when creatine phosphate concentrations can be used that are much larger than the concentrations of MgATP. This is because the diffusion coefficient of creatine phosphate is expected to be not more than about twice the diffusion coefficient of MgATP, such that only larger concentration gradients will provide more efficient diffusion and backing-up of MgATP.

As shown by Yu and Podolsky (1990), even poor substrates, e.g. ATP-analogues that are only slowly hydrolysed, may develop relevant gradients in fibre bundles; this has been observed when titrating Mg adenosine 5′-O-(3-thio triphosphate) (MgATPγS) into skinned fibres (Kraft *et al.* 1996).

As a result, for some approaches, e.g. high-sensitivity ATPase measurements or 2D-X-ray diffraction studies, where the amount of contractile material of one single fibre is too small, the use of fibre bundles is questionable and certainly rather limited, especially at room temperature. Due to the diffusion limitations, it might instead become necessary to use an array of single fibres, i.e. several single fibres that are mounted to the equipment in parallel, so that the diffusion situation for each fibre is the same as for one isolated fibre (Kraft *et al.* 1996).

2.7 Force–pCa relation

Early measurements of force versus free [Ca^{++}] in skinned fibres were performed by Hellam and Podolsky (1969) on mechanically skinned fibres of the frog (*Rana pipiens*). They observed that, especially at low [Ca^{++}], the rate of force development was much slower than expected from diffusion of Ca ethylene glycol-bis(β-aminoethylether)N, N, N', N'-tetraacetic acid (CaEGTA) into the skinned fibres. This was attributed to Ca^{++}-uptake by the sarcoplasmic reticulum (SR) at low activation levels. Residual Ca^{++}-accumulation by the SR or its vesicles, which is observed even when fibres are treated with detergents like Triton X-100, may result in quite sizeable variations in active tension with time, particularly at partial activation levels. This variation in active force can be reduced by the addition of 10–20 mM caffeine to the bathing solution.

Another problem in recording reproducible force–pCa relations is limited fibre stability, specifically at high levels of Ca^{++}-activation, as noticed by Hellam and Podolsky (1969). Progressive disorder at higher activation levels can result in artificially low levels of active force at saturating [Ca^{++}], while force levels at partial activation appear too high. Thus, limited structural stability has to be ruled out when analysing and interpreting force–pCa relations, e.g. by fitting the Hill equation to the data in order to extract information about Ca^{++}-sensitivity (pCa for half-maximum active force) and the slope of force–pCa relations.

Another factor affecting force–pCa relations is hysteresis observed when comparing data recorded upon stepwise increase in [Ca^{++}] versus stepwise decrease in [Ca^{++}] (e.g. Ridgeway *et al.* 1983). Such effects have been interpreted as the result of increased Ca^{++}-binding to troponin-C when force-generating cross-bridges are attached to actin. However, similar larger force responses can be recorded at low Ca^{++}-levels after fibres have been cycled through full activation followed by full relaxation. Stepwise increase in [Ca^{++}] after such an activation/relaxation cycle, again results in larger force response at low activation levels. This might point to the possibility that part of the observed hysteresis may originate from changes in the level of phosphorylation of the regulatory light chain that was found to significantly increase the observed isometric force at low activation levels (Persechini *et al.* 1985).

It is widely accepted practice to analyse and interpret force–pCa relations in terms of Ca^{++}-sensitivity (pCa for half-maximum force) and co-operativity (slope of force–pCa relations) described by the Hill coefficient (for a review, see Rüegg 1992). While Ca^{++}-sensitivity is usually thought to be determined by

Ca^{++}-affinity to troponin-C, the Hill coefficient is generally thought to reflect co-operativity either due to nearest-neighbour interactions among regulated units of the thin filament or resulting from strong cross-bridge attachment increasing Ca^{++}-binding to troponin-C (Bremel and Weber 1972; Weber and Murray 1973; Fuchs 1977; Güth and Potter 1987).

It has been demonstrated that, in addition to such factors, changes in cross-bridge cycling kinetics can have pronounced effects on the observed Ca^{++}-sensitivity and apparent co-operativity, that go beyond possible changes in affinity of Ca^{++} for troponin-C, changes in nearest-neighbour interactions of regulated units of the actin filament, or increased Ca^{++}-binding to troponin-C upon strong cross-bridge attachment (Brenner 1988). Phosphorylation of the regulatory light chain has been demonstrated to be one example of such changes in Ca^{++}-sensitivity and co-operativity via cross-bridge cycling kinetics (Brenner and Morano 1990; Sweeney and Stull 1990). Thus, straightforward interpretation of changes in observed Ca^{++}-sensitivity and co-operativity, in terms of Ca^{++}-affinity of troponin-C and co-operativity within the regulatory system, respectively, may be rather misleading. Instead, detailed interpretation of these parameters requires analysis of cross-bridge cycling kinetics and of Ca^{++}-binding to troponin-C.

2.8 Force–velocity relations and V_{max}

Load-controlled releases (isotonic releases)

For recording the force–velocity relationships of skinned fibres, isotonic releases have been established. During isometric steady-state contraction, contraction conditions are suddenly switched to a constant load level that is less than the maximum isometric force (Fig. 2.2a). After a period of adjustment ('isotonic transients'; Podolsky 1960), steady shortening is reached during which the speed of shortening slows down somewhat, however, as the fibre length becomes shorter.

Due to limited fibre stability, either one such isotonic release has been imposed for each activation/relaxation cycle (e.g. Podolsky and Teichholz 1970) or a multiple-step-release protocol has been employed (Julian 1971), in which a step-wise reduction in load is applied so that several points of a force–velocity relationship can be recorded in each activation/relaxation cycle. In these protocols, fibres are relaxed when fibre length has reached a pre-set length and active force starts to redevelop.

Alternatively, a protocol in which fibres are cycled between isometric steady-state contraction and periods of unloaded or lightly loaded shortening, followed by re-stretch of the fibres to their initial isometric length (Brenner 1983), can be used to record a large number of data points within one activation/relaxation cycle. To avoid structural deterioration, each isotonic period during which fibres are significantly loaded ($P/P_0 > 0.10$), has to be followed by one or several periods of unloaded shortening.

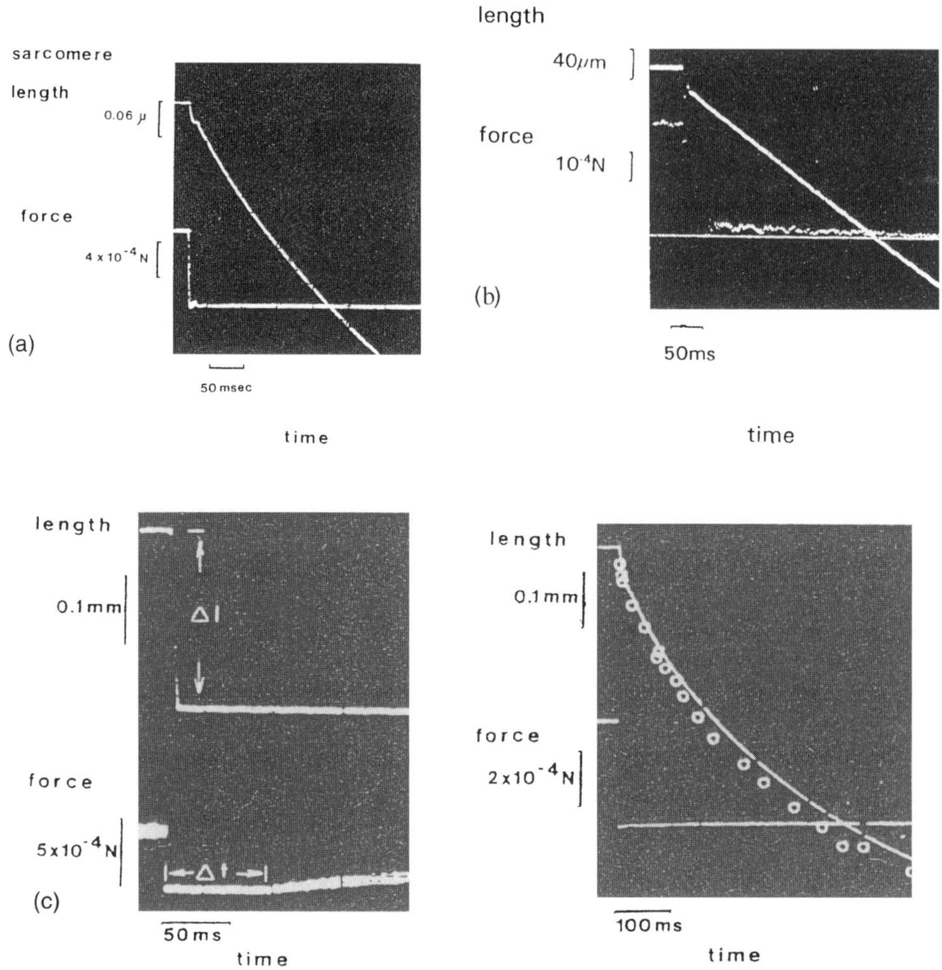

Fig. 2.2 Techniques for recording fibre shortening. (a) Isotonic release (load-controlled release). Note that the length trace is not straight but curved (decrease in shortening velocity as shortening proceeds). (b) Isovelocity release with imposed initial step to rapidly drop tension to zero. Note that in this protocol the load imposed on the fibre is continuously decreasing. (c) Slack-test analysis: *left panel*, experimental protocol; Δl = amplitude of imposed length stop, Δt = time to take up the imposed slack by free shortening; *right panel*, time course of active shortening (*open circles*) data points obtained for imposed length steps of different amplitude (free fibre shortening). The continuous curve was recorded from the same fibre but with the isotonic release technique.

Isovelocity releases

As a different approach, instead of controlling the load imposed on a fibre, the time course of fibre length observed in isotonic releases can be mimicked by imposing a stepwise length release followed by a ramp of constant velocity (Fig 2.2b).

With such a manoeuver, force rapidly decays during the initial step and remains at a nearly constant level, depending on the speed of the imposed ramp-shaped release. With progressive constant-velocity shortening, the force level, however, decays slowly towards lower levels, equivalent to the reduction in shortening velocity of load- controlled isotonic releases. It should be noted that the multiple-step release technique used by Julian (1971) was similar to such isovelocity releases, with initial stepwise length change. Thus, shortening traces in the records of Julian (1971) are rather straight but tension slowly decreases during the different periods of shortening.

Slacktest

For only recording the maximum speed of unloaded fibre shortening, the 'slack test' (Hill 1970) is widely used. Fibres are suddenly released to different pre-set fibre lengths and the time interval is measured from the release until force starts to re-develop. This is the time needed to take up the imposed slack by free shortening. Plotting time against amplitude of the imposed length release provides the time course of unloaded (free) fibre shortening (Fig. 2.2c). Again, the plots reveal a progressive decrease in velocity of unloaded shortening, i.e. the plots are curved so that it becomes difficult to define a unique value for shortening velocity.

One serious shortcoming of this approach is the loss of any diffraction signal or marker image when fibres go slack upon release to the different pre-set lengths. Thus, no length signal can be obtained that properly reflects the sarcomere length behaviour. The slack test data are, therefore, limited by unknown effects from end-compliance and fibre deterioration, which may go unnoticed.

Problems with measurements of fibre shortening

One limitation of measurements of speed of active contraction is the rather pronounced sensitivity to structural disorder. Such disorder results in:

(1) progressive increase in the initial instantaneous length change that coincides with the unloading;

(2) much slower active shortening subsequent to the instantaneous length change.

This is especially critical at higher temperatures (e.g. 20 °C), consequently without careful control of structural order, data on shortening velocity at higher temperatures tend to reveal much too low velocities active shortening.

A second problem, encountered when analysing the velocity of fibre shortening, is the fact that all techniques reveal a non-constant speed of shortening, i.e. as shortening proceeds, the speed of shortening continuously decreases as fibre length becomes shorter. Moreover, this decrease in shortening velocity is highly dependent upon experimental conditions. For example, this decrease in shortening velocity becomes much more pronounced when reducing the level of Ca^{++}-activation. In analogy, during constant-speed length changes (isovelocity contraction; Fig. 2.2b), the continuous decrease in active force as shortening proceeds is also more pronounced at reduced levels of Ca^{++}-activation.

In earlier work (e.g. Podolsky and Teichholz 1970; Julian 1971) this problem was not considered when studying the effects of Ca^{++} on V_{max} and on the force–velocity relationship. According to Podolsky and Teichholz (1970) neither V_{max} nor the shape of the force–velocity relationship recorded from mechanically skinned frog fibres are affected by Ca^{++}-activation. From these data it was concluded that Ca^{++} has no effect on cross-bridge kinetics. In contrast, Julian (1971), using chemically skinned fibres, concluded that Ca^{++} affects both, V_{max} and the shape of the force–velocity relationship. It was later proposed that this controversy arises from the progressive reduction of shortening velocity as the fibre keeps shortening, since this effect is more pronounced at low activation levels. While the initial speed of fibre shortening appears to be Ca^{++}-independent, shortening velocity after increasing distance of active shortening becomes more and more Ca^{++}-dependent (Brenner 1980, 1986*a*). While Podolsky and Teichholz have recorded only one isotonic release during each activation/relaxation cycle, thus recording speed of shortening soon after the instantaneous length change, i.e. after only a small distance of fibre shortening, Julian (1971) has imposed three step releases in which the low-load data were recorded after high-load data. Consequently all data at intermediate and especially at low-load levels were recorded after quite substantial fibre shortening. The shortening traces in Julian's work were of the isovelocity type (i.e. load decreased continuously). The actual loads, however, were not corrected for the decrease that had already occurred during high-load and intermediate-load shortening. Correcting for this, force–velocity relationships derived for saturating activation and partial activation essentially superimpose (Brenner 1980, 1986*a*).

Different possible mechanisms have been proposed for the continuous reduction in shortening velocity as fibres keep shortening:

1. An internal load resulting from elastic deformation of, for example, cytoskeletal structures that provide re-extension of fibres upon relaxation (Brenner 1986*a*).
2. Slowly cycling cross-bridges that accumulate at low Ca^{++}-activation levels generating an internal load (Moss 1986), or dissociation of Ca^{++} from the actin filaments upon reduction in fibre length (Ridgway and Gordon 1984).

The basis for the continuous decrease in shortening velocity in isotonic releases (or, correspondingly, the decrease in force in isovelocity releases) is still unclear and may arise from several of the already mentioned sources. Nevertheless, it is necessary to find suitable ways to analyse shortening velocity, so that these effects can be accounted for. Otherwise, when probing for factors that might affect maximum speed of unloaded shortening at the cross-bridge level, apparent changes in (unloaded) shortening velocity, which only result from differences in the continuous decrease in shortening velocity, may be mistaken as evidence for changes in those cross-bridge kinetics that are thought to limit V_{max} (e.g. g_2, the rate constant for detachment of negatively strained cross-bridges in the concept of A.F. Huxley 1957).

Analysis of isotonic shortening

Two different approaches for analysing the non-linear length changes have been proposed. For isotonic releases, traces can be linearized by plotting the natural logarithm of the instantaneous velocity against fibre length (sarcomere length; Brenner 1986*a*). This means that sarcomere length against time is well described by a logarithmic-type function. On this basis the initial shortening velocity, can easily be derived (Brenner 1980, 1986*a*), while technically, due to inertia of the lever system, the initial parts of the shortening phase cannot be recorded precisely. Detailed analysis has suggested that the slowing down of shortening velocity can be accounted for by an internal load that increases linearly with the extent of fibre shortening. Thus shortening velocity is continuously reduced as expected from the force–velocity relationship, since cross-bridges experience a progressively increasing load.

For the slack test analysis, Moss (1986) proposed an approximation of the observed plots of length change against time with two straight lines: one for the initial part, one for the later part. With this approximation, Moss (1986) found that the initial part of active shortening appears to be Ca^{++}-independent. This approximation, however, has some limitations in the difficulty of unambiguously defining two linear sections, especially in view of the fact that two such different phases are not seen in the continuous length traces recorded during isotonic releases. Therefore, care must be taken to ensure that the approximation of two straight sections does not introduce apparent changes in the initial speed of shortening, which in reality only result from such approximation. It is therefore recommended that the appropriate logarithmic function is fitted to the slack test data, to obtain the initial speed of shortening.

The issue of whether Ca^{++} affects the force–velocity relationship and/or V_{max} is still not completely agreed upon, but it is beyond the scope of this chapter to resolve the issue. Instead, this controversy serves as an illustration of how analysis of the same data in different ways may create controversial interpretation.

2.9 Fibre stiffness

Fibre stiffness is usually measured by imposing small length changes (stretches or releases) upon one end of a muscle fibre. Fibre stiffness is thereby defined as the slope when force is plotted against the imposed length change (Fig. 2.3). Explicitly, stiffness measurements are performed in three different ways:

1. By imposing stepwise length changes of different amplitude and plotting the extreme in the force response (maximum for stretches, minimum for releases) against the amplitude of the imposed stepwise length change (Fig. 2.3a; Huxley and Simmons 1973).

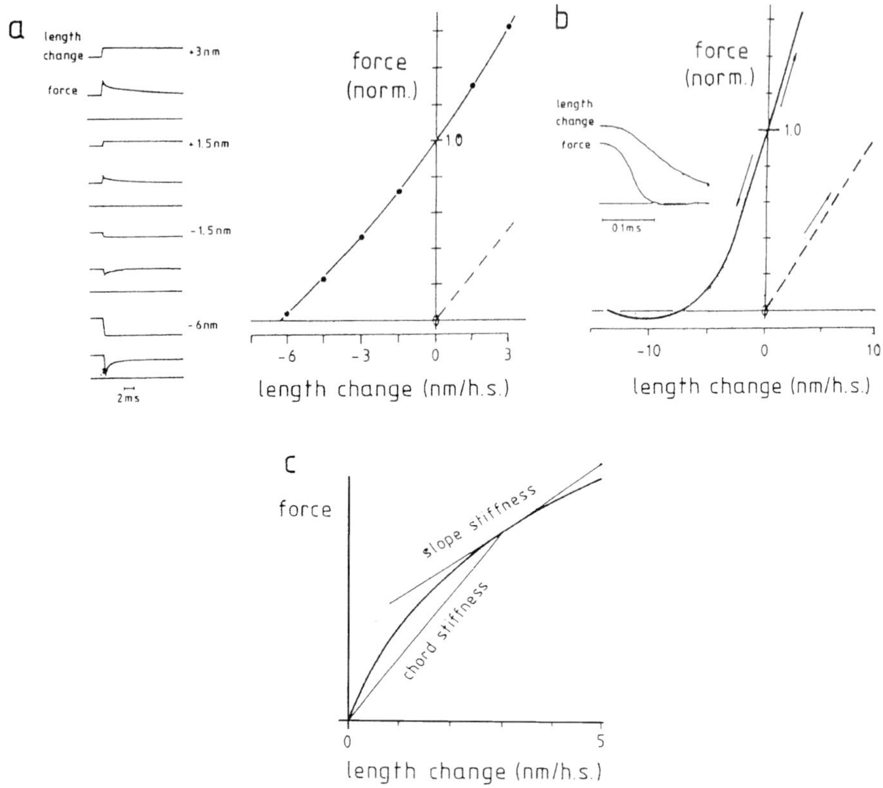

Fig. 2.3 Measurements of fibre stiffness. (a) Fibre stiffness derived from the force response to stepwise length releases of different amplitude, and plot of extreme values of the tension responses (maxima for stretches, minima for releases) versus amplitude of imposed length change ('T_1-plot', cf. Ford *et al.* 1977). Dashed line illustrates plot obtained in fibres without steady active tension, e.g. a relaxed fibre. (b) T-plot obtained by plotting force versus length change recorded during a length release. Dashed line as in (a). (c) Definition of fibre stiffness for non-linear plots of force versus imposed length change (h.s. = half-sarcomere). (Modified from Brenner 1990.)

2. Force is continuously recorded during imposed length changes (e.g. ramp-shaped stretches or releases). Force is then plotted against the imposed length change recorded during a single stretch or release (Fig. 2.3b; Ford *et al.* 1977).
3. Small-amplitude sinusoidal length changes (500 Hz to some 2 kHz) are continuously superimposed, usually on overall fibre length (e.g. Julian and Sollins 1975).

Plots of force against imposed length change are mostly non-linear, such that the term 'stiffness' has to be defined precisely; e.g. slope stiffness versus chord stiffness (cf. Fig. 2.3c). Moreover, for almost all experimental conditions it has been found that plots of force against imposed length change, and thus, both 'slope stiffness' and 'chord stiffness', depend on the speed of the imposed length change (e.g. Huxley and Simmons 1973; Ford *et al.* 1977; Brenner *et al.* 1982, 1986); i.e. for almost all experimental conditions, there is not a unique value for fibre stiffness. For this reason, it is useful to call fibre stiffness that has been recorded at a certain speed of the applied length change 'apparent' fibre stiffness.

For interpretation of the observed fibre stiffness, several technical points need to be considered. A length change applied to one end of a fibre may not be distributed homogeneously along the entire length of the fibre. This might be due to structural inhomogeneities along the fibre (e.g. end compliances) or when the rise time of the applied length change is comparable to, or even shorter than, the propagation time for such a length perturbation (Schoenberg *et al.* 1974). Any inhomogeneity in the distribution of the applied length change, however, will lead to non-parallel behaviour between total fibre length and relative filament sliding at the sarcomere level. Thus, to determine fibre stiffness for interpretation in terms of cross-bridge properties, the observed change in force has to be correlated with filament sliding, i.e. with the length change at the sarcomere level. Consequently, for meaningful stiffness measurement, the actual change in sarcomere length has to be recorded by some of the above-mentioned approaches (e.g. diffraction-based techniques, monitoring of markers). To minimize delays between force signal and sarcomere length signal, due to limited propagation velocity of the imposed length change, the change in sarcomere length has to be monitored at a point as close as technically possible to the force transducer. Thus, segment-length monitoring techniques may have some limitations if the segment length is rather large. The use of diffraction techniques requires coping with the domain-organization of the striation pattern. As already discussed, this can be achieved by averaging the length response of several repetitive measurements using different incidence angles between laser beam and fibre (cf. Brenner 1991; Fig. 2.1b) or white light diffraction (Goldman 1987) as long as the relevant range of necessary incidence angles is covered by this technique.

The stiffness–speed relationship

Aside from these more technical limitations, for interpretation of stiffness data, the dependence of apparent fibre stiffness on the speed of the applied length change has to be considered. To provide a comprehensive characterization of fibre stiffness, including such 'speed-dependence', it is necessary to determine the stiffness–speed relationship, i.e. the dependence of apparent fibre stiffness on the speed of the applied length change (Fig. 2.4; Brenner *et al.* 1986; Brenner 1990). In fact, it was found that the stiffness–speed relationship represents a distinct feature of different experimental conditions/different cross-bridge states (cf. Brenner 1990).

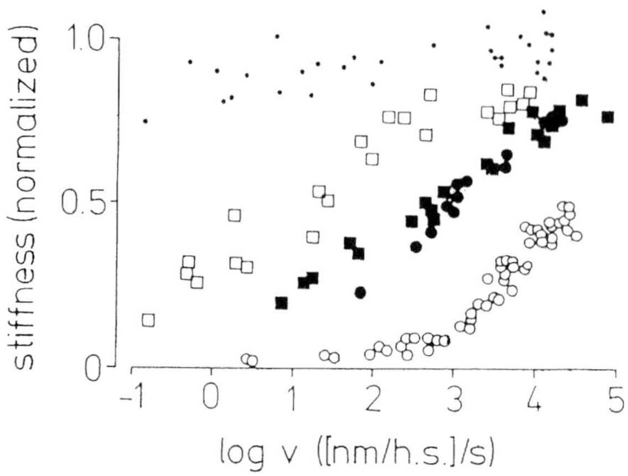

Fig. 2.4 Stiffness–speed relationship (plots of apparent fibre stiffness versus speed of imposed length change) recorded from single skinned fibres under different experimental conditions: (*closed spots*) rigor conditions; (*open squares*) 4 mM Mg pyrophosphate; (*closed squares*) isometric contraction (stretch); (*closed circles*) isometric contraction (release); (*open circles*) relaxed fibre, ionic strength 20 mM. Ionic strength for all other conditions 170 mM.

To reproducibly construct the stiffness–speed relationship in spite of the non-linear plots of force against length change (Fig. 2.3c), it was proposed to plot the chord stiffness for a displacement of, for example, 0.5 or 1 nm per half sarcomere, depending on the noise of the signal (Fig. 2.3c). Stiffness–speed relationships derived in this way for different experimental conditions are shown in Fig. 2.4.

There is one technical aspect that needs to be considered before interpreting stiffness–speed relationships in terms of cross-bridge properties. Even using laser diffraction techniques (or other techniques to monitor length changes at the sarcomere level) there appears to be a limitation for speed dependence of fibre stiffness at very high speeds of stretch, especially when fibre stiffness is high (e.g. rigor conditions). In several tests we observed a speed dependent fibre stiffness in rigor when fibres were not firmly attached to the motor or force transducer. This speed-dependence, seen at high speeds of stretch, disappeared (or was at least much reduced) after fibres had been remounted (own unpublished results). The actual basis for this observation is not completely clear. It seems possible that cytoskeletal structures, which provide lateral connectivity between myofibrils, might transmit forces from those parts of the fibre cross-section that are not firmly attached to the force transducer. However, due to visco-elastic properties of such cytoskeletal structures, the transmitted force decays or, equivalently, becomes smaller for slower stretches.

One process/factor that can introduce the observed dependence of apparent fibre stiffness on the speed of the imposed length change is the possibility of reversible cross-bridge dissociation and reassociation from and to actin (Brenner *et al.* 1982, 1986; Schoenberg *et al.* 1984; Schoenberg 1985, 1988). As has been demonstrated by model calculations (Schoenberg *et al.* 1984; Schoenberg 1985; Brenner 1990), dynamic actin interaction of myosin heads, i.e. reversible dissociation and reassociation, can generate plots of force against imposed length change and a dependence of apparent fibre stiffness on speed of stretch, just as seen experimentally. It has also been demonstrated that, assuming reversible actin interaction as the basis for the observed stiffness–speed relationship, the position of a stiffness–speed relationship along the abscissa provides information about the rate constant for cross-bridge dissociation from actin. If it is possible to derive information about the fraction of attached cross-bridges (e.g. from X-ray diffraction or under some conditions from stiffness data, see below), knowledge of the rate constant for dissociation allows an estimation of the rate constant for attachment to actin (cf. Schoenberg 1985; Brenner 1990, for a detailed discussion).

It should be noted, however, that under many experimental conditions, only a part of the stiffness–speed relationship can be recorded experimentally, i.e. under most experimental conditions the speed of the applied length changes are not fast enough to reach a stiffness level that is independent of further increases of the speed of stretch (cf. Fig. 2.4). As a result, only a rough estimate of the rate constants for dissociation is usually possible (Schoenberg 1985, 1988; Brenner 1990).

The width of the experimentally observed stiffness–speed relationship is rather wide. It has been demonstrated that this could be accounted for by the fact that attached cross-bridges experience different amount of strain due to the mismatch between actin and myosin filament periodicities. Such differences in cross-bridge strain are expected to result in different rate constants for dissociation and reassociation of myosin heads from and to actin (Hill 1974).

A second factor that can result in a dependence of chord stiffness on speed of the imposed length changes is redistribution of cross-bridges among different attached (e.g. force-generating) states. Such reactions, however, can be differentiated from reversible actin interactions by inspection of the plots of force against imposed length change recorded during ramp-shaped length changes. Redistribution among attached states of different average strains (e.g. different force-generating states) has been shown to generate a rather characteristic shape of such plots (Ford *et al.* 1977; Kuhn *et al.* 1979; van Kaam *et al.* 1984; Brenner 1991; see next paragraph).

T-plots and possible implications

Plots of force response against applied length change recorded during ramp-shaped releases that were imposed during isometric steady state contraction have been introduced (Ford *et al.* 1977; Kuhn *et al.* 1979) as an alternative approach to the study of isometric tension transients in response to stepwise length releases

(Huxley and Simmons 1973; Ford *et al.* 1977). Such 'T-plots' (cf. Kuhn *et al.* 1979) were later also studied for ramp-shaped stretches imposed during isometric steady-state contraction (van Kaam *et al.* 1984; Brenner 1991). It was noticed that T-plots recorded during stretches and releases do not show the kind of symmetric behaviour that is expected from the model of Huxley and Simmons (1971; cf. van Kaam *et al.* 1984; Brenner 1991). Moreover, such T-plots reveal changes in their initial slope when recorded during ramps of different speed (Brenner 1991). Based on model calculations it was proposed that recording of T-plots during ramps of different speed allows differentiation between effects arising from reversible dissociation and reassociation of cross-bridges from and to actin, and effects arising from re-equilibration among different force-generating cross-bridge states (Brenner 1991; Brenner *et al.* 1995), i.e. between two processes that are difficult to differentiate in isometric tension transients in response to stepwise length changes.

Since rapid equilibration among different force-generating states is essentially absent during stretches (Brenner 1991), stiffness–speed relationships that are only affected by dissociation and reassociation of force-generating cross-bridges from and to actin may be constructed from ramp-shaped stretched of different speed (Brenner 1986*a*; cf. Fig. 2.4). Again, apparent fibre stiffness changes over a rather wide range of stretch velocities and a speed-independent stiffness is not quite reached even for the highest stretch velocities attained.

In conclusion, from plots of force against imposed length change recorded during ramp-shaped stretches and releases (Fig. 2.3c), we have demonstrated that for essentially all conditions studied, fibre stiffness is not single valued. Instead, we have to derive the stiffness–speed relationship, i.e. the relationship between observed fibre stiffness and speed of the applied ramp-shaped length change. We have also seen the need and feasibility of differentiating between effects on fibre stiffness originating from reversible dissociation and reassociation of cross-bridges from and to actin, and effects arising from re-equilibration of different force-generating states in activated fibres.

Using stepwise length changes to derive fibre stiffness from the T_1-plots (extreme force level against amplitude of imposed step; Fig. 2.3a), it will be necessary to account for effects arising from dissociation and reassociation of cross-bridges from and to actin and to separate such effects from re-equilibration of different force-generating states in activated fibres. Variation of the time for completing the step and analysis of the quick tension recovery in the isometric transients might be the way to achieve this, although it appears difficult since both dissociation/reassociation from and to actin, as well as re-equilibration of different force-generating states, are thought to contribute to quick tension recovery (Eisenberg *et al.* 1980; Chen and Brenner 1993). Thus, it remains to be shown how such characterization could be achieved when using stepwise length changes.

Similarly, for application of small sinusoidal length oscillations it is necessary to at least record the tension response for different frequencies, and to differentiate between in-phase stiffness and out-of-phase stiffness (Goldman *et al.* 1982, 1984). Again, more systematic studies are necessary to characterize and differentiate between different factors contributing to the observed fibre stiffness.

2.10 Interpretation of fibre stiffness in terms of cross-bridge attachment

In interpreting fibre stiffness in terms of cross-bridge attachment, several central assumptions are usually made:

(1) Fibre stiffness essentially represents cross-bridge stiffness, i.e. actin and myosin filaments, to a first approximation, are essentially rigid.

(2) Observed fibre stiffness directly reflects the fraction of attached cross-bridges, i.e. no detachment/re-attachment of cross-bridges or re-equilibration among different force-generating states is considered.

(3) Cross-bridge stiffness is the same for all cross-bridge states.

(4) In rigor, essentially all cross-bridges are attached to actin.

On the basis of these assumptions, the ratio of observed fibre stiffness over rigor stiffness is quite widely used to estimate the fraction of attached cross-bridges. Such estimates, however, have to be treated with care, since most of these four assumptions may in fact be violated quite significantly.

Filament compliance

Ever since the work of Huxley and Simmons (1973) and Ford *et al.* (1981) it has been widely assumed that observed fibre stiffness, if either segment-length control or suitable diffraction techniques had been used, essentially represents cross-bridge stiffness, i.e. it was thought that filament compliance contributes only little to observed fibre stiffness, at most some 20% (Ford *et al.* 1981). This was confirmed by studies of Julian and Morgan (1981), and Bagni *et al.* (1990) who examined changes in stiffness of frog fibres when sarcomere length was changed between 2.0 and 2.2 μm. Over this range, only the 'free' length of the actin filaments, i.e. the part from the end of the overlap zone to the Z-line, is changing while active force remains constant.

The first to raise questions about the assumed 'inextensibility' of thin filaments was Sugi and his co-workers, who stretched skinned fibres in rigor state and demonstrated extensibility of thin filaments electron microscopically (Suzuki and Sugi 1983, 1984; Suzuki *et al.* 1993). They also estimated the thin filament compliance and showed that it can not be ignored in the fibre stiffness measurements. This was confirmed by direct measurement of single thin filament compliance (Kojima *et al.* 1994) and time-resolved X-ray diffraction studies (Wakabayashi *et al.* 1994; H.E. Huxley *et al.* 1994).

These studies suggested not only a quite significant compliance of the actin filaments but also within the myosin filaments. It was proposed that filament compliance is well within the range of cross-bridge compliance (Huxley *et al.* 1994), or might even represent the dominant compliance in stiffness measurements (Wakabayashi *et al.* 1994). Based on stiffness measurements in rigor fibres, it was confirmed that about 50% of fibre stiffness might reside in the myofilaments (Higuchi *et al.* 1995).

Several estimates have been represented about how filament compliance will affect the plot of fibre stiffness against active force (Goldman and Huxley 1994; Mijailovich *et al.* 1996). It was demonstrated that, depending on the extent of filament compliance, the plots of fibre stiffness against force will be curved, thus providing yet another interpretation for the apparent lead of stiffness over force in, for example, the rising phase of a tetanus.

The issue of how much fibre stiffness is affected by filament compliance is not yet settled. Proper controls are necessary before interpreting fibre stiffness in terms of cross-bridge attachment. For instance, careful analysis of plots of fibre stiffness against active force might be one such way, especially when the force level is changed by different interventions, like different degrees of Ca^{++}-activation, changes in temperature, or application of positive inotropic drugs (cf. Kraft and Brenner 1997).

Contribution of dissociation/reassociation or equilibration among different states

The assumption that the observed fibre stiffness is not reduced by factors like dissociation/re-association or equilibration among different cross-bridge states can only be verified when fibre stiffness has been recorded over a wide range of stretch/release velocities and stiffness is found independent of changes in stretch/release velocity.

Analogous controls have to be performed when applying stepwise length changes or sinusoidal length perturbations. Otherwise, changes in fibre stiffness could simply result from changes in, for example, dissociation kinetics of cross-bridges that shift the stiffness–speed relationship along the abscissa, while the speed independent stiffness, and thus possibly the fraction of attached cross-bridges, might be unchanged.

Cross-bridge stiffness in different states

The assumption that stiffness of a cross-bridge is the same for all conditions (e.g. for all nucleotide states) is difficult to assess. However, probing the component of fibre stiffness in radial direction by resistance of the myofilament lattice to osmotic compression, it was found that this radial component has distinctly different magnitudes for the various nucleotide states of the cross-bridge (Xu *et al.* 1993; Brenner *et al.* 1996). Thus, it seems quite possible that the axial component of cross-bridge stiffness also varies quite significantly for different nucleotide states. It is hoped that mechanical measurements on individual myosin molecules will provide the answer to this question, once approaches have been developed that allow to cope with compliances outside the myosin head in these single molecule assays.

The question of cross-bridge attachment in rigor has been addressed in several studies (Cooke and Franks 1980; Cooke and Thomas 1980; Lovell and Harrington 1981). It is usually thought that, at least in rabbit skeletal muscle, basically all cross-bridge are attached under rigor conditions.

As pointed out elsewhere (Brenner 1990), estimates of the fraction of attached cross-bridges needed to determine actin affinity of cross-bridges, if possible at all, are useful only over a rather limited range, i.e. over the range from some 10% attachment to 90% attachment, corresponding to a binding constant in muscle of some 0.1–9.0. Any binding constant outside this range would be unrealistic to estimate from the small additional change in the fraction of attached cross-bridges.

2.11 Analysis of cross-bridge cycling kinetics

To form a complete understanding of cross-bridge action in muscle, detailed characterization of the kinetics of the various reaction steps within the cross-bridge cycle is essential. How mechanical studies can be used to characterize actin binding kinetics for the various cross-bridge states was demonstrated in the discussion of possible interpretations of stiffness–speed relationships.

The next step in characterizing cross-bridge kinetics is the systematic analysis of cross-bridge turnover. Biochemical studies in solution, as well as mechanical, biochemical and structural studies in skinned fibres, have suggested that in muscle, as in solution, two main groups of cross-bridge states can be distinguished: a group of weak-binding (non-activating) states of the myosin head, and a group of strong-binding (activating) states, i.e. states in which the myosin heads, upon attachment to regulated actin, can activate the regulated actin filament (Taylor 1979; Eisenberg and Greene 1980; Eisenberg and Hill 1985; Hibberd and Trentham 1986). The solution and fibre studies further suggested that active turnover of myosin heads, can be considered as cycling between these two groups of cross-bridge states (e.g. Taylor 1979; Eisenberg and Greene 1980; Eisenberg and Hill 1985; Hibberd and Trentham 1986).

In a systematic analysis of cross-bridge cycling kinetics, the first step is characterization of turnover (and thus distribution) of myosin heads between these two main groups of cross-bridge states, and changes in turnover (distribution) upon different interventions, e.g. changing from isometric to isotonic contraction conditions and vice versa. This analysis will provide the framework for analysis of kinetics of individual reaction steps. Such a framework is necessary to recognize and account for possible changes in the 'background' distribution, i.e. the distribution of myosin heads between these two main groups of states during protocols imposed on fibres for analysis of individual reaction steps. Such changes in background distribution are readily missed if analysis of individual reaction steps in based upon 'reduced' kinetics schemes, i.e. schemes that only account for a few reaction steps of the whole cycle. One typical example is the quick tension recovery in response to stepwise length changes or the equivalent, T-plots recorded during ramp-shaped length changes. These are often discussed only in terms of redistribution among different force-generating cross-bridge states (e.g. Huxley and Simmons 1971, 1973; Kuhn *et al.* 1979; Brenner 1991), while possible transitions from and to the weak-binding states are ignored, in spite of evidence (experimental and modelling) for such transitions, at least under some conditions (e.g. Kawai 1978; Eisenberg *et al.* 1980; Chen and Brenner 1993).

Consideration of active cross-bridge turnover as cycling between two groups of cross-bridge states is an extension of the two-state model of A.F. Huxley (1957). The single force-generating state is replaced by a group of force-generating states, and the single detached state is replaced by a group of weak-binding states. Note that due to low actin affinity, weakly attached myosin heads do not stay attached over appreciable periods of time when strained, i.e. exerting force onto the myofilaments (Hill 1974). Active turnover between two groups of cross-bridge states can still be characterized by two main rate constants which, in accordance to the proposal of A.F. Huxley (1957) are often called f_{app} (for the transition from the group of weak-binding states to the strong-binding states) and g_{app} for the transition from the strong-binding states back to the weak binding states via release of ADP and binding of a new MgATP-molecule. Note that f_{app} and g_{app} take on different meanings in a scheme with more than one attached and one detached state. Both f_{app} and g_{app} are effective composite rate constants for the transitions between the group of weak-binding (non-force-generating) and the strong-binding (force-generating) states. Thus, these rate constants are not actual association and dissociation rate constants of weakly bound or strongly bound intermediates. For a complete discussion, we also have to keep in mind that the transitions between the two main groups of cross-bridge states are, in principle, reversible reactions and that the reverse rate constants, f^-_{app} and g^-_{app} can become significant, at least under certain experimental conditions (e.g. f^-_{app} at high concentrations of inorganic phosphate, since the reaction sequence described by f^-_{app} involves rebinding of inorganic phosphate; e.g. Taylor 1979; Eisenberg and Hill 1985; Hibberd and Trentham 1986).

Steady-state parameters

One essential steady-state parameter for analysing overall cross-bridge cycling kinetics is the measurement of ATPase activity in skinned fibres. For measurements of the steady-state ATP-hydrolysis, several approaches have been developed:

1. Measurement of products of ATP-hydrolysis, e.g. inorganic phosphate (Schädler 1967) of ADP (Kushmerick and Krasner 1982; Krasner and Kushmerick 1983; Ferenczi *et al.* 1984).

2. Coupled enzyme reaction systems in which rephosphorylation of ADP is coupled to a reporter system. One approach is rephosphorylation of ADP by the phosphoenolpyruvate/pyruvate kinase system and coupling of this reaction to the reduction of pyruvate to lactate by lactate dehydrogenase using NADH H^+. The change in absorption (or fluorescence) when NADH is oxidized is directly proportional to the hydrolysis of ATP (e.g. Takashi and Putnam 1979; Glyn and Sleep 1985).

In the first approach (monitoring the hydrolysis product ADP), no ATP-backup is possible. Thus, this method might appear inferior to the coupled enzyme system for this reason. However, in the coupled enzyme system, the concentration of phosphoenolpyruvate used is usually 2 mM (e.g. Glyn and Sleep 1985), consequently, in the presence of millimolar MgATP concentrations, the ATP-backup does not much improve possible gradients in [MgATP] inside the fibres generated by hydrolysis of MgATP (cf. Yu and Podolsky 1990). An advantage of the coupled enzyme system over the recording of ADP or phosphate, however, is the possibility of continuous monitoring of the ATPase activity (e.g. Glyn and Sleep 1985). Limited enzyme activity and diffusion of products, however, limit the time resolution of this approach.

To correlate the observed ATP hydrolysis with the ATPase activity of the contractile apparatus, other sources of non-cross-bridge ATPase activities have to be ruled out (e.g. ATPase activity of vesicles of the sarcoplasmic reticulum, mitochondrial ATPase activity). To do so, inhibitors have been introduced (e.g. dibucaine and sodium azide; cf. Krasner and Kushmerick 1983), or ATPase activity in full filament overlap is compared with ATPase activity recorded when fibres were stretched beyond overlap (cf. Ferenczi *et al.* 1984). Assuming that stretching beyond overlap does not affect the non-actomyosin ATPase activities, the observed difference in ATPase activity is thought to represent the actomyosin ATPase. When correlating ATPase activity with active force at different levels of Ca^{++}-activation, incomplete relaxation might result in non-parallel behaviour, i.e. at low force levels the ATPase activity is higher than expected if ATPase and force change in proportion. To minimize this problem, in our experiments, we only use fibres that have been kept in skinning solution for at most 2 days (cf. Kraft and Brenner 1997).

Pre-steady-state parameters

For analysis of cross-bridge cycling kinetics, e.g. the magnitude of f_{app} and g_{app}, measurements of steady-state parameters only (e.g. force, fibre stiffness, ATPase activity, etc.) are not sufficient to arrive at unique conclusions about the magnitude of these rate constants or their changes when experimental conditions are modified. Instead, pre-steady-state parameters are necessary to complement steady-state data. One approach to obtaining pre-steady-state kinetic data about overall cross-bridge cycling kinetics from mechanical interventions is analysis of the time course of 'force re-development' (Brenner 1985*b*).

The protocol for recording of the rate constant of force re-development is illustrated in Fig. 2.5. An isometrically contracting fibre is suddenly unloaded, so that unloaded (or nearly unloaded) isotonic contraction conditions are established. After some 100–200 msec of isotonic contraction, the fibre is suddenly re-stretched to its initial isometric sarcomere length (to re-establish initial overlap conditions). After a short tension spike, active isometric force starts to re-develop from a low but non-zero level. Parallel recording of stiffness reveals that during low-load shortening, stiffness is low and re-increases in parallel with active force during the period of force re-development. Essentially the same time course of re-increase in force is observed when fibres are held at constant length (isometric conditions) without re-stretch to the initial sarcomere length, although after active shortening over $>5\%$ of initial fibre length, an increasing deficit in force re-development is seen (Brenner and Eisenberg 1986). As is also shown in Fig. 2.5, the time course of force re-development is closely fitted by a single exponential function.

To observe this behaviour, a well-preserved sarcomeric structure is essential. As the striation pattern deteriorates, the transient force spike becomes larger and does not decay as rapidly to low levels but stays rather high, thus obscuring the period of force re-development. This appears to be due to overstretched sarcomeres with reduced contribution to active force and active shortening, i.e. deterioration of force re-development usually goes in parallel with an increased

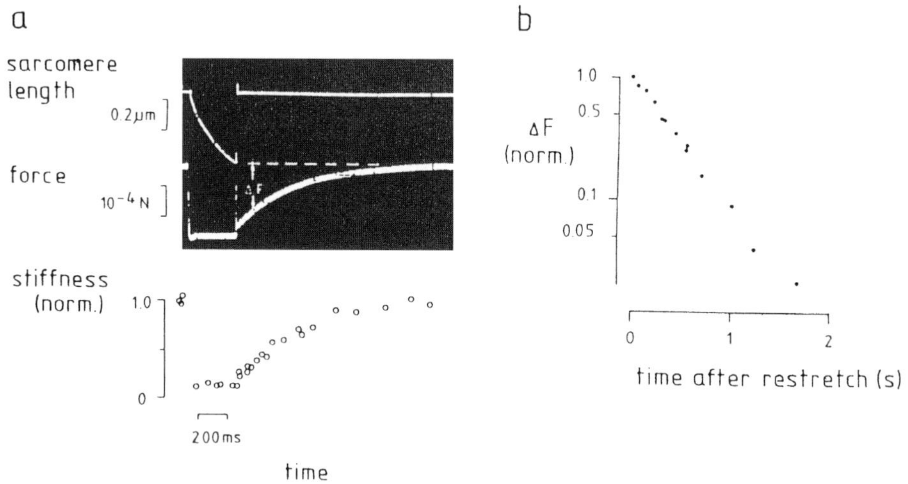

Fig. 2.5 Experimental protocol for recording of force re-development (a) Force, sarcomere length, and fibre stiffness versus time. Maximum Ca^{++}-activation, 5 °C, 170 mM ionic strength. Stiffness is apparent fibre stiffness recorded during stretches. Speed of stretches 10^3 to 5×10^3 (nm/half-sarcomere) s^{-1}. (b) Time course of force re-development. Difference between instantaneous force and isometric force level (ΔF) plotted versus time after re-stretch. ΔF normalized to ΔF immediately after the re-stretch. (Modified from Brenner 1990.)

amplitude of the instantaneous length change and reduced active shortening in isotonic releases.

Also important is careful control of sarcomere length during the period of force re-development in order to ensure isometric conditions. This is because the time course of force re-development was found to be rather sensitive to changes in sarcomere length during re-development of active tension (non-isometric conditions). Thus, some kind of sarcomere-length-controlled feedback system is required. Fixation of fibre ends according to Chase and Kushmerick (1988) helps to reduce changes in sarcomere length during force re-development, such that sarcomere-length-controlled feedback can readily be applied even during maximum Ca^{++}-activation.

It has been proposed that the rate constant of force re-development represents the redistribution of cross-bridges among non-force-generating (weak-binding) and force-generating (strong-binding) states after cross-bridges have been shifted toward non-force-generating (weak-binding) states during the period of high-speed shortening (Brenner and Eisenberg 1986; Brenner 1988). Retardation of force re-development by reactivation of thin filaments that might have become deactivated to some extent during high-speed shortening (Millar and Homsher 1990) or during the re-stretch manoeuver (Swatz and Moss 1992), was shown to be unlikely by several controls. These controls include measurement of force re-development after shortening under substantial load (Brenner 1988) up to the analysis of the kinetics of changes in thin filament activation upon sudden changes in strong cross-bridge attachment (Brenner *et al.* 1997). This analysis revealed that for all conditions studied, changes in thin filament activation, if detectable at all, are more than an order of magnitude faster than force re-development.

Thus, assuming that force re-development represents the time course of redistribution of cross-bridges among non-force-generating and force-generating states under isometric conditions, the rate constant of force re-development is expected to represent, in principle, the sum of the rate constants that determine the transitions between non-force-generating states and force-generating states, i.e. $f_{app} + g_{app}$, if the reverse rate constants can be ignored (e.g. if inorganic phosphate is minimized by sucrose and sucrose phosphorylase; cf. Pate and Cooke 1989).

One limitation might appear to be the fact that a whole range of rate constants is expected for cross-bridge cycling kinetics, due to different strains experienced by individual attached cross-bridges in the myofilament lattice (mismatch between thin- and thick-filament periodicities). Such a wide spectrum of rate constants is expected to result in substantial deviations from a single exponential time course (cf. model of Huxley 1957; Eisenberg *et al.* 1980). The observed close fit by a single exponential function is unexpected on the basis of the models of Huxley (1957) or Eisenberg *et al.* (1980), and differs in that respect from the time course of phase 4 of the isometric transients (own unpublished results). This was proposed to suggest that under isometric conditions, the strain distribution of cross-bridges might be rather narrow (cf. model in Brenner 1990), e.g. due to the presence of binding sites every 55 Å instead of the some 380 Å assumed in the models of Eisenberg *et al.* (1980) or Huxley (1957). After a stepwise release, a wider distribution of cross-bridge strain is expected from the presence of attached cross-bridges that became unstrained, while newly attaching cross-bridges experience the strain present prior to release. Nevertheless, while detailed characterization of strain-dependence of cross-bridge turnover kinetics requires further investigation, analysis of the time course of force re-development appears to represent a useful approach to obtain the necessary pre-steady-state kinetic data for deriving overall kinetics of active cross-bridge cycling.

Time course of force generation in response to release of ATP from caged-ATP in the presence of Ca^{++}

Using release of ATP from an inactive precursor (caged-ATP) Goldman *et al.* (1984) recorded the time course of force generation upon release of ATP into a rigor fibre in the presence of high Ca^{++}. The time course was later shown to be essentially identical to the time course of force re-development observed with the protocol described above (Burton *et al.* 1989). Thus, the release of caged compounds can also be used to obtain pre-steady-state kinetic data to analyse cross-bridge cycling kinetics. Details on experiments with caged compounds, however, are covered in another chapter of this volume.

Deriving overall kinetics of cross-bridge cycling

For deriving the apparent rate constants for active cross-bridge cycling, f_{app} and g_{app}, the following assumptions are made and should be considered as limitations when interpreting these rate constants:

1. During active contraction, cross-bridges are assumed to cycle between two groups of states of the myosin head. Weak-binding states, in which cross-bridges cannot make significant contribution to force generation (non-force-generating states), and the strong-binding states, in which, when attached, cross-bridges can generate active tension (force-generating states).
2. At least under isometric conditions, cross-bridges experience only a narrow distribution of different strain, i.e. to a first approximation, cross-bridges are kinetically homogenous.
3. For the conditions of interest, the reverse rate constants of f_{app} and g_{app} (i.e. f^-_{app} and g^-_{app}) can be neglected (low inorganic phosphate, isometric conditions where the drop of free energy upon return to the weak-binding states is assumed to be large; Huxley 1957 and all related concepts).
4. Contribution to active force is similar for all states within a group (e.g. insignificant for all weak-binding states) or else, the states within a group that contribute differently are either in rapid equilibrium on the time-scale of active turnover (e.g. strong-binding states) or are not significantly occupied during steady-state contraction.

With these assumptions, the following relations apply (Huxley 1957; Woledge *et al.* 1985; Brenner 1986*b*, 1988). Isometric force is given by:

$$F = n \cdot \bar{F} \cdot f_{\mathrm{app}} / (f_{\mathrm{app}} + g_{\mathrm{app}}),$$

where n is the number of actively cycling cross-bridges per half-sarcomere, $\bar{F}$ is the mean force generated by a cross-bridge in the force-generating states. $\bar{F}$ will depend on the amount of force generated in each force-generating state and the distribution among these states. For a procedure to distinguish the two components, see Kraft and Brenner (1997).

Assuming hydrolysis of one ATP-molecule per cross-bridge per cycle, the isometric ATPase of a fibre, measured as production of ADP or inorganic phosphate per second is:

$$ATPase = n \cdot b \cdot f_{\mathrm{app}} \cdot g_{\mathrm{app}} / (f_{\mathrm{app}} + g_{\mathrm{app}}),$$

where b is the number of half-sarcomeres in the fibre.

Assuming that f^{-}_{app} and g^{-}_{app} are not significant, the rate constant of force re-development is:

$$k_{\mathrm{redev}} = f_{\mathrm{app}} + g_{\mathrm{app}}.$$

Note that these relations are independent of the details of a kinetic scheme, e.g. the exact number of states specified in the kinetic scheme, as long as the kinetic scheme is consistent with the above stated four general assumptions, i.e. definition and estimates of the apparent rate constants are not based upon oversimplification into a two-state model. In fact most, if not all, of the common kinetic schemes satisfy the above stated assumptions (if more than a single binding site per 380 Å is assumed on the actin filament). The detailed dependence of f_{app} and g_{app} on the various rate constants and equilibrium constants, however, is characteristic for each of the commonly used schemes.

One frequently used parameter to estimate effects of various interventions on cross-bridge cycling kinetics, specifically on g_{app}, is the tension cost, i.e. the ratio of ATPase to force. This is given by:

$$tension\ cost = b \cdot g_{\mathrm{app}} / \bar{F}.$$

Note that changes in tension cost observed on the same fibre (b = constant) for different experimental conditions, may originate from both changes in g_{app} and/or $\bar{F}$. Thus, measurements of ATPase and force alone do not permit differentiation.

Based on these relationships, attempts have been made to estimate the magnitude of the main apparent rate constants for cross-bridge cycling, f_{app} and g_{app}, by either including effects of changes in substrate concentration (Brenner 1986*b*) or temperature (Kraft *et al.* 1995*a*) on force, force re-development, and ATPase activity. Furthermore, work is in progress to characterize the overall cycling kinetics for unloaded, high-speed shortening, in order to solve the still open, yet central questions, as to what fraction of cross-bridges contribute to force generation under isometric conditions, and to what extent high-speed isotonic shortening affects cross-bridge cycling kinetics.

Kinetic schemes and cross-bridge models

Development of kinetic schemes and cross-bridge models by correlating mechanical, biochemical, and structural data that were obtained from both solution and skinned fibre data have been outlined in an earlier review (Brenner 1990) to which the interested reader is referred. One point, however, may be reiterated here. For construction of kinetic schemes, it is strongly recommended to allow, in a qualitative but visually obvious way, for changes in free energy across various reaction steps, e.g. by using the concept proposed in the earlier review and demonstrated in the kinetic scheme shown in Fig. 2.6 (cf. Brenner 1990).

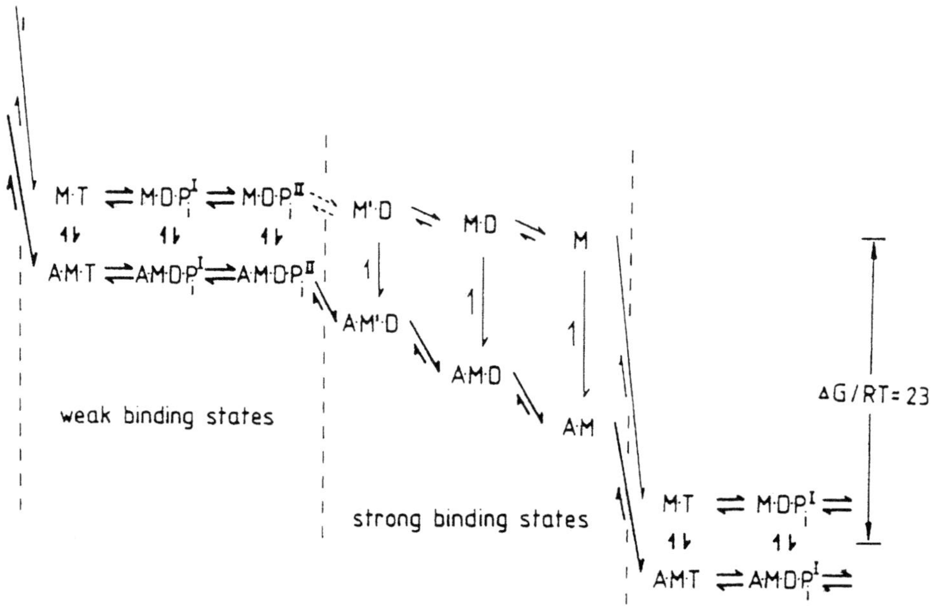

Fig. 2.6 Kinetic scheme of actomyosin ATPase: (*Heavy arrows*) indicate the main pathway. A = actin, M = myosin, D = ADP, T = ATP, P_i = inorganic phosphate. The relative magnitude of the arrows, as usual, illustrates the change in free energy across this reaction step. In addition, the free energy of each state is qualitatively illustrated by its relative vertical position. During one complete cycle a cross-bridge passes from the upper left to the lower right. During this pass the free energy of the system drops by the free energy of ATP-hydrolysis ($\Delta G/RT \approx 23$; cf. Eisenberg *et al.* 1980). (Modified from Brenner 1990.)

Using this principle to directly visualize changes in free energy, although only in a qualitative manner, helps to avoid confusing reversible reactions (during which no change in free energy occurs and thus no net work can be done) with 'cycles' in which the cross-bridge is assumed to do some work. For instance, a cross-bridge detaching from the state AMD to MD⇔M′D and coming back to AMD (through whichever way) has passed through a (series of) reversible reaction(s) ending at the same free energy level. Therefore, there is no way that this cross-bridge has done any net work. If the cross-bridge is supposed to have done some net work, the AMD state in which it ends up has to have a lower free energy level, and thus has to be specified explicitly as a different state in a kinetic scheme of the type shown in Fig. 2.6.

The kinetic scheme illustrated in Fig. 2.6 thus helps to illustrate the fundamental difference between, for example, reversible cross-bridge attachment and multiple power strokes. With reversible detachment/reattachment (e.g. AMD⇔MD⇔M′D) a cross-bridge ends up in the same attached state (AMD) and thus has not generated any net work. In contrast, during multiple power strokes (Lombardi *et al.* 1992), cross-bridges are assumed to generate some net work. Thus, if the above reaction, directly or through additional intermediates, is thought to be one of several power strokes, the AMD-state reached the second time has to be different from the first. Although it still has the same nucleotide bound (ADP) its free energy level ought to be lower and thus this state has to be specified as a separate state.

Multiple step-binding (Marston and Taylor 1980; Geeves *et al.* 1984; Taylor 1991) has not been included in this scheme to limit complexity. Two-step attachment would require specification of attached intermediates between the detached and attached states specified here.

References

Ahnert-Hilger, G., Mach, W., Föhr, K.J. and Gratzl, M. (1989) Poration by alpha-toxin and streptolysin O: an approach to analyse intracellular processes. *Methods Cell Biol.* **31**, 63–90.

Bagni, M.A., Cecchi, G., Colomo, F. and Poggesi, C. (1990) Tension and stiffness of frog muscle fibres at full filament overlap. *J. Muscle Res. Cell Motil.* **11**, 371–7.

Bremel, R.D. and Weber, A. (1972) Cooperation within actin filament in vertebrate skeletal muscle. *Nature (New Biol.)* **238**, 97–101.

Brenner, B. (1980) Effect of free sarcoplasmic Ca^{2+}-concentration on maximum unloaded shortening velocity: measurements on single glycerinated rabbit psoas muscle fibres. *J. Muscle Res. Cell Motil.* **1**, 409–28.

Brenner, B. (1988) Effect of Ca^{2+} on cross-bridge turnover kinetics in skinned single rabbit psoas fibers: implications for regulation of muscle contraction. *Proc. Natl. Acad. Sci. USA* **85**, 3265–9.

Brenner, B. (1983) Technique for stabilizing the striation pattern in maximally calcium-activated skinned rabbit psoas fibers. *Biophys. J.* **41**, 99–102.

Brenner, B. (1985*a*) Sarcomeric domain organization within single skinned rabbit psoas fibers and its effects on laser light diffraction patterns. *Biophys. J.* **48**, 967–82.

Brenner, B. (1985*b*) Correlation between the cross-bridge cycle in muscle and the actomyosin ATPase in solution. *J. Muscle Res. Cell Motil.* **6**, 659–64.

Brenner, B. (1986*a*) The necessity of using two parameters to describe isotonic shortening velocity of muscle tissues: the effect of various interventions upon initial shortening velocity (v_i) and curvature (*b*). *Basic Res. Cardiol.* **81**, 54–69.

Brenner, B. (1986*b*) The cross-bridge cycle in muscle. Mechanical, biochemical and structural studies on single skinned rabbit psoas fibers to characterize cross-bridge kinetics in muscle for correlation with the actomyosin ATPase in solution. *Basic Res. Cardiol.* **81**, (1), 1–15.

Brenner, B. (1990) Muscle mechanics and biochemical kinetics. In *Molecular mechanism of muscular contraction* (ed. John Squire), pp. 77–149. Macmillan Press.

Brenner, B. (1991) Rapid dissociation and reassociation of actomyosin cross-bridges during force generation: a newly observed facet of cross-bridge action in muscle. *Proc. Natl. Acad. Sci. USA* **88**, 10 490–4.

Brenner, B. and Eisenberg, E. (1986) The rate of force generation in muscle: correlation with the actomyosin ATPase in solution. *Proc. Natl. Acad. Sci. USA* **83**, 3542–6.

Brenner, B. and Morano, I. (1990) Effects of myosin light chain phosphorylation on isometric force and cross-bridge turnover kinetics. *Pflüg. Arch.* **415**, R73.

Brenner, B. and Yu, L.C. (1993*a*) Evidence for structural changes in crossbridges during force generation. *Adv. Exp. Med. Biol.* **332**, 461–7.

Brenner, B. and Yu, L.C. (1993*b*) Structural changes in the actomyosin cross-bridges associated with force generation. *Proc. Natl. Acad. Sci. USA* **90**, 5252–6.

Brenner, B., Chalovich, J.M., and Yu, L.C. (1995) Distinct molecular processes associated with isometric force generation and with rapid tension recovery after quick release. *Biophys. J.* **68**, 106s–11s.

Brenner, B., Kraft, T. and Chalovich, J.M. (1997) Kinetics of thin filament activation probed by the fluorescence of IANBD-labeled troponin-I. *Biophys. J.* **72**, A127.

Brenner, B., Xu, S., Chalovich, J.M. and Yu, L.C. (1996) Radial equilibrium lengths of actomyosin cross-bridges in muscle. *Biophys. J.* **71**, 2751–8.

Brenner, B., Chalovich, J.M., Greene, L.E., Eisenberg, E. and Schoenberg, M. (1986) Stiffness of skinned rabbit psoas fibers in MgATP and $MgPP_i$ solution. *Biophys. J.* **50**, 685–91.

Brenner, B., Schoenberg, M., Chalovich, J.M., Greene, L.E. and Eisenberg, E. (1982) Evidence for cross-bridge attachment in relaxed muscle at low ionic strength. *Proc. Natl. Acad. Sci. USA* **79**, 7288–91.

Burton, K., Irving, M. and Sleep, J. (1989) The rate of force recovery after rapid shortening and restretch is equal to that after photolysis of caged ATP at 10 °C in rabbit psoas muscle. *Biophys. J.* **55**, 10a.

Chase, P.B. and Kushmerick, M.J. (1988) Effects of pH on contraction of rabbit fast and slow skeletal muscle fibers. *Biophys. J.* **53**, 935–46.

Chen, Y. and Brenner, B. (1993) On the regeneration of the actin–myosin power stroke in contracting muscle. *Proc. Natl. Acad. Sci. USA* **90**, 5148–52.

Cleworth, D.R. and Edman, K.A.P. (1972) Changes in sarcomere length during isometric tension development in frog skeletal muscle. *J. Physiol. (Lond.)* **227**, 1–17.

Cooke, R. and Franks, K. (1980) All myosin heads form bonds with actin in rabbit rigor skeletal muscle. *Biochemistry* **19**, 2265–9.

Cooke, R. and Thomas, D.D. (1980) Spin label studies of the structure and dynamics of glycerinated muscle fibers applications. *Fed. Proc.* **39**, 1962.

Eastwood, A.B., Wood, D.S., Bock, K.L. and Sorenson, E.E. (1979) Chemically skinned mammalian skeletal muscle. I. the structure of skinned rabbit psoas. *Tissue Cell* **11**, 553–6.

Edman, K.A.P. and Reggiani, C. (1984) Redistribution of sarcomere length during isometric contraction of frog muscle fibres and its relation to tension creep. *J. Physiol. (Lond.)* **351**, 169–98.

Eisenberg, E. and Greene, L.E. (1980) The relation of muscle biochemistry to muscle physiology. *Ann. Rev. Physiol.* **42**, 293–309.

Eisenberg, E. and Hill, T.L. (1985) Muscular contraction and free energy transduction in biological systems. *Science N.Y.* **227**, 999–1006.

Eisenberg, E., Hill, T.L. and Chen, Y. (1980) Cross-bridge model of muscle contraction. *Quantitative Analysis. Biophys. J.* **29**, 195–227.

Ferenczi, M.E., Homsher, E. and Trentham, D.R. (1984) The kinetics of magnesium adenosine triphosphate cleavage in skinned muscle fibres of the rabbit. *J. Physiol. (Lond.)* **352**, 575–99.

Ford, L.E., Huxley, A.F. and Simmons, R.M. (1977) Tension responses to sudden length change in stimulated frog muscle fibres near slack length. *J. Physiol. (Lond.)* **269**, 441–515.

Ford, L.E., Huxley, A.F. and Simmons, R.M. (1981) The relation between stiffness and filament overlap in stimulated frog muscle fibres. *J. Physiol. (Lond.)* **311**, 219–49.

Fuchs, F. (1977) The binding of calcium to glycerinated muscle fibers in rigor. The effect of filament overlap. *Biochim. Biophys. Acta* **491**, 523–31.

Geeves, M.A., Goody, R.S. and Gutfreund, H. (1984) Kinetics of acto-S1 interaction as a guide to a model for the crossbridge cycle. *J. Muscle Res. Cell Motil.* **5**, 351–61.

Glyn, H. and Sleep, J. (1985) Dependence of adenosine triphosphatase activity of rabbit psoas muscle fibres and myofibrils on substrate concentration. *J. Physiol. (Lond.)* **365**, 259–76.

Goldman, Y. (1987) Measurement of sarcomere shortening in skinned fibers from frog muscle by white light diffraction. *Biophys. J.* **52**, 57–68.

Goldman, Y.E. and Huxley, A.F. (1994) Actin compliance: are you pulling my chain? *Biophys. J.* **67**, 2131–6.

Goldman, Y.E., Hibberd, M.G. and Trentham, D.R. (1984) Relaxation of rabbit psoas muscle fibres from rigor by photochemical generation of Adenosine-5′-triphosphate. *J. Physiol. (Lond.)* **354**, 577–604.

Goldman, Y.E., Hibberd, M.G., McCray, J.A. and Trentham, D.R. (1982) Relaxation of muscle fibres by photolysis of caged ATP. *Nature* **300**, 701–5.

Gordon, A.M., Huxley, A.F. and Julian, F.J. (1966) Tension development in highly stretched vertebrate muscle fibres. *J. Physiol. (Lond.)* **184**, 143–69.

Güth, K. and Potter, J.D. (1987) Effect of rigor and cycling crossbridges on the structure of troponin-C and on the calcium affinity of the calcium specific regulatory sites in skinned rabbit psoas fibers. *J. Biol. Chem.* **262**, 13 627–35.

Hellam, D.C. and Podolsky, J.R. (1969) Force measurements in skinned muscle fibres. *J. Physiol. (Lond.)* **200**, 807–19.

Hibberd, M.G. and Trentham, D.R. (1986) Relationships between chemical and mechanical events during muscular contraction. *Ann. Rev. Biophys. Biophys. Chem.* **15**, 119–61.

Higuchi, H., Yanagida, T. and Goldman, Y.E. (1995) Compliance of thin filaments in skinned fibers of rabbit skeletal muscle. *Biophys. J.* **69**, 1000–10.

Hill, A.V. (1970) *First and last experiments in muscle mechanics*. Cambridge University Press.

Hill, T.L. (1974) Theoretical formalism for the sliding filament model of contraction of striated muscle. Part I. *Prog. Biophys. Mol. Biol.* **28**, 267–340.

Huxley, A.F. (1957) Muscle structure and theories of contraction. *Prog. Biophys. Biophys. Chem.* **7**, 255–318.

Huxley, A.F. and Simmons, R.M. (1971) Proposed mechanism of force generation in striated muscle. *Nature* **233**, 533–8.

Huxley, A.F. and Simmons, R.M. (1973) Mechanical transients and the origin of muscular force. Cold Spring Harb. *Symp. Quant. Biol.* **37**, 669–80.

Huxley, A.F., Lombardi, V. and Peachey, L.D. (1981) A system for fast recording of longitudinal displacement of a striated muscle fibre. *J. Physiol. (Lond.)* **317**, 12–13P.

Huxley, H.E., Stewart, A., Sosa, H. and Irving, T. (1994) X-ray diffraction measurements of the extensibility of actin and myosin filaments in contracting muscle. *Biophys. J.* **67**, 2411–21.

Iizuka, K., Ikebe, M., Somlyo, A.V. and Somlyo, A.P. (1994) Introduction of high molecular weight (IgG) proteins into receptor coupled, permeabilized smooth muscle. *Cell Calcium* **16**, 431–45.

Jung, D.W.G., Blange, T., de Graaf, H. and Treijtel, B.W. (1988) Weakly attached cross-bridges in relaxed frog muscle fibers. *Biophys. J.* **55**, 605–19.

Julian, F.J. (1971) The effect of calcium on the force–velocity relation of briefly glycerinated frog muscle fibres. *J. Physiol. (Lond.)* **218**, 117–45.

Julian, F.J. and Morgan, D.L. (1981) Tension, stiffness, unloaded shortening speed and potentiation of frog muscle fibres at sarcomere lengths below optimum. *J. Physiol. (Lond.)* **319**, 205–17.

Julian, F.J. and Sollins, M.R. (1975) Variation of muscle stiffness with force at increasing speeds of shortening. *J. Gen. Physiol.* **66**, 287–302.

Julian, F.J., Sollins, M.R. and Moss, R.L. (1978) Sarcomere length non-uniformity in relation to tetanic responses of stretched skeletal muscle fibers. *Proc. R. Soc. London Ser. B* **200**, 109–16.

Kawai, M. (1978) Head rotation or dissociation? A study of exponential rate processes in chemically skinned muscle fibers when MgATP concentration is changed. *Biophys. J.* **22**, 97–103.

Kobayashi, S., Kitazawa, T., Somlyo, A.V. and Somlyo, A.P. (1989) Cytosolic heparin inhibits muscarinic and alpha-adrenergic Ca^{2+} release in smooth muscle. Physiological role of inositol 1,4,5-triphosphate in pharmacomechanical coupling. *J. Biol. Chem.* **264**, 17 997–8004.

Kojima, H., Ishijima, A. and Yanagida, T. (1994) Direct measurement of stiffness of single actin filaments with and without tropomyosin using *in vitro* nano-manipulation. *Proc. Natl. Acad. Sci. USA* **91**, 12 962–6.

Kraft, T. and Brenner, B. (1997) Force-enhancement without changes in cross-bridge turnover kinetics: the effect of EMD 57033. *Biophys. J.* **72**, 272–82.

Kraft, T., Becker, F., Regel, G. and Brenner, B. (1995*a*) Cross-bridge kinetics in single fibers of human soleus muscle. *Biophys. J.* **68**, A72.

Kraft, T., Chalovich, J.M., Yu, L.C. and Brenner, B. (1995*b*). Parallel inhibition of active force and relaxed fiber stiffness by caldesmon fragments at physiological ionic strength and temperature conditions: additional evidence that weak cross-bridge binding to actin is an essential intermediate for force generation. *Biophys. J.* **68**, 2404–18.

Kraft, T., Xu, S., Brenner, B. and Yu, L.C. (1996) Does lowering ionic strength change weak cross-bridge binding from a non-stereospecific to a stereospecific conformation? *Biophys. J.* **70**, A290.

Kraft, T., Messerli, M., Rothen-Rutishauser, B., Perriard, J.-C., Wallimann, T. and Brenner, B. (1995*c*) Equilibration and exchange of fluorescently labeled molecules in skinned skeletal muscle fibers visualized by confocal microscopy. *Biophys. J.* **69**, 1246–58.

Krasner, B.H. and Kushmerick, M.J. (1983) Tension and ATPase rate in steady-state contractions of rabbit soleus fiber segments. *Am. J. Physiol.* **245**, C405–14.

Kuhn, H.J., Güth, K., Drexler, B., Berberich, W. and Rüegg, J.C. (1979) Investigation of the temperature dependence of the cross bridge parameters for attachment, force generation and detachment deduced from mechano-chemical studies in glycerinated single fibres from the dorsal longitudinal muscle of Lethocerus maximus. *Biophys. Struct. Mech.* **6**, 1–29.

Kushmerick, M.J. and Krasner, B.H. (1982) Force and ATPase rate in skinned skeletal muscle fibers. *Fed. Proc.* **41**, 2232–7.

Leung, A.F. (1983) Light diffractometry for determining the sarcomere length of striated muscle: an evaluation. *J. Muscle Res. Cell Motil.* **4**, 473–84.

Lombardi, V., Piazzesi, G. and Linari, M. (1992) Rapid regeneration of the actin–myosin power stroke in contracting muscle. *Nature* (*Lond.*) **255**, 638–41.

Lovell, S.J. and Harrington, W.F. (1981) Measurement of the fraction of myosin heads bound to actin in rabbit skeletal myofibrils in rigor. *J. Mol. Biol.* **149**, 659–74.

Magid, A. and Reedy, M.K. (1980) X-ray diffraction observations of chemically skinned frog skeletal muscle processed by an improved method. *Biophys. J.* **30**, 27–40.

Marston, S.B. and Taylor, E.W. (1980) Comparison of the myosin and actomyosin ATPase mechanism of the four types of vertebrate muscles. *J. Mol. Biol.* **139**, 573–600.

Mijailovich, S.M., Fredberg, J.J. and Butler, J.P. (1996) On the theory of muscle contraction: filament extensibility and the development of isometric force and stiffness. *Biophys. J.* **71**, 1475–84.

Millar, N.C. and Homsher, E. (1990) The effect of phosphate and calcium on force generation in glycerinated rabbit skeletal muscle fibers *J. Biol. Chem.* **265**, 20 234–40.

Moss. R.L. (1986) Effects on shortening velocity of rabbit skeletal muscle due to variations in the level of thin-filament activation. *J. Physiol.* (*Lond.*) **377**, 487–505.

Nishimura, J., Kolber, M. and van Breemen, C. (1988) Norepinephrine and GTP-gamma-S increase myofilament Ca^{2+} sensitivity in alpha-toxin permeabilized arterial smooth muscle. *Biochem. Biophys. Res. Commun.* **157**, 677–83.

Pate, E. and Cooke, R. (1989) Addition of phosphate to active muscle fibers probes actomyosin states within the power stoke. *Pflügers Arch.* **414**, 73–81.

Persechini, A., Stull, J.T. and Cooke, R. (1985) The effect of myosin phosphorylation on the contractile properties of skinned rabbit skeletal muscle fibers. *J. Biol. Chem.* **260**, 7951–4.

Pfitzer, G. (1996) Permeabilized smooth muscle. In *Biochemistry of smooth muscle contraction* (M. Barany, ed.), pp. 191–9. Academic Press.

Podolsky, R.J. (1960) Kinetics of muscular contraction: the approach to the steady state. *Nature* **188**, 666–8.

Podolsky, R.J. and Teichholz, L.E. (1970) The relation between calcium and contraction kinetics in skinned muscle fibres. *J. Physiol.* (*Lond.*) **211**, 19–35.

Ridgway, E.B. and Gordon, A.M. (1984) Muscle calcium transient effect of post-stimulus length changes in single fibers. *G. Gen Physiol.* **83**, 75–103.

Ridgeway, E.B., Gordon, A.M. and Martyn, D.A. (1983) Hysteresis in the force–calcium relationship in muscle. *Science* **219**, 1075–7.

Rüdel, R. and Zite-Ferenczy, F. (1979) Interpretation of light diffraction by cross-striated muscle as bragg reflexion of light by the lattice of contractile proteins. *J. Physiol.* (*Lond.*) **290**, 317–30.

Rüegg, J.C. (1992) *Calcium in muscle contraction* (2nd edn). Springer, Berlin, Heidelberg.

Schädler, M. (1967) Proportionale Aktivierung von ATPase-Aktivität and Kontraktionsspannung durch Calciumionen in isolierten contractilen Stukturen verschiedener Muskelarten. *Pflüg. Arch.* **296**, 70–90.

Schoenberg, M. (1985) Equilibrium muscle cross-bridge behavior: theoretical considerations. *Biophys. J.* **48**, 467–75.

Schoenberg, M. (1988) Characterization of the myosin adenosine triphosphate (M.ATP) cross-bridge in rabbit and frog skeletal muscle fibers. *Biophys. J.* **54**, 135–48.

Schoenberg, M., Wells, J. and Podolsky, R.J. (1974) Muscle compliance and the longitudinal transmission of mechanical impulses. *J. Gen. Physiol.* **64**, 623–42.

Schoenberg, M., Brenner, B., Chalovich, J.M., Greene, L.E. and Eisenberg, E. (1984) Cross-bridge attachment in relaxed muscle. In *Contractile mechanisms in muscle* (G.H. Pollack and H. Sugi, eds), pp. 269–84. Plenum Press, New York.

Stienen, G.J.M., Güth, K. and Rüegg, J.C. (1983) Force and force transients in skeletal muscle fibres of the frog skinned by freeze-drying. *Pflügers Arch.* **397**, 272–6.

Sundell, C.L., Peachey, L.D. and Goldman, Y.E. (1984) Interpretation of laser light diffraction patterns from single muscle fibers. *Biophys. J.* **45**, 102a.

Suzuki, S., Oshimi, Y. and Sugi, H. (1993) Freeze-fracture studies on the cross-bridge angle distribution at various states and the thin filament stiffness in single skinned frog muscle fibers. *J. Electron Microsc.* **42**, 107–16.

Suzuki, S. and Sugi, H. (1983) Extensibility of the myofilaments in vertebrate skeletal muscle as revealed by stretching rigor muscle fibers. *J. Gen. Physiol.* **81**, 531–46.

Swartz, D.R. and Moss, R.L. (1992) Influence of a strong-binding myosin analogue on calcium-sensitive mechanical properties of skinned skeletal muscle fibers. *J. Biol. Chem.* **267**, 20 497–506.

Szent-Györgyi, A. (1949) Free-energy relations and contractions of actomyosin. *Biol. Bull.* **96**, 140–61.

Takashi, R. and Putnam, S. (1979) A fluorimetic method for continuously assaying ATPase: application to small specimens of glycerol-extracted muscle fibres. *Anal. Biochem.* **92**, 375–82.

Tameyasu, T., Ishide, N. and Pollack, G.H. (1982) Discrete sarcomere length distribution in skeletal muscle. *Biophys. J.* **37**, 489–92.

Taylor, E.W. (1979) Mechanism of actomyosin ATPase and the problem of muscle contraction. *CRC Crit. Rev. Biochem.* **6**, 103–64.

Taylor, E.W. (1991) Kinetic studies on the association and dissociation of myosin subfragment 1 and actin. *J. Biol. Chem.* **266**, 294–302.

van Kaam, F.A.M., deBeer, E.L., Stienen, G.J.M. and Blange, T. (1984) The influence of velocity of length change on tension development in skeletal muscle: model calculations and experimental results. *J. Biomechanics* **17**, 501–11.

Wakabayashi, K., Sugimoto, Y., Tanaka, H., Ueno, Y., Takezawa, Y. and Amemiya, Y. (1994) X-ray diffraction evidence for the extensibility of actin and myosin filaments during muscle contraction. *Biophys. J.* **67**, 2422–35.

Weber, A. (1951) Muskelkontraktion and Modellkontraktion. *Biochim. Biophys. Acta* **7**, 214–24.

Weber, A. and Murray, J.M. (1973) Molecular control mechanisms in muscle contraction. *Physiol. Rev.* **53**, 612–73.

Weber, H.H. and Portzehl, H. (1954) The transference of the muscle energy in the contraction cycle. *Prog. Biophys. Mol. Biol.* **4**, 60–111.

Weber, A. and Weber, H.H. (1951) Zur Thermodynamik der Kontraktion des Fasermodells. *Biochim. Biophys. Acta* **7**, 339–48.

Winegrad, S. (1971) Studies of cardiac muscle with high permeability to calcium produced by treatment with ethylenediaminetetraacetic acid. *J. Gen. Physiol.* **58**, 71–93.

Woledge, R.C., Curtin, N.A. and Homsher, E. (1985) *Energetic aspects of muscle contraction.* Academic Press, Orlando, FL.

Xu, S., Brenner, B. and Yu, L.C. (1993) State-dependent radial elasticity of attached cross-bridges in single skinned fibres of rabbit psoas muscle. *J. Physiol.* (*Lond.*) **461**, 283–99.

Yu, L.C. and Podolsky, R.J. (1990) Equatorial X-ray diffraction studies of single skinned muscle fibers. In *Molecular mechanisms of muscular contraction* (J.M. Squire, ed.), pp. 265–286. Macmillan Press, London.

Zite-Ferenczy, F., Häberle, K.D., Wilke, W. and Rüdel, R. (1984) Light diffraction of skeletal muscle explained with the Ewald formalism. *Biophys. J.* **45**, 10a.

3 The use of caged compounds

Carl Morris and Earl Homsher

3.1 Introduction

Studies of the mechanism of ATP hydrolysis by myosin, heavy meromyosin (HMM), myosin subfragment 1 (S1), and acto-S1 have shown that the mechanism involves four sequential steps: ATP binding, ATP cleavage, Pi release, and ADP release (Goldman 1987; Taylor 1979). The same basic mechanism seems to be used by all molecular motors—myosins, kinesins, and dyneins (Hackney 1996; Howard 1996)—involved in chemo-mechanical transduction. Binding studies of actin to myosin have revealed that myosin binds to actin in two general ways: strongly bound states (in which the equilibrium constant for binding is $> 10^5\,\mathrm{M}^{-1}$) and weakly bound states in which the equilibrium constant is $< 10^5\,\mathrm{M}^{-1}$ (Chalovich 1992). Whether or not the actin molecule is strongly bound to the myosin depends on the substrates or products bound to the myosin. A general and over-simplified mechanism is shown in Fig. 3.1 and is represented by six steps. In this figure, strongly bound species appear in the upper line and weakly bound species in the lower line. In the equation, actin is represented by A and the S1 portion of myosin by M. Table 3.1 lists the rate constants for each of these labelled steps (1–6) using the designation k_{+i} for the ith rate constant going to the right, k_{-i} for the ith rate constant going to the left, and K_i for the equilibrium constant (k_{+i}/k_{-i}) of the ith step.

Table 3.1 shows that the rates of steps 1 and 2 are fast (the pseudo-first order rate constant, $K_1 k_{+2}$, for ATP-induced AM dissociation is about $1 \times 10^6\,\mathrm{M}^{-1}\,\mathrm{s}^{-1}$), indicating that at physiological MgATP concentrations, the rate of dissociation will be $> 5 \times 10^3\,\mathrm{s}^{-1}$. The equilibrium constant for ATP binding is large and provides the energy to promote actomyosin dissociation (White and Taylor 1976).

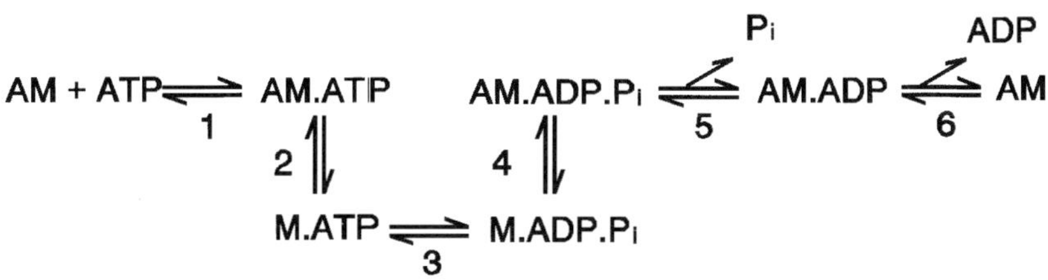

Fig. 3.1 The reaction mechanism for acto-S1 ATPase (see text for details).

Table 3.1 Rate constants for ATPase reaction steps of solution biochemistry (acto-S1 and acto-HMM and myofibrils) and skinned muscle fibres (rabbit psoas fibres)

Rate constants (units)	Temp (°C)	Condition (pH)	Ionic strength (mM)	Acto-S1 and acto-HMM	Myofibril native	Myofibril X-linked	Rabbit psoas fibres
	5[11]	7.0	<50	0.9	0.6	–	–
	10[11]	7.0	<20	1.5	–	–	–
K_1k_{+2}	13[7]	7.0	200	–	–	–	0.2–1.0*
$M^{-1}s^{-1}$	20[11]	7.0	50	2.7	3.0	–	–
$(\times 10^6)$	20[10,11]	7.0	150	1.0	1.1	–	–
	10[11]	7.0	<20	40.0	55.0	–	–
	13[5]	7.0	200	–	–	–	60.0*
$k_{+3}+k_{-3}$	15[3]	7.0	100	55.0	–	–	–
(s^{-1})	20[11]	7.0	<20	85.0	140.0	–	–
	20[10,11]	7.0	100	125.0	140.0	–	–
	20[1]	8.0	150	250.0	–	–	–
	10[11]	7.0	<20	1.5	1.2	–	–
K_3	13[5]	7.0	200	–	–	–	5.0*
(unitless)	20[11]	7.0	<20	2.4	2.8	–	–
	20[1]	8.0	150	9.0	–	–	–
	10[4]	7.0	200	–	–	–	20.0*
k_{+5}	10[9]	7.0	200	–	–	–	$>100.0^{\dagger}$
(s^{-1})	20[11]	7.0	100	140.0	–	–	–
	20[13]	7.0	10–50	325.0–200.0	–	–	–
k_{+6}	10[4]	7.0	200	–	–	–	<10.0*
(s^{-1})	10[9]	7.0	200	–	–	–	$>20.0^{\dagger}$
	5[8]	7.4	<20	1.3	–	–	–
	5[2,8]	7.4	150–200	–	1.7	0.85	1.0*
	10[8]	7.4	<20	2.4	–	–	–
k_{cat}	10[8]	7.4	150	–	3.4	1.3	–
(s^{-1})	10[12]	7.0	150	–	–	–	0.8*
	13[5a]	7.0	200	–	–	–	1.9*
	15[6,8]	7.0	150–200	–	4.2	1.8	1.8*
	20[8,14]	7.0	150–200	–	8.2	3.5	3.2*
	20[8]	7.4	<50	15.0	–	–	–

* Isometric contraction
† Isovelocity contraction
References for data: (1) Bagshaw and Trentham 1973; (2) Brenner 1988; (3) Chock *et al.* 1979; (4) Dantzig *et al.* 1992; (5) Ferenczi and Spenser 1986; (5a) Ferenczi *et al.* 1984; (6) Glyn and Sleep 1985; (7) Goldman *et al.* 1984*a*; (8) Herrmann *et al.* 1994; (9) Homsher *et al.* 1997; (10) Johnson and Taylor 1978; (11) Ma and Taylor 1994; (12) Pate *et al.* 1993; (13) Taylor 1991; (14) Webb *et al.* 1986

The ATP cleavage step in which water oxygen is incorporated to the gamma-O-P bond, occurs during step 3 in an equilibrium reaction which is readily reversible and rapid, and dependent on temperature and ionic strength (see data in Table 3.1). The rates of these reactions indicate that at 20 °C the rate of ATP cleavage by actomyosin at physiological MgATP will occur at a rate of about $150\,s^{-1}$. The re-binding of actin to myosin in step 4 is rapid and dependent on the actin concentration (Brenner *et al.* 1982). In the fibre lattice the effective actin concentration near the S1 head is estimated to be about 1 M (Eisenberg *et al.* 1980).

In solution there are no mechanical forces constraining the attachment and motion of the actin to S1.ADP.Pi. The release of Pi is thought to be the rate limiting step for the ATP hydrolysis cycle, and is estimated to be $> 100\,s^{-1}$ in a process known to be reversible (Chalovich *et al.* 1981; Ma and Taylor 1994; Webb *et al.* 1986). Despite this, a large free energy production (> 10 kJ/mole) accompanies the Pi release, and leads to the suggestion that the Pi release step is associated with the 'power stroke'—that step in the cross-bridge ATPase cycle in which the cross-bridge applies mechanical work to the external world (by the development of force or movement of the thick and thin filaments with respect to each other). The Pi release step is reversible because ^{32}P inorganic phosphate (Pi) can be incorporated into unlabelled ATP when added to a solution of actomyosin (Bagshaw and Trentham 1973; Bowater and Sleep 1988; Sleep and Hutton 1980). Further there is ^{18}O water exchange with the oxygen molecules of the gamma-Pi of ATP (Webb *et al.* 1986). Release of ADP occurs subsequent to Pi release and involves a step which it essentially irreversible (Sleep and Hutton 1980). The release of ADP from AM.ADP in solution is thought to be quite rapid ($\sim 300\,s^{-1}$) in the absence of strain on the cross-bridge (see Table 3.1). As discussed below, there is good evidence that the ADP release is greatly slowed when stress or strain is applied to the cross-bridge. Accurate measurements of the post-hydrolytic steps in solution are approximate because they are fast. However, understanding how fast and to what extent these individual steps occur *in the fibre*, and how they are modified by mechanical, biochemical, and genetic differences are central to our understanding of muscle function and regulation.

3.2 The use of caged compounds to study contraction

A variety of different methodologies have been used to probe the kinetics of cross-bridge transitions, e.g. rapid force or displacement transients, transients associated with X-ray diffraction, ATP hydrolysis by contracting muscle fibres, temperature jumps, and pressure jumps (Fortune *et al.* 1994; Goldman *et al.* 1987; Huxley and Faruqi 1983; Huxley and Simmons 1973; Irving *et al.* 1992; Knight *et al.* 1993). However, these types of approaches are limited in that they do not target specific steps of the cross-bridge mechanism and, as such, produce data which are difficult to interpret. A major advance in our understanding of the mechanism of ATP hydrolysis occurred with the application of transient kinetic analysis to the study of isolated proteins in solution (Lymn and Taylor 1971; Trentham *et al.* 1976). However, it was only in the last decade that this technique became available to probe specific steps of the fibre ATPase cycle using caged-ATP, caged-ADP, and caged-Pi (Goldman 1987; Homsher and Millar 1990). The term 'caged' refers to the covalent bonding of a 1-(2-nitrophenylethyl) group (NPE) or a 3′,5′ dimethyoxybenzoin (DMB) ester to phosphate or the terminal phosphate moiety of ATP or ADP. These covalently attached cages reduce the binding of ATP, ADP, or Pi to the active site in the myosin molecule and thus render a physiologically inert compound. For experiments, one of these compounds is added to the muscle fibre lattice (by diffusing the molecule into a mechanically or chemically skinned muscle fibre or by micro-injecting the molecule into the fibre or solution) with no significant effects on the muscle force or shortening. After equilibration the muscle volume is exposed to a pulse of light produced by a laser or a flash-lamp (at 320–350 nm using > 5 mJ of energy per cm length of muscle fibre; see Goldman (1986) and McCray and Trentham (1989) for excellent discussions of the technical details of these procedures. The goal is to cleave the caging compound photolytically from the substrate or product and thus rapidly (in several ms) increase the substrate or product concentration. One can thus provoke a synchronized physiological or biochemical response in the fibre in a fashion comparable to transient kinetic studies in solutions. There are a number of factors that should be considered if this strategy is to be used successfully, and we briefly consider some of them and discuss avoidable pitfalls.

1. Effective use of this methodology depends on producing a known amount of product at a specific point in time. To do this one must be sure that the number of photons produced is both reproducible and sufficient to induce photolysis in each flash. It is, therefore, imperative that the light output at the relevant wave lengths (320–350 nm) be frequently measured to assure that the extent of photolysis is as expected. Frequency doubled dye or ruby lasers are often used for photogeneration and the efficiency of the doubling is dependent on temperature, so frequent measurement of the secondary beam (or if a flash-lamp is being used, the exciting beam) is important. It is also important to test the uniformity of the laser flash using laser-flash alignment paper and an energy meter to obtain a consistent burn pattern. Similarly, if the beam is modified or shaped in any way, it is important to place laser-alignment paper at the site to be illuminated, to assure that the beam is striking the target and has the desired shape. If a flash-lamp is being used, the lamp should be put in the repetitive mode and the shaped image at the target visually examined. With lasers one must avoid striking the t-clips used to mechanically attach the fibre to the force and displacement transducers, because when light strikes a t-clip a great deal of mechanical noise is produced, which makes accurate force recordings impossible. This restriction complicates experiments in muscles whose length is changing with time.

2. Measurement of the extent of photolysis is done by recovery of the bathing solution and analysis by HPLC (Dantzig *et al.* 1991; Dantzig *et al.* 1992; Goldman *et al.* 1984*a*). Although time-consuming and involving some inherent limitations on accuracy, this approach assures that the change in [ATP], [ADP], or [Pi] is as anticipated. Alternatively, if a change of contractile behaviour (e.g. force, shortening velocity, stiffness) can be correlated with a specific change in concentration separate from the parameter studied, that correlation can be used as a measure of the photolytic conversion.
3. The methodology assumes that the caged compound itself does not alter the behaviour of the muscle. However, it has been observed that incubation of a fibre in caged-ATP and caged-Pi inhibits isometric force to a variable extent (Dantzig *et al.* 1992; Ferenczi *et al.* 1984; Ferenczi and Spencer 1986; Goldman *et al.* 1984*a*; Goldman *et al.* 1984*b*). This effect may be produced by spontaneous decay of the compound, the formation of contaminant Pi, contamination by agents involved in the synthesis of the caged moiety or by-products of spontaneous photolysis, or direct binding of the caged compounds to the proteins. The refinement of the synthesis of caged compounds (Walker *et al.* 1988), involving the aqueous partitioning of product from an ether phase, brought a major reduction in the inhibitory effect of caged-ATP and caged-Pi on muscle force. Furthermore, it is good practice to column-purify caged compounds by preparative HPLC followed by passage through a DE-52 column with gradient elution (from 10 mM to 1000 mM triethylammonium bicarbonate) followed by thorough desalting by rotary evaporation (Walker *et al.* 1988). After storage of caged compounds at $-20\,^{\circ}$C for several months, column purification is essential and reduces contaminant photolytes to trace values ($<1\%$). Recently, Thirwell *et al.* (1995) have shown that in 100 μM ATP (without an ATP regenerating system), 2 mM NPE caged-ATP reduced unloaded shortening velocity by $\sim 30\%$. It would be useful to determine the caged-ATP effects at 5 mM ATP *and* a regenerating system to better evaluate possible binding of caged-ATP to the active site in the S1. Thus it may be that caged ATP can bind to the active site in myosin (albeit 1000-fold less strongly than ATP). Caged-Pi does not affect the unloaded shortening velocity in skinned muscle fibres.
4. By-products of photolysis represent a serious problem in obtaining reliable data (Kaplan *et al.* 1978). Millimolar changes in ATP or Pi are often desired, and with each mole of product generated from NPE-caged compounds, one mole of protons and one mole of 2-nitrosoacetophenone are generated. This latter agent reacts with SH groups so proteins can be protected from SH modification by inclusion of generous amounts ($\sim$30 mM) of DTT or glutathione at the site of photolysis. Similarly high concentrations ($\sim$100 mM and an appropriate pK) of H^+ buffers are added to prevent significant changes in pH subsequent to photolysis. This aspect is particularly important when

EGTA-containing calcium buffers are present. If one is photogenerating relatively small amounts of ATP and it is not possible to provide pH and SH buffering agents (e.g. at intracellular locations), experiments in which neutral products (such as AMP) can be generated in amounts comparable to those of ATP, Pi, or ADP to control for the effects of photolytic by-products should be performed.

5. Photogeneration of ATP and ADP from NPE-cages is relatively slow ($\sim 120\,s^{-1}$) and is not a truly instantaneous step increase in substrate or product concentration (Goldman *et al.* 1984*a*; Goldman *et al.* 1984*b*). Recently, Trentham's group has synthesized a DMB-caged ATP which is photolysed at a rate of $\sim 8 \times 10^4\,s^{-1}$ and has been used in the study of actomyosin dissociation in skinned muscle fibres (see below; also Thirwell *et al.* 1994). While this is a welcome development, DMB-caged ATP has two drawbacks. First it is not very stable, spontaneously forming about 1% ADP every 40 h. If the initial concentration of the DMB-caged ATP is 10 mM, then about 100 μM ADP will be present in these solutions a day and a half after purification. This contamination is circumvented by the use of apyrase to convert contaminant ADP and ATP to AMP and Pi (Thirwell *et al.* 1994). Second, both the quantum yield (0.3) and extinction coefficient ($170\,M^{-1}\,cm^{-1}$) for photolysis of DMB-ATP are two to four times less than those for NPE-caged ATP, thus effectively limiting the extent of conversion per flash to < 1 mM.

3.3 Studies of the ATPase mechanism in fibres using caged-ATP

ATP binding and cross-bridge dissociation

In a seminal series of studies, Goldman and his colleagues characterized contractile behaviour following photogenerated ATP in muscle fibres in the presence and absence of calcium (Goldman *et al.* 1984*a*; Goldman *et al.* 1984*b*) and at various concentrations of Pi (Hibberd *et al.* 1985). These experiments showed that the decay of force and stiffness after photogeneration of ATP in the absence of Ca^{2+} from rigor followed a complicated time course. There was an initial decline in force, followed by a plateau or rise, and ending with a further decay of force having a fast and slow phase. These results were explained by an initial detachment of cross-bridges after ATP binding, followed by a rise or maintenance of force as the detached cross-bridges hydrolysed ATP and reattached to the thin filaments, activated by the presence of the remaining rigor cross-bridges. As progressively more cross-bridges bound ATP and detached, force fell and the thin filaments were deactivated. Plots of the rate of force decline versus the concentration of ATP photogenerated revealed that the apparent second-order rate constant for ATP binding was $0.2\text{–}1 \times 10^6\,M^{-1}\,s^{-1}$ (Goldman *et al.* 1984*a*).

Subsequent studies have basically confirmed these rates and interpretation under a variety of conditions (Emoto *et al.* 1995; Horiuti *et al.* 1993; Horiuti *et al.* 1994). It has also been shown that exposure of fibres to caged-ATP solutions containing apyrase essentially eliminated the plateau or rise in force following ATP photogeneration. This result suggests that the plateau was produced by the presence of contaminant ADP bound to the cross-bridges, as such ATP binding to the cross-bridge is slowed by the dissociation of the ADP from the cross-bridge, as earlier shown by Dantzig *et al.* (1991). Additional studies using DMB-caged ATP showed a two to three-fold more rapid decline in the rate of the cross-bridge dissociation than when NPE-caged ATP was used. Thus the apparent second-order rate constant for ATP binding to fibre cross-bridges may be as high as $0.4\text{–}3 \times 10^6\,M^{-1}s^{-1}$. This result compares well with the $1\text{–}3 \times 10^6\,M^{-1}s^{-1}$ seen in purified acto-S1 or myofibrillar studies (see Table 3.1). These results suggest that the geometric constraints of the filament lattice exercise little effect on the binding of ATP to cross-bridges. Application of positive strain to the cross-bridges in fibres slightly increased the rate of ATP binding (Dantzig *et al.* 1991). Negatively strained cross-bridges, however, detach six to eight times faster than positively strained cross-bridges (Dantzig *et al.* 1991).

Cross-bridge detachment following photogeneration of ATP has also been monitored by using X-ray diffraction to observe the decline in the ratio of the equatorial $I_{11}:I_{10}$ reflections (Poole *et al.* 1991). This ratio change correlates to the movement of mass away from the thin filament. In this case there is a decline in the $I_{11}:I_{10}$ ratio with a half-time of 1 ms ($690\,s^{-1}$) when 2 mM ATP is photogenerated. This corresponds to an apparent second-order rate constant of $3.5 \times 10^5\,M^{-1}s^{-1}$ and is within the range reported above.

ATP cleavage

Ferenczi and his colleagues (Ferenczi *et al.* 1984; Ferenczi and Spencer 1986) have examined the cleavage step of ATP hydrolysis in fibres using caged-^{3}H-ATP photolysis in 'flash 'n smash' experiments. Glycerinated fibres were incubated in caged-^{3}H-ATP in the presence and absence of calcium, flashed to photogenerate ATP, and freeze-clamped between copper blocks at $-180\,°C$ to stop the ATPase reactions. The fibre contents were extracted and the extent of photolysis and the amount of ADP formed were determined by scintillation counting of the HPLC-separated nucleotides. In these studies, the ATP hydrolysis exhibited a burst of about one mole of ADP formed/mole of S1 in the fibre which was followed by a much slower steady-state rate of ATP splitting. The steady-state ATPase in the presence of calcium was $\sim 2\,s^{-1}$ and in the absence of calcium decreased by more than 10-fold. This pattern, a rapid burst followed by a steady-state rate, was similar to that seen in isolated acto-S1 and myofibrillar systems (see Table 3.1). Subsequent experiments showed that the rate of ATP cleavage in fibres was about $60\,s^{-1}$ and had an equilibrium constant of 5, which also fits with studies of isolated proteins and myofibrils (see Table 3.1).

Effects of products on tension transients following ATP photogeneration

In other experiments, Goldman and his colleagues (Hibberd *et al.* 1985) examined the effects of Pi concentration on the rate of relaxation at pCa > 8 or force development at pCa 4 following the photogeneration of ATP. At pCa > 8, the final rate of force decline was $18\,s^{-1}$ (at 20 °C) and rose to $46\,s^{-1}$ at 10 mM Pi. At pCa 4 force rose at a rate of $76\,s^{-1}$ in the absence of Pi and at $140\,s^{-1}$ in 10 mM Pi, despite the fact that the force produced in the latter case was reduced. These results were interpreted in terms of the model given in Fig. 3.1. Hibberd *et al.* 1985 reasoned that if force generation is associated with the release of Pi, then addition of Pi would increase the rate of the reaction from AM.ADP to AM.ADP.Pi, and the rate of force production should be the sum of the forward and backward rates for this step ($k_{+5} + k_{-5}$[Pi]). Similarly, reversal of step 5 provides a pathway out of the force generating state during relaxation (besides flux through the AM state) and thus increasing Pi might increase the rate of relaxation. A variety of additional experiments by others have confirmed these conclusions (see below). The steady-state rate of ATP hydrolysis (k_{cat}, obtained by dividing the rate of ATP hydrolysis in moles/s per unit mass of muscle by the number of moles of S1 heads per unit mass of muscle) in muscle fibres in the isometric case is only about $3\,s^{-1}$ at 20 °C (see Table 3.1). Because the rate of force development is $76\,s^{-1}$, this result suggests that the rate-limiting step for the cross-bridge cycle must follow the release of Pi.

Dantzig *et al.* (1991) performed experiments in which ATP was photogenerated in the presence of various concentrations of MgADP. These experiments showed that force decayed in four different phases. The data was interpreted as indicating that MgADP bound to the rigor cross-bridge with an apparent second-order rate constant of $2\text{–}5 \times 10^5\,M^{-1}\,s^{-1}$ and that the rate of ADP release from the positively strained cross-bridges was about $35\,s^{-1}$, while that from negatively strained cross-bridges ranged from 77 to $400\,s^{-1}$. Because these rates are much greater than the steady-state ATPase (k_{cat}, Table 3.1), the results imply the existence of an AM′.ADP state whose isomerization to AM.ADP is rate limiting and irreversible for the isometric cycle (Sleep and Hutton 1980).

3.4 Studies of the fibre ATPase mechanism using caged-Pi

Isometric contractions

The fact that the elevation of Pi produces a decline in isometric force is of considerable theoretical and practical importance. The hypothesis that the power stroke is associated with the Pi release stems from the fact that the free energy for the Pi release step is large (White and Taylor 1976). Further, as described above, Hibberd *et al.* 1985 have suggested that it was the Pi release that resulted in the power stroke. Finally, the work of Cooke and Pate (Cooke *et al.* 1988; Cooke and Pate 1985; Pate and Cooke 1989*a*,*b*) examined both the ability to develop force and shorten at various [Pi], and produced the suggestion (and a cross-bridge model) that was consistent with the notion of a force generating Pi release. The practical importance of this idea is that during strenuous exercise the myoplasmic [Pi] increases, which will inhibit force production and in part accounts for muscle fatigue. This cross-bridge hypothesis makes several predictions: if the power stroke occurs as the result of Pi release, then:

(1) force decline following a step increase in [Pi] should begin immediately after the [Pi] jump;

(2) the rate of force decline following a step increase in [Pi] should be exponential;

(3) the rate of force decline following a step increase in [Pi] should be linearly dependent on the [Pi].

Tests of these predictions were made by incubating skinned muscle fibres in a solution containing caged-Pi and MgATP at pCa 4 (Dantzig *et al.* 1992). The force was allowed to rise to an isometric steady-state and then 1 mM Pi was photogenerated in the fibre lattice. The results of this experiment are shown in Fig. 3.2 (trace a). Force does decline exponentially at a rate of about $22\,s^{-1}$ at 2 mM final Pi concentration (there is about 1 mM Pi contamination in the solutions prior to the flash). However, there is a 2–5 ms lag after Pi photogeneration during which force does not change and which is inconsistent with the model.

When the [Pi] is raised to examine the dependence of the phosphate transient rate (k_{Pi}), the increase in k_{Pi} was not linear with [Pi] but exhibited hyperbolic behaviour and a plateau near $100\,s^{-1}$ at high [Pi] at 10 °C. An interpretation which accounts quantitatively for this behaviour is the idea that the force-generating power stroke occurs during isomerization of the strongly bound AM.ADP.Pi state and is stabilized by the release of Pi (in a rapid equilibrium dissociation) immediately after the power stroke, as seen in Fig. 3.3. Similar conclusions, using very different experimental approaches, have been arrived at by Geeves and his collaborators (Fortune *et al.* 1991; Fortune *et al.* 1994) and Kawai and his collaborators (Kawai *et al.* 1987; Kawai and Halvorson 1991). The force generating isomerization itself (k_{+f}) is very temperature dependent having a Q_{10} of about 4, while its reverse step (k_{-f}) is little affected by temperature (Dantzig *et al.* 1992). It should be emphasized that the phosphate transient rate (k_{Pi}) is not the same as the rate of force redevelopment following a rapid shortening and restretch cycle (k_{tr}) using the protocol of Brenner (1988). The phosphate transient is at least twice as fast as k_{tr} in cases in which both have been measured in the same fibre under identical conditions (Millar and Homsher 1990; Regnier *et al.* 1995).

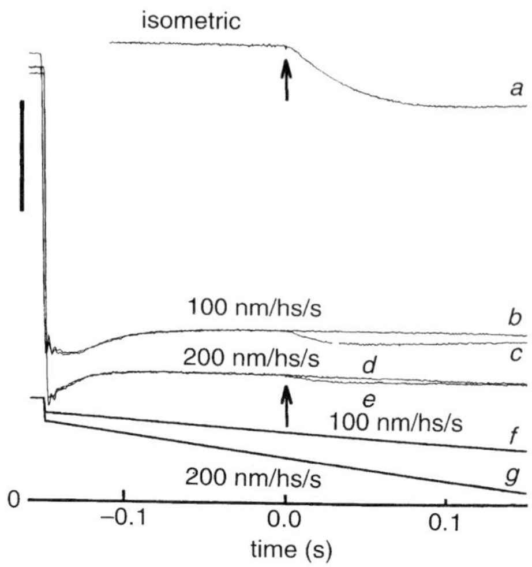

Fig. 3.2 The time course of force during isometric or steady-state shortening before and after the photogeneration of Pi. Record (*a*) is the isometric control; record (*b*) is for a muscle suddenly released by 48.4 μm (1.4% of its initial length), followed by shortening at 100 nm/half-sarcomere/s; record (*c*) is as (*b*), but at zero time, 1 mM Pi is photogenerated; record (*d*) is a muscle shortening at 200 nm/half-sarcomere/s following a sudden release of 71 μm (1.9% of initial muscle length). The *upward arrow* indicates the photogeneration of Pi. The *vertical bar* to the *left* corresponds to a force of 30.4 kN/m^2 for force and 342 μm for the displacement record. The *horizontal line* at the *bottom* of the figure corresponds to zero force and is a time-scale in which the time is given in seconds. Fibre initial length, 3.8 mm at sarcomere spacing = 2.8 μm, cross-sectional area = 3,612 μm^2, isometric force 134 kN/m^2. k_{Pi} for record *a* is 22.7 s^{-1}, for record *c* is 56.6 s^{-1}, and for record *e* is 61.9 s^{-1}.

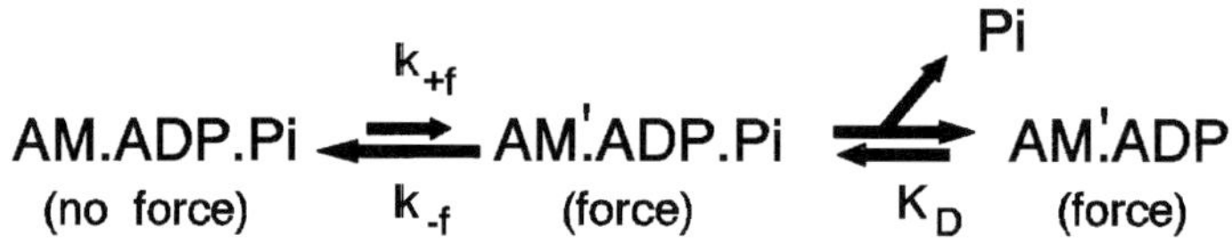

Fig. 3.3 A two-step mechanism for the force generating isomerization (k_{+f} and its reversal, k_{-f}) and the Pi dissociation (K_d is the dissociation constant, here about 10 mM. The typical values for k_{+f} and k_{-f} at 10 °C would be 20 s^{-1} and 80 s^{-1}. The ' after AM indicates that the cross-bridge has gone through its power stroke.

The behaviour of k_{Pi} has been examined in a number of different conditions. First, in slow twitch muscle fibres k_{Pi} is reduced 25–100-fold as compared to that in fast twitch muscle fibres. In slow twitch fibres of the rabbit, k_{Pi} is so slow that the force-generating isomerization itself may be the rate-limiting step for the cross-bridge cycle (Millar and Homsher 1992). The effect pCa on the k_{Pi} has been studied independently by two groups (Millar and Homsher 1990; Walker *et al.* 1992). In these studies, muscle fibres developed force to a steady-state at various pCa, Pi was photogenerated, and the k_{Pi} was measured. Although differing slightly in details, both groups observed that k_{Pi} is little if at all affected by pCa in fast twitch fibres, even though force ranged from 0.1 to 1.0 of its maximal value and k_{tr} changed almost 25-fold. The implication of this result is that the force-generating isomerization is not directly controlled by pCa, and that another step, perhaps strong binding of the cross-bridge to the thin filament prior to force generation, is the step regulated by $[Ca^{2+}]$. This idea has gained experimental support from studies of the effects of butanedione monoxime (BDM) in which force, shortening, k_{tr}, and k_{Pi} have been measured (Regnier *et al.* 1995). Walker's group (Araujo and Walker 1996) has extended their measurements to studies of the calcium dependency of k_{Pi} in cardiac tissue, where pCa does seem to modulate k_{Pi}, although this may also occur through steps immediately before the force-generating isomerization.

Isotonic contractions

Studies of the effects of shortening on the Pi transient have been made at 10 °C (see Fig. 3.2, traces b–e) (Homsher *et al.* 1997). In these studies, the muscle was allowed to contract isometrically in the presence of caged-Pi to a steady-state. Then the muscle was brought to steady-state shortening by a small quick release followed by isovelocity shortening. Pi was photogenerated during isovelocity shortening and the tension transient measured. Figure 3.2 illustrates the observed changes in the Pi transient during shortening:

1. The amplitude of the Pi transient is reduced by the steady-state shortening. This implies that shortening depletes the AM.ADP population in the contracting muscle. This result could occur if the reduction in strain on the cross-bridge during shortening increased k_{+6}, increased k_{-f}, or both.
2. The rate of the Pi transient k_{Pi} increases linearly and markedly with shortening. At a shortening velocity of 300 nm/half-sarcomere/s (~15% of the unloaded shortening velocity, V_{max}, at 10 °C) k_{Pi} is 110 s^{-1}. Extrapolating these data indicates that at V_{max}, k_{Pi} would approach 500 s^{-1}.
3. At a fixed shortening velocity of 100 nm/half-sarcomere/s, k_{Pi} was measured as the [Pi] was varied. k_{Pi} increases with [Pi] in a fashion similar to that in the isometric case, but at each Pi concentration, it is about 35 s^{-1} greater than the isometric value.

These observations have been interpreted as indicating that reduction of strain increases k_{+f} and that the k_d for Pi binding to the AM.ADP state is unaffected by strain, remaining at about 10 mM in the presence or absence of shortening. It has been found that minor modifications of a three-state strain-dependent model proposed by Dantzig *et al.* (1992) to explain the isometric results, accounts for these results observed during shortening. Experiments similar to those above should be performed on other types of muscle.

3.5 Studies of the ATPase mechanism using caged-ADP

Cooke and Pate (1985) showed that ADP increases force production in the isometric case by binding to force exerting AM cross-bridges and competitively inhibiting dissociation of the cross-bridges by ATP. Only a few studies of the effects of photogeneration of ADP on force have been made (Lacktis and Homsher 1987; Lu *et al.* 1991; Lu *et al.* 1993). In these studies, an ATP regenerating system is omitted from the bathing solution because it would modify the ADP perturbation in a time-dependent fashion. However, omission of an ATP regenerating system produces an atypical cross-bridge distribution and a gradient of ADP through the fibre cross-section complicates analysis. In their initial work on the effects of photogenerated ADP, Lacktis and Homsher (1987) showed that the rate of force rise subsequent to ADP photogeneration ($10\,s^{-1}$) is essentially independent of ADP concentration and interpreted it as indicating that ADP bound to the AM cross-bridge and the rate of force rise was produced by the flux of cross-bridges into the AM state. The rate of force rise (k_{ADP}) would be set by the rate-limiting step of the cross-bridge cycle (presumably the isomerization of the AM′.ADP state to the AM.ADP state). Work by Lu *et al.* (1993) extended and better defined the fibre response to step increases in ADP. They showed that the rate of ADP transient, k_{ADP} was only weakly dependent on ADP. Also, increasing [ATP] increased k_{ADP} from about $8\,s^{-1}$ at 2 mM ATP to $10\,s^{-1}$ at 8 mM ATP for a photorelease of 0.5 mM ADP. Elevation of the [Pi] increased k_{ADP} asymptotically from $10\,s^{-1}$ at 0 mM Pi to about $27\,s^{-1}$ at 20 mM Pi. While the former two results are difficult to explain in the terms of the cross-bridge model described above, the latter effect suggests that elevating Pi allows the system to approach the steady-state more rapidly. An increase of ADP in the fibre lattice may form an AM.ADP intermediate normally occurring in the cross-bridge cycle, but it could also represent a branching path which is *not* part of the normal cross-bridge cycle. There is an intriguing report too that k_{ADP} is also modulated by pCa in such a way that the greater the pCa, the slower is the rise of force following a step increase in [ADP] (Lu *et al.* 1991).

3.6 Summary

The view emerging from this series of studies is that the binding, attachment and detachment of M.ATP and M.ADP.Pi to actin, and the cleavage of ATP by S1 heads is similar in isolated proteins and in the organized filament lattice. The release of Pi and ADP appear, however, to be very fast (>100–$500\,s^{-1}$) in the absence of strain (as would be the case in isolated proteins and native myofibrils in solution). Consequently, k_{cat} in such systems at 20 °C will be 15–25 s^{-1} (see Table 3.1). In fibres contracting isometrically and in cross-linked myofibrils, the release of Pi is much slower than in solution and can be resolved as two steps: a slow isomerization whose equilibrium favours the unstrained state, followed by a release of Pi which pulls the equilibrium to the right. This force-generating isomerization rate is strain-dependent, increasing as cross-bridge strain is *reduced*. The release of ADP is also very slow in the strained cross-bridge and k_{+6} must be $< 10\,s^{-1}$ at 10 °C in the isometric case to account for the observed k_{cat}. This step too must be increased during shortening (and the concomitant reduction in cross-bridge strain) to account for k_{Pi} and the increased rate of ATP hydrolysis accompanying shortening and the performance of work.

3.7 Studies using caged compounds to synchronize the cross-bridges

Three series of studies deserve mention in this regard. First there are the studies of ATP hydrolysis during shortening (Higuchi and Goldman 1991; Higuchi and Goldman 1995; Wada *et al.* 1996; Yamada *et al.* 1993). In these studies, fibres in rigor in a caged-^{3}H-ATP solution and in an oil bath were put under a strain, ^{3}H-ATP was photogenerated in them and the fibres were allowed to shorten at selected velocities. By measuring the amount of ATP photogenerated and hydrolysed per myosin head, the distance shortened, and the average stiffness, it was possible to estimate the distance a cross-bridge moved per ATP split (interaction distance). There is a problem with this type of approach in that as shortening must take place in the absence of an ATP regenerating system, there will be a build-up of ADP, which will slow cross-bridge detachment. Nevertheless, Huguchi and Goldman (1991, 1995) and Sugi's group (Wada *et al.* 1996; Yamada *et al.* 1993) have taken pains to conduct the appropriate series of controls so that the results must be taken seriously. The results indicate that the interaction distance increases with decreasing force from about 20 nm at $0.46P_0$ to > 40 nm at $0.07P_0$. These results seem inconsistent with a simple cross-bridge model in which the cross-bridge attaches, moves the thin filament 7–12 nm, then detaches and in the cycle consumes a single ATP. Higuchi and Goldman (1995) have suggested two possible explanations for the result:

1. The cross-bridge does perform a power stroke (exerting positive force) for 12 nm then (at high velocities) is dragged into a negative force-bearing region (which need not be a linear extrapolation of the behaviour at positive strains) and remains attached for an additional distance of about 28 nm before detaching.

2. The cross-bridge may exert multiple power strokes per ATP hydrolysed. In this case, a mechanism needs to be worked out to distribute the energy from the ATP hydrolysis over multiple power strokes.

The recent data from optical trap studies of molecular motors seem to argue against this possibility and favour interpretation 1 (Finer *et al.* 1994; Molloy *et al.* 1995; Svoboda and Block 1994). These experiments also emphasize the central role of stiffness measurements, and their attendant uncertainties, play in the interpretation of this type of data.

Photogeneration of ATP has also been used in a number of other studies to synchronize cross-bridges and examine various aspects of the cross-bridge cycle (Berger *et al.* 1995; Ostap *et al.* 1993; Tanner *et al.* 1992). Photogeneration of ATP has been used in the study of cross-bridge behaviour in fibres whose SH-1 sulfhydryl group (Cys 707) has been covalently reacted with electron spin resonance probes. While the basic result from these studies is that cross-bridge behaviour is not grossly distorted by such labelling, the authors speculate about the extent to which these studies actually resolve the behaviour of the *covalently* modified cross-bridges. The reason for our skepticism is that recent studies from the Reisler laboratory (Bobkov *et al.* 1997) suggest that covalent modification of Cys 707 completely inactivates the cross-bridge, so that the mechanical behaviour observed in these earlier studies may have been produced by the unmodified cross-bridges, while the probe monitored the behaviour of the non-cycling modified cross-bridges.

Recently, an exciting new development has made it possible to monitor the time course of Pi release from the muscle fibre or myofibrils. The periplasmic phosphate binding protein (PBP) of *Escherichia coli* binds Pi tightly ($K_d \sim 0.1\,\mu M$), rapidly ($1.4 \times 10^8\,M^{-1}\,s^{-1}$) and with high specificity. The crystal structure has been solved to high resolution (0.17 nm) and the binding of Pi to the protein involves 12 hydrogen bonds with no salt linkages, covalent bonds or hydrophobic interactions (Luecke and Quiocho 1990). Brune *et al.* (1994) utilized the phosphate binding function of the PBP to generate a fluorescent probe to directly measure Pi release by acto-S1. They mutated the PBP to incorporate a reactive cysteine residue for attachment of a synthesized fluorescent compound. The fluorescent probe, N-[2-(1-maleimidyl)-ethyl]-7-(diethylamino) coumarin-3-carboxamide (MDCC), was reacted with the mutated residue A197C. Upon addition of Pi, the fluorescence intensity of the labelled protein, excited at 425 nm and measured at 464 nm, increased five-fold. A 'Pi mop', 7-methylguanosine and purine nucleoside phosphorylase, was used to remove any contaminant Pi from the solutions. Single turnover and multiple turnover stopped flow experiments were conducted in the presence of the MDCC–PBP and the Pi mop. No Pi burst was observed using MDCC–PBP to directly measure Pi release at 5–22 °C. This strongly suggests that the rate-limiting step for acto-S1 in solution is Pi release or a step preceding it. The Pi release rate was calculated using a four-step model and available data for ATP binding to acto-S1, the ATP cleavage, and k_{cat} and was estimated to be $45\,s^{-1}$ at 22 °C. Work by Lionne *et al.* (1995) in calcium-activated myofibrils at 4 °C showed that there is no Pi burst using MDCC–PBP, while quench flow measurement of the time course of ATP hydrolysis indicated an initial cleavage of 0.84 ± 0.08 mol Pi/mol myosin S1 head at a rate of $20 \pm 2\,s^{-1}$. This result indicates the Pi burst is the formation of AM.ADP.Pi rather than release of free Pi and therefore a build-up of the AM.ADP.Pi state. Cross-linked myofibrils also showed a burst by acid quench-flow ATP splitting but not with the MDCC–PBP. Ferenczi *et al.* (1995) and He *et al.* (1997) used the PBP to measure the rate of Pi release from skinned rabbit psoas muscle fibres at 20 °C. Relaxed fibres (pCa 9) exhibited an ATPase rate (Pi release rate) of $0.24 \pm 0.03\,s^{-1}$. Unexpected behaviour was, however, observed at pCa 4 when the muscle fibre developed force. The fibres were incubated with 600 μM MDCC–PBP and 5 mM NPE-caged-ATP, and when 1 mM ATP was photogenerated the time course of Pi release was best described as an exponential Pi release rate at a rate of $37\,s^{-1}$. Comparable behaviour has been seen in fibres incubated in 2 mM MDCC–PBP. Photogeneration of ATP in this case produced a Pi release that could be approximated by an exponential release of Pi over 500 ms (amounting to 4 moles of Pi for each cross-bridge head) followed by a steady-state Pi release rate of roughly $3\,s^{-1}$. This result has at least two possibilities:

1. At the start of isometric contraction the cross-bridge heads cycle more rapidly and over 500 ms slow to the steady-state rate for reasons that are unknown (He *et al.* 1997).
2. The development of force involves significant internal shortening and work performance in the skinned fibre which obligates an increased rate of ATP hydrolysis.

It will be important to monitor the time course of ATP hydrolysis, Pi release, and sarcomere spacing in fibres in similar experiments to better define the reasons for these surprising results.

Acknowledgement

This work was supported by the National Institutes of Health grant AR 30988.

References

Araujo, A. and Walker, J.W. (1996) Phosphate release and force generation in cardiac myocytes investigated with caged phosphate and caged calcium. *Biophys. J.* **70**, 2316–26.

Bagshaw, C. and Trentham, D.R. (1973) The reversibility of adenosine triphosphate cleavage by myosin. *Biochem. J.* **133**, 323–8.

Berger, C.L., Craik, J.S., Trentham, D.R., Corrie, J.E.T. and Goldman, Y.E. (1995) Fluorescence polarization from isomers tetramethylrhodamine at SH-1 in rabbit psoas muscle fibers. *Biophys. J.* **68**, 78s–80s.

Bobkov, A.A., Bobkova, E.A., Homsher, E. and Reisler, E. (1997) Activation of regulated actin by SH1-modified myosin subfragment 1. *Biochemistry* **36**, 7733–8.

Bowater, R. and Sleep, J. (1988) Demembranted muscle fibers catalyze a more rapid exchange between phosphate and adenosine triphosphate than actomyosin subfragment 1. *Biochemistry* **27**, 5314–23.

Brenner, B. (1988) Effect of Ca^{2+} on crossbridge turnover kinetics in skinned single rabbit psoas fibers: implications for regulation of muscle contraction. *Proc. Natl. Acad. Sci. USA* **85**, 3265–9.

Brenner, B., Schoenberg, M., Chalovich, J.M., Greene, L.E. and Eisenberg, E. (1982) Evidence for crossbridge attachment in relaxed muscle at low ionic strength. *Proc. Natl. Acad. Sci. USA* **79**, 88–91.

Brune, M., Hunter, J.L., Corrie, J.E.T. and Webb, M.R. (1994) The direct real time measurement of rapid inorganic phosphate release using a novel fluorescent probe and its application to actomyosin subfragment 1 ATPase. *Biochemistry* **33**, 8262–71.

Chalovich, J. (1992) Actin mediated regulation of muscle contraction. *Pharmacol. Ther.* **55**, 95–148.

Chalovich, J., Chock, B. and Eisenberg, E. (1981) Mechanism of action of troponin–tropomyosin: inhibition of actomyosin ATPase without inhibition of myosin binding to actin. *J. Biol. Chem.* **256**, 575–8.

Chock, S., Chock, P. and Eisenberg, E. (1979) The mechanism of skeletal muscle myosin ATPase. *J. Biol. Chem.* **254**, 3236–43.

Cooke, R. and Pate, E. (1985) The effects of ADP and phosphate on the contraction of muscle fibers. *Biophys. J.* **48**, 789–98.

Cooke, R., Franks, K., Luciani, G.B. and Pate, E. (1988) The inhibition of rabbit skeletal muscle contraction by hydrogen ions and phosphate. *J. Physiol.* **395**, 77–97.

Dantzig, J.A., Hibberd, M.G., Trentham, D.R. and Goldman, Y.E. (1991) Crossbridge kinetics in the presence of MgADP investigated by photolysis of caged ATP in rabbit psoas muscle fibres. *J. Physiol.* **432**, 639–80.

Dantzig, J.A., Goldman, Y.E., Millar, N.C., Lacktis, J. and Homsher, E. (1992) Reversal of the crossbridge force-generating transition by photogeneration of phosphate in rabbit psoas muscle fibres. *J. Physiol.* **451**, 247–78.

Eisenberg, E., Hill, T. and Chen, Y. (1980) Cross-bridge model of muscle contraction. *Biophys. J.* **29**, 195–227.

Emoto, Y., Horiuti, K., Tawada, K. and Yamada, K. (1995) Tension relaxation induced by pulse photolysis of caged ATP in partially crosslinked fibers from rabbit psoas muscle. *Proc. Natl. Acad. Sci. USA* **92**, 1461–4.

Ferenczi, M.A. and Spencer, I.C. (1986) The elementary steps of the actomyosin ATPase in muscle fibres studied with caged ATP. *Adv. Exp. Med. and Biol.* **22**, 181–8.

Ferenczi, M.A., Homsher, E. and Trentham, D.R. (1984) The kinetics of magnesium adenosine triphosphate cleavage in skinned muscle fibres of the rabbit. *J. Physiol.* **352**, 575–99.

Ferenczi, M.A., He, Z.-H., Chillingworth, R.K., Brune, M., Corrie, J.E.T., Trentham, D.R. and Webb, M.R. (1995) A new method for the time-resolved measurement of phosphate release in permeabilized muscle fibers *Biophys. J.* **68**, 191s–3s.

Finer, J.T., Simmons, R.M. and Spudich, J.A. (1994) Single myosin molecule mechanics: piconewton forces and nanometer steps. *Nature* **368**, 113–9.

Fortune, N.S., Geeves, M.A. and Ranatunga, K.W. (1991) Tension responses to rapid pressure release in glycerinated rabbit muscle fibers. *Proc. Natl. Acad. Sci. USA* **88**, 7323–7.

Fortune, N., Geeves, M. and Ranatunga, K. (1994) Contractile activation and force generation in skinned rabbit muscle fibers: effects of hydrostatic pressure. *J. Physiol.* **474**, 283–90.

Glyn, H. and Sleep, J. (1985) Dependence of adenosine triphosphatase activity of rabbit psoas muscle fibres and myofibrils on substrate concentration. *J. Physiol.* **365**, 259–76.

Goldman, Y.E. (1986) Laser pulsed release of ATP and other optical methods in the study of muscle contraction. In *Optical methods in cell physiology*, (eds Dewell, D. and Salzberg, B.M.), pp. 397–415. Wiley, New York.

Goldman, Y.E. (1987) Kinetics of the actomyosin ATPase in muscle fibers. *Ann. Rev. Physiol.* **49**, 637–54.

Goldman, Y.E., Hibberd, M.G. and Trentham, D.R. (1984*a*) Relaxation of rabbit psoas muscle fibres from rigor by photochemical generation of adenosine-5′-triphosphate. *J. Physiol.* **354**, 577–604.

Goldman, Y.E., Hibberd, M.G. and Trentham, D.R. (1984*b*) Initiation of active contraction by photogeneration of adenosine-5′-triphosphate in rabbit psoas muscle fibres. *J. Physiol.* **354**, 605–24.

Goldman, Y.E., McCray, J.A. and Ranatunga, K.W. (1987) Transient tension changes initiated by laser temperature jumps in rabbit psoas muscle fibers. *J. Physiol.* **392**, 71–95.

Hackney, D.D. (1996) The kinetic cycles of myosin, kinesin, and dynein. *Ann. Rev. Physiol.* **58**, 731–50.

He, Z.-H., Chillingworth, R.K., Brune, M., Corrie, J.E.T., Trentham, D.R., Webb, M.R. and Ferenczi, M.A. (1997) ATPase kinetics on activation of rabbit and frog permeabilized isometric muscle fibers: a real time phosphate assay. *J. Physiol.* **501.1**, 125–48.

Herrmann, C., Lionne, C., Travers, F. and Barman, T. (1994) Correlation of actoS1, myofibrillar, and muscle fiber ATPases. *Biochemistry* **33**, 4148–54.

Hibberd M.G., Dantzig, J.A., Trentham, D.R. and Goldman, Y.E. (1985) Phosphate release and force generation in skeletal muscle fibers. *Science* **228**, 1317–9.

Higuchi, H. and Goldman, Y.E. (1991) Sliding distance between actin and myosin filaments per ATP molecule hydrolysed in skinned muscle fibres. *Nature* **352**, 352–4.

Higuchi, H. and Goldman, Y.E. (1995) Sliding distance per ATP molecule hydrolyzed by myosin heads during isotonic shortening of skinned muscle fibers. *Biophys. J.* **69**, 1491–507.

Homsher, E. and Millar, N.C. (1990) Caged compounds and striated muscle contraction. *Ann. Rev. Physiol.* **52**, 875–96.

Homsher, E., Lacktis, J. and Regnier, M. (1997) Strain-dependent modulation of phosphate transients in rabbit skeletal muscle fibers. *Biophys. J.* **72**, 1780–91.

Horiuti, K., Kagawa, K. and Yamada, K. (1994) Transient contraction of muscle fibers on photorelease of ATP at intermediate concentrations of Ca^{2+}. *Biophys. J.* **67**, 1925–32.

Horiuti, K., Sakoda, T. and Yamada, K. (1993) Mechanical response to photolytic ATP pulses of skinned muscle fibres pre-activated with a small pulse of ATP. *J. Mus. Res. Cell Motil.* **14**, 335–40.

Howard, J. (1996) The movement of kinesin along microtubules. *Ann. Rev. Physiol.* **58**, 703–29.

Huxley, H.E. and Faruqi, A.R. (1983) Time resolved X-ray diffraction studies on vertebrate striated muscle. *Ann. Rev. Biophys. Bioeng.* **12**, 381–417.

Huxley, A.F. and Simmons, R.M. (1973) Mechanical transients and the origin of muscula force. *Cold Spring Harbor Symp. Quant. Biol.* **37**, 669–80.

Irving, M., Lombardi, V., Piazzesi, G. and Ferenczi, M. (1992) Myosin head movements are synchronous with the elementary force-generating process in muscle. *Nature* **357**, 156–8.

Johnson, K.A. and Taylor, E.W. (1978) Intermediate states of subfragment 1 and actosubfragment 1 ATPase: reevalution of the mechanism. *Biochemistry* **17**, 3432.

Kaplan, J.H., Forbush, B.I. and Hoffman, J.F. (1978) Rapid photolytic release of adenosine 5′-triphosphate from a protected analogue: utilization by the Na : K pump of human red blood cell ghosts. *Biochemistry* **17**, 1929–35.

Kawai, M. and Halvorson, H.R. (1991) Two step mechanism of phosphate release and the mechanism of force generation in chemically skinned fibers of rabbit psoas muscle. *Biophys. J.* **59**, 329–42.

Kawai, M., Guth, K., Winnikes, K., Haist, C. and Ruegg, J.C. (1987) The effect of inorganic phosphate on ATP hydrolysis rate and the tension transients in chemically skinned rabbit psoas fibers. *Pflügers Arch.* **408**, 1–9.

Knight, P., Fortune, N. and Geeves, M. (1993) Effects of pressure on equatorial X-ray diffraction from skeletal muscle fibers. *Biophys. J.* **65**, 638–51.

Lacktis, J.W. and Homsher, E. (1987) The force response to photogeneration of ADP in isometrically contracting glycerinated rabbit psoas muscle fibers. *Biophys. J.* **51**, 475a (abstract)

Lionne, C., Brune, M., Webb, M.R., Travers, F. and Barman, T. (1995) Time resolved measurements show that phosphate release is the rate limiting step of myofibrillar ATPase *FEBS Letters* **364**, 59–62.

Lu, Z., Moss, R.L. and Walker, J.W. (1993) Tension transients initiated by photogeneration of MgADP in skinned skeletal muscle fibers. *J. Gen. Physiol.* **101**, 867–88.

Lu, Z., Walker, J.W. and Moss, R.L. (1991) Effects of Ca^{2+} and Pi on tension transients induced by photogeneration of ADP in skinned rabbit psoas fibers. *Biophys. J.* **59**, 417a (abstract).

Luecke, H. and Quiocho, F.A. (1990) High specificity of a phosphate transport protein determined by hydrogen bonds. *Nature* **347**, 402–6.

Lymn, R.W. and Taylor, E.W. (1971) Mechanism of adenosine triphosphate hydrolysis by actomyosin. *Biochemistry* **10**, 4617–24.

Ma, Y. and Taylor, E.W. (1994) Kinetic mechanism of myofibril ATPase. *Biophys. J.* **66**, 1542–53.

McCray, J.A. and Trentham, D.R. (1989) Properties and uses of photoreactive caged compounds. *Ann. Rev. Biophys. Chem.* **18**, 239–70.

Millar, N.C. and Homsher, E. (1990) The effect of phosphate and calcium on force generation in glycerinated rabbit skeletal muscle fibers. *J. Biol. Chem.* **265**, 20234–40.

Millar, N.C. and Homsher, E. (1992) Kinetics of force generation and phosphate release in skinned rabbit soleus muscle fibers. *Am. J. Physiol.* **262**, C1239–45.

Molloy, J.E., Burns, J.E., Kendrick-Jones, J., Tregear, R.T. and White, D.C.S. (1995) Movement and force produced by a single myosin head. *Nature* **378**, 209–12.

Ostap, E.M., White, H.D. and Thomas, D.D. (1993) Transient detection of spin-labelled myosin subfragment 1 conformational states during ATP hydrolysis. *Biochemistry* **32**, 6712–20.

Pate, E. and Cooke, R. (1989*a*) Addition of phosphate to active muscle fibers probes actomyosin states within the powerstroke. *Pflügers Arch.* **414**, 73–81.

Pate, E. and Cooke, R. (1989*b*) A model of crossbridge action: the effects of ATP, ADP, and Pi. *J. Mus. Res. Cell Motil.* **10**, 181–96.

Pate, E., Franks-Skiba, K., White, H. and Cooke, R. (1993) The use of differing nucleotides to investigate crossbridge kinetics. *J. Biol. Chem.* **268**, 10046–53.

Poole, K.J.V., Maeda, Y., Rapp, G. and Goody, R.S. (1991) Dynamic X-ray diffraction measurements following photolytic relaxation and activation of skinned rabbit psoas fibres. *Adv. Biophys.* **27**, 63–75.

Regnier, M., Morris, C. and Homsher, E. (1995) Regulation of the cross-bridge transition from a weakly to strongly bound stake in skinned muscle fibers. *Am. J. Physiol.* **269**, C1532–9.

Sleep, J.A. and Hutton, R.L. (1980) Exchange between inorganic phosphate and adenosine 5′-triphosphate in the medium by actomyosin subfragment 1. *Biochemistry* **19**, 1276–83.

Svoboda, K. and Block, S. (1994) Force and velocity measured for single kinesin molecules. *Cell* **77**, 773–8.

Tanner, J.W., Thomas, D.D. and Goldman, Y.E. (1992) Transients in orientation of a fluorescent crossbridge probe following photolysis of caged nucleotides in skeletal muscle fibres. *J. Mol. Biol.* **223**, 185–203.

Taylor, E.W. (1979) Mechanism of actomyosin ATPase and the problem of muscular contraction. *Crit. Rev. Biochem. Mol. Biol.* **6**, 103.

Taylor, E.W. (1991) Kinetic studies on the association and dissociation of myosin subfragment 1 and actin. *J. Biol. Chem.* **266**, 294–302.

Thirwell, H., Corrie, J.E.T., Reid, G.P., Trentham, D.R. and Ferenczi, M.A. (1994) Kinetics of relaxation from rigor of permeabilized fast-twitch skeletal fibers from the rabbit using a novel caged ATP and apyrase. *Biophys. J.* **67**, 2436–47.

Thirwell, H., Sleep, J.A. and Ferenczi, M.A. (1995) Inhibition of unloaded shortening velocity in permeabilized muscle fibers by caged ATP compounds. *J. Mus. Res. Cell Motil.* **16**, 131–7.

Trentham, D.R., Eccleston, J. and Bagshaw, C.R. (1976) Kinetic analysis of ATPase mechanisms. *Q. Rev. Biophys.* **9**, 217–81.

Trybus, K.M. and Taylor, E.W. (1982) Transient kinetics of adenosine 5′-diphosphate and adenosine 5′(β,γ-imidotriphosphate) binding to subfragment 1 and actosubfragment 1. *Biochemistry* **21**, 1284–94.

Wada, H., Yamada, T. and Sugi, H. (1996) Effect of inorganic phosphate and ADP on the myofilament sliding induced by laser flash photolysis of caged ATP. *Biochim. Biophys. Acta* **1274**, 89–93.

Walker, J.W., Lu, Z. and Moss, R.L. (1992) Effects of Ca^{2+} on the kinetics of phosphate release in skeletal muscle. *J. Biol. Chem.* **276**, 2459–66.

Walker, J.W., Reid, G.P., McCray, J.A. and Trentham, D.R. (1988) Photolabile 1-(2-nitrophenyl)ethyl phosphate esters of adenine nucleotide analogues. Synthesis and mechanism of photolysis. *J. Am. Chem. Soc.* **110**, 7170–7.

Webb, M.R., Hibberd, M.G., Goldman, Y.E. and Trentham, D.R. (1986) Oxygen exchange between Pi in the medium and water during ATP hydrolysis mediated by skinned fibers from rabbit skeletal muscle. *J. Biol. Chem.* **261**, 15557–64.

White, H.D. and Taylor, E.W. (1976) Energetics and mechanism of actomyosin ATPase. *Biochemistry* **15**, 5818–26.

Yamada, T., Abe, O., Kobayashi, T. and Sugi, H. (1993) Myofilament sliding per ATP molecule in rabbit muscle fibres studied using laser flash photolysis of caged ATP. *J. Physiol.* **466**, 229–43.

4 The use of fluorescent nucleotides

Clive R. Bagshaw

4.1 Introduction

Fluorescence is one of the most sensitive forms of spectroscopy available. In the last decade, detection has been achieved by several groups at the single molecule level (Funatsu *et al.* 1995; Peck *et al.* 1989; Sase *et al.* 1995). Furthermore, fluorescence has good time resolution because the associated electronic events occur on a nanosecond time-scale and are thus many orders of magnitude faster than the cross-bridge cycle time of muscle contraction. However, within the basic components necessary for motility (i.e. actin, myosin and MgATP), the only naturally occurring moieties, which display significant fluorescence emission, are the aromatic amino acids, tryptophan and tyrosine. The majority of fluorescence techniques applied to study muscle proteins, therefore, involve the use of probe molecules. This chapter focuses on probes that are combined with nucleotides to make fluorescent ATP analogues.

A major drawback of fluorescence, compared with other forms of spectroscopy, is that the structural information that can be gleaned is limited. Changes in excitation and emission spectra when a fluorescent ATP analogue binds to actin or myosin may indicate that the active site is more hydrophobic compared with bulk solvent, but it is difficult to extract any quantitative data. Nevertheless, such a change may present a very sensitive way of quantifying the kinetics and thermodynamics of the binding process. Thus one of the most fruitful applications of fluorescent analogues has been in the analysis of the elementary steps of the myosin ATPase reaction (Bagshaw *et al.* 1972; Rosenfeld and Taylor 1984; Smith and White 1985*a*; Woodward *et al.* 1991). Quantitative structural information can be extracted in favourable cases by examining fluorescence resonance energy transfer (FRET) between fluorescent probes (dos Remedios *et al.* 1987). However, the method has a range of between 2 and 7 nm because the probes themselves may be up to 1 nm in length and the efficiency of transfer falls off rapidly with distance.

The environment of the bound nucleotide might be defined in an empirical way by comparison with solvent mixtures of water and alcohols, which give rise to comparable spectral shifts. Another procedure for defining the degree to which the bound state is protected from the solvent is to add a fluorescence quenching agent, such as acrylamide, which reduces the signal from accessible fluorophores

(Ando *et al.* 1982; Franks-Siba *et al.* 1995). Collisional quenching coefficients can be evaluated, which provide a quantitative estimate as to the exposure of the fluorophore.

The fluorescence process involves an initial photon absorption process, which is complete in around 10^{-15} s, to give an excited state (the singlet state). The excited state involves a rearrangement of the electron distribution. The excited molecule quickly (about 10^{-12} s) loses some energy as heat to populate the lowest vibrational level of the excited state. The excited molecule can undergo a number of fates at this point. Usually further energy is lost as heat by collision with surrounding molecules to re-establish the ground state. An alternative route is the loss of energy by photon emission, which gives rise to the phenomenon of fluorescence. Rarer fates of the excited state are bond cleavage (photodecomposition) or further electronic rearrangement to a long-lived triplet state, which may ultimately emit a photon in the process termed phosphorescence. In general, the latter two routes are problematic for fluorescence measurements, but they can be exploited in their own right to reveal useful information. A good introduction to fluorescence methodology is given by Lakowicz (1983), while specific applications to contractile systems are discussed by Cooke (1982).

The ratio of the number of photons emitted to those initially absorbed is termed the fluorescence quantum yield. For practical purposes, compounds having quantum yields above 0.1 would be regarded as strongly fluorescent, while those with a quantum yield less that 0.01 would be weakly or non-fluorescent. The fact that the excited state normally loses some energy as heat results in the emitted photons having less energy, and hence a longer wavelength, than those absorbed (the Stokes shift). Typically this shift is a few tens of nanometres.

The return to the ground state from the excited singlet state typically takes several nanoseconds. This time-scale is slow compared with the rotational motion of small molecules in solution, but comparable to rotations of small proteins or protein domains. If follows, therefore, that the emission from free fluorescent nucleotides is likely to be completely randomized (i.e. depolarized), whereas when bound to myosin the fluorescence emission is likely to retain orientation information. In favourable cases, this property can be used to characterize the orientation of the fluorophore, and hence a protein domain relative to sample axis (e.g. axis of a muscle fibre, Yanagida 1981). Time-resolved studies can provide information about the rotational motion of protein domains (Rosenfeld *et al.* 1994).

The rationale for designing nucleotide analogues for the study of myosin has evolved over three decades and the emphasis has shifted in response to the problems posed. Initially, modifications of the adenosine, ribose and phosphate moieties were made to map out the important interacting groups in the active site (Tonomura 1972; Yount 1975). Such studies have shown that myosin is rather non-discriminating in its substrates, although the rates of the elementary processes may be adversely effected so as to diminish actin activation and coupling to filament sliding. The hydrolysis step itself can be drastically slowed by incorporating sulphur into the terminal phosphate group (ATPγS) or stopped completely with a bridging nitrogen (AMPPNP). A different rationale applies to the introduction of spectroscopic reporter groups where, in general, the aim is to synthesize an analogue whose interaction with myosin is perturbed to a minimal extent compared with ATP. Such reporter groups could be chromophores, fluorophores, esr spin labels or nmr labels (e.g. fluorine). As outlined above, the particular advantages of fluorescent reporters relate to their exquisite sensitivity.

Absorption, and hence fluorescence, in the accessible u.v. and visible regions usually arise from aromatic compounds with easily delocalized electrons. As a rule of thumb, the greater the degree of conjugation, the more red-shifted the absorption and fluorescence will be. Fluorescence properties are, however, difficult to predict from theory. Early ATP analogues were based on minimal changes to the adenosine ring (which itself shows no significant fluorescence) to make fluorescent derivatives. Modification to the ribose moiety, in contrast, required the introduction of an extrinsic fluorophore which, at first sight, did not seem such an attractive route. The requirements of the first analogues to be prepared were that their excitation should be clear of protein absorption (i.e. > 300 nm), while their emission should be clear of protein emission and ideally match the peak sensitivity of photomultiplier tubes (typically around 400 nm). Such analogues (e.g. ε-ATP, mant-ATP, formycin triphosphate) have been widely used to study the kinetics of actomyosin in solution and within muscle fibres. More recently, in response to single filament and single molecule studies, new analogues have been developed with excitation in the visible region to match the transmission characteristics of conventional optical microscopes and standard laser sources. These analogues naturally demand larger conjugated ring systems and hence a greater risk to perturbations to their kinetics of interaction with myosin. On the other hand, the further the excitation is red-shifted the less the problem of photodegradation of the protein components as well as the probe itself.

Apart from its absolute sensitivity, fluorescence emission intensity is also very sensitive to the local environment. It is for this reason that fluorescent ATP analogues are often able to meet the conflicting demand of a large perturbation of the reporter signal with minimal effect on the system under study (i.e. the actomyosin ATPase kinetics). It is also desirable that the bound states give a larger signal than the free nucleotide, so that the measurement is not swamped by the background emission. This requirement is also often met because the bound state is protected from quenching by collisions from solvent water. On the other hand, with experiments involving FRET, it is desirable that the fluorophores are not too sensitive to environmental changes because of the effect of errors introduced by small differences in fractional occupation of sites or conformation, on the estimated intensities with and without the energy acceptor.

In this review, the fluorescent nucleotide analogues that have been used to study actomyosin systems will be discussed. Emphasis will be focused on the myosin nucleotide site involved in energy transduction in muscle contraction. However, fluorescent analogues have also been substituted at the nucleotide site of actin. More recently other motor proteins, such as kinesin, have been characterized with the aid of fluorescent analogues (Sadhu and Taylor 1992). The interaction of ATP analogues with kinases is also of interest because many experiments require the use of an ATP regenerating system to maintain the nucleotide triphosphate concentration. In addition, an enzymic route may provide a convenient route for synthesis in laboratories not geared to non-aqueous chemical methods. As far as regenerating systems are concerned, creatine kinase may be preferable to pyruvate kinase because pyruvate substrates can act as fluorescence quenching agents (Ando and Miyata 1983; Nagano and Yanagida 1984).

4.2 Methods and results

Nucleotide analogues

Fluorescent derivatives have been prepared by modifications to the adenosine, ribose and phosphate moieties of ATP. Syntheses will be described for the nucleoside triphosphates. The corresponding diphosphates can be prepared in an analogous way starting from ADP. However, in many cases, it is easier to prepare the nucleoside diphosphates, as required, by hydrolysis of the triphosphate using catalytic amounts of myosin. Nucleoside diphosphates have a tendency to disproportionate on storage to a mixture of mono- and tri-phosphates and, therefore, critical studies require freshly prepared samples.

The systematic nomenclature of nucleotide analogues can become unwieldy. To complicate matters further, at least two numbering systems exist for the adenine ring. However, the one most widely used in biological texts labels the carbon atoms, which are shared by the 5- and 6-membered rings, as C^4 and C^5, so that attachment to the ribose is via N^9. Substitutions on the ribose moiety are distinguished from adenosine derivatives using a prime ($'$). Derivatization of the ribose hydroxyl groups usually results in a mixture of $2'$ and $3'$ substituents, denoted $2'(3')$, which may be resolved by chromatography, although some derivatives quickly re-equilibrate to give a mixture of isomers (Rensland *et al.* 1991). The generally accepted trivial names, denoted in brackets in the following section, will be used elsewhere in this review to denote specific analogues. Their chemical structures are given in Fig. 4.1 and a summary of their fluorescence properties in Table 4.1.

Some of these analogues are commercially available, while most may be synthesized in a one- or two-step reaction. A generally useful purification procedure for all of these analogues is ion-exchange chromatography on diethylaminoethyl (DEAE) cellulose or other support. Nucleotides may be eluted using

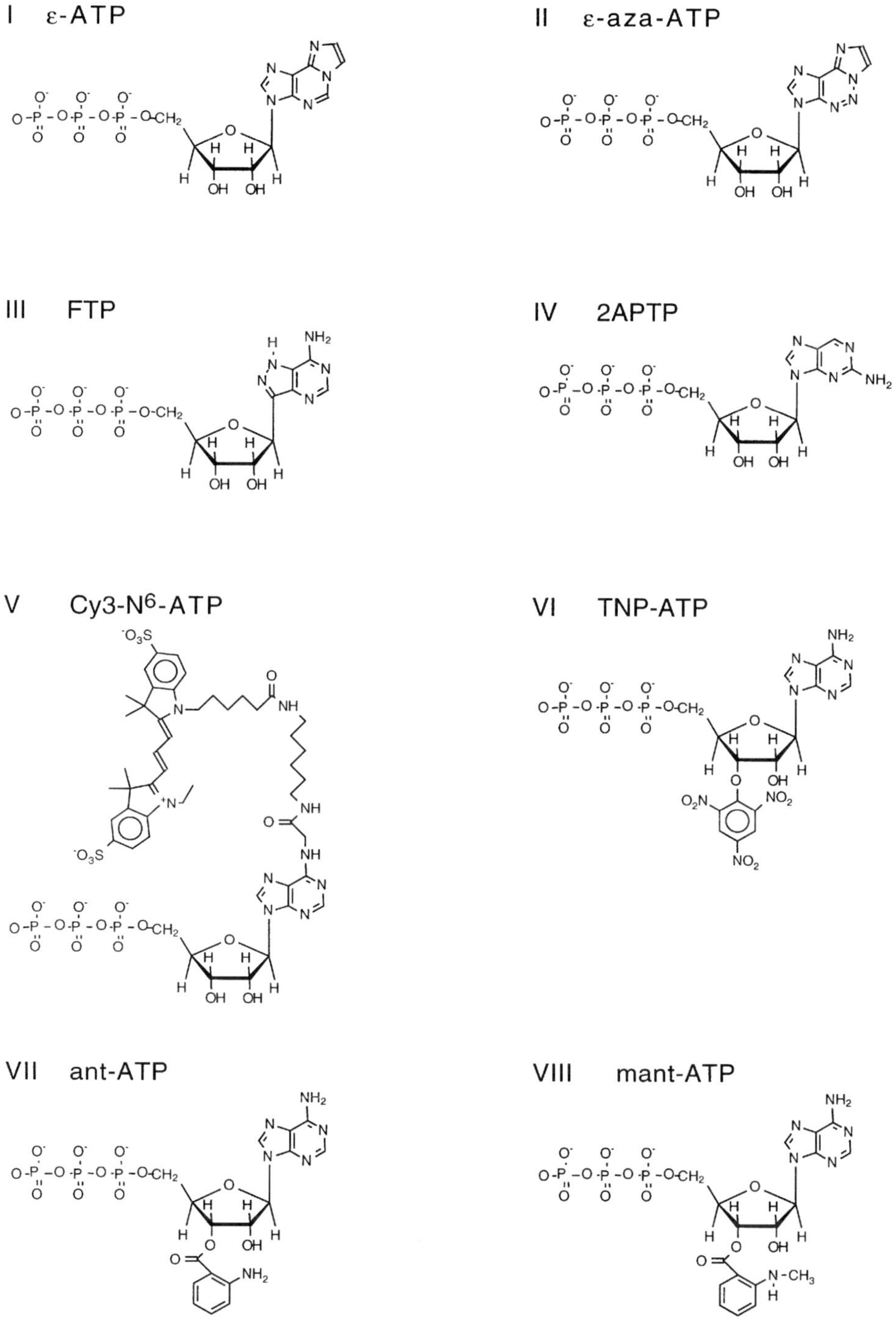

Fig. 4.1 Fluorescent ATP analogues. Ribose derivatives are shown as their 3′ positional isomers, whereas synthesis normally produces the 2′ derivative as well.

IX DEDA-ATP

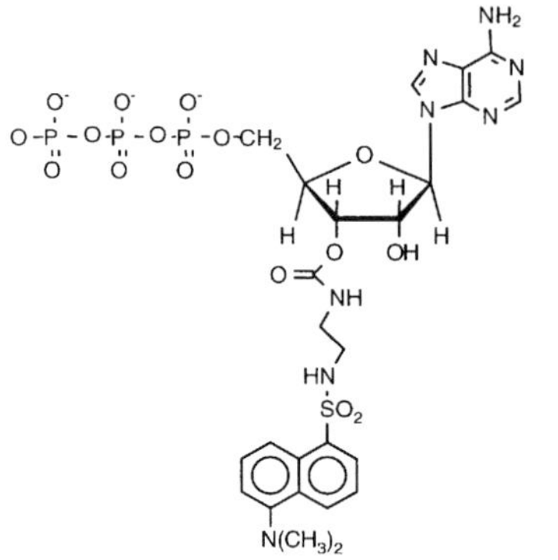

X FEDA-ATP

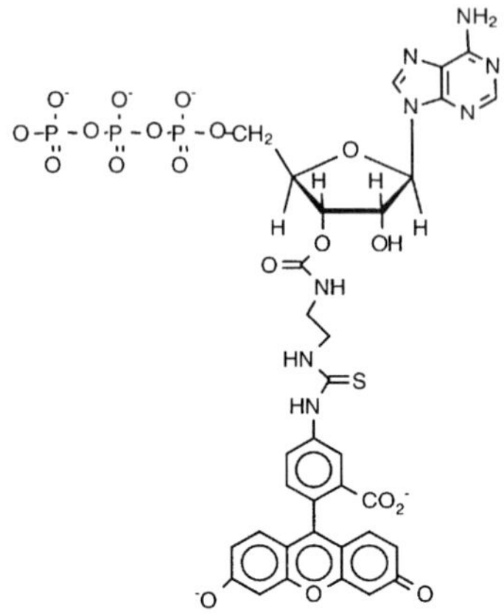

Xa caged-FEDA ATP

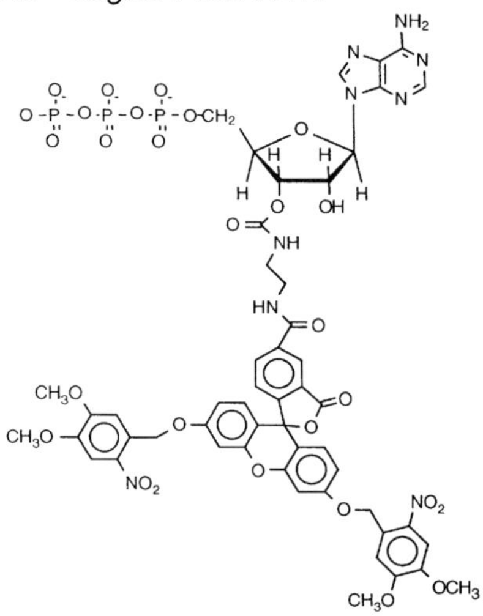

XI REDA-ATP

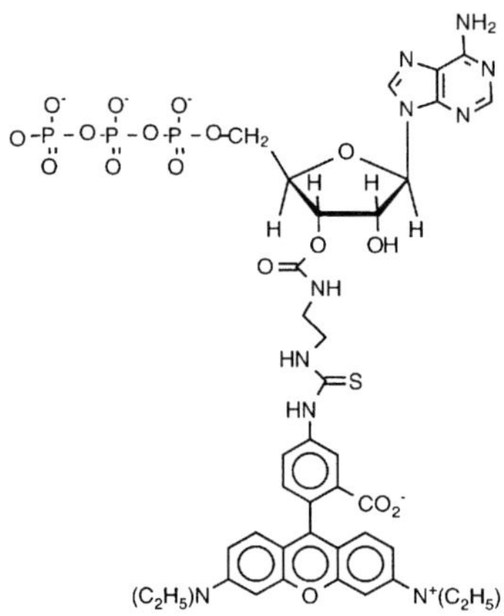

Fig. 4.1 (*Continued*).

a 0.05–0.8 M gradient of the volatile buffer, triethylamine bicarbonate (TEAB). The latter is prepared by bubbling CO_2 (e.g. from a flask containing dry ice) through a 1 M solution of triethylamine, on ice, until the pH drops to 7.5. The triethylamine should be fresh (or freshly distilled) to reduce secondary amines and the TEAB buffer kept at 0–4 °C and used within a few days. Following elution from the column, the TEAB is removed by rotary evaporation at 35 °C and the nucleotides recovered by several cycles of rotary evaporation with methanol.

XII Cy3-EDA-ATP

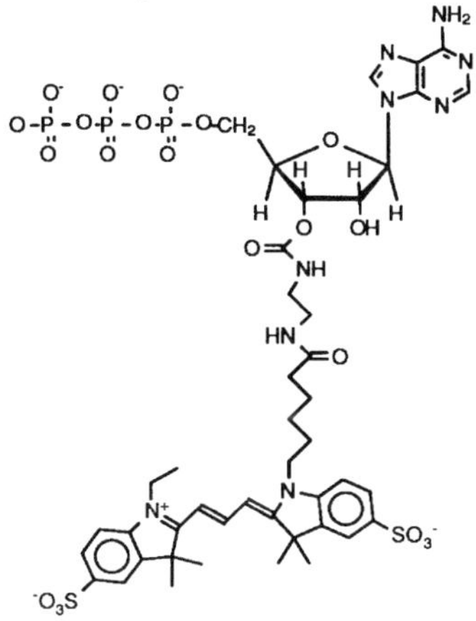

XIII Cy5-EDA-ATP

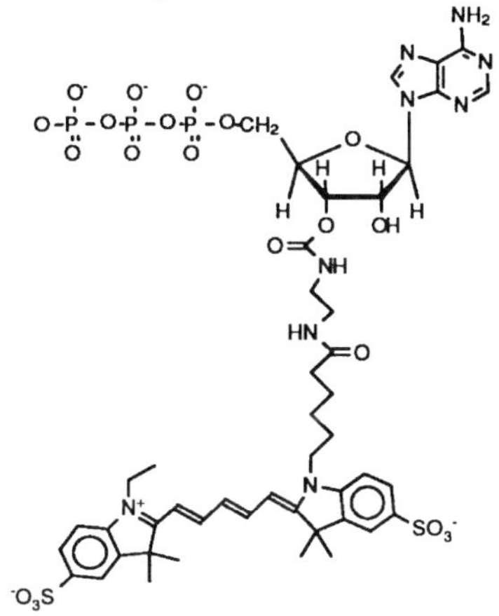

XIV FMP-PNP

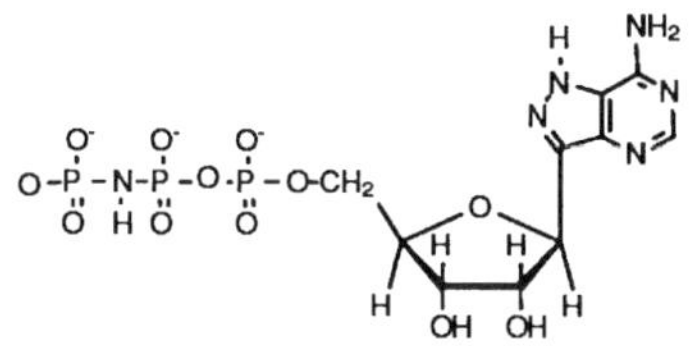

XV DTP

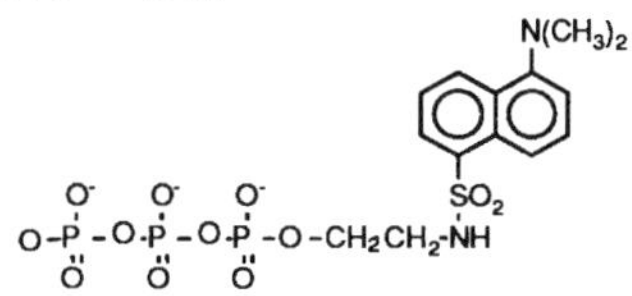

Fig. 4.1 (*Continued*).

Adenosine modified analogues

I 1-N^6-ethenoadenosine 5′ triphosphate (ε-ATP). The preparation of ε-ATP is described by Secrist *et al.* (1972) and involves the condensation of chloroacetaldehyde with ATP to introduce a third ring by linking the N^1 and 6NH groups of adenosine. ε-ATP is commercially available from Molecular Probes Inc. (Eugene, Oregon) and the Sigma Chemical Company. The absorption spectrum overlaps with protein absorption but, at neutral pH, ε-ATP has a shoulder extending out to 360 nm, allowing excitation well clear of protein fluorescence (Table 4.1). It is a substrate for adenylate kinase and pyruvate kinase, showing about 80% the V_{max} and similar (to within a factor of 2) K_m for substrate (Secrist *et al.* 1972), thus allowing the use of regenerating systems. It is also a substrate for myosin and shows a similar V_{max} and K_m to ATP, however, more detailed kinetic studies show that the hydrolysis step is slowed significantly and consequently, actin activation is perturbed (Rosenfeld and Taylor 1984). On direct excitation (325 nm) there is a small (10%) quench in the nucleotide fluorescence when ε-ATP binds to myosin

Table 4.1 Spectroscopic characteristics of fluorescent ATP analogues

Analogue	References	λ_{ex} (nm)	λ_{em} (nm)	ε_{max} ($M^{-1} \cdot cm^{-1}$)	Φ	τ (ns)	RF
ε-ATP	a, b	275 (310)	415	6000	0.59	26.0	0.9*
ε-aza-ATP	c, d	354 (365)	485	1530			3.0
FTP	e, f	295 (313)	340	9500	0.054	<1.0	5.0
2AP-TP	e, g	303	370	7100	0.063	7.0	0.95
TNP-ATP	h	408 470	540	26 400 18 500			†
ant-ATP	i	332	428	4700	0.14		
mant-ATP	i, j	350	446	5800	0.22	4.0	2.5
DEDA-ATP	j	330	578	4350	0.089	3.6	3.2
FEDA-ATP	k	495 (436)	520	78 000	0.9	4.2	0.4
REDA-ATP	k, l	553 (546)	580	106 000	0.3	1.7	0.55
Cy3-EDA-ATP	k, m	552	568	150 000	0.03		1.3
Cy5-EDA-ATP	m	648	665	215 000	0.1		0.93
DTP	n	320	534	5410			1.6

Sources: (a) Secrist *et al.* 1972; (b) Rosenfeld and Taylor 1984; (c) Miyata and Asai 1982; (d) Smith and White 1985*a*; (e) Ward *et al.* 1969*a*; (f) Jackson and Bagshaw 1988; (g) Bagshaw 1974; (h) Hiratsuka and Uchida 1973; (i) Hiratsuka 1983; (j) Cremo *et al.* 1990; (k) Conibear *et al.* 1996; (l) Chang and Cheung 1992; (m) Mujumdar *et al.* 1993; (n) Onodera and Yagi 1971.
* A fluorescence enhancement seen if ε-ATP is excited via tryptophan or free ε-ATP is quenched with acrylamide.
† A large enhancement observed when TNP-ATP binds to secondary site.

Absorption (excitation) and emission wavelengths, molar absorption coefficients (ε_{max}), quantum yield (Φ) and fluorescence lifetime (τ) are given for the free fluorophores in aqueous solvents (at pH 7.0 where available). λ_{ex} values in brackets refer to recommended (non-peak) settings to avoid overlapping protein absorption and/or to match the output from a Hg arc lamp.

RF represents the fluorescence of the nucleotide when bound to myosin as its predominant steady-state intermediate or trapped form relative to the fluorescence of the free nucleotide. These numbers are approximate because they depend on the bandwidth of the emission detector relative to any accompanying spectral shifts, as well as on solvent conditions

(Yanagida 1981) but energy is transferred from the protein tryptophan residues to give enhanced emission when excited at 290 nm (Onishi *et al.* 1973). The binding process has also been followed by the quench in protein fluorescence by Garland and Cheung (1979), while other workers have exploited the selective quenching of free ε-ATP fluorescence by acrylamide to provide a spectroscopic signal (Ando *et al.* 1982; Rosenfeld and Taylor 1984).

II 1-N^6-etheno-2-aza-adenosine 5′ triphosphate (ε-aza-ATP). The preparation of ε-aza-ATP is described by Yip and Tsou (1973) and Tsou *et al.* (1974). The monophosphate, ε-aza-AMP is available commercially from Molecular Probes. ε-aza-ATP has the advantage over ε-ATP of giving a large (three-fold) enhancement and shift in fluorescence emission on binding to myosin, without the need for selective quenching of free fluorophore by acrylamide (Miyata and Asai 1982). This property was exploited to show that myosin S1 crystals were able to bind ε-aza-ATP without shattering, and thus demonstrate that no large conformational rearrangement of the motor domain occurs (White *et al.* 1986). Solution studies indicate, as for ε-ATP, some rate constants for the actin-activated myosin ATPase are perturbed (Smith and White 1985*a*), but the analogue does support shortening of glycerinated fibres at a rate of about a quarter of that for ATP (Pate *et al.* 1993).

III Formycin 5′ triphosphate (FTP). Formycin is a naturally occuring nucleoside antibiotic, which may be chemically phosphorylated to give a fluorescent ATP analogue, FTP, as described by Ward *et al.* (1969*b*). Formycin A was commercially available from the Sigma Chemical Company but unfortunately supplies have now been exhausted. Formycin B lacks the 6 amino group and, therefore, is a poorer analogue of adenosine. Only formycin A derivatives will be discussed here. Formycin nucleotides are good substrates for adenylate kinase and pyruvate kinase and hence FTP can be made enzymically from FMP (Rossomando *et al.* 1981). The spectroscopic properties of FTP are described by Ward *et al.* (1969*a*,*b*), where it is also shown to be a good substrate for many enzymes. FTP retains the same chemical substituents as ATP on its 6-membered ring, so that interactions should be perturbed minimally. Interestingly, however, polyformycin nucleotides showed anomalous base pairing properties, possibly because the pyrazolopyrimidine ring is folded back over the ribose rather than extending in the required anti conformation (Ward *et al.* 1969*b*). The carbon–carbon glycosyl bond linking the ring to the ribose is slightly longer than the corresponding carbon–nitrogen bond of ATP. X-ray crystallography shows formycin adopts a syn or intermediate conformation in a number of derivatives (Giranda *et al.* 1988; Koyama *et al.* 1966; Prusiner *et al.* 1973). However, the barrier to rotation is not large, so that FTP is a substrate for enzymes requiring the anti-conformation, including the myosin ATPase (Bagshaw *et al.* 1972).

The bias towards the syn conformation may account for the slower association rates of FTP with myosin compared with ATP. However, once bound, the kinetics of the ATPase are mimicked closely and the large fluorescent enhancement of the analogue has found particular use for characterizing the trapped states associated with regulated myosins, such as those from scallop adductor muscle (Ankrett *et al.* 1991; Jackson and Bagshaw 1988), smooth muscle (Cross *et al.* 1988) and non-muscle myosins (Citi *et al.* 1989). Although the absorption of formycin overlaps with that of proteins, it extends further to the red and can be selectively excited by the 313 nm Hg line of an arc lamp.

IV 2 aminopurine 5′ triophosphate (2APTP). The preparation and spectroscopic properties of 2APTP are described by Ward *et al.* (1969*a*). The small quench in nucleotide fluorescence on binding to myosin did not appear favourable for full kinetic characterization (Bagshaw *et al.* 1972; Bagshaw 1974).

V N^6-((6-(N-Cy3)-aminohexyl)carbamoylmethyl)-5′-adenosine triphosphate (Cy3-N^6-ATP). This analogue was the first to be detected, when bound to myosin, at the single molecule level (Funatsu *et al.* 1995). It was prepared in a one-step synthesis from commercially available precursors: the succinimidyl ester of a carbocyanine dye (Cy3.29-OSu; Biological Detection Systems Inc., Pittsburgh, Pa) and N^6-[(6-aminohexyl)carbamoylmethyl]-5′-adenosine triphosphate. Although it appears a reasonable substrate for the myosin ATPase, the kinetics have not been reported in detail. The favourable fluorescence properties of the Cy3 dye (Mujumdar *et al.* 1993; Southwick *et al.* 1990) are probably better exploited in the ribose derivative, Cy3-EDA-ATP, described below.

Ribose modified analogues

VI 2′(3′)-O-(2,4,6-trinitrophenyl) adenosine 5′-triphosphate (TNP-ATP). TNP-ATP was among the first fluorescent derivatives to be prepared by attaching a fluorophore to the free 2′(3′) ribose hydroxyl groups. Its synthesis and initial characterization are described by Hiratsuka and Uchida (1973) and Hiratsuka (1976). At neutral pH, the trinitrophenyl group bridges both 2′ and 3′ hydroxyl groups to form a so-called Miesenheimer compound. Under acid conditions (pK 5.1), the TNP group resides on either the 2′ or 3′ hydroxyl groups. Although a substrate for myosin, Moss and Trentham (1983) reported two complications in its use as a fluorescent probe:

(1) a secondary binding site for the TNP-adenosine moiety appeared to be responsible for the major fluorescence enhancement observed on binding the TNP-ATP;

(2) TNP-ATP appeared to react slowly with the myosin, probably at the SH1 cysteine residue, to form a covalent adduct.

While conditions could be achieved where TNP-nucleotides were bound predominantly at the active site (Moss and Trentham 1983), this analogue has not found wide application in contractile systems.

VII 2′(3′)-O-(anthraniloyl) adenosine 5′-triphosphate (ant-ATP). The preparation and initial characterization of ant-ATP was described alongside the related mant-ATP by Hiratsuka (1983). However, the red-shifted spectra and higher quantum yield of the latter has made it the more favoured analogue in subsequent studies with myosin. In general, however, ant-ATP could be used in many of the applications described for mant-ATP below.

VIII 2′(3′)-O-(N-methylanthraniloyl) adenosine 5′-triphosphate (mant-ATP). Mant-ATP, first synthesized by Hiratsuka (1983), has proved to be one of the most useful analogues for solution studies of actomyosin and other motor proteins. It is prepared in a one-step reaction of the adenosine nucleotide with N-methyl isatoic anhydride under mild aqueous conditions. As with other ribose derivatives, synthesis results in a mixture of 2′ and 3′ isomers. They may be separated by HPLC but re-equilibration occurs with a half-time of about 10 minutes at neutral pH (cf. guanosine analogues, Rensland *et al.* 1991). Both isomers appear to be good substrates for myosin, but to avert problems of a heterogeneous substrate, the 2′ deoxy derivative was prepared (Woodward *et al.* 1991). Mant-ATP is a poor substrate for pyruvate kinase, but is much better with creatine kinase, which may therefore find use as a regenerative enzyme (Hitasuka 1983). On binding to myosin there is a two- to three-fold enhancement in fluorescence. An even greater enhancement can be achieved by exciting the bound mant nucleotide at 280 nm via energy transfer from protein tryptophan residues (Woodward *et al.* 1991).

IX 2′(3′)-O-[N-[2[[[5-(dimethylamino) naphthyl] sulphonyl] amino] ethyl] carbamoyl]-adenosine 5′-triphosphate (DEDA-ATP). The introduction of a fluorophore via an ethylene diamine linker on the ribose ring, as described by Cremo *et al.* (1990), established the route for a number of useful analogues. In the original preparation, the dansyl group was first linked to ethylene diamine to produce DEDA and then this was coupled to ATP. However, by linking ethylene diamine to ATP to produce EDA-ATP as a common intermediate, a whole range of amine-reactive fluorophores can be introduced in the second step (Fig. 4.2). The carbamoyl group, which couples the ethylene diamine to the ribose, is stable to alkaline conditions and exchange between the 2′ and 3′ is slow at neutral pH, so that isomers can be separated.

The dansyl group was introduced by Cremo *et al.* (1990) to produce DEDA-ADP, which showed a fluorescence enhancement on binding to myosin and a further increase in quantum yield when trapped as a vanadate complex or by thiol cross-linking. DEDA-ATP is a good substrate for myosin S1, although a full characterization has not been published (Woodward *et al.* 1991).

X 2′(3′)-O-[N-[2-[3-[5-fluoresceinyl] thioureido] ethyl] carbamoyl]-adenosine 5′ triphosphate (FEDA-ATP). The ethylene diamine linker strategy allowed Eccleston (personal communication) to introduce fluorescein and rhodamine into ATP to produce analogues, which were substrates for myosin. Both 2′ and 3′ FEDA-ATP are reasonable analogues, as judged by their kinetics of interaction with myosin (Fig. 4.3). Compared with the O-acyl link in mant-ATP, the carbamoyl link in EDA-based derivatives is relatively stable and the positional isomers do not interconvert on the minutes time-scale. The favourable fluorescence properties of fluorescein for optical microscopy were exploited by Sowerby *et al.* (1993) to follow ATP turnover at the level of single myosin filaments. Using Fluorescein isothiocyanate (FITC) as a reactant gives a thiourea link with the free amine of EDA-ATP, while the potentially more stable amide link can be achieved using fluorescein succinimidyl esters (Fig. 4.2). The latter was

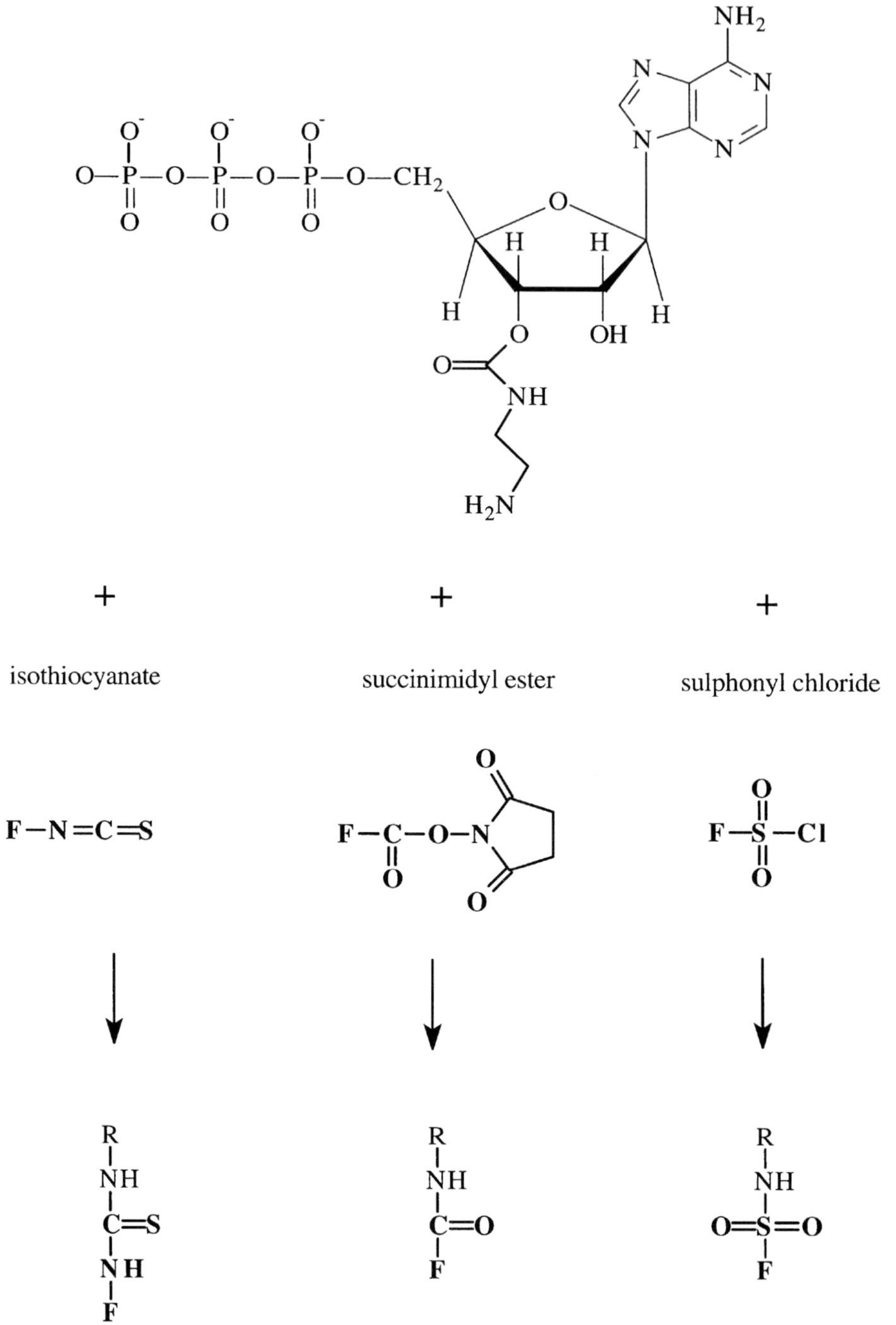

Fig. 4.2 Synthetic routes of fluorescent analogues based on EDA-ATP. In the products shown, R represents the EDA-ATP moiety and F a fluorophore. The preferred route for derivatization is via the succinimidyl ester, which is a highly reactive reagent, while the amide link in the product is stable against subsequent hydrolysis.

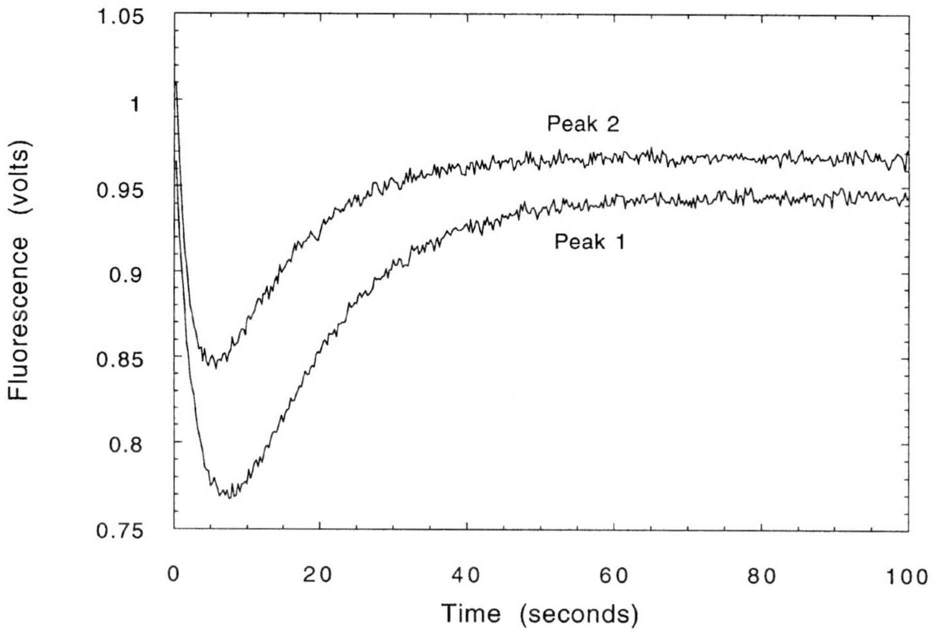

Fig. 4.3 Fluorescent stopped flow records showing the turnover of 2′ and 3′ FEDA-ATP by myosin S1. A FEDA-ATP preparation comprising mixed 2′(3′) isomers was separated by HPLC on a SAX column. Two nucleotide triphosphate peaks were eluted in a 2:1 absorbance ratio which, by analogy with other EDA-derivatives (Cremo *et al.* 1990), probably correspond to the 3′ and 2′ isomers, respectively. The two fractions were diluted to 50 nM FEDA-ATP in a reaction buffer comprising 40 mM NaCl, 1 mM $MgCl_2$ and 20 mM TES at pH 7.5 and 20 °C, and then mixed with 1 μM rabbit skeletal myosin S1 (syringe concentrations) in a stopped-flow apparatus. The fluorescein fluorescence quench corresponds to the binding phase and the recovery to the rate-limiting product release step, which was similar for both isomers ($0.07\,s^{-1}$ and $0.09\,s^{-1}$ for Peak 1 and 2, respectively). Records courtesy of P.B. Conibear and C.K. Seehra (cf. Conibear *et al.* 1996).

used to prepare a caged FEDA-ATP (Fig. 4.1, Xa) in which the fluorescein was locked into its non-fluorescent lactone state by photolabile dimethoxynitrobenzyl groups (Jeffreys *et al.* 1995). Exposure of caged FEDA-ATP to a pulse of 350 nm light releases fluorescent FEDA-ATP. Such an analogue might prove useful in microscopy, where local activation of caged FEDA-ATP avoids having a high background fluorescence signal from the analogue in bulk solution. Fluorescein is well suited for excitation by the 488 nm line of an Argon ion laser.

XI 2′(3′)-O-[N-[2-[3-[5(6)-tetraethylrhodamine] thioureido] ethyl] carbamoyl]-adenosine 5′ triphosphate (REDA-ATP). REDA-ATP was synthesized in a similar manner to FEDA-ATP by reacting rhodamine isothiocyanate RITC (5/6 mixed isomers) with EDA-ATP. Rhodamine dyes comprise a family of compounds of which the rhodamine B present in RITC refers to the tetraethyl derivative. The kinetics of interaction with myosin have not been fully characterized but, were measured, they are similar to those of FEDA-ATP (Conibear *et al.* 1996). Rhodamine shows better photostability than fluorescein and its reduced quantum yield is not a disadvantage when measuring fluorescence against a high background of free nucleotide. These properties have been exploited in the microscopy studies of Conibear and Bagshaw (1996).

XII 2′(3′)-O-[N-[2-[(Cy3) amido] ethyl] carbamoyl]-adenosine 5′ triphosphate (Cy3-EDA-ATP). Following the demonstration of the favourable fluorescence properties of carbocyanine dye, Cy3, for single molecule detection by Funatsu *et al.* (1995), several groups prepared the corresponding Cy3-EDA-ATP analogue by reacting Cy3.29-OSu succinimidyl ester (Mujumdar *et al.* 1993; Biological Detection Systems Inc.) with EDA-ATP (Conibear *et al.* 1996; Oiwa *et al.* 1996; Saito *et al.* 1995). As with FEDA-ATP, the rate constants for interaction of Cy3-EDA-ATP with myosin and actomyosin are within a factor of 2 to 4 of those of ATP itself (Conibear *et al.* 1996; Eccleston *et al.* 1996). Cy3-EDA-ATP, however, has the advantage of giving a 30% enhancement in fluorescence on binding to myosin compared with a 45% quench found with FEDA-ATP, thus facilitating detection of the bound states.

Cy3 analogues, like rhodamine derivatives, can be excited by the 514.5 nm line of an argon ion laser, but a better match is provided by a frequency-doubled YAG laser at 532 nm. Green HeNe lasers at 543.5 nm are presently only available at low power (up to 1.5 mW) but this may be sufficient for some microscopy experiments.

XIII 2′(3′)-O-[N-[2-[(Cy5) amido] ethyl] carbamoyl]-adenosine 5′ triphosphate (Cy5-EDA-ATP). The corresponding Cy5-EDA-ATP analogue has been prepared and characterized by Oiwa *et al.* (1996) and Eccleston *et al.* (1996). On binding to myosin it shows a 7% quench in fluorescence (J.F. Eccleston, personal communication) and the emission in the far-red makes detection less sensitive with standard photomultiplier tubes compared with emission from Cy3. Cy5 analogues can be excited by red HeNe lasers at 633 nm.

Phosphate modified analogues

Introduction of a fluorescent group into the phosphate moiety of ATP is likely to be very detrimental to biological function as the triphosphate group is buried in the active site of myosin (Fisher *et al.* 1995*a*,*b*). Derivatives such as adenosine 5′-triphosphoro-γ-1-(5-sulphonic acid) naphthylamidate (Yarbrough *et al.* 1979) are therefore unlikely to be suitable for probing contractile systems. On the other hand, a nucleotide analogue may be required which is not hydrolysed but will still bind with high affinity. All of the above fluorescent derivatives could be prepared in conjunction with thio- or imido-phosphate modifications, and a limited number of such analogues have been reported. Thus ε-AMPPNP was prepared by Yanagida (1981) for orientation studies on muscle fibres from AMPPNP using the method of Secrist *et al.* (1972). Mant-AMPPNP has been prepared chemically from mant-ATP by Rosenfeld *et al.* (1994), while the formycin derivative, FMPPNP (also abbreviated to p[NH]ppF) was prepared enzymically by Jackson and Bagshaw (1988), using the amino acid-activating reaction to exchange the PNP group with the terminal β, γ groups of ATP (Fig. 4.1, XIV). In this reaction, a crude extract of amino acid-activating enzymes (E), obtain by centrifuging disrupted yeast cells at 100 000 g, was added to a mixture of FTP, sodium imidodiphosphate and 20 amino acids (a.a) to catalyse the following reactions:

$$\text{FTP} + \text{a}\cdot\text{a} + \text{E} \leftrightarrow \text{E-FMP-a}\cdot\text{a} + \text{PPi},$$

$$\text{E-FMP-a}\cdot\text{a} + \text{PNP} \leftrightarrow \text{E} + \text{a}\cdot\text{a} + \text{FMPPNP}.$$

After a 1-h incubation, the mixture is boiled and any remaining FTP converted to FMP by myosin and adenylate kinase, and the FMPPNP is purified by ion-exchange chromatography.

The rationale for using imido analogues is to generate a non-hydrolysable species, which corresponds to the M.ATP state. However, there remains a considerable degree of uncertainty as to what state bound AMPPNP actually mimics (see Solution Kinetics section). Better analogues for the M.ATP state are based on thiophosphate derivatives, which have a rate-limiting hydrolysis step (Bagshaw *et al.* 1974; Moore and Lohman 1994).

In other situations, it is desirable to mimic the M.ADP.Pi state but with a longer lifetime, so that more complete characterization is possible. In these cases, fluorescent nucleotide diphosphates may be trapped as their vanadate or aluminium fluoride complexes (Conibear *et al.* 1996; Cremo *et al.* 1990; Hiratsuka 1984). The corresponding beryllium fluoride complex is proposed to mimic a long-lived M.ATP state (Fisher *et al.* 1995*a,b*).

Fluorescent analogues have also been prepared in which the γ-phosphate group is caged with a photolabile group (cf. the caged fluorescein analogue Xa described above, in which the fluorophore is caged). For example, ε-aza-ATP has been caged with a 2-nitrobenzyl group to give an intramolecularly quenched species (McCray and Trentham 1989). On flash photolysis, the fluorescence of the ε-aza-adenosine increases as the caging group is released at the rate limited by the decay of the *aci*-nitro anion. By using a brief laser pulse, the fluorophore escapes severe photodecomposition even when the photolysis wavelength (typically around 350 nm) overlaps with the excitation spectrum. Compounds caged on their γ-phosphate bind only weakly to myosin and therefore can be diffused into muscle fibres and activated within milliseconds, as required.

In principle, fluorescent probes for the ATPase sites can be introduced by substitution of the Mg^{2+} ion for a fluorescent Tb^{3+} ion. While this has been exploited for the slowly exchanging metal-nucleotide site of actin (Barden and dos Remedios 1984), the method would be difficult to use under steady-state hydrolysis conditions of the myosin ATPase, where mM concentrations of free metal ion are desirable.

Non-nucleotide analogues

The broad specificity of myosin allows the use of non-nucleotides as substrates (Wang *et al.* 1993), including fluorescent derivatives. Thus, Onodera and Yagi (1971) showed 2-(dansylamino)ethyl triphosphate (DTP; Fig. 4.1, XV) was a substrate for heavy meromyosin and showed a 60% enhancement on binding during the steady-state hydrolysis reaction.

Solution kinetics

The development of fluorescent ATP analogues to study the elementary steps of the myosin ATPase in the late 1960s and early 1970s coincided with the characterization of the changes of the absorption and emission of the intrinsic protein tryptophan fluorescence, which allowed characterization of the interactions with the natural ATP substrate (Bagshaw *et al.* 1972; Morita 1967; Werber *et al.* 1972). Rather than making the extrinsic probe approach redundant, this situation allows a detailed comparison of the kinetics of interaction of fluorescent ATP analogues with ATP on interaction myosin to establish the degree to which the probe is perturbing the system. Once characterized, fluorescent ATP analogues can then be used to follow kinetics in preparations where changes in tryptophan fluorescence are masked by other protein components (e.g. actin and other myofibrillar proteins). In principle, analogues could reveal additional steps in the myosin ATPase, which are spectroscopically silent with respect to tryptophan. However, there remains an ambiguity as to whether such a process is intrinsic to the myosin ATPase or is an artefact for the particular analogue under study. Fluorescent analogues also provide an important tool for myosin isoforms, which show only small perturbations in the tryptophan signal or where proteins are available in small amounts, so that the increased sensitivity of probes may prove valuable (Ritchie *et al.* 1993).

Myosin is anomalous compared with 'typical' enzymes because it forms a particularly long-lived product complex, which limits the overall ATPase rate in the steady-state. A poor substrate, which is not hydrolysed rapidly, may actually show an increased turnover rate in the steady-state if the product complex is not as stable as with ATP itself. Thus V_{max} is not a reliable parameter. A further complication with steady-state measurements is that the K_m for MgATP is so low ($< 10^{-7}$ M) that it may be difficult to achieve the requirement of $[M] \ll [ATP]$ in steady-state assays and thus reported K_m values should be viewed critically. Comparison of V_{max} and K_m data for ATP analogues with myosin alone from literature data is, therefore, not always informative. The actin-activated rates provide a more stringent test of an analogue, but these are also not without problems because the degree of activation is extremely sensitive to ionic strength and comparisons across data sets in the literature are difficult.

Transient kinetic measurements with myosin are generally more informative than steady-state kinetics, but these are not without difficulties. Stopped-flow methods provide a convenient way for resolving reaction on the millisecond time-scale (Eccleston 1987; White 1982). At low nucleotide concentrations, ATP itself, and most ATP analogues, show a linear dependence on the apparent binding rate constant on nucleotide concentration, yielding second-order rate constants in the region of 10^5 to $10^6\,M^{-1}\cdot s^{-1}$. This value provides a useful comparison, in that any analogue which fails to match that for ATP (1 to $2\times10^6\,M^{-1}\cdot s^{-1}$) clearly is not recognized with the same efficacy as ATP itself. However, this number represents the combination of several rate constants because the binding process is at least two-step and part of the fluorescence change may also reflect the hydrolysis step. Thus, at high nucelotide concentrations, the observed rate constant of the fluorescence change may become independent of concentration, reflecting a first-order step. Indeed, more than one transition may be revealed, yielding biphasic profiles. However, interpretation is not straightforward because such biphasicity may represent heterogeneity in the protein preparation, heterogeneity of the ATP analogue or genuine multi-step reactions (Rosenfeld and Taylor 1984; Smith and White 1985*a*).

Nevertheless, it is desirable that comparisons between ATP and its analogues are made at high nucleotide concentrations, where at least part of the spectroscopic change may become independent of nucleotide concentration and resolvable by stopped-flow methods. In practice, the high background signal from the excess free analogue reduces the relative change in fluorescence on binding and sets an upper limit for the nucleotide concentration, which can then be investigated. There are a number of 'tricks' that can be tried to push the limit to its maximum. Clearly, increasing the protein concentration will increase the signal from the bound state and aid its detection. However, the overall absorbance of the protein and nucleotide may ultimately limit the fluorescence detection due to a large inner filter effect, which attenuates the excitation light. Using front-faced optics may help in this regard (Smith and White 1985*b*).

Quenching agents, such as acrylamide, which selectively reduce the fluorescence from the free fluorescent nucleotide, have proved valuable in the case of ε-ATP (Ando *et al.* 1982; Rosenfeld and Taylor 1984). However, agents such as the iodide ion are generally unsatisfactory as they perturb the ionic strength. Unfortunately acrylamide does not quench all fluorophores and, in the case of ribose-modified analogues, the myosin active site might not be expected to contribute much protection against quenching (mant-ATP is an exception, see next section).

Another route, applicable to fluorophores that are excited by light in the 320–360 nm range, is to selectively excite the bound fluorophore via protein fluorescence (Onishi *et al.* 1973; Woodward *et al.* 1991). Thus, if the free fluorophore is not efficiently excited at 280 nm, this wavelength may be used to excite tryptophan, which will transfer its energy to a nearby fluorophore by FRET (see next section). As a result, an enhancement in emission will be observed on binding to myosin.

Alternatively, the reaction can be monitored by the quench in tryptophan emission rather than analogue emission (Garland and Cheung 1979).

In any event, it makes sense to determine the optimal wavelengths that give maximum change in signal before carrying out a detailed kinetic study. Although rapid scanning instruments are available, stopped-flow fluorescence measurements are usually recorded at single excitation and emission wavelengths to maximize signal-to-noise. While the amplitudes of transients can be compared for various combinations of excitation and emission wavelengths, a more efficient way of optimizing the signal is to compare the excitation and emission spectra in a scanning spectrofluorimeter of the free nucleotide with that trapped at the myosin active site as the BeF_x or AlF_4 complex (Fisher *et al.* 1995*a*,*b*). Vanadate trapping is best avoided in such a study as it can interfere with fluorescence in the u.v. region and also cause photolytic cleavage under strong illumination.

Direct comparison of rate constants for the elementary steps of the myosin and actomyosin ATPase for different fluorescent nucleotides in the literature is practically impossible owing to the range of buffer conditions (pH, ionic strength and temperature) used and the specific myosin isoform and subfragment studied. However, many papers do provide a comparison of the kinetics of a particular analogue with ATP and ADP itself under the same conditions. It is, therefore, more appropriate to discuss the properties of individual analogues rather than tabulating data that may invite dubious comparisons.

Garland and Cheung (1979) have suggested that nucleotide binding to rabbit skeletal myosin subfragment 1 involves three distinct steps (apart from hydrolysis), on the basis of deviation from a simple two-step mechanism (which predicts a hyperbolic dependence of the observed pseudo first-order rate constant on [nucleotide] or an unusually low initial collision rate constant) and on the marked difference in the maximum first-order rate constant recorded at high [nucleotide] for ATP and ADP compared with ε-ATP and ε-ADP:

$$\mathrm{M + NTP \rightleftharpoons M.NTP \rightleftharpoons M^{\dagger}NTP \rightleftharpoons M^{*}NTP.}$$

On the other hand, Rosenfeld and Taylor (1984) found that the binding of ε-ATP and ε-ADP was biphasic and both phases were concentration-dependent leading to two first-order transitions at saturating concentrations. They modelled this in terms of a branched pathway for nucleotide binding. More significanlty, they showed that the rate constants for both fluorescence phases were an order of magnitude faster that the hydrolysis step for ε-ATP. Thus the hydrolysis step for ε-ATP was around $5\,\mathrm{s}^{-1}$ under conditions where ATP hydrolysis was $\geqslant 50\,\mathrm{s}^{-1}$. Furthermore, the magnitude of the phosphate burst suggested that the hydrolysis equilibrium constant was in favour of the M·ε-ATP state. Following hydrolysis, the apparent isomerization limiting phosphate release was three times faster for ε-ATP ($0.2\,\mathrm{s}^{-1}$) than that for ATP ($0.06\,\mathrm{s}^{-1}$). Given the unfavourable equilibrium of the preceding hydrolysis step, the actual rate constant could be more than ten times that for ATP. The subsequent ε-ADP dissociation was comparable to that for ADP at $1\,s^{-1}$.

The perturbed hydrolysis step was reflected in the actin interaction kinetics. While ε-ATP was practically as efficient at dissociating acto-S1 as ATP itself (both in rate and extent), the slow hydrolysis rate constant of the former limited the subsequent activation of product release by actin. Indeed, as the actin concentration was increased, the steady-state ATPase showed a biphasic response with a limited (four- to five-fold) activation, followed by inhibition as the degree of association of the actin exceed about 0.4. Overall, ε-nucleotides might be considered good analogues for the binding steps, but poor analogues for mimicking the characteristic rapid hydrolysis and the subsequent rate-limiting isomerization controlling phosphate release.

The related ε-aza-nucleotides also show complicated saturation kinetics on binding, indicative of multi-step binding mechanisms. However, the data did not allow an unambiguous assignment to a linear or branched scheme (Smith and White 1985*a*). The slower maximum rate constant for actin-activated ATPase and the marked inhibition of this rate at high [actin] in the case of the subfragment 1 A2 isozyme, suggests that hydrolysis may be considerably slower than that for ATP (White *et al.* 1993). This is supported by the finding that the release of phosphate from the A.M.ADP.Pi species, as monitored with a fluorescence Pi indicator (Webb and White 1996), is biphasic, with the slow phase attributed to the A.M.ε-aza-ATP → A.M.ε-aza-ADP.Pi step being ten times slower than that for ATP itself (H.D. White, personal communication). On the other hand, the steady-state rate and single turnover analysis of ε-aza-ATP in the absence of actin, indicate that the isomerization that limits phosphate release occurs at the same rate as that for ATP, although this may be partly coincidential if the hydrolysis equilibrium is perturbed.

In the case of formycin nucleotides, the slow apparent second-order association rate constant (about 25 times slower than ATP) did not allow saturation to be achieved and, therefore, multiple steps in the binding mechanism could not be resolved using these analogues (Jackson and Bagshaw 1988). The slow binding may be due to an initially unfavourable weak association of the collision complex due to the predominance of the syn conformation. However, once bound, the hydrolysis step is rapid and the bound products complex represents the predominant steady-state complex. The close similarity of the product release rate constants, together with a large fluorescence enhancement, has made formycin analogues valuable probes for the myosin-linked regulatory systems found in molluscan myosins, vertebrate smooth muscles and cellular myosins (Ankrett *et al.* 1991; Citi *et al.* 1989; Cross *et al.* 1988; Jackson and Bagshaw 1988).

A feature of all these preparations is that the 'off-state' at low Ca^{2+} results from inhibition of the ATPase through the action of the light chain subunits. If these subunits or the crucial neck region of the heavy chain are removed or damaged in some way, then the ATPase becomes activated to its high default value. As a result, most preparations contain a small fraction of permanently active molecules, which dominant the steady-state ATPase in the absence of Ca^{2+}. By adding near stoichiometric amounts of nucleotide triphosphate, a single turnover of the ATPase can be monitored, which provides a direct estimate of both the fraction of regulated and unregulated molecules present and their specific ATPase activities. While this can be done discontinuously using radioactive ATP in combination with a physical separation technique (Cross *et al.* 1986; Wells and Bagshaw 1985), it is much more convenient to use a fluorescent analogue such as FTP. Two levels of nucleotide trapping have been characterized. With heavy meromyosin and myosin molecules in filaments, removal of Ca^{2+} slows down product release from the M.ADP.Pi state by up to 50-fold (Jackson and Bagshaw 1988). On the other hand, with soluble myosin the tail has the capacity to fold back on itself and interact with the neck region, which is accompanied by a further 10-fold reduction in the product release step (Ankrett *et al.* 1991; Cross *et al.* 1988). This compact form of myosin has a higher sedimentation coefficient (10S) compared with the conventional straight conformation (6S) and may represent an inert storage form of myosin, which exists prior to filament assembly in developing muscle or cellular reorganization. More recently, mant-ATP has been used to probe this conformational state (see next section and Rosenfeld *et al.* 1994).

All aspects considered, mant-ATP offers the best fluorescent analogue for solution studies. The apparent second-order association rate constant with myosin is similar to that of ATP, while the plateau rate constant, controlling the hydrolysis step, is also a good match (Woodward *et al.* 1991). As monitored by the Pi linked assay (Webb and White 1996), the acto-S1 hydrolysis step is about two-fold slower than that for ATP (H.D. White, personal communication). However, the mant-ADP dissociation rate constant is about 10-fold slower than that of ADP, and this is reflected in its tighter binding under equilibrium conditions. Mant-ADP dissociation from the ternary acto-S1.mant-ADP complex revealed biphasic kinetics suggesting two acto-S1.nucleotide diphosphate states.

Analogues designed for studies using optical microscopy have exploited fluorophores with high molar absorption coefficients and fluorescence excitation in the visible region, well beyond potential energy transfer from tryptophan. As a result, they are not ideal for conventional solution studies, where optical pathlengths exceed 1 mm, because they are difficult to characterize at concentrations above 100 μM. Nevertheless, transient kinetic studies with bulky fluorophores attached via an ethylenediamine linker to the 2′(3′) ribose hydroxyls (EDA derivatives), indicate that their rate constants for interaction with myosin and actomyosin are perturbed less than four-fold (Conibear *et al.* 1996; Eccleston *et al.* 1996).

Analogues based on imidophosphate derivatives (derivatives of AMPPNP), generally show apparent second-order association rate constants that are an order of magnitude slower than that of the corresponding nucleotide triphosphate (Bagshaw *et al.* 1972; Jackson and Bagshaw 1988). This parameter, together with their relatively high dissociation rate constant, suggests they are not ideal analogues of the M.ATP state (Trentham *et al.* 1976). Although less widely used and more difficult to synthesize, analogues based on ATPγS may be better because the apparent association rate of the latter is comparable to that of ATP (Bagshaw *et al.* 1974).

Actomyosin structural studies

The sensitivity of fluorescence emission to the environment often poses a problem in interpretation in structural terms. That is, large intensity changes may reflect relatively minor local structural events. However, with appropriate methodology, questions that might be tackled include:

(1) is the fluorophore exposed to or protected from the solvent?

(2) does the fluorophore have independent motion when bound to a protein or protein domain?

(3) is the fluorophore within about 6 nm of a second fluorophore at another location?

In addition, fluorophores can be incorporated into affinity labels so that active site residues can be identified without the need for radioactive tracers.

Solvent exposure can be assessed by adding agents that quench fluorescence (Lakowicz 1983). Two basic mechanisms of quenching exist. In the case of static (ground state) quenching, the agent and fluorophore pre-combine to form a species with reduced fluorescence. Collisional quenching, however, refers to when the agent combines with the excited state of the fluorophore to provide a radiationless decay pathway and hence reduces both the fluorescence lifetime and intensity. Their contributions can be ascertained from a Stern-Volmer plot in which fluorescence quenching is plotted as a function of quenching agent concentration $[Q]$:

$$F_0/F=(1+K_D\cdot[Q])(1+K_S\cdot[Q]),$$

where F_0 is the fluorescence intensity in the absence of quencher, F the observed fluorescence in the presence of quencher, K_D is the dynamic quenching constant and K_S the static quenching constant. Static quenching leads to an upward curvature of the plot F_0/F against $[Q]$, while multiple sites lead to downward curvature. Ambiguities may be resolved by measuring the effect of the agent of fluorescence lifetime (τ), which only effects collisional quenching:

$$1/\tau=(1/\tau_0)+k_q\cdot[Q],$$

where k_q is the bimolecular quenching constant ($=K_D/\tau_0$ where τ_0 is the lifetime in the absence of quencher).

Using this approach, Rosenfeld and Taylor (1984) showed that ε-nucleotides were protected from acrylamide in the solvent in the M.ε-ADP.Pi and M.ε-ADP complexes to a similar degree, but both indicated heterogeneity of sites, suggesting two conformational states for each complex. Thus the Stern-Volmer quenching constant K_D for the free nucleotides of $57\,M^{-1}$ is reduced to $10\,M^{-1}$ and $1\,M^{-1}$ for the bound states.

In the light of the subsequent crystal structures of myosin (Fisher *et al.* 1995*a,b*), it is to be expected that fluorophores based on adenosine derivatives are protected from solvent, provided the fluorophore is not so bulky as to prevent it binding to the active site. However, it is somewhat surprising that the mant group attached to the ribose moiety in mant-nucleotides is also protected from quenching reasonably efficiently when bound to skeletal myosin S1 beacuse the hydroxyl groups, which are derivatized, face the solvent. Cremo *et al.* (1990) reported a reduction in K_D from $2.4\,M^{-1}$ to $0.2\,M^{-1}$ on binding mant-ADP. Comparable values for the bound species were found by Franks-Skiba and Cooke (1995).

The discussion following the Franks-Skiba and Cooke (1995) paper illustrates the uncertainty as to what extent protection would be afforded when a fluorophore associates with the protein by one face. There is also the possibility of protection being dependent on local structure, which varies between myosin isoforms. The *Dictyostelium* myosin head domain, for which nucleotide binding has been defined by X-ray crystallography, has a shorter linker region between the 25 kD and 50 kD fragments, which lies at the opening of the nucleotide pocket. Rosenfeld *et al.* (1994) found mant-nucleotides were somewhat less protected in smooth muscle myosin S1 compared with that from skeletal muscle.

Binding of the fluorophore in a protein pocket or association with the surface residues will change its rotational properties and, therefore, fluorescence depolarization measurements should report on this event. In the case of myosin, most studies to explore protein domain movements have been performed by attaching probes covalently to reactive cysteine or lysine residues in the heavy or light chain subunits. However, provided conditions can be established where the nucleotide is predominantly bound at the active site, there is no reason why rotational information cannot be obtained using fluorescent nucleotides. Nucleotide can be trapped using the phosphate analogues AlF_4 and BeF_x mentioned previously. This approach was used by Rosenfeld *et al.* (1994) to demonstrate a reduction in head mobility of smooth muscle myosin, as revealed by the time resolved anisotropy of bound mant-ADP.BeF_x, when the molecule folded up from its 6S to 10S conformation. This is in line with electron microscopy studies that indicate the heads become locked in a predominantly 'heads-down' conformation.

Quantitative distance information can be obtained by fluorescence resonance energy transfer (FRET) measurements (dos Remedios *et al.* 1987; Lakowicz 1983). This involves the dipolar interaction between the excited state of one fluorophore (the donor), which exchanges energy to a second chromophore, or fluorophore (the acceptor), whose absorption spectrum overlaps with the emission spectrum of the donor. As a result, the donors emission intensity is reduced and the acceptor, if fluorescent, emits at its characteristic longer wavelength. The efficiency of transfer depends on the sixth power of the distance between the donor and acceptor. However, this process is also sensitive to the relative orientation of the emission and excitation dipoles, a factor which often is poorly characterized. If both fluorophores are mobile (as indicated by a low polarization or anisotropy value) then all relative orientations are explored and an appropriate average orientation factor can be calculated. More often than not, bound fluorophores show constrained motion—most often rapid motion within a restricted cone—which is more difficult to characterize in practice. The other main practical difficulties are labelling the sites specifically and near stoichiometrically, and in controlling for changes in fluorescence intensity which are due to conformation changes as a consequence of the label.

The procedure for FRET measurements firstly involves selection of suitable donor–acceptor pairs and selective labelling of the two sites. In the context of this chapter, one site would involve the myosin or actin nucleotide binding site using the fluorescent ATP analogues discussed above. The donor fluorescence emission yield (Φ) or lifetime (τ) is then measured in the absence ($_{\rm d}$) and presence ($_{\rm da}$) of the acceptor and the efficiency of FRET estimated from:

$$E = (1 - \tau_{\rm da}/\tau_{\rm d}) = (1 - \Phi_{\rm da}/\Phi_{\rm d}).$$

To allow for conformational effects it is useful to use a dummy acceptor with similar properties but lacking absorption in the region of the donor emission. The need for near stoichiometric labelling becomes apparent from the ambiguity introduced by substoichiometric labelling. Thus 50% labelling at each site could give anywhere between 0 and 100% interacting pairs, depending on any synergistic or antagonistic interactions during labelling.

The theoretical distance, R_0 (in nm) which would give 50% reduction in the measured Φ or τ can be calculated for pairs of donors and acceptors from the expression:

$$R_0 = 9.79 \times 10^2 (\kappa^2 \cdot n^{-4} \cdot \Phi_{\rm d} \cdot J)^{1/6},$$

where κ^2 is the orientation factor and can take on a value between 0 and 4 (for freely mobile fluorophores $\kappa^2 = 2/3$), n is the refractive index of the medium, $\Phi_{\rm d}$ is the quantum yield of the donor, most readily determined by comparison with a known standard having similar spectral properties, and J is the overlap integral of the donor emission spectrum and the acceptor absorption spectrum. From R_0 the actual distance of separation, r is calculated from:

$$E = R_0^6/(R_0^6 + r^6).$$

Clearly there are many potential uncertainties in the estimated r, added to which there may be further problems in the actomyosin system of FRET occurring between heads rather than within heads and between different G-actin monomers within an F-actin filament. The review by dos Remedios *et al.* (1987) gives a very comprehensive summary of FRET measurements on the actin and myosin, involving ten distinct sites. Moreover, this work precedes knowledge of the high-resolution crystal structures of actin (Kabsch *et al.* 1990) and myosin S1 (Fisher *et al.* 1995*a*,*b*; Rayment *et al.* 1993*a*) and therefore provides a true blind test of the methodology. Table 4.2 compares the distances estimated between the myosin active site and actin nucleotide site with other covalently attached probes by FRET and their approximate separations based on the crystal structures. Agreement is as good as might be expected in many cases. Clearly, having a crystal structure as a reference gives more confidence to FRET measurements. Nevertheless, without such a structure, dos Remedios *et al.* (1987) were able to identify inconsistent pairs (e.g. actin nucleotide site to myosin lysine 83), by triangulating several sets of data. While FRET is grossly inferior to crystallography for distance measurements of detailed structure, it has the potential advantage of being able to measure changes in distance in solution with good time resolution. However, there remains a problem of interpretation as to whether any observed change in energy transfer efficiency reflects a change in r or κ.

Another technique that has provided structural information prior to the availability of a crystal structure is affinity labelling. Even now that high resolution structures are available, labelling studies may complement the information by identifying interactions between flexible domains. Affinity labelling of myosin, using nucleotide analogues, proved difficult until methods were developed to trap the nucleotide specifically at the active site in an unreactive form so that non-specifically bound material could be removed. The bound nucleotide affinity label could then be photoactivated to explore nearby active site amino acids without a high background of non-specific labelling. Trapping was achieved either by cross-linking the SH1/SH2 reactive thiols with pPDM (*N*,*N*′-*p*-phenylenedimaleide) or using vanadate or other phosphate analogues.

Cremo and Yount (1987) used photoaffinity labels based on ε-ADP in which a benzoylbenzoyl group (Bz_2) was coupled to the 2′ and 3′ ribose hydroxyls. On trapping at the active site by pDPM cross-linking, photoactivation resulted in the ε-nucleotide being incorporated into the 50 K fragment of the heavy chain. Hiratsuka (1985) found that the 50 K fragment was preferentially labelled when incubated with mant-ATP and a photosensitizer, flavine mononucleotide (FMN). These results are in line with the crystal strucutre of myosin S1, which show that groups attached to the ribose 2′ and 3′ hydroxyls could contact the 50 K fragment, in contrast to the adenine moiety that is buried in the N terminal 25 K fragment (Fisher *et al.* 1995*a*,*b*).

Measurements in myofibrils and fibres

Although myofibrils and muscle fibres scatter more light than proteins in solution, the discrimination and sensitivity of fluorescence allows measurements to be made without severe problems in terms of detection capabilities. However, difficulties may arise owing to the very high local concentration of myosin sites leading to local

Table 4.2 Comparison of distances between sites based on FRET measurements with fluorescent nucleotide analogues with the crystal structures of actin, myosin S1 and the acto-S1 complex

Donor site	Donor	Acceptor site	Acceptor	FRET distance (nm)	Crystal distance (nm)
Myosin SH1 cys 705	IAEDANS	Myosin nucleotide (ribose)	TNP-ADP	2.9–3.9	2.7
Myosin nucleotide (adenine)	ε-ADP	Myosin SH1/SH2	nitro-benzene derivatives	2.3–2.6	1.6–2.7
Myosin ELC cys 177	IAEDANS	Myosin nucleotide (ribose)	TNP-ADP	5.5	≈5
Myosin nucleotide (adenine)	ε-ADP	Myosin lys 83	TNP	4.2	2.8
Myosin nucleotide (adenine)	ε-ADP	F-actin nucleotide (ribose)	TNP-ADP	6.2	5.8
F-actin nucleotide (adenine)	ε-ADP	Myosin nucleotide (ribose)	TNP-ADP	7.7	5.8
F-actin cys 374	IAEDANS	Myosin nucleotide (adenine)	TNP-ADP	5.1	5.3
Myosin SH1 cys 705	IAEDANS	F-actin nucleotide (ribose)	TNP-ADP	6.1	6.7
F-actin nucleotide (adenine)	ε-ADP	Myosin SH1 cys 705	IAF	7.2	6.7
F-actin nucleotide (adenine)	ε-ADP	Myosin lys 83	TNP	4.5	8.3
G-actin nucleotide (adenine)	ε-ADP	G-actin cys 374	IANBD	3.3–3.5	2.7
G-actin nucleotide (adenine)	ε-ADP	G-actin cys 10	DDPM	3.9	1.9
G-actin nucleotide (adenine)	FTP	G-actin Tyr 69	Dansyl-Cl	2.1	0.7

FRET distances were obtained from the review of dos Remedios *et al.* (1987), which should be consulted for original references. Crystallographic data were obtained from Rayment *et al.* (1993*a*) for S1, Kabsch *et al.* (1990) for actin and Rayment *et al.* (1993*b*) for the acto-S1 complex, and analysed by A. Gulick and P.A. Kuhlman who measured the distances between the points of attachment of the fluorophores to the nucleotide or amino-acid side chains. Thus the fluorescence dipoles may be up to 0.5 nm from this position. Note that the major discrepancy with the distance between the actin nucleotide site and myosin lys 83 was previously identified because it was internally inconsistent with the fluorescence data (dos Remedios *et al.* 1987).
Abbreviations: IAF, iodoacetamido fluorescein;
IANBD, 4-[N-(iodoacetoxy)ethyl-N-methyl]amino-7-nitrobenz-2-oxa-1,3diazole;
DDPM, N-(4-dimethylamino-3,5-dinitrophenyl) maleimide; TNP, trinitrophenyl;
IAEDANS, N-iodoacetyl-N′-(1-sulpho-5-naphthyl)ethylenediamine. For others see Fig. 4.1

substrate depletion and product build-up within the myofibril lattice. Myofibrils provide a useful intermediate preparation between homogenous solutions and single muscle fibres, in which diffusional lags are not quite as severe. Nevertheless, there remains a problem of controlling myofibril length under activating conditions so as to maintain their ordered structure. A number of strategies have been employed in an attempt to retain the myofibrillar structure.

By working at high nucleotide concentrations under relaxing conditions, no significant shortening occurs. Using this strategy, Myburgh *et al.* (1995) were able to measure the turnover rate of the myosin ATPase in myofibrils by following the displacement of ATP by mant-ATP and considered that its value was about two-fold lower than that of isolated myosin under the same conditions. Generally, when using fluorescent analogues, it is easier to follow the displacement of the fluorescent nucleotide by a large excess of ATP, but this gives the characteristic turnover rate for the fluorescent analogue (k'_{to}). In the case of mant-ATP, the turnover rate constant is so similar to that of ATP, that Myburgh *et al.* (1995) followed the partial displacement of ATP by a limited concentration of mant-ATP to obtain the rate constant common to both nucleotides. The theoretical expression for the observed rate constant k_{obs}, is a function of both the association (k_{on}) and turnover rate constants (k_{to}) of both nucleotides:

$$k_{obs} = k'_{to} \cdot k_{on} \cdot [\text{ATP}]/(k'_{on} \cdot [\text{mant-ATP}] + k_{on} \cdot [\text{ATP}]) + k_{to} \cdot k'_{on} \cdot [\text{mant-ATP}]/(k'_{on} \cdot [\text{mant-ATP}] + k_{on} \cdot [\text{ATP}]).$$

The independence of the observed rate constant with a change in the concentration of the displacing nucleotide, argued the procedure of Myburgh *et al.* (1995), was valid. More generally, increasing the concentration of the displacing nucleotide gives, at its limit, the true turnover rate of the displaced nucleotide.

In contrast to these studies, Smith and White (1985*b*) found that single turnovers of ε-aza-ATP, examined in a stopped-flow apparatus, were practically identical for cardiac myofibrils compared with acto-S1 under relaxing (non-activating) conditions.

Under activating conditions with myofibrils, either measurements have to be made within about 100 ms during the shortening phase or some means of supporting the tension is required. The latter can be achieved, in principle, by working at low nucleotide, where a significant fraction of the heads remain in rigor, or by limited chemical cross-linking to hold the myofibrilar lattice in register (Herrmann *et al.* 1993). A better method, but more tedious to apply routinely, is to attach the myofibrils to glass needles. This method has the advantage that the force and velocity may be quantified by using needles of different compliance (Anazawa *et al.* 1992; Chaen *et al.* 1996). In the latter study, myofibrils were first incubated with a low (0.2 μM) concentration of REDA-ATP, which selectively stained the A band owing to the local high concentration of myosin heads (Fig. 4.4b). On release of excess ATP by flash photolysis of caged-ATP, the

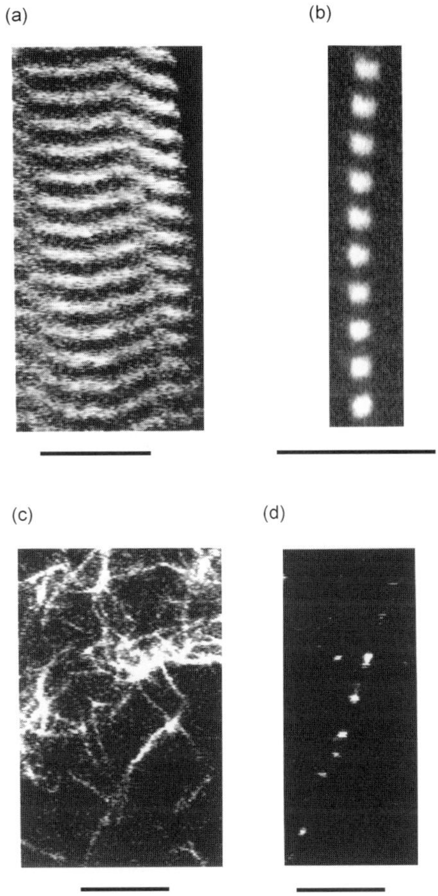

Fig. 4.4 Fluorescent ATP analogues bound to muscle preparations (a) Rabbit psoas muscle fibres were incubated with 1 μM REDA-ATP to show selective staining of the A-band. (b) A single rabbit psoas muscle myofibril was mounted between two glass needles and incubated with 200 nM REDA-ATP and 0.1 mM caged-ATP. Upon flash photolysis the fluorescence from the A-band disappeared over several seconds (Chaen *et al.* 1996). (c) FEDA-ADP was trapped as its AlF_4 complex by reacting 2 μM rabbit myosin with 10 μM FEDA-ATP in the presence of 0.1 mM $AlCl_3$ and 5 mM NaF in 0.5 M NaCl. After 2 h incubation, synthetic rabbit myosin filaments were made by dialysing the mixture against 0.12 M NaCl, 20 mM imidazole at pH 6.8, and examined by fluorescence microscopy. (d) Native thick filaments were isolated from the anterior byssus retractor muscle of *Mytilus edulis* and incubated with REDA-ATP in the absence of Ca^{2+}. At 2 nM nucleotide concentrations, fluorescent streaks were observed corresponding to the position of single thick filaments, while at 0.5 nM nucleotide the fluorescence was punctate, as shown, suggestive of single molecule labelling or local regions of myosin under co-operative control (Miyata *et al.* 1996). Scale bars = 10 μm.

Parts (a) and (b) were obtained courtesy of S. Chaen and I. Shirakawa using a confocal microscope with an Arg ion laser at 514.5 nm. (c) Was recorded by P.B. Conibear using a conventional epifluorescence microscope and Hg arc illumination, while (d) was obtained by H. Miyata, S. Chaen and colleagues using the prism-based TIRF method in conjunction with a frequency-doubled YAG laser at 532 nm.

REDA-nucleotide was displaced over a period of seconds. Interestingly the rate constant was an order of magnitude faster than that expected for myosin alone, even under relaxing conditions, probably because the high local concentration of myosin heads within the myofibril lattice (about 240 μM, Bagshaw 1993) caused local depletion of REDA-ATP and a build-up of REDA-ADP. As a result, the predominant steady-state complex was likely to be M.REDA-ADP, which dissociates with a rate constant of around $0.8\,s^{-1}$ (Conibear and Bagshaw 1996). At a higher initial concentration of REDA-ATP of 5 μM, the displacement rate constant ($0.2\,s^{-1}$) approached that expected for the M.REDA-ADP.Pi complex. However, under these conditions the background fluorescence increased significantly, so that the A-band contrast was lower, necessitating the use of confocal microscopy for imaging. With these baseline values established for relaxed myofibrils, the effect of actin-activation under isometric or auxotonic contraction could be investigated.

Muscle fibre studies with ATP analogues require the fibre or fibre bundle to be permeabilized. The simplest method is to glycerinate the preparation. Using such an approach, Yanagida (1981) investigated the binding of ε-nucleotides to rabbit skeletal muscle and measured the polarization along and across the fibre axis of the bound nucleotide by microspectrofluorimetry. Under each condition, four polarization measurements were made, corresponding to parallel and perpendicular polarized light measured parallel and perpendicular to the fibre axis. These parameters were modelled on a helical array to calculate the average angle of the absorption and emission dipole and their angular spread. In relaxed fibres, or those stretched beyond filament overlap, the fluorescence polarization indicated that the ε-nucleotides were oriented randomly. On the other hand, under conditions where significant numbers of the myosin heads were attached, a preferred orientation was observed. At partial occupancy with nucleotide, these angles were similar for both the A.M.ε-ATP and A.M.ε-ADP states, indicating that the active site does not change angle significantly during the phosphate release step conventionally associated with tension generation. This conclusion has also been reached with probes attached covalently at the reactive SH1 cysteine and indicates that any angle change might be restricted to more distal regions of the head (i.e. the neck region bearing the light chains). As the ε-nucleotide concentration neared saturation, the polarization data did not fit a single class of sites, possibly reflecting the dissociation or differential interaction of the second head with the actin filament.

The emission intensity of ε-nucleotides, when bound to myosin subfragments, does not change significantly ($< 10\%$), so that intensity measurements alone are not informative as to the bound chemical state (Yanagida 1981). It does, however, allow quantitation of the bound fluorescence from the increased signal intensity compared with a control sample in which excess ATP is present to block nucleotide binding to the active site. These studies confirmed that the myosin active site concentration in a skeletal muscle fibre is around 170 to 270 μM. Variations can arise from the fibre swelling after the skinning procedure.

In a related study, Nagano and Yanagida (1984) showed the fluorescence of ε-aza-ATP was much more sensitive to the environment giving around 12-fold enhancement on binding to myosin at optimal wavelength settings. Furthermore, the bound nucleotide during the steady-state hydrolysis (M.ε-aza-ATP/M.ε-aza-ADP.Pi) was more than twice as fluorescent as the bound ADP state (M.ε-aza-ADP). With this information, attempts were made to perturb the relative populations of these states by allowing the muscle to generate tension against high or low loads. While ε-aza-ATP supported an isometric tension comparable to that of ATP, little change in fluorescence intensity was found amongst various conditions, suggesting A.M.ε-aza-ATP/A.M.ε-aza-ADP.Pi remained the predominant chemical state of the attached cross-bridges. Although ε-aza-ATP appeared a good analogue as far as tension development was concerned, ten times higher concentrations of the analogue were required to reach the maximum compared with ATP. This is in line with the reduced apparent second-order association constant for ε-aza-ATP, compared with ATP as monitored by dissociation of acto-S1 (White *et al.* 1993).

Muscle fibres may also be readily visualized with the fluorescent nucleotides specifically developed for microscopy (Fig. 4.4a).

Microscopy and *in vitro* motility assays

With the advent of the *in vitro* motility assay to follow directly the relative sliding of actin filaments over myosin heads (Kron and Spudich 1986), attention has focused on the coupling to ATP hydrolysis in an attempt to address the step-size problem (Burton 1992; Harada *et al.* 1990; Uyeda *et al.* 1990). Initially ATP turnover within the flow cell was determined by measuring Pi production after several minutes by the bulk preparation (Harada *et al.* 1990; Kron *et al.* 1991). However, it would be desirable to measure ATP turnover by the filaments actually under observation to get a direct measure of the ATP utilized per μm of sliding. One approach to this is to use fluorescent ATP analogues as substrates and determine their rate of turnover or exchange at the myosin active site by high power light microscopy. So far, this has been reported for myosin preparations in the absence of actin.

The first attempts at imaging myosin filaments through mant-nucleotide binding showed the potential of this technique but revealed deficiencies in fluorophores that were initially developed for solution studies (Bagshaw *et al.* 1992). Excitation in the near u.v. is possible in microscopes with neofluar objectives but nevertheless transmission is poor and there is a high background fluorescence from other components, such as glass and immersion oil. Furthermore, although mant-nucleotides are reasonably stable at light intensities used in standard fluorimeters, the light intensity required for microscopy is considerably higher and caused extensive photobleaching. For this reason analogues were synthesized that contained more suitable fluorophores, which had been developed for light microscopy. The generally desirable features to achieve sensitivity at the single molecule level are a high molar absorption coefficient, high quantum yield and high photostability.

Sowerby *et al.* (1993) showed that clam thick filaments containing several thousand myosin sites could be readily imaged in the presence of low concentrations (50 nM) of FEDA-ATP. Contrast is obtained because the local concentration of myosin heads is much higher than the free nucleotide concentration and the FEDA-ATP concentration used is close to its K_m giving around 50% saturation of the active sites. The FEDA-ATP turnover rate was measured directly by adding a large excess of ATP and following the displacement kinetics with photomultiplier detection. Subsequently, this assay was modified by Conibear and Bagshaw (1996) for REDA-ATP, which has better photostability. Time resolution was improved by using flash photolysis of caged-ATP to effect displacement and sensitivity increased by using total internal reflectance fluorescence microscopy to reject signal from the free REDA-ATP in the bulk solution. In setting up filament preparations for microscopy, it is useful to trap the fluorescent nucleotide at the active site in a stable manner, as has been done with FEDA-ADP.AlF_4 complex (Fig. 4.4c), so that free nucleotide can be washed out to give better contrast. Studies of assemblies of myosin heads in filaments or tracks is important as it can be combined with the actin filament sliding assay in an attempt to determine the number of ATP molecules turned over during active sliding.

On the other hand, it is useful to extend the sensitivity to the single molecule level, so that single force and displacement events can be correlated with single turnovers of the actomyosin ATPase. To this end, Funatsu *et al.* (1995) achieved the impressive feat of imaging single molecules of Cy3-N^6-ATP while being turned over by immobilized S1. This was done by using a very low concentration (10 nM) of fluorescent nucleotide, so that fluorescence spots could only be imaged when immobilized on binding to the S1 active site. The lifetime of this state was around 17 s, corresponding to a turnover rate of $0.059\,s^{-1}$. Free nucleotide moved too rapidly to be imaged, but its contribution to the background signal was minimized using total internal reflectance microscopy. Other measures were also needed to reduced the background to allow single molecule detection, including use of sharp cut-off filters, a selected low fluorescence objective lens, low fluorescence immersion oil, and quartz coverslips. Further technical points are described by Sase *et al.* (1995), who achieved single molecule detection with tetra methylrhodamine-labelled actin molecules. In addition to the above, they recommend for epifluorescence set-ups, that the dichroic holder is modified to let the small proportion of unreflected excitation light to exit through the back of the holder, rather than scatter within it. From these results it is likely that REDA-ATP will also be suitable for single molecule ATPase studies (Fig. 4.4d; Miyata *et al.* 1996).

The limiting background signal in most fluorescence measurements is that resulting from scattering. Rayleigh scattering causes no change in the wavelength and hence is minimized by use of high quality barrier filters and reduction of dust from optical components and solutions. Raman scattering is generally much weaker but the scattered light is shifted and may overlap with fluorescence emission wavelengths. The predominant source of Raman scattering in biological samples is from the solvent water and hence generally cannot be removed. However, with judicious choice of excitation wavelength and the emission band width, it is possible to minimize the overlap with the desired fluorescence signal. To a reasonable approximation, the wavelength of the Raman peak (in nm) can be calculated from the relationship $\lambda_R = 1.28 \cdot \lambda_{ex} - 49.5$. It is worth having a graph relating the Raman scatter peak to the excitation wavelength to hand by any fluorimeter because such peaks are often the source of initial confusion in excitation and emission spectra.

Conventional actin sliding *in vitro* motility assays provide a functional test for ATP analogues. Mant-ATP supports sliding at about half the velocity of that with ATP (Bagshaw *et al.* 1992), while FTP and ε-ATP show about a quarter of the velocity (Shimizu *et al.* 1991). With Cy3 analogues the background fluorescence swamps that of the label on the actin, so only μM nucleotide concentrations can be explored. While Cy3-EDA-ATP and Cy5-EDA-ATP support movement comparable to that for low ATP concentrations, it is difficult to perform a quantitative comparison (Oiwa *et al.* 1996). It is preferable to use assays where fluorescence does not interfere. At 0.5 mM Cy5-EDA-ATP, the maximum shortening velocity of permeabilized rabbit fast-twitch fibres was 10% that for ATP (Eccleston *et al.* 1996).

4.3 Problems and limitations

In the above sections, some of the problems and limitations of fluorescence-based assays have already been addressed. It is appropriate to summarize these. The paramount factor for most fluorescent analogues is, how well does the analogue mimic ATP? Physiological tests, such as isometric tension and maximum shortening velocity, are probably the best to use (Pate *et al.* 1993). *In vitro* motility assays may be complicated by the analogue flourescence, although other assays, such as bead sliding, should be unaffected (see Chapter 1, Section 1.7). Ideally an analogue will perform within a factor of two of that for ATP under the same conditions. Solution kinetics can sometimes identify a specific difference, such as reduced association or hydrolysis rate constants. In this context it is appropriate to note that when fluorophores are attached to the ribose there is a possibility that they may associate intramolecularly with the adenine base, and thus present unfavourable conformation for binding. Rensland *et al.* (1991) found that the 2′ and 3′ isomers of mant-GTP had a different quantum yield, which may be a result of the 2′ isomer showing more extensive association. Likewise, Cremo and Yount (1987) suggested that the 2′ isomer of the benzoylbenzoyl (Bz_2) photoaffinity label derivative of ε-ADP caused static fluorescence quenching, presumably by intramolecular association with the adenine ring. However, no such difference in quantum yield has been reported for mant-adenosine nucleotides. Molecular modelling of FEDA-ATP also revealed the possibility of intramolecular association of the fluorescein and adenine groups but the measured fluorescence lifetime of 4.2 ns, is very similar for that of fluorescein alone, suggesting internal quenching is not extensive (Conibear *et al.* 1996). These complications also raise another difficulty in that molar absorption coefficients may be affected on derivatization, leading to uncertainities in concentration.

Whatever the source of analogues, the problem of analogue purity and heterogeneity also have to be considered. Many fluorophores are available only in small quantities and hence might not be subject to extensive characterization. Furthermore, in some derivatives there is a possibility of positional isomers (e.g. most commercial RITC sources are mixtures of 5 and 6 derivatives and hence 2′(3) REDA-ATP made from these will comprise four isomers). Given that no sample is 100% pure, a more practical question to address is, what is the consequence of an impurity at a given level? Thus a fluorescent ATP analogue may be contaminated with corresponding fluorescent di- and monophosphates, the free fluorophore and ATP or another reagent nucleotide. These can be added to the assay in turn to see how sensitive the measured parameters are. In some cases the assay can be devised to minimize the effects of contamination or heterogeneity problems. Single turnover kinetic measurements (Fig. 4.3) are particularly valuable because by using a large excess of myosin active sites, any turnover of contaminant nucleotides are likely to remain silent, while any differential turnover of positional isomers will be resolved. A similar argument applies to FRET measurements, where steady-state flourescence emission intensities may represent a complicated weighted average, while direct measurements of the donor lifetime in the presence and absence of acceptor will show any significant heterogeneity.

Positional isomers have a particularly severe effect on orientation studies using fluorescence polarization because information is obtained from just four measurements and have to be analysed in terms of a predominant orientation and spread. Thus, mixed states may give an average orientation angle that is applicable to none of them. This is in contrast to epr studies where the spin label probe provides angularly resolved information (Thomas 1987). However, fluorescence measurements are more than an order of magnitude more sensitive, permitting single fibre measurements (Cooke 1982).

Fluorescence and epr spectroscopy also provide alternative methods for measuring rotational motion of proteins or protein domains on the μs time scale (Thomas 1987). Given that fluorescence lifetimes are usually a few or few tens of ns, the time-resolved fluorescence depolarization techniques tend to run out of signal for motions longer than 1 μs. Saturation transfer epr, however, extends the magnetic resonance technique to the ms time-scale. In principle, phosphorescence anisotropy could report on slower motions by optical means, but the choice of labels is limited.

The problem of photodecomposition with respect to ATP analogues has come more to the fore with the increased use of microscope based-assays. It is particularly severe in the case of motility assays because photodecomposition results not only in loss of signal but also possible generation of free radicals, which can cause cross-links between proteins (Knight and Parsons 1991). Only a small fraction of cross-links need form between actin and myosin to bring motility to a halt. Oxygen scavenging systems may help but, on the other hand, the flavin moiety in glucose oxidase may contribute to background fluorescence and itself is another potential source of free radicals. However, addition of thiols is generally prudent. The ability to initiate motility using flash photolysis (Conibear and Bagshaw 1996) shows that high light intensities and motility are not mutually exclusive situations. In general, it is preferable to work with analogues which fluoresce at the red end of the spectrum to minimize the energy of the excitation light. However, this decision is a compromise because the detector sensitivity may not be well matched, thus requiring higher excitation light intensities to give a comparable signal.

4.4 Future prospects

The approach that is likely to show the most development over the next few years is developing ATP analogues suitable for optical microscopy. In turn, microscopies that restrict the volume of sample under study to allow the kinetics of single or small numbers of molecules to be followed will be further advanced.

There are a number of ways to restrict the region under observation and reject the signal from the bulk free fluorophore, some of which have been eluded to above. For surface assays, the total internal reflectance fluorescence (TIRF) method is particularly suitable. Here, the excitation light is delivered to the glass/water interface at a glancing angle, so that it undergoes total internal reflection. Fluorescence is, however, induced in molecules close to the interface, which are excited by the evanescent field. The field strength typically decays exponentially with a half-penetration depth of the order of 100 nm. Thus, molecules in the bulk solution, which may be several tens of μm thick, are not excited. Usually the light source is a laser, which is more readily collimated than an arc lamp, and the beam is introduced via a prism in optical contact with the glass surface (Axelrod 1989; Conibear *et al.* 1995; Funatsu *et al.* 1995). However, it is also possible to introduce the light around the periphery of the back focal plane of a 1.4 N.A. oil immersion objective lens, so that TIRF is induced by epi-illumination by rays above the critical angle (typically around 61° for a glass/water interface; Stout and Axelrod 1989). This method can be used with an arc lamp, although it is difficult to direct the light path as well as with a laser (Conibear and Bagshaw 1996).

The standard prism method for TIRF demands that the objective lens be focused through the sample on to the opposite face of the aqueous layer. For high-powered objective lenses (60–100 ×) necessary for *in vitro* assays, this limits the sample thickness to a maximum of about 50 μm and there is some loss in image quality. The advantages of achieving TIRF through the objective lens include focusing on the near face for optimal correction of lens aberrations, no limit to sample thickness, ease of scanning the field without the need to readjust the prism position, free access to the opposite face for other apparatus (e.g. photolysis light source) and the need for reduced laser powers (< 1 mW). On the other hand, the field of view that is illuminated is restricted to around 30 μm diameter, there is more potential for stray light entering the bulk solution from spurious reflections within the objective and there is less control over the precise penetration depth owing to the range of angles of the incident light.

The major difficulty with TIRF is achieving a perfectly optical flat surface to generate the evanescent field because even 1% scattered light will cause increased fluorescence in the bulk solution. Glass surfaces may be cleaned by brief exposure to HF (Todd *et al.* 1988), while freshly cleaved mica provides an alternative substrate. Depending on the protein involved, surfaces may need coating to allow it to adhere. In general, nitrocellulose is to be avoided owing to its high background fluorescence. Polylysine is better in this regard (Sase *et al.* 1995). Imperfections in coating and the protein itself can lead to scatter of the excitation light, thus degrading the theoretical contrast enhancement. A promising extension of the TIRF method is to coat the surface with metallic silver to induce surface plasmon resonance. This reduces the penetration depth to around 10 nm but gives a stronger field in the region of the immobilized proteins (Saito *et al.* 1996).

Another way to obtain necessary z resolution to reduce background fluorescence is by confocal microscopy (Pawley 1995). Depending on the pin hole size, the optical section is of the order of 1 μm, somewhat greater than that of TIRF. On the other hand, confocal microscopy is not restricted to a surface and hence it is more suitable for preparations held under mechanical restraint in the bulk solution (Chaen *et al.* 1996). A wide variety of scanning mechanisms exist, some of which allow data to be collected at standard video rates or faster (Pawley 1995). An important consideration in confocal microscopy is the laser power. An excellent discussion of optimizing laser power and choice of fluorophore for confocal microscopy is given by Tsien and Waggoner (1995), although many of the points they raise have general significance. A useful rule to note is that under any particular condition, the quantum yield for photodecomposition together with the quantum yield for fluorescence defines a fixed number of fluorescence photons that can be emitted by a compound before it ultimately degrades (e.g. around 30 000 photons for fluorescein). The excitation light intensity must be chosen so that the experiment is complete before a significant amount of photodecomposition occurs. On the other hand, using too low a light intensity is wasting potentially available signal. In confocal microscopy, the light intensity at the sample is potentially very high because the excitation beam is brought to a diffraction limited spot of the order of 0.2 μm diameter. When such high excitation intensities are used it is possible to achieve optical saturation, where few molecules are left in the ground state and the intensity of the emission becomes one of diminishing returns. Worse still is the problem that any longer-lived triplet states are more readily saturated than the desired singlet state and the former become a kinetic sink if the sample is illuminated continuously. This effect can be reduced in confocal microscopy by minimizing the dwell time, i.e. the time that each region is illuminated, to less than the triplet-state lifetime (e.g. to around 1 μs). In this way the small amount triplet-state species that forms has time to relax before the same region is scanned again.

The confocal principle is also exploited in fluorescence correlation spectroscopy (Elson and Magde 1974; Rigler 1995). Here a non-scanning pin-hole arrangement allows illumination of a volume element as small as $0.4 \times 0.4 \times 2\,\mu m^3 = 0.32$ fl. Thus if the fluorophore concentration is 10 nM, then on average there will be two molecules in the volume element. However, this number will fluctuate as molecules diffuse into and out of the volume element. Analysis of such fluctuations can give the diffusion constant of the fluorophore. If the diffusion is restricted by binding to a large protein, then the fluctuation profile will be perturbed. Equilibrium binding constants can be determined from such data. If the diffusion time is lengthened by taking a larger volume element and/or examining fluorophores attached to a large protein then, in principle, fluctuations in chemical states may be revealed if they involve a change in fluorescence quantum yield. The recent commercial availability of a confocal fluorescence correlation microscope will no doubt lead to further exploitation in the field of biochemical kinetics at the single molecule level.

A third way of inducing fluorescence in a small volume is through dual photon excitation (Denk *et al.* 1995). Here very high intensities of long wavelength light are used to excite a fluorophore by the unlikely event of two photons being absorbed simultaneously. The two practical advantages of the method are that fluorescence is only excited efficiently at the point of focus of the light because the probability of two-photon absorption depends on the square of its instantaneous intensity. This gives the method good z resolution because the intensity of the focused beam, and hence of the emission, falls off as the fourth power in the z direction. Outside the region of focus there is little photodamage by one-photon excitation as the long wavelengths used (typically 700–1050 nm) are generally not absorbed by the fluorophore. However, the method requires high-power pulsed lasers and its use has not yet become routine.

In vitro motility assays can be used to address the two key conditions of muscle activity: isotonic and isometric contraction. In the former case the interest is to determine how many ATP molecules are consumed per μm of sliding in order to define the interaction distance. Currently, indirect estimates of this number are in conflict (Burton 1992; Harada *et al.* 1990; Uyeda *et al.* 1990). However, standard *in vitro* motility assays using myosin-coated surfaces in conjunction with fluorescent ATP analogues would be insufficient to answer this question because the number of heads that interact with the actin filaments would be swamped by the background myosin. Instead some mechanism of limiting the number of heads is required, such as the use of heavy meromyosin (HMM) tracks (Toyoshima *et al.* 1989) or myosin/light meromyosin co-filaments (Saito *et al.* 1994). In order to study isometric contraction a method of restraining the actin filament from moving is required, such as the use of glass needles or an optical trap (Finer *et al.* 1994). Here the interest is to determine the coincidence of single force transients with single molecule ATP turnovers on a sparse surface of myosin heads, as has already been attempted for single kinesin turnovers (Funatsu *et al.* 1996).

Apart from *in vitro* motility assays, the other major new applications of fluorescent ATP analogues in the field of muscle and motility are likely to involve the characterization of novel myosin isoforms and other motor proteins. There are now at least 12 classes of myosin characterized from sequence data (Sellers *et al.* 1996). Many of these are not available in protein form but if and when they become expressed, quantities are likely to be limited. Therefore, fluorescent analogues will be vital to characterize their ATPase activity. Other motor proteins are worthy of study not only in their own right but because basic mechanisms are likely to be shared with muscle proteins. In this regard it is useful to note kinesin, like myosin is relatively non-specific and uses fluorescent analogues as substrates, whereas dynein is much more selective and therefore will be difficult to probe (Shimizu *et al.* 1991). Mant-nucleotides have been used for transient state kinetic analysis of the kinesin and ncd ATPases (Lockhart *et al.* 1995; Sadhu and Taylor 1992).

As far as new analogues are concerned, the possibilities are expanding as new fluorophores are developed. Haugland (1996) lists over a dozen classes of amine-reactive probes that could be used to derivatize EDA-ATP to generate novel ATP analogues (e.g. BODIPY, oregon green, rhodamine green). There is likely to be much development on this front, driven in particular by the requirements of fluorescent nucleotides for automated nucleic acid sequencing at the single molecule level.

Acknowledgements

I am grateful for discussions and comments from Drs. John Eccleston, Howard White, Paul Conibear, Philip Kuhlman and Shigeru Chaen. The experimental results shown in the figures came from research supported by the Japanese Society for the Promotion of Science, The Wellcome Trust and the Human Frontier Science Program.

References

Anazawa, T., Yasuda, K. and Ishiwata, S. (1992) Spontaneous oscillation of tension and sarcomere length in skeletal myofibrils: microscopic measurement and analysis. *Biophys. J.* **61**, 1099–108.

Ando, T. and Miyata, H. (1983) Pyruvate as a fluorescence quencher: a new spectroscopic assay for pyruvate reactions. *Anal. Biochem.* **129**, 170–5.

Ando, T., Duke, J.A., Tonomura, Y. and Morales, M.F. (1982) Spectroscopic isolation of ES complexes of myosin subfragment-1 ATPase by fluorescence quenching. *Biochem. Biophys. Res. Comm.* **109**, 1–6.

Ankrett, R.J., Rowe, A.J., Cross, R.A., Kendrick-Jones, J. and Bagshaw, C.R. (1991) A folded (10S) conformer of myosin from a striated muscle and its implications for regulation of ATPase activity. *J. Mol. Biol.* **217**, 323–35.

Axelrod, D. (1989) Total internal reflection fluorescence microscopy. *Meth. Cell Biol.* **30**, 245–70.

Bagshaw, C.R. (1974) The adenosine triphosphatase of myosin. PhD thesis. University of Bristol.

Bagshaw, C.R. (1993) *Muscle contraction* (2nd edn). Chapman & Hall, London.

Bagshaw, C.R., Eccleston, J.F., Trentham, D.R., Yates, D.W. and Goody, R.S. (1972) Transient kinetic studies of the Mg^{2+}-dependent ATPase of myosin and its proteolytic subfragments. *Cold Spring Harbor Symp. Quant. Biol.* **37**, 127–35.

Bagshaw, C.R., Eccleston, J.F., Eckstein, F., Goody, R.S., Gutfreund, H. and Trentham, D.R. (1974) The Mg^{2+}-dependent adenosine triphosphatase of myosin; two-step processes of ATP association and ADP dissociation. *Biochem. J.* **141**, 351–64.

Bagshaw, C.R., Matuska, M. and Spudich, J.A. (1992) Measurement of ATP turnover by single, immobilized myosin filaments. *J. Muscle Res. Cell Motil.* **13**, 266–7.

Barden, J.A. and dos Remedios, C.G. (1984) The environment of the high-affinity cation binding site on actin and the separation between cation and ATP sites as revealed by proton NMR and fluorescence spectroscopy. *J. Biochem.* (*Tokyo*) **96**, 913–21.

Burton, K. (1992) The myosin step size: estimates from motility assays and shortening muscle. *J. Muscle Res. Cell Motil.* **13**, 590–607.

Chaen, S., Shirakawa, I., Bagshaw, C.R. and Sugi, H. (1996) Studies of ATP turnover within contracting muscle myofibrils using a fluorescent ATP analogue. *J. Muscle Res. Cell Motil.* **17**, 280.

Chang, T.-L. and Cheung, H.C. (1992) Solvent effects on the photoisomerisation rates of the zwitterionic and cationic forms of rhodamine B in protic solvents. *J. Phys. Chem.* **96**, 4874–8.

Citi, S., Cross, R.A., Bagshaw, C.R. and Kendrick-Jones, J. (1989) Parallel modulation of brush border myosin conformation and enzyme activity induced by monoclonal antibodies. *J. Cell Biol.* **109**, 549–56.

Conibear, P.B. and Bagshaw, C.R. (1996) Measurement of nucleotide exchange kinetics with single myosin filaments using flash photolysis. *FEBS Lett.* **380**, 13–16.

Conibear, P.B., Seehra, C.K., Bagshaw, C.R. and Gingell, D. (1995) Observation of ATP turnover during *in vitro* motility assays. *Biochem. Soc. Trans.* (*London*) **23**, 400S.

Conibear, P.B., Jeffreys, D.S., Seehra, C.K., Eaton, R.J. and Bagshaw, C.R. (1996) Kinetic and spectroscopic characterization of fluorescent ribose-modified ATP analogs upon interaction with skeletal muscle myosin subfragment 1. *Biochemistry* **35**, 2299–308.

Cooke, R. (1982) Fluorescence as a probe of contractile systems. *Meth. Enzymol.* **85**, 574–93.

Cremo, C.R. and Yount, R.G. (1987) 2′-Deoxy-3′-O-(4-benzoylbenzoyl)- and 3′(2′)-O-(4-benzoylbenzoyl)-1,N6-ethenoadenosine 5′-diphosphate, fluorescent photoaffinity analogues of adenosine 5′-diphosphate. Synthesis, characterization and interaction with myosin subfragment 1. *Biochemistry* **26**, 7524–34.

Cremo, C.R., Neuron, J.M. and Yount, R.G. (1990) Interaction of myosin subfragment 1 with fluorescent ribose-modified nucleotides: a comparison of vanadate trapping and SH1-SH2 cross-linking. *Biochemistry* **29**, 3309–19.

Cross, R.A., Cross, K.E. and Sobieszek, A. (1986) ATP-linked monomer-polymer equilibrium of smooth muscle myosin: the free folded monomer traps ADP.Pi. *EMBO J.* **5**, 2637–41.

Cross, R.A., Jackson, A.P., Citi, S., Kendrick-Jones, J. and Bagshaw, C.R. (1988) Active site trapping of nucleotides by smooth and non-muscle myosins. *J. Mol. Biol.* **203**, 173–81.

Denk, W., Piston, D.W. and Webb, W.W. (1995) Two-photon molecular excitation in laser-scanning microscopy. In *Handbook of biological confocal microscopy* (ed. Pawley, J.B.), pp. 445–58. Plenum Press, New York.

dos Remedios, C.G., Miki, M. and Barden, J.A. (1987) Fluorescence resonance energy transfer measurements of distances in actin and myosin. A critical evaluation. *J. Muscle Res. Cell Motil.* **8**, 97–117.

Eccleston, J.F. (1987) Stopped-flow spectrophotometric techniques. In *Spectrophotometry and spectrofluorimetry: a practical approach* (ed. Harris, D.A. and Bashford, C.L.), pp. 137–64, IRL Press, Oxford.

Eccleston, J.F., Oiwa, K., Ferenczi, M.A., Anson, M., Corrie, J.E.T., Yamada, A. *et al.* (1996) Ribose-linked sulfoindocyanine conjugates of ATP: Cy3-EDA-ATP and Cy5-EDA-ATP. *Biophys. J.* **70**, A159.

Elson, E. and Magde, D. (1974) Fluorescence correlation spectroscopy. I. Conceptual basis and theory. *Biopolymers* **13**, 1–27.

Finer, J.T., Simmons, R.M. and Spudich, J.A. (1994) Single myosin molecule mechanics: piconewton forces and nanometre steps. *Nature (London)* **368**, 113–9.

Fisher, A.J., Smith, C.A., Thoden, J., Smith, R., Sutoh, K., Holden, H.M. *et al.* (1995*a*) Structural studies of myosin : nucleotide complexes: a revised model for the molecular basis of muscle contraction. *Biophys. J.* **68**, 19s–28s.

Fisher, A.J., Smith, C.A., Thoden, J., Smith, R. Sutoh, K., Holden, H.M. *et al.* (1995*b*) X-ray structure of the myosin motor domain of *Dictyostelium discoideum* complexed with MgADP.BeF_x and MgADP.AlFA. *Biochemistry* **34**, 8960–72.

Franks-Skiba, K. and Cooke, R. (1995) The conformation of the active site of myosin probed using mant-nucleotides. *Biophys. J.* **68**, 142s–9s.

Funatsu, T., Harada, Y., Tokunaga, M., Saito, K. and Yanagida, T. (1995) Imaging of single fluorescent molecules and individual ATP turnovers by single myosin molecules in aqueous solution. *Nature (London)* **374**, 555–9.

Funatsu, T., Harada, Y., Higuchi, H., Tokonaga, M., Saito, K., Vale, R. *et al.* (1996) Single molecule imaging of motion and ATPase reaction of motor proteins. *Biophys. J.* **70**, A6.

Garland, F. and Cheung, H.C. (1979) Fluorescence stopped-flow study of the mechanism of nucleotide binding to myosin subfragment 1. *Biochemistry* **18**, 5281–9.

Giranda, V.L., Berman, H.M. and Schramm, V.L. (1988) Crystal structure of formycin 5′-phosphate: an explanation for its tight binding to AMP deaminase. *Biochemistry* **27**, 5813–8.

Harada, Y., Sakurada, K., Aoki, T., Thomas, D.D. and Yanagida, T. (1990) Mechanochemical coupling in actomyosin energy transduction studied by *in vitro* movement assay. *J. Mol. Biol.* **216**, 49–68.

Haugland, R.P. (1996) *Handbook of fluorescent probes and research chemicals* (6th edn). Molecular Probes Inc., USA.

Herrmann, C., Sleep, J., Chaussepied, P., Travers, F. and Barman, T. (1993) A structural and kinetic study on myofibrils prevented from shortening by chemical cross-linking. *Biochemistry* **32**, 7255–63.

Hiratsuka, T. (1976) Fluorescence properties of 2′ (or 3′)-O-(2,4,6-trinitrophenyl) adenosine 5′-triphosphate and its use in the study of binding to heavy meromyosin ATPase. *Biochem. Biophys. Acta* **453**, 293–7.

Hiratsuka, T. (1983) New ribose-modified fluorescent analogs of adenine and guanine nucleotides available as substrates for various enzymes. *Biochem. Biophys. Acta* **742**, 496–508.

Hiratsuka, T. (1984) Affinity labeling of the myosin ATPase with ribose-modified fluorescent nucleotides and vanadate. *J. Biochem.* (*Tokyo*) **96**, 147–54.

Hiratsuka, T. (1985) Photosensitized direct cross-linking of fluorescent analogs of ATP to the adenine recognition domain in myosin ATPase. *J. Biochem.* (*Tokyo*) **97**, 71–8.

Hiratsuka, T. and Uchida, K. (1973) Preparation and properties of 2′(or 3′)-O-(2,4,6-trinitrophenyl) adenosine 5′-triphosphate, and analog of adenosine 5′-triphosphate. *Biochem. Biophys. Acta* **320**, 635–47.

Jackson, A.P. and Bagshaw, C.R. (1988) Kinetic trapping of intermediates of the scallop heavy meromyosin adenosine triphosphatase reaction revealed by formycin nucleotides. *Biochem. J.* **251**, 527–40.

Jeffreys, D.S., Eaton, R.J. and Bagshaw, C.R. (1995) Caged FEDA-ATP: a new tool in the measurement of ATP turnover during the *in vitro* motility assay. *Biochem. Soc. Trans.* (*London*) **23**, 401S.

Kabsch, W., Mannherz, H.G., Suck, D., Pai, E.F. and Holmes, K.C. (1990) Atomic structure of the actin : DNase 1 complex. *Nature* **347**, 37–44.

Knight, P. and Parsons, N. (1991) Modification of myofibrils by fluorophore-induced photo-oxidation. *J. Muscle Res. Cell Motil.* **12**, 183–91.

Koyama, G., Maeda, K., Umezawa, H. and Iitaka, V. (1966) The structural studies of formycin and formycin B. *Tetrahedron Lett.* **6**, 597–602.

Kron, S.J. and Spudich, J.A. (1986) Fluorescent actin filaments move on myosin fixed to a glass surface. *Proc. Nat. Acad. Sci.* (*USA*) **83**, 6272–6.

Kron, S.J., Toyoshima, Y.Y., Uyeda, T.Q.P. and Spudich, J.A. (1991) Assays for actin sliding movement over myosin-coated surfaces. *Meth. Enzymol.* **196**, 399–416.

Lakowicz, J.R. (1983) *Principles of fluorescence spectroscopy*. Plenum Press, New York and London.

Lockhart, A., Cross, R.A. and McKillop, D. (1995) ADP release is the rate-limiting step of the MT activated ATPase of non-claret disjunctional and kinesin. *FEBS Lett.* **368**, 531–5.

McCray, J.A. and Trentham, D.R. (1989) Properties and uses of photoreactive caged compounds. *Ann. Rev. Biophys. Biophys. Chem.* **18**, 239–70.

Miyata, H. and Asai, H. (1982) 1,N^6-etheno-2-aza-adenosine triphosphate: enormous increase in fluorescence induced by its binding to heavy meromyosin and estimation of the kinetic parameters. *Biochem. Biophys. Res. Comm.* **105**, 296–302.

Miyata, H., Chaen, S., Sase, I., Kinosita, K. and Sugi, H. (1996) An application of evanescent excitation for visualisation of dynamic aspects of protein-ligand interactions. *J. Muscle Res. Cell Motil.* **17**, 280–1.

Moore, K. and Lohman, T.M. (1994) Kinetic mechanism of adenine nucleotide binding to and hydrolysis by the *Escherischia coli* Rep monomer. Use of fluorescent nucleotide analogues. *Biochemistry* **33**, 14 550–64.

Morita, F. (1967) Interaction of heavy meromyosin with substrate. *J. Biol. Chem.* **242**, 4501–6.

Moss, D.J. and Trentham, D.R. (1983) Distance measurement between the active site and cysteine-177 of the alkali one light chain of subfragment 1 from rabbit skeletal muscle. *Biochemistry* **22**, 5261–70.

Mujumdar, R.B., Ernst, L.A., Mujumdar, S.R., Lewis, C.J. and Waggoner, A.S. (1993) Cyanine dye labeling reagents: sulphoindocyanine succinimidyl esters. *Bioconj. Chem.* **4**, 105–11.

Myburgh, K.H., Franks-Siba, K. and Cooke, R. (1995) Nucleotide turnover rate measured in fully relaxed rabbit skeletal muscle myofibrils. *J. Gen. Physiol.* **106**, 957–73.

Nagano, H. and Yanagida, T. (1984) Predominant attached state of myosin cross-bridges during contraction and relaxation at low ionic strength. *J. Mol. Biol.* **177**, 769–85.

Oiwa, K., Anson, M., Yamada, A., Eccleston, J.F., Corrie, J.E.T., Ferenczi, M.A. *et al.* (1996) Microscopic observations of Cy3-EDA-ATP and Cy5-EDA-ATP binding to myosin filaments *in vitro*. *Biophys. J.* **70**, A159.

Onishi, H., Ohtsuka, E., Ikehara, M. and Tonomura, Y. (1973) Energy transfer from tryptophan residues to a fluorescent ATP analog, 1 N6-ethenoadenosine triphosphate bound to H-meromyosin. *J. Biochem.* (*Tokyo*) **74**, 435–50.

Onodera, M. and Yagi, K. (1971) Synthesis of 2-(dansylamino)ethyl triphosphate and its properties as a fluorescent substrate of heavy meromyosin-ATPase. *Biochem. Biophys. Acta* **253**, 254–65.

Pate, E., Franks-Siba, K., White, H. and Cooke, R. (1993) The use of differing nucleotides to investigate cross-bridge kinetics. *J. Biol. Chem.* **268**, 10 046–53.

Pawley, J.B. (ed.) (1995) *Handbook of biological confocal microscopy*) (2nd edn). Plenum Press, New York.

Peck, K., Stryer, L., Glazer, A.N. and Mathies, R.A. (1989) Single-molecule fluorescence detection: autocorrelation criterion and experimental realization with phycoerythrin. *Proc. Nat. Acad. Sci.* (*USA*) **86**, 4087–91.

Prusiner, P., Brennan, T. and Sundaralingam, M. (1973) Crystal structure and molecular conformation of formycin monohydrates. Possible origin of the anomolous circular dichroic spectra in formycin mono- and polynucleotides. *Biochemistry* **12**, 1196–202.

Rayment, I., Rypniewski, W.R., Schmidt-Bäse, K., Smith, R., Tomchick, D.R., Benning, M.M. *et al.* (1993*a*) Three-dimensional structure of myosin subfragment 1: a molecular motor. *Science* **261**, 50–8.

Rayment, I., Holden, H.M., Whittaker, M., Yohn, C.B., Lorenz, M., Holmes, K.C. *et al.* (1993*b*) Structure of the actin-myosin complex and its implications for muscle contraction. *Science* **261**, 58–65.

Rensland, H., Lautwein, A., Wittinghofer, A. and Goody, R.S. (1991) Is there a rate-limiting step before GTP cleavage by H-ras p21? *Biochemistry* **30**, 11 181–5.

Rigler, R. (1995) Fluorescence correlations, single molecules detection and large number screening—applications in biotechnology. *J. Biotechnology* **41**, 177–86.

Ritchie, M.D., Geeves, M.A., Woodward, S.K.A. and Manstein, D.J. (1993) Kinetic characterization of a cytoplasmic myosin motor domain expreseed in *Dictylostelium discoideum*. *Proc. Natl. Acad. Sci.* (*USA*) **90**, 8619–23.

Rosenfeld, S.S. and Taylor, E.W. (1984) Reactions of 1-N^6-ethenoadenosine nucleotides with myosin subfragment 1 and acto-subfragment 1 of skeletal and smooth muscle. *J. Biol. Chem.* **259**, 11 920–9.

Rosenfeld, S.S., Xing, J., Rener, B., Lebowitz, J., Kar, S. and Cheung, H.C. (1994) Structural and kinetic studies of the 10S↔6S transition in smooth muscle myosin. *J. Biol. Chem.* **269**, 30 187–94.

Rossomando, E.F., Jahngen, J. and Eccleston, J.F. (1981) A fluorometric high-performance liquid chromatographic assay procedure for several enzymatic activities using formycin analogs of adenosine 5′-mono, 5′-tri and cyclic 3′5′ monophosphates as substrates. *Anal. Biochem.* **116**, 80–8.

Sadhu, A. and Taylor, E.W. (1992) A kinetic study of the kinesin ATPase. *J. Biol. Chem.* **267**, 11 352–9.

Saito, K., Aoki, T., Aoki, T. and Yanagida, T. (1994) Movement of single myosin filaments and myosin step size on an actin filament suspended in solution by a laser trap. *Biophys. J.* **66**, 769–77.

Saito, K., Tokunaga, M., Yokota, H. and Yanagida, T. (1995) ATPase reaction during sliding of actin filaments. *Biophysics* (*Japanese Biophys. Soc.*) **35**, S208.

Saito, K., Yokota, H. and Yanagida, T. (1996) Imaging single protein molecules by surface plasmon resonance. *Biophys. J.* **70**, A430.

Sase, I., Miyata, H., Corrie, J.E.T., Craik, J. and Kinosita, K. (1995) Real time imaging of single fluorophores on moving actin with an epifluorescence microscope. *Biophys. J.* **69**, 323–8.

Secrist, J.A., Barrio, J.R., Leonard, N.J. and Weber, G. (1972) Fluorescent modification of adenosine-containing coenzymes. Biological activities and spectroscopic properties. *Biochemistry* **11**, 3499–506.

Sellers, J.R., Goodson, H.V. and Wang, F. (1996) A myosin family reunion. *J. Muscle Res. Cell Motil.* **7**, 7–22.

Shimizu, T., Furusawa, K., Ohashi, S., Toyoshima, Y.Y., Okuno, M., Malik, F. *et al.* (1991) Nucleotide specificity of the enzymatic and motile activities of dynein, kinesin and heavy meromyosin. *J. Cell. Biol.* **112**, 1189–97.

Smith, S.J. and White, H.D. (1985*a*) Kinetic mechanism of 1-N^6-etheno-2-aza-ATP hydrolysis by bovine ventricular myosin subfragment 1 and actomyosin subfragment 1: the nucleotide binding steps. *J. Biol. Chem.* **260**, 15 146–55.

Smith, S.J. and White, H.D. (1985*b*) Kinetic mechanism of 1-N^6-etheno-2-aza-ATP and 1-N^6-etheno-2-aza-ADP binding to bovine ventricular actomyosin-S1 and myofibrils. *J. Biol. Chem.* **260**, 15 156–62.

Southwick, P.L., Ernst, L.A., Tauriello, E.W., Parker, S.R., Mujumdar, R.B., Mujumdar, S.R. *et al.* (1990) Cyanine dye labeling reagents—carbomethylindocyanine succinimidyl esters. *Cytometry* **11**, 418–30.

Sowerby, A.J., Seehra, C.K., Lee, M. and Bagshaw, C.R. (1993) Turnover of fluorescent nucleotide triphosphates by isolated immobilized myosin filaments: transient kinetics on the zeptomole scale. *J. Mol. Biol.* **234**, 114–23.

Stout, A.L. and Axelrod, D. (1989) Evanescent field excitation of fluorescence by epi-illumination. *Applied Optics* **28**, 5237–42.

Thomas, D.D. (1987) Spectroscopic probes of muscle crossbridge rotation. *Ann Rev. Physiol.* **49**, 691–709.

Todd, I., Mellor, J.S. and Gingell, D. (1988) Mapping cell-glass contacts of *Dictyostelium amoeba* by total internal reflection aqueous fluorescence overcomes a basic ambiguity of interference reflection microscopy. *J. Cell Sci.* **89**, 107–14.

Tonomura, Y. (1972) *Muscle proteins, muscle contraction and cation transport*. University of Tokyo Press.

Toyoshima, Y.Y., Toyoshima, C. and Spudich, J.A. (1989) Bidirectional movement of actin filaments along tracks of myosin heads. *Nature* **341**, 154–6.

Trentham, D.R., Eccleston, J.F. and Bagshaw, C.R. (1976) Kinetic analysis of ATPase mechanisms. *Quart. Rev. Biophys.* **9**, 217–81.

Tsien, R.Y. and Waggoner, A. (1995) Fluorophores for confocal microscopy: photophysics and photochemistry. In *Handbook of biological confocal microscopy* (ed. Pawley, J.B.), pp. 267–79. Plenum Press, New York.

Tsou, K.C., Yip, K.F., Miller, E.E. and Lo, K.W. (1974) Synthesis of 1,N^6-etheno-2-aza-adenosine (2-aza-e-adenosine): a new cytotoxic fluorescent nucleotide. *Nucleic Acid Res.* **1**, 531–47.

Uyeda, T.Q.P., Kron, S.J. and Spudich, J.A. (1990) Myosin step size: estimation from slow sliding movement of actin over low densities of heavy meromyosin. *J. Mol. Biol.* **214**, 699–710.

Wang, D., Pate, E., Cooke, R. and Yount, R.G. (1993) Synthesis of non-nucleotide ATP analogues and characterization of their chemomechanical interactions with muscle fibres. *J. Muscle Res. Cell Motil.* **14**, 484–97.

Ward, D.C., Reich, E. and Stryer, L. (1969*a*) Fluorescence studies of nucleotides and polynucleotides. *J. Biol. Chem.* **244**, 1228–37.

Ward, D.C., Cerami, A., Reich, E., Acs, G. and Altwerger, L. (1969*b*) Biochemical studies of the nucleotide analogue, formycin. *J. Biol. Chem.* **244**, 3243–50.

Webb, M.R. and White, H.D. (1996) Kinetics of ATP cleavage and phosphate release steps by associated rabbit skeletal actomyosin using a novel fluorescent phosphate probe. *Biophys. J.* **70**, A40.

Wells, C. and Bagshaw, C.R. (1985) Calcium regulation of molluscan myosin ATPase in the absence of actin. *Nature* (*London*) **313**, 696–7.

Werber, M.M., Szent-Györgyi, A.G. and Fasman, G.D. (1972) Fluorescence studies of heavy meromyosin-substrate interaction. *Biochemistry* **11**, 2872–83.

White, H.D. (1982) Special instrumentation and techniques for kinetic studies of contractile systems. *Meth. Enzymol.* **85**, 698–708.

White, H.D., Smith, S.J. and Rayment, I. (1986) Myosin S1 crystals bind 1,N^6-etheno-2-aza ATP. *Biophys. J.* **49**, 45a.

White, H.D., Belknap, B. and Jiang, W. (1993) Kinetics of binding and hydrolysis of a series of nucleoside triphosphates by actomyosin-S1: relationship between solution rate constants and properties of muscle fibers. *J. Biol. Chem.* **268**, 10039–45.

Woodward, S.K.A., Eccleston, J.F. and Geeves, M.A. (1991) Kinetics of interaction of 2′(3′)-O-(N-methylanthraniloyl)-ATP with myosin subfragment 1 and actomyosin subfragment 1: characterization of two Acto.S1.ADP complexes. *Biochemistry* **30**, 422–30.

Yanagida, T. (1981) Angles of nucleotide bound to cross-bridges in glycerinated muscle fiber at various concentrations of e-ATP, e-ADP and e-AMPPNP detected by polarised fluorescence. *J. Mol. Biol.* **146**, 539–60.

Yarbrough, L.R., Schlageck, J.G. and Baughman, M. (1979) Synthesis and properties of fluorescent nucleotide substrates for DNA-dependent RNA polymerase. *J. Biol. Chem.* **254**, 12 069–73.

Yip, K.F. and Tsou, K.C. (1973) Synthesis of fluorescent adenosine derivatives. *Tetrahedron Lett.* **53**, 3087–90.

Yount, R.G. (1975) ATP analogs. *Adv. Enzymol.* **43**, 1–56.

5 The use of peptide mimetics

Harald Kögler and J. Caspar Rüegg

5.1 Introduction

Muscle contraction totally depends on the two proteins actin and myosin. However, if they were the only proteins to exist in muscle, the ability to contract would be completely useless. It is the additional presence of the various regulatory proteins that enables the nervous system to determine when and where muscle contraction or relaxation is required for the benefit of the whole organism. Hence, in muscle cells a variety of protein–protein interactions occur in an orchestrated way, resulting in a mechanical response that is precisely controlled by the neural and/or humoral output of the central nervous system. To better understand not only the physiology but also the pathology of muscle contraction, it is vital to gain detailed insight into the nature of these protein–protein interactions. It is crucial, therefore, to identify the specific protein-recognition sites—if possible to the level of single amino acid residues—in order to be able to deduce the mechanisms of interaction.

In this respect, for many years peptide approaches have been fruitful in unravelling mechanisms of muscle contraction and regulation. Peptides with a structure corresponding to that of a specific protein recognition site may be used as inhibitors to compete for protein–protein interactions in order to probe the functional importance of these interaction sites (Rüegg 1990). Such peptides can serve not only as competitors but also as mimetics that imitate the action of the parent protein from which they are derived.

The use of peptides or proteolytic fragments to identify functionally important segments of proteins originates from the 1950s and 1960s when Szent-Györgyi (1953) and Mueller and Perry (1962) reported that after short and prolonged tryptic digestion of myosin, its actin binding and ATPase properties were associated with heavy mero-myosin (HMM) and myosin subfragment-1 (S1), respectively. Atassi (1975) pioneered the use of peptide competition when he demonstrated that the recognition of myoglobin by a myoglobin antibody could effectively be inhibited by peptide fragments derived from the antigenic sites. Only when the amino acid sequence was identical to that of the antigenic site, was there an inhibition of the antibody–antigen recognition by peptide competition.

Subsequently, a great number of studies were carried out using protein fragments obtained by chemical or limited enzymatic proteolysis as mimics or competitors of protein–protein interactions. However, this approach was limited by the location of the cleavage sites, which did not allow examination of every region of interest along the polypeptide chain of a protein (Trayer *et al.* 1991). Therefore, it was a milestone when Talbot and Hodges (1979), for the first time in muscle research, employed solid phase peptide synthesis to produce a 20-residue synthetic peptide corresponding to the inhibitory region of fast skeletal muscle troponin-I. This new method opened up the possibility of preparing specific isolated structural motifs of a polypeptide chain, which would not have been accessible to proteolysis due to lack of specific cleavage sites. Furthermore, peptide synthesis offered the additional advantage that single residues could be systematically substituted in order to study the biological relevance of single specified amino acid residues (Van Eyk and Hodges 1993).

In this chapter, we will highlight some of the recent advances in muscle physiology and biochemistry, which have been brought about using proteolytic fragments or synthetic peptides as tools to probe the molecular interactions of actin, myosin, troponins, caldesmon and other proteins involved in muscle regulation and contraction. The presentation of these results will show the advantages of this method, but also point out some of the methodological pitfalls that have to be kept in mind when working with peptides. This will lead to a discussion of the problems and limitations of this powerful approach.

5.2 Examples of application

Actin and myosin peptides

Actin peptides

The zero-length cross-linking studies of Sutoh (1982) using 1-ethyl-3-[3-(dimethylamino)propyl]carbodiimide (EDC) to covalently cross-link actin and myosin, revealed that the highly negatively charged N-terminal segment of actin (residues 1–12) participates in the binding of both the 20 K and 50 K fragments of S1 heavy chain. Consequently, Levine and Moir (1985) as well as Moir *et al.* (1987) used a proteolytic fragment of actin to confirm this result: specific signals in the ^{1}H-nuclear magnetic resonance (NMR) spectrum arising from the N-terminal cyanogen bromide (CNBr) fragment of actin (residues 1–44) were perturbed upon titration with S1, which is indicative of close interaction.

Additional evidence for the functional importance of the N-terminus of actin was presented by DasGupta *et al.* (1990). Their immunological study showed that F(ab) fragments of antibodies raised against residues 1–7 of actin, inhibited the S1-induced polymerization of G-actin to F-actin filaments. In the three-dimensional structure of actin as determined by X-ray crystallography (Kabsch *et al.* 1990), this N-terminal segment was located to the outermost surface of the actin monomer, which identified it as a candidate readily accessible for cross-bridge binding. Thus, Van Eyk *et al.* (1991) synthesized an actin peptide 1–28 and three truncated analogues to examine their effects on S1 and HMM Mg^{2+} ATPase activity. Actin 1–28 increased the S1 Mg^{2+}ATPase activity to 150% of the level found in the absence of peptide. A peptide actin 8–20 was considerably less potent, whereas peptides 1–14 and 18–28 displayed intermediate activating potency. Complex formation of the peptide 1–28 with S1 could be demonstrated by high performance size-exclusion chromatography, where the peptide was found to be present in the S1 containing fractions (Van Eyk and Hodges 1991). Kögler *et al.* (1991) examined the CNBr fragment 1–44 of actin with respect to its effects on S1 and acto S1 Mg^{2+}ATPase activity and demonstrated that it effectively activated the S1 Mg^{2+}ATPase. On the other hand, actin 1–44 inhibited the actin-activated S1 Mg^{2+}ATPase, presumably competing with actin for S1 binding. In contrast, a synthetic peptide actin 1–18 did not show any of the effects elicited by the CNBr fragment 1–44 in S1 and acto-S1 Mg^{2+}ATPase assays.

From these peptide investigations it is possible to conclude that the region covalently cross-linked to S1 by EDC treatment, indeed represents a functionally important actin-S1 interaction site. The differential effects of the various peptides employed in these studies provide additional information about the process of protein–protein interaction: obviously, peptides shorter than residues 1–28 show only minor biological activity, if at all. Thus, the actin segment 1–28 appears to be the minimum sequence required for a functionally relevant actin-S1 interaction. The higher activity of the CNBr fragment 1–44, especially the inhibition of acto-S1 Mg^{2+}ATPase which was not observed with actin 1–28, suggests that actin 1–44 represents a binding site more closely resembling the one found in native actin than does the smaller peptide. One possible explanation is that the bigger fragment might contain more information about secondary or tertiary structure, which would allow it to fit more precisely to the corresponding binding site on S1 than the smaller peptide.

Actin peptides were also used to investigate the actin–troponin-I interaction. Levine *et al.* (1988) performed ^{1}H-NMR spectroscopy on CNBr fragments of actin and troponin-I (TnI) to registrate spectral perturbations resulting from complex formation. When the CNBr fragment 1–44 of actin was titrated with the CNBr fragment 96–116 of skeletal TnI the effect induced by complex formation was closely identical to the effects observed for F-actin and TnI 96–116 indicating that this segment of actin is also involved in TnI binding when present as part of the intact molecule. With a synthetic actin peptide 1–18, spectral perturbations could be detected as well, which were, however, distinguishable from the effects exerted by actin 1–44. A probable hypothesis is the presence of a second interaction site, located within segment 19–44 of actin. Consistently, Van Eyk *et al.* (1991) have demonstrated that actin 1–28 showed much stronger binding to a TnI-affinity high performance liquid chromatography (HPLC) column than actin 1–14 or 18–28. Actin 8–20 did not bind at all.

In conclusion, peptide studies have made an important contribution to our present understanding of the N-terminal actin segment acting as a flip-flop switch with alternative binding sites on S1 and TnI, which is a major step in the process of activation/inactivation in striated muscles.

Myosin heavy chain peptides

After Mornet *et al.* (1981) had shown that actin can be covalently cross-linked to both the 50 K and 20 K tryptic fragments of the S1 heavy chain, considerable effort was made using peptides to locate more closely possible actin-binding sites on the myosin heavy chain. Sutoh (1983) demonstrated that the 20-residue N-terminal CNBr-cleavage product of the 20 K tryptic fragment of S1 participated in actin binding. Also, Katoh *et al.* (1985) employed CNBr-cleavage of the 20 K tryptic S1 fragment and pointed out that all CNBr fragments that retained F-actin binding ability contained both reactive thiol groups of the heavy chain, denoted as SH1 and SH2 (residues Cys^{705} and Cys^{695} of the rabbit skeletal myosin heavy chain, respectively). Thus, Suzuki *et al.* (1990) synthesized several peptides spanning the region around these two thiol groups. They observed that a heptapeptide I-R-I-C^{707}-R-K-G, covering the segment closely around SH1, in micromolar concentrations inhibited the acto-S1 Mg^{2+}ATPase in solution at low ionic strength. S1 Mg^{2+}ATPase activity in the absence of actin was not affected by the peptide, which indicated that the peptide specifically interfered with the process of activation of S1 ATPase by F-actin. However, the peptide did not compete with S1 for actin binding, since kinetic analysis revealed that with increasing peptide concentrations, only V_m decreased, leaving K_m essentially unchanged. Acto-S1 was not dissociated by the peptide even when the acto-S1 Mg^{2+}ATPase was inhibited completely (Morita *et al.* 1991).

Kraft *et al.* (1991) reported that a synthetic dekapeptide 697–706, partially overlapping with the heptapeptide referred to above, after addition to single skinned rabbit psoas fibres resulted in a decrease in relaxed fibre stiffness at low ionic strength. Equatorial X-ray diffraction studies revealed that this was accompanied by a reduction of the intensity ratio $I_{11}:I_{10}$ indicating weakening of cross-bridge attachment. Peptides corresponding to the neighbouring segments 706–715 and 716–725, respectively, exerted no significant effects.

Several synthetic peptides spanning the region around Cys^{705} were examined by Keane *et al.* (1990) with regard to F-actin binding, effects on acto-S1 Mg^{2+}ATPase activity in solution, and isometric force development in skinned rabbit psoas fibres. Peptides containing all or a portion of the residues 690–725 acted as mixed inhibitors of the acto-S1 Mg^{2+}ATPase, whereas those from just outside 690–725 had no inhibitory effect. In skinned rabbit psoas fibres, all peptides from the region 690–725 inhibited isometric force at maximally activating Ca^{2+} concentration. However, at submaximally activating Ca^{2+} concentrations, the peptides' effects on isometric tension were not uniform: those derived from the N-terminal end of this region (residues 690–715) increased tension concentration dependently between 20 and 100 μM peptide, thus acting as calcium sensitizers. Peptides from the C-terminal end (residues 716–728), on the other hand, displayed no calcium-sensitizing activity. Finally, a peptide containing almost the whole segment (residues 701–728) inhibited force at all Ca^{2+} concentrations examined.

Rüegg *et al.* (1993) reported that two myosin heavy chain peptides (residues 687–716 and 707–717, respectively) in skinned porcine cardiac fibres exerted an effect on isometric tension very similar to the one obtained if the ATP concentration was lowered: under both conditions, much lower Ca^{2+} concentrations were required to induce a given submaximal force. At 10 μM ATP, the peptides had no additional activating effect. These findings were taken to mean that the peptides possibly mimicked the permission of co-operative activation of the thin filament, which is thought to be relevant at low ATP concentration, which induces the formation of rigor cross-bridges.

However, there remained some uncertainty towards the specificity of the peptide effects summarized above for the peptides corresponding to the region around SH1 and SH2 of the myosin heavy chain. As already mentioned, Morita *et al.* (1991) observed that the inhibition of acto-S1 Mg^{2+}ATPase evoked by their heptapeptide containing SH1 was not brought about by a competitive mechanism. Chase *et al.* (1991*a*) have suggested that the primary factor responsible for the inhibitory effect of these peptides was their cationic nature rather than some specific information encoded in the amino acid sequence. They observed that a scrambled heptapeptide, comprising the same amino acids as the one employed by Morita *et al.* (1991), arranged in random order, was equally effective in inhibiting isometric tension in skinned fibres as had been the original heptapeptide. A Cys/Pro-mutated heptapeptide, resulting in the sequence I-R-I-***P***-R-K-G, had no substantially altered (mean effective concentration) EC_{50} for inhibition of isometric tension in skinned fibres, as compared to the native sequence, even though the proline residue is expected to disrupt any induced structure within the peptide (Chase *et al.* 1991*b*). Re-evaluation of several studies performed on SH1 peptides has revealed that the most pronounced inhibitory peptide effects have been exerted by the peptides carrying the highest positive net charge. Cartoux *et al.* (1992) have reported that up to six molecules of the native, as well as the scrambled, heptapeptide were bound per actin monomer in co-sedimentation assays followed by SDS-polyacrylamide gel electrophoresis, which makes the specificity of the peptide effects questionable. And indeed, when the three-dimensional structure of S1 and its complex with actin was determined using X-ray crystallography (Rayment *et al.* 1993*a*,*b*), SH1 was located a distance of 4–5 nm from the actin-binding site of S1, which ruled out the possibility that this segment may interact with actin when present as part of the parent protein.

The question remains of how the peptides from the region around SH1 exerted their dramatic effects in solution and in skinned fibres if such overwhelming evidence argues against a specific interaction with actin. It is tempting to speculate that non-specific binding of the positively charged peptides to the cluster of negatively charged residues at the N-terminus of actin might occur. However, Cartoux *et al.* (1992) found no significant competition between the SH1 peptide and F(ab) fragments of an antibody raised against actin 1–7. They concluded that the peptide effects were mediated through sites on actin other than the N-terminus.

An alternative target zone for the peptides along the actin sequence has not yet been presented to our knowledge. Therefore, at the moment the mechanism of action of the peptides derived from the segment around SH1 of the myosin heavy chain remains obscure.

An interesting alternative peptide approach for investigating a possible actin-binding site on myosin was chosen by Chaussepied and Morales (1988). They designed an 'antipeptide' fitting the segment 633–642 of the myosin heavy chain. The amino acid sequence of this segment is G^{633}-G-K-K-G-G-K-K-K-G^{642}. It forms the junctional loop between the 50 K and 20 K tryptic fragments of S1. The antipeptide C-G-G-D-D-G-G-D-D-D-G provided attractive coulombic interactions with the target segment of the myosin heavy chain. An additional cysteinyl residue was introduced N-terminally to provide the opportunity of introducing fluorescence or spin labels. The peptide bound specifically to S1, since it was demonstrated that the peptide's Asp side-chain carboxyl groups and the target segment's Lys side-chain amino groups could be covalently cross-linked by EDC. The S1 Mg^{2+}ATPase activity was not affected by the antipeptide, indicating that residues 633–642 do not form part of the ATPase site of myosin. However, actin-activation of S1 Mg^{2+}ATPase was greatly inhibited in the presence of the antipeptide. The K_m for acto-S1 Mg^{2+}ATPase activity was shifted towards higher actin concentrations, whereas V_m remained unchanged. In contrast to native S1, antipeptide-S1 did not induce polymerization of G-actin to F-actin (Chaussepied and Kasprzak 1989). Consistently, complex formation between antipeptide-S1 and G-actin was not detected. The authors have suggested that the N-terminus of actin might be a possible binding site for the negatively charged segment 633–642 of the myosin heavy chain. This was confirmed by the structural investigations of Rayment *et al.* (1993*a*,*b*), who concluded from their X-ray crystallographic data that the 50 K–20 K junctional loop binds to the N-terminus of actin as an initial step of myosin cross-bridge attachment to actin.

An additional binding site of the myosin heavy chain to actin is represented by the flexible loop 405–415, which is located within the upper domain of the 50 K fragment and, according to the model of Rayment *et al.* (1993*a*), extends towards the actin filament to form contact with residues 332–334 of actin. A physiologically relevant function of this loop region was demonstrated by Kögler *et al.* (1996), who observed that a synthetic peptide, comprising residues 400–417 of the skeletal myosin heavy chain, exerted a calcium sensitizing effect by enhancing isometric force production in porcine cardiac skinned fibres at submaximally activating Ca^{2+} concentrations.

Myosin light chain peptides

The covalent cross-linking study of Sutoh (1982) revealed that besides the complex formation between actin and the myosin heavy chain, EDC also cross-linked actin to the A1 myosin light chain. The A2 light chain, however, does not form a complex with actin. Thus, several studies were performed using peptides corresponding to segments from the N-terminal extension specific for the A1 light chain, which is not present in the A2 light chain and which appeared to most likely represent the reason for the differential binding behaviour of the two myosin light chain isoforms towards actin. This N-terminal segment in fast-twitch skeletal muscle has a length of 41 residues. Henry *et al.* (1985) treated fast twitch skeletal muscle A1 light chain with the protease thrombin, and isolated from the digest a 76-residue N-terminal fragment containing the 41-residue N-terminal tail. This fragment bound specifically to a F-actin–tropomyosin affinity column. ^{1}H-NMR spectroscopy revealed strong interactions between F-actin and the 76-residue fragment, the most prominent difference spectrum signals originating from the unusual N-terminal residue α-N-trimethylalanine. Titration of F-actin with the C-terminal light chain fragment, however, exerted no prominent change in the ^{1}H-NMR spectrum. Pepsin digestion of the A1 light chain yields a smaller N-terminal fragment comprising residues 1–37 (Trayer *et al.* 1987). When this peptide fragment was titrated with actin to which a spin label had been coupled at Cys^{374}, ^{1}H-NMR spectroscopy showed changes characteristic for specific interactions between the peptide and actin's C-terminus. Trayer *et al.* (1991) also examined a series of synthetic peptides corresponding to several segments of the N-terminal tail of the A1 light chain: in these experiments binding of peptides 1–10 and 1–32 of the A1 light chain to actin was detected by ^{1}H-NMR, whereas peptides 1–3 and 8–31 did not bind to actin.

Morano *et al.* (1995) examined several synthetic peptides derived from segments within the N-terminal tail of the human cardiac ventricular light chain 1. A peptide residue 5–14 enhanced force production in skinned human ventricular muscle fibres. The ED_{50} for this activation was found to be in the picomolar range. A scrambled peptide comprised of the same amino acids also slightly activated force production, at concentrations 10^3 times higher than the native peptide. Less effective than the native peptide 5–14, were truncated fragment residues 5–8 and 5–10, which required much higher concentrations to enhance force production to a degree comparable with the peptide 5–14. Morano *et al.* (1995) also raised a monoclonal antibody against this peptide, which activated isometric force in skinned ventricular muscle fibres equally as well as the peptide itself. Interestingly, the peptide 5–14 was effective in strikingly low concentrations, despite the fact that the light chain's unusual N-terminal α-N-trimethylalanine residue, found by Henry *et al.* (1985) and Trayer *et al.* (1987) to be a site of major importance for the interaction between light chain and actin, was not present in the peptide examined by Morano *et al.* (1995).

Troponin peptides

Troponin-I peptides

The most extensive use of peptides in the whole field of muscle research was made with respect to TnI. Both its interactions with actin and troponin-C (TnC) have been probed by peptides corresponding to various segments of the protein, especially from the so-called inhibitory region. The small size of the protein has recently allowed production of TnI species modified at specified residues by site-directed mutagenesis. This is of great importance, since it offers the opportunity to cross-check results obtained using peptides and thus to evaluate the usefulness of the peptide approach.

Troponin-I inhibitory region

Troponin-I is the inhibitory subunit of the troponin complex. It exerts its inhibitory effect on muscle contraction by tightly binding to actin in the absence of calcium. When Ca^{2+} ions are released from their intracellular stores, binding of Ca^{2+} to TnC alters the TnC–TnI interaction, resulting in the relief of TnI inhibition of the actin–myosin interaction. A major role both in inhibiting actin–myosin interaction, as well as in binding to TnC, is played by the inhibitory region of TnI.

Syska *et al.* (1976) demonstrated that a CNBr fragment 96–116 of rabbit fast skeletal TnI retained the parent molecule's ability to inhibit the actomyosin Mg^{2+}ATPase in the presence of tropomyosin. The 20-residue peptide comprising residues 98–117 of TnI synthesized by Talbot and Hodges (1979), was the first synthetic peptide to show biological activity in muscle research. Like both native TnI and the corresponding CNBr fragment 96–116, this peptide potentiated the inhibitory effect of tropomyosin on actomyosin Mg^{2+}ATPase. Stepwise deletion of single amino acid residues from the N-terminus and the C-terminus of the 20-residue peptide revealed that the segment essential for inhibition of the actomyosin Mg^{2+}ATPase could be narrowed to a peptide as little as the nine residues 105–114 of TnI (Talbot and Hodges 1981*a*), which were denoted as TnI inhibitory peptide. Deletion of further residues greatly impaired the inhibitory potency of the peptide.

Talbot and Hodges (1981*b*) constructed peptide analogues of the inhibitory region of skeletal (G^{104}-K-F-K-R-P-P-L-R-R-V-R-A^{116}) and cardiac TnI (G^{137}-K-F-K-R-P-T-L-R-L-R-V-R-A^{150}), and of hybrids of the two sequences. Replacement of Pro^{110} by threonine, as well as the mere insertion of an arginine residue between residues 112 and 113 (including a misalignment of the consecutive residues), resulted in no significant change of the inhibitory effect on actomyosin Mg^{2+}ATPase. However, if the residue inserted was a leucine, as is the case in cTnI, the inhibitory activity of the peptide was markedly decreased. In a similar approach, Van Eyk and Hodges (1988) synthesized a series of 'mutant' analogues to the inhibitory peptide in which one residue after the other was replaced by a glycine. All these TnI inhibitory peptide analogues inhibited the acto-S1 Mg^{2+}-ATPase but were quantitatively less effective than the native peptide, indicating that every single residue of the inhibitory segment to some extent contributed to the biological activity.

The TnI inhibitory segment not only binds to actin but also to TnC, as was shown by Cachia *et al.* (1983). They synthesized an inhibitory peptide analogue in which a spin label was coupled to Phe^{106}, and monitored TnC binding of this analogue using ^{1}H-NMR spectroscopy. The inhibitory peptide interacted strongly with TnC, both in the absence and presence of Ca^{2+}, with signals from the same residues being perturbed under both conditions. TnI inhibitory peptide binding to TnC was observed to increase the Ca^{2+} affinity of TnC by Rüegg *et al.* (1989*b*). This was accompanied by an increase in the number of helical residues within TnC as detected by circular dichroism. The inhibitory peptide competed with TnI for TnC binding when added to the preformed complex of TnI/TnC.

Rüegg *et al.* (1989*b*) were the first to demonstrate that the inhibitory peptide not only affected actomyosin Mg^{2+}ATPase activity in solution but also exerted effects on force production in skinned cardiac and skeletal muscle fibres: after addition of inhibitory peptide the pCa–tension relationship was markedly shifted to the right indicating a desensitization of the myofilaments to Ca^{2+}. Isometric force at maximally activating Ca^{2+} concentrations was also diminished (Rüegg *et al.* 1991). The replacement of Pro^{109} by Gly completely abolished the inhibition of force production elicited by the native peptide. Interestingly, a Leu^{111}Gly replacement resulted in a profound inhibition of force in skeletal muscle fibres, yet activated cardiac muscle fibres. A similar observation was reported by Strauss *et al.* (1993) for a Phe^{106}Gly replacement which converted the peptide from strongly inhibitory to strongly stimulatory in cardiac skinned fibres.

Van Eyk *et al.* (1993) removed endogenous TnI from skinned cardiac muscle fibres using treatment with vanadate (Strauss *et al.* 1992*b*). In these TnI-depleted fibres, which contracted Ca^{2+} independently, dose-dependent relaxation could be achieved by addition of TnI inhibitory peptide. Subsequent addition of TnC restored the fibres' ability of contracting in the presence of Ca^{2+}. Thus, the inhibitory peptide, after reconstitution into TnI-depleted skinned cardiac muscle fibres, was able to switch between binding sites on TnC and actin–tropomyosin, depending on the Ca^{2+} concentration. Interestingly, the inhibition of Ca^{2+}-independent force in TnI-depleted skinned fibres exerted by the skeletal inhibitory peptide (residues 104–115, Strauss *et al.* 1993), as well as the corresponding cardiac inhibitory peptide (residues 137–148, Barth *et al.* 1995), was antagonized by the Ca^{2+}-sensitizing compound EMD 53998.

The cardiac inhibitory peptide (residues 137–148), when loaded into isolated intact cardiomyocytes, upon lowering of the ATP concentration by metabolic poisoning, prevented the entry of these cells into rigor and the concomitant rise in intracellular [Ca^{2+}] (Duffy and Cobbold 1995). However, as much as 5 mM peptide was necessary to achieve this protective effect, probably due to the high affinity of rigor cross-bridges towards actin, which required the competitor to be present at such a high concentration. Stapleton *et al.* (1995) have reported that a corresponding scrambled peptide failed to protect cardiomyocytes against rigor contraction.

As quoted above, several 'mutant' peptide studies were aimed at investigating the importance of single amino acid residues within the inhibitory segment of TnI. All of these studies made use of the advantage that solid phase peptide synthesis allows to place any desired amino acid at any desired position within the peptide.

Thus, it is fairly easy to substitute a 'mutant' residue for a specified residue of the native sequence. A loss of biological activity in the 'mutant' as compared to the native peptide can then be attributed to the single amino acid replacement and the degree of weakening of the biological activity is considered to be correlated with the physiological relevance of this specified residue of the native sequence.

The studies of Zhang *et al.* (1996) and Strauss *et al.* (1996), however, have revealed that results obtained with 'mutant' peptides need to be interpreted carefully. After extraction of endogenous TnI from skinned cardiac muscle fibres, Strauss *et al.* (1996) reconstituted these preparations with skeletal muscle TnI mutants $Lys^{105}Gly$, $Leu^{111}Gly$, $Pro^{110}Thr$, and $Pro^{110}Gly$. The $Pro^{110}Thr$ mutant, as had been expected from the 'mutant' peptide investigation of Talbot and Hodges (1981*b*), had no significantly different effect from intact wild types sTnI. However, the $Pro^{110}Gly$ mutant exerted no significant effect either, despite the observation of Van Eyk and Hodges (1988) that the corresponding 'mutant' peptide had been considerably less effective than the native peptide in inhibiting acto-S1 Mg^{2+}ATPase. The 'mutant' peptide $Leu^{111}Gly$ when added to TnI-depleted cardiac skinned fibres had been observed to exert no inhibiting but an activating effect on force production (Rüegg *et al.* 1991). The corresponding mutant protein had no significant effect. On the other hand, the mutant protein $Lys^{105}Gly$ to a significantly lesser degree inhibited force when reconstituted into TnI-depleted skinned cardiac fibres, as compared with wild type TnI, which was consistent with the finding of Van Eyk and Hodges (1988) that the corresponding 'mutant' peptide's inhibitory potency was greatly diminished. Zhang *et al.* (1996) examined the inhibition of actomyosin Mg^{2+}ATPase in solution occurring after addition of mutant sTnI and the relief of this inhibition upon addition of TnC and Ca^{2+}. Mutations within the basic C-terminal segment of the inhibitory region were shown to impair only weakly the inhibitory potency of TnI, which is in contradiction to the 'mutant' peptide studies of Talbot and Hodges (1981*b*) and Van Eyk and Hodges (1988), who considered the basic C-terminal residues of the inhibitory region to be essential for its biological function.

Thus, when interpreting results obtained using 'mutant' peptides, one should bear in mind that the substitution of a single residue within the inhibitory peptide will probably disrupt the peptide's structure to a much higher degree than if the same substitution is carried out in the intact protein, where the entire three-dimensional structure stabilizes the corresponding segment. Additionally, protein domains other than the inhibitory region of TnI may contribute to the biological activity of the intact molecule and may, therefore, partly compensate for any structural and functional alteration of the inhibitory region (Strauss *et al.* 1996).

Miscellaneous Troponin-I peptides

The peptide approach has also proved to be useful with regard to functionally important domains of TnI different from the inhibitory region. The identification of the protein-kinase A-dependent phosphorylation sites on cardiac TnI, for example, was greatly promoted by the peptide work of Solaro *et al.* (1976) who demonstrated that after treatment of isolated perfused hearts with adrenaline, phosphate incorporation almost exclusively occurred within the N-terminal CNBr fragment containing residues 1–48 of cTnI. Consistently, the site of phosphorylation was located by Mittmann *et al.* (1990) to the residues Ser^{23} and Ser^{24}.

Binding of the N-terminal region of skeletal TnI to TnC was examined by Ngai and Hodges (1992), who synthesized peptides corresponding to residues 1–40, 10–40, and 20–40 of sTnI. None of the peptides inhibited acto-S1-tropomyosin Mg^{2+}ATPase or bound to actin–tropomyosin indicating that the N-terminal region of sTnI does not interact with actin. However, disinhibition of acto-S1-tropomyosin-TnI Mg^{2+}ATPase by addition of TnC was blocked by the peptides. STnI 1–40 was found to bind to TnC Ca^{2+} dependently.

In summary, peptide investigations have played an important role in promoting our understanding of TnI and its interactions with actin–tropomyosin as well as TnC.

Troponin-C peptides

Peptide investigations using analogues of specific segments of TnC have focused on two aspects of TnC function that characterize the central role of this protein in the Ca^{2+}-dependent regulation of contraction: interaction with TnI and binding of Ca^{2+}. Weeks and Perry (1978) identified a CNBr fragment 83–134 of skeletal TnC that was able to form a Ca^{2+}-dependent complex with skeletal TnI. Moreover, the fragment, which comprised relevant parts of the C-terminal domain of TnC, neutralized the inhibition of desensitized actomyosin Mg^{2+}ATPase by TnI. This neutralization of TnI-induced inhibition did not occur in the presence of excess ethylene glycol-bis(β-aminoethylether)*N*,*N*,*N*′,*N*′-tetraacetic acid (EGTA). A similar study was carried out by Grabarek *et al.* (1981): various TnC fragments obtained by proteolysis using trypsin, thrombin, and CNBr were added to actomyosin inhibited by the presence of tropomyosin, TnI and troponin-T (TnT). Three of the fragments examined, containing residues 1–120, 1–100, and 89–159, respectively, like the parent protein reversed the inhibition of actomyosin Mg^{2+}ATPase activity. It is interesting to note that fragment 1–100, which contained only the low-affinity Ca^{2+}-specific sites of TnC, and fragment 89–159, which contained only the high-affinity Ca^{2+}/Mg^{2+} sites of TnC, were equally effective. The only sequence that both fragments shared, residues 89–100, belonged to the N-terminal helix of calcium-binding site III. Furthermore, a remarkable observation was that the fragment 89–159, containing only the Ca^{2+}/Mg^{2+} sites, which are not considered to be involved in regulation, could substitute for TnC in conferring Ca^{2+} sensitivity to TnI-inhibited actomyosin.

As a second aspect of TnC function, the structure of the Ca^{2+}-binding sites has been studied in detail with the use of peptides. Reid *et al.* (1980) observed that a 12-residue synthetic peptide, corresponding to the calcium-binding loop III of skeletal TnC, bound Ca^{2+} with low affinity. The solution structure of this peptide in ^{1}H-NMR experiments was very similar to that of the corresponding calcium-binding loop in the crystal structure of TnC (Marsden *et al.* 1989). When the C-terminal helix of site III was added, the resulting 21-residue loop–helix segment bound Ca^{2+} with markedly increased affinity. Circular dichroism examination revealed that if the medium containing the peptide was rendered hydrophobic, considerable helix formation occurred. In an aqueous medium, the peptide possessed very little secondary structure in the absence of Ca^{2+}. However, addition of Ca^{2+} induced the development of secondary structure (Reid *et al.* 1981). The development of secondary structure was slightly increased in a 26-residue peptide containing five additional residues of the N-terminal helix; however, the Ca^{2+} affinity was still low. Eight further N-terminal residues conferred a reasonable Ca^{2+}-affinity on the peptide; the resulting 34-residue peptide showed a considerable development of secondary structure in aqueous medium even in the absence of Ca^{2+}.

Shaw *et al.* (1990) used ^{1}H-NMR spectroscopy to study the changes elicited by the addition of Ca^{2+} to a solution of the 34-residue peptide representing Ca^{2+}-binding site III. The difference spectra indicated that upon binding Ca^{2+} a symmetric peptide dimer similar in tertiary structure to the 70-residue C-terminal domain of TnC was formed. Comparable behaviour was found by Kay *et al.* (1991) when examining a 39-residue proteolytic fragment containing Ca^{2+}-binding site IV of skeletal TnC, which formed a stable dimer of helix–loop–helix motifs related by a two-fold axis of symmetry. Two synthetic 34-residue peptides encompassing Ca^{2+}-binding sites III and IV did not interact with each other in Ca^{2+}-free solution. However, when Ca^{2+} was added, a heterodimer was again formed (Shaw *et al.* 1992).

Taken together, the peptide studies of the TnC calcium-binding sites demonstrate that, even in relatively short peptide segments, considerable information is encoded about secondary, tertiary and even quaternary structure of the parent proteins, which can be recalled under favourable conditions.

Peptide mimetics of smooth muscle regulatory proteins

An essential step in contractile activation of smooth muscle involves phosphorylation of the regulatory myosin light chain (MLC_{20}) by the calcium/calmodulin-dependent myosin light chain kinase (MLCK) (Itoh *et al.* 1989). Indeed, smooth muscle contraction is often preceeded by myosin light chain phosphorylation, but contraction without light chain phosphorylation has also been observed more recently. Thus, phosphorylation-independent mechanisms for regulation of smooth muscle contraction have been proposed, and these may involve the regulatory proteins calponin and caldesmon (for review see Marston and Huber 1995). In recent years, synthetic peptide approaches applied to permeabilized smooth muscle fibres have been found useful in unravelling the relative contributions of myosin light chain kinase, calponin and caldesmon in smooth muscle activation. For instance, specific inhibition of MLCK and of myosin phosphatase (Gailly *et al.* 1996) by peptide inhibitors should be a very promising approach in elucidating the role of MLC phosphorylation in smooth muscle contraction and relaxation.

MLC and MLCK peptides

One of the most potent inhibitors of MLCK known is the nonapeptide Lys-Lys-Arg-Ala-Ala-Arg-Ala-Thr-Ser-NH_2, which has an IC_{50} value of only 14 μM (Pearson *et al.* 1986). The peptide is a competitive inhibitor because it is based on the sequence of the substrate of myosin light chain kinase, i.e. residues Lys^{11}–Ser^{19} of the smooth muscle myosin light chain MLC_{20}. Interestingly, the hexapeptide Lys^{11}–Arg^{16} is nearly as potent (IC_{50} 27 μM, Hunt *et al.* 1989). Therefore, both peptides might be used to probe the role of myosin light chain phosphorylation in regulating contractility in skinned smooth muscle. Surprisingly, however, Strauss *et al.* (1992*a*) found that these peptides did not alter the extent of myosin light chain phosphorylation or contractile tension during steady-state force generation of permeabilized guinea-pig *Taenia coli*. The reason was that the peptide did not only inhibit myosin light chain kinase activity, but also myosin phosphatase activity thereby 'freezing' myosin light chain phosphorylation at an intermediate level. The latter finding may not be so surprising since myosin light chains are, of course, also substrates of myosin light chain phosphatase. Interestingly, the peptide also inhibited unloaded shortening of the skinned *Taenia coli* fibres. This effect is unspecific, as Cai *et al.* (1995) have found that a 13-amino acid peptide, mimicking the C-terminal region of the smooth muscle myosin isoform SM_1, also reduced the unloaded shortening velocity of skinned guinea-pig *Taenia coli*. Obviously, there would be a need for more specific inhibitors. One of these inhibitors, SM_1, is based on the so-called pseudosubstrate sequence of myosin light chain kinase (cf. Itoh *et al.* 1989).

The pseudosubstrate sequence of myosin light chain kinase contains residues 480–501. This sequence is partly similar to that of the myosin light chain residues, which are the substrate of MLCK. Thus, the pseudosubstrate sequence competes with the light chain for the substrate-binding active site of the enzyme. During calcium activation, this intramolecular peptide competition is relieved as the pseudosubstrate sequence interacts with calcium/calmodulin, rather than with the active site of the enzyme. The interaction of MLCK with calmodulin involves a stretch of 20 residues (MLCK 493–512) that partially overlap with the pseudosubstrate sequence. This overlapping part of the molecule may, therefore, be regarded as a dynamic flip-flop mechanism switching as it were calcium dependently between calmodulin-binding and active site inhibitory states (Hardie 1988; Kemp *et al.* 1987). The C-terminally located pseudosubstrate sequence may, however, be removed from the enzyme by proteolysis (Ikebe *et al.* 1987). In this way, a truncated myosin light chain kinase denoted as IMLCK is obtained, which is no longer inhibited in the absence of calcium/calmodulin, but is constitutively active. IMLCK can, however, be inhibited by a synthetic analogue of the pseudosubstrate sequence (denoted SM_1, cf. Ikebe *et al.* 1987) containing the residues AKKLSKORMKKYMARRKWGKTG. This inhibitory MLCK peptide obviously competes with the natural substrate (the myosin light chains) for the catalytic site of IMLCK (Kargacin *et al.* 1990).

By micro-injecting the SM_1 peptide and/or IMLCK, Itoh *et al.* (1989) were able to show that myosin light chain phosphorylation is both necessary and sufficient to induce smooth muscle contraction. Thus, micro-injection of IMLCK induced—without increase in intracellular Ca^{2+} concentration—within 20 s a contraction of smooth muscle cells which could be completely inhibited when SM_1 peptide was also injected (Fig. 5.1). Remarkably, this contraction could not be inhibited by another MLCK-derived peptide denoted as RS_{20}. This peptide is based on the calmodulin-binding sequence containing amino acid residues 493–512 of smooth muscle myosin light chain kinase and has the sequence ARRKWQKTGHAV-RAIGRLSS (Lukas *et al.* 1986). It is noteworthy that both the IMLCK-induced contraction and its inhibition by the MLCK inhibitory peptides occurred at basal levels of free calcium. In contrast, high K^+ solution caused a rise in the concentration of the intracelluar free calcium that was bound to calmodulin, thereby activating myosin light chain kinase and inducing contraction. Significantly, this contraction was inhibited by both the RS_{20} calmodulin-binding peptide and the SM_1 peptide. Calcium-induced contractions of permeabilized smooth muscle cells (Kargacin *et al.* 1990), as well as skinned fibres (Rüegg *et al.* 1989*a*), were also inhibited by the RS_{20} peptide in micromolar concentrations, and this inhibitory effect was competitively antagonized by calmodulin. In conjunction, these results show in a most direct manner that, 'activation of the calmodulin myosin light chain kinase pathway is both necessary and sufficient to trigger contraction of smooth muscle' (Itoh *et al.* 1989).

Given the high affinity of calmodulin for RS_{20} (KD 10^{-9} M), this peptide can also be used to buffer and stabilize the free calmodulin concentration (Rüegg

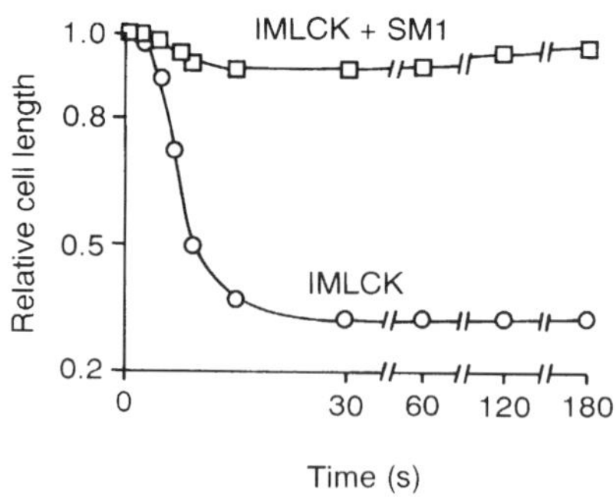

Fig. 5.1 Effects of intracellular injection of IMLCK on the cell length of a single smooth muscle cell in the presence of (*square*) and absence of (*circles*) the SM_1 inhibitory peptide (adapted from Fig. 3 in Itoh *et al.* 1989).

et al. 1989). In this way it could be shown that free calmodulin activates force in the same range (nanomolar concentrations) in which it also activates myosin light chain kinase. On the other hand, in skinned fibres the calmodulin concentration required to induce maximal contraction was far in excess of the RS_{20} peptide. The reason is that other calmodulin-binding proteins, such as caldesmon and calponin, will presumably bind calcium/calmodulin. Obviously, these binding sites must be saturated under conditions of maximal contractile activation. On the other hand, it is noteworthy that the calmodulin-binding peptide RS_{20} may induce full relaxation, while myosin light chain phosphorylation is only decreased to intermediate values. In conjunction, these findings suggest that mechanisms other than myosin light chain phosphorylation are also involved in controlling the contractile state of smooth muscle.

Caldesmon peptides

Caldesmon is a thin filament actin-binding protein, which inhibits actomyosin ATPase *in vitro*. At high concentrations of calcium, caldesmon binds to calmodulin rather than actin, so that its inhibitory action on actin–myosin interaction is derepressed (for review see Marston and Huber 1995). Szpacenko *et al.* (1985) have shown that caldesmon relaxes skinned fibres of permeabilized smooth muscle without causing myosin light chain dephosphorylation; but clearly this relaxing effect could also have been due to a reduction in free calmodulin concentration by the addition of the caldesmon. Progress in the molecular understanding of caldesmon action was subsequently obtained by the use of caldesmon fragments and synthetic peptides. Thus, Pfitzer *et al.* (1993) have shown that caldesmon inhibited the force production of chemically skinned gizzard fibres, which were activated at basal levels of Ca^{2+} by the constitutively active myosin light chain kinase (IMLCK), and this effect could be mimicked by the C-terminal actin-binding fragment of caldesmon. The N-terminal myosin-binding fragment, on the other hand, had no inhibitory effect. Inhibition caused by the actin-binding fragment was not observed at high levels of myosin light chain phosphorylation and it was also reversed by calcium/calmodulin at high calcium concentrations in these experiments. The mode of action of caldesmon was presumably similar to that observed in skinned skeletal muscle fibres. Here, force production could be completely inhibited by the actin-binding fragment of caldesmon, which seemed to block the attachment of myosin cross-bridges to actin (Brenner *et al.* 1991; Kraft *et al.* 1995). In conjunction, these studies support the concept that caldesmon causes relaxation of smooth muscle by binding to actin in a calcium/calmodulin-dependent manner.

Caldesmon binds to the same N-terminal region of actin as TnI (actin residues 1–40), but not to the same residues; it also binds to an additional site that is further along the peptide chain in the region where S1-binding has been demonstrated (Levine *et al.* 1990). To locate more precisely the actin-binding and actomyosin ATPase inhibitory sequence, Mezgueldi *et al.* (1994) made use of proteolytic fragments, as well as of several synthetic overlapping peptides covering the sequence Asn^{675}–Lys^{729} of caldesmon, and they were able to show that the stretch of residues 692–722 played a major role in binding to actin, as well as in the inhibition of actomyosin ATPase. Thus, a peptide spanning this sequence bound to actin and this was competitively inhibited by an antibody directed to the sequence 693–721 of smooth muscle caldesmon. In addition, there is another actin-binding site, as well as a calmodulin-binding site, within the stretch of residues ranging from Trp^{659} to Phe^{665} (Zhan *et al.* 1991). These authors pinpointed the calmodulin-binding site to residues Trp^{659}–Phe^{665} and then synthetized an 18-residue peptide containing this site, as well as an actin-binding site (containing residues 651–667 of caldesmon) with an additional cysteine at the C-terminus. This peptide (GVRNIKSMWEKGNVFSSC) bound to F-actin and it could be displaced from actin by calcium/calmodulin in much the same way as native caldesmon but, unlike caldesmon, the peptide had no inhibitory effect on actomyosin ATPase. When the peptide was infused into β-escin skinned fibres from arteries, force was increased (Katsuyama *et al.* 1992); these findings were taken to mean that the peptide competed with endogenous caldesmon for the binding site on actin, thereby releaving its inhibitory action. In conjunction, all these findings support the idea that caldesmon may play a major role in the calcium/calmodulin-dependent regulation of smooth muscle tone. However, as mentioned already, it would, not impair force development at high levels of myosin light chain phosphorylation (cf. Pfitzer *et al.* 1993), which seems to override thin filament regulation.

Calponin peptides

Calponin is a smooth muscle regulatory protein that binds to the thin filaments, thereby inhibiting actin–myosin interaction and, hence, the ATPase activity of actomyosin and contraction (Obara *et al.* 1996; Winder and Walsh 1990). Disinhibition of actomyosin ATPase occurs *in vitro* when calponin is phosphorylated by protein kinase C. Thus dephosphorylated, but not phosphorylated, calponin will inhibit protein kinase C induced contraction of permeabilized smooth muscle myocytes (Horowitz *et al.* 1996). Obviously, the phosphorylatable serine residue 175 seems to be critical for the inhibitory action of calponin, since a mutant protein in which serine 175 has been replaced by alanine was unable to cause inhibition. This residue is within the actin-binding and ATPase inhibitory domain of calponin, which comprises a 38-residue stretch, Ala^{145}–Tyr^{182} (Mezgueldi *et al.* 1992, 1995). Recently, the interaction of this domain with actin, tropomyosin and calcium-binding proteins was studied in great detail, employing a series of overlapping synthetic peptides spanning various parts of this region (Mezgueldi *et al.* 1995). While some synthetic peptides of that region inhibited the actin-activated myosin ATPase activity, other peptides displayed only calponin's capacity to recognize various calcium-dependent proteins and tropomyosin or actin. Based on these findings, Horowitz and colleagues (1996) designed a 30-residue peptide

(LQMGTNKFASQQGMTAYGTRRHLYDPKLGC) that comprised part of the actin-binding site (Ala^{145}–Tyr^{182}) but did not contain the 19 residues Ala^{145}–Ile^{163}, which are essential for inhibition of the actomyosin-ATPase according to Mezgueldi *et al.* (1995, 1996). Hence, the peptide was unable to inhibit actin–myosin interaction, while still retaining calponin's capacity to bind to actin. This synthetic peptide corresponding to the sequence Leu^{166}–Gly^{194} of calponin, but containing an additional cysteine, was found to be useful in competing with endogenous dephosphorylated calponin and displacing it from its actin-binding sites.

Horowitz and colleagues then demonstrated that 50–100 μM peptide induced a contraction of aortic smooth muscle cells that were permeabilized with saponin and immersed into an ATP-salt solution containing ATP and EGTA to reduce the Ca^{2+} level to below 10^{-8} M. On the other hand, a scrambled peptide containing the same number of residues, but in a disordered manner, had no effect. The effect was dose-dependent (Fig. 5.2). Thus, the specific effect observed with the peptide mimetic of the actin-binding region was taken to indicate a derepression of calponin-induced inhibition by peptide competition. A shorter peptide from that region (Phe^{173}–Arg^{185}) was much less effective and less specific (Itoh *et al.* 1995), as it produced a comparatively small increase of force in β-escin permeabilized skinned fibres from smooth muscle that were precontracted at low levels of Ca^{2+}.

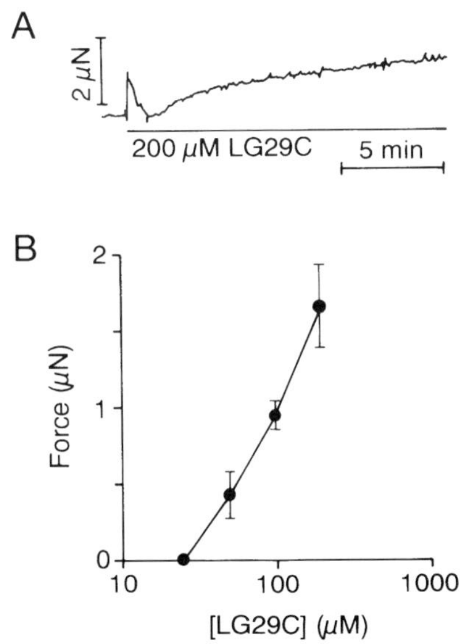

Fig 5.2 Effect of a calponin peptide (LG29C) on unstimulated single smooth muscle cell at pCa 9 (adapted from Fig. 4 in Horowitz *et al.* 1996). (A) Time course of contraction following injection (causing an initial mechanical artefact) of the peptide. (B) Dependence of isometric force on the peptide concentration.

Interestingly, the peptide used by Horowitz and colleagues (1996) produced a contraction that was similar in magnitude to that elicited in permeabilized cells by the calcium-independent ε isoform of protein kinase C that was found to cause phosphorylation of calponin and its translocation away from the thin filaments to the cell membrane. In both the peptide-induced contraction, as well as in the protein kinase C-induced contraction, contractile activation is obviously brought about at basal Ca^{2+} levels by dissociating the inhibitory calponin from the actin filaments. If it were possible to introduce peptides (or peptide mimetics) into intact cells, peptides or bio-organic peptide mimetics of calponin (and caldesmon) might offer the prospect of very selective pharmacological intervention into the regulatory mechanism of smooth muscle contraction and relaxation.

Miscellaneous

Peptide mimetic approaches have also been useful to investigate the mechanism of excitation–contraction coupling in skeletal muscle. It is well known that during muscle activation an action potential invades the T-tubular system, thereby causing calcium release from the sarcoplasmic reticulum. In studying this process, the interface between the T-system and the terminal cisternae of the sarcoplasmic reticulum is a focus of great interest. It is now clear that two calcium channel proteins are involved: one residing in the T-tubular membrane, the so-called dihydropyridine receptor (DHP receptor), and one located at the membrane of the terminal cisternae, the calcium release channel, also known as ryanodine receptor (RYA receptor). When the T-tubular membrane is depolarized, the DHP receptor senses the voltage change and transmits a signal to the RYA receptor, but the nature of this signal is still not yet clear (for review see Rüegg 1992). It is generally assumed, though, that the two proteins interact directly, rather than indirectly, by means of other proteins, e.g. triadin. Indeed, recent biochemical studies show that the DHP receptor may directly interact with ryanodine receptors *in vitro*, thereby forming a supramolecular complex as reviewed by Melzer *et al.* (1995). However, the site of this interaction is not known, but may well involve the 'effector loop' linking transmembrane domain 2 and 3 of the DHP receptor (Tanabe *et al.* 1990).

Lu *et al.* (1994) have shown that a 61-residue peptide spanning this loop sequence (Glu^{666}–Glu^{726}) could interact with the RYA receptor, thereby enhancing its affinity for ryanodine and also increasing the open probability of this calcium channel. Skeletal muscle ryanodine receptors were incorporated into planar lipid bilayer separating two compartments of a plastic chamber filled with saline. By means of a current amplifier connected to two electrodes inserted into the compartments and suitably applied potentials, an ionic current could be measured across the bilayer due to the opening of a pore protein when Ca^{2+} (or Ba^{2+}) ions were present in the experimental chambers. Most interestingly, the open probability of the calcium release channel increased from 0.02 to 0.7 when the peptide spanning the DHP receptor loop was added to the experimental chamber.

However, in comparable experiments where cardiac ryanodine receptor was incorporated into planar lipid bilayer, its open probability was not enhanced. More recently, Lu *et al.* (1995) showed that serine 687 is essential for a functional interaction of the peptide with the RYA receptor. Thus, an effector loop peptide in which serine 687 was either phosphorylated by cAMP-dependent protein kinase or mutated into alanine could bind to, but not activate, the RYA receptor. Binding of the phosphorylated peptide could, however, be competitively inhibited by the unphosphorylated peptide. In conjunction, these studies suggest that the effector loop region of the DHP receptor functionally interacts with skeletal, but not cardiac, calcium release channels, in direct support of the hypothesis of the mechanical coupling between the T system and SR in skeletal muscle. These studies also suggest that phosphorylation of serine 687 by cAMP-dependent protein kinase may modulate the interaction of RYA receptor and DHP receptor in excitation–contraction coupling.

In cardiac muscle cells, the DHP receptor is identical to the L-type Ca^{2+} channel, which is the predominant gate for the entry of Ca^{2+} ions into the cell. The open probability of the channel and, hence, an important factor determining the intracellular Ca^{2+} concentration, is modulated by β-adrenergic receptor stimulation via the adenylyl cyclase–cyclic adenosine-3′,5′-monophosphate–cAMP-dependent protein kinase pathway. Recently, an interesting peptide approach has been of great use in identifying a serine residue of the α_{1C}-subunit of the cardiac L-type Ca^{2+} channel as a potential site of functional modulation by cAMP-dependent phosphorylation, namely Ser^{1928} (Perets *et al.* 1996). Native and mutant channel proteins were expressed in *Xenopus* oocytes where, as a well-known phenomenon, PKA treatment does not enhance the Ca^{2+} current. This is possibly due to an abnormally high level of basal phosphorylation, which cannot be further increased by addition of PKA. The PKA-inhibiting peptide (PKI) of Fernandez *et al.* (1991) was used to inhibit the cAMP-dependent protein kinase resulting in a decreased Ca^{2+} current. In the mutant channel proteins, several of the serine residues of the α_{1C}-subunit had been replaced by alanine. The mutant carrying a Ser^{1928}Ala replacement prior to treatment with PKI displayed a decreased basal Ca^{2+} current as compared to wild-type protein and mutants with other serines being replaced. In oocytes expressing the Ser^{1928}Ala mutant α_{1C}-subunit, PKI treatment did not reduce the Ca^{2+} current, which was markedly depressed in the presence of wild-type and all other mutants. Thus, it was concluded that cAMP-dependent phosphorylation of Ser^{1928} may well be responsible for the increased L-type Ca^{2+} current observed under conditions of β-adrenergic receptor stimulation.

Besides EC coupling, other aspects of signal transduction, such as G protein receptor interaction, have been studied by peptide approaches. For instance, a C-terminal 11-residue peptide of the stimulatory G protein α-subunit, namely α-384–394, was found to compete with G protein receptor interaction, thereby selectively blocking β-adrenergic stimulation of adenylate cyclase in saponin-permeabilized cells (Hamm 1991).

5.3 Problems and limitations

In the preceding sections, the usefulness of different peptide approaches has been exemplified by the diversity of important results that have been obtained in muscle research using peptides. Some specific problems and limitations have already been addressed. One more general limitation arises from the fact that in most cases protein–protein interactions involve more than just one segment of the respective polypeptide chains. In the case of the actin–myosin interaction, for example, several actin-binding sites are arranged over a region comprising more than 200 residues of the myosin heavy chain (Rayment *et al.* 1993*a*). It is not possible to synthetically produce a peptide covering this segment completely. This apparent limitation of the peptide approach, however, may also be interpreted as one of its biggest advantages, since it allows examination of each of these sites functionally isolated from the remainder of the molecule. Thus, it is possible to obtain information about the relative importance of the isolated segments with respect to each other.

On the following pages we will summarize some specific critical points which have to be taken into consideration for the work with peptides.

Solution structure and bound structure of peptides

The conformation of a protein domain is stabilized by the whole three-dimensional structure of the protein. A specified segment, when present as part of the protein, is involved in elements of secondary structure, which are arranged three-dimensionally to determine the appearance of the whole molecule. On the other hand, when the same segment has been isolated from the stabilizing remaining parts of the protein, it should not be presumed that it adopts the same structure in solution as in the parent protein. Rather, the peptide may exist in a variety of random conformations that are in rapid equilibrium in solution. If the peptide is derived from a specific protein recognition site, only one of these conformations may bind with high affinity and presumably the fraction of total peptide presenting in the appropriate conformation will be small. Based on this assumption, the ratio of the concentrations of peptide in the conformation binding with high affinity and of alternate conformations, will be virtually equal to the ratio of the apparent stability constant of the peptide and the real stability constant (corresponding to the association constant of the peptide/protein complex). Any intervention that stabilizes the peptide in its binding conformation can therefore be expected to enhance the apparent affinity constant, i.e. it decreases the EC_{50} values. If the different peptide conformations are in rapid equilibrium in solution, the binding process will cause the peptide to accumulate in the bound state, which is equivalent to the high affinity conformation. In other words, interaction with the target segment transforms the peptides from a random coil into a highly ordered structure by a kind of selection mechanism exerted by the target segment. This mechanism is generally referred to as 'induced fit'.

In various studies, synthetic peptides and proteolytic fragments have been shown to display considerable three-dimensional structure also in the absence of a target protein. Miller *et al.* (1987) produced polyclonal site-specific antibodies by immunizing rabbits against a synthetic peptide spanning the N-terminal residues 1–7 of actin from skeletal muscle. Obviously, when inducing antibody production in the host's lymphocytes, the seven-residue peptide must have adopted a conformation closely resembling the one present in intact actin, since the antibodies specifically bound to an immunoaffinity chromatography column with G-actin covalently coupled to sepharose. DasGupta *et al.* (1988, 1990) have demonstrated that F(ab) fragments derived from these antibodies completely inhibited actin-activation of S1 Mg^{2+}ATPase, as well as the S1-induced polymerization of G-actin, by inhibiting the binding of these proteins to one another.

If even a peptide as small as actin 1–7 possesses sufficient structure to induce specific antibody production, it will not be surprising that larger peptides have been shown to adopt distinct conformations. ^{1}H-NMR experiments carried out by Marsden *et al.* (1989) revealed that a 12-residue peptide corresponding to the Ca^{2+}-binding loop III of TnC had a solution structure very similar to that determined as part of the crystal structure of TnC. The extensive peptide work on the Ca^{2+}-binding sites III and IV of TnC by Shaw *et al.* (1990, 1992) and Kay *et al.* (1991) have clearly proved that within the amino acid sequence of peptides, enough structural information was encoded for the peptides to aggregate specifically into dimers closely identical in structure to domains of the intact molecule.

Purity and specificity

Great care must be taken when working with peptides to ensure that they are properly purified. Contaminants of different kinds may exert effects, which must not be mistakenly attributed to specific peptide action. Chase and co-workers discovered that the inhibition of maximum Ca^{2+}-activated force in skinned psoas muscle fibres, seemingly exerted by a peptide from the SH1 region of myosin (Chase and Kushmerick 1989) after HPLC purification of the peptide, was greatly altered (Chase *et al.* 1990). Consequently they attributed most of the inhibitory activity observed beforehand to an unspecified contaminant. This contaminant later was identified as aluminium fluoride, which was an interesting discovery, since it allowed characterization of aluminium fluoride as an analogue of inorganic phosphate binding tightly to and dissociating slowly from S1 (Chase *et al.* 1994).

Of special importance is the pH of peptide solutions. Most peptide synthesis protocols use hydrofluoric acid for cleavage of the peptide from the resin and it is impossible to completely get rid of these acid equivalents during the subsequent process of purification. Additionally, trifluoroacetic acid from purification is often still present in minor amounts as a counterion to the peptide (Van Eyk and Hodges 1993). Thus, when dissolving freeze-dried peptide, sufficient buffering capacity should be provided with the solvent and the resulting solution must be carefully checked for and, if necessary, adjusted for pH. Both buffering capacity and pH of the peptide solution must be taken into consideration when carrying out control experiments in the absence of peptide.

In Section 5.2 we exemplified the difficulties of interpreting the specificity of peptide results when discussing the peptides corresponding to the region around SH1 of the myosin heavy chain. Several studies have demonstrated peptide effects, which were attributed to specific structural information encoded in the amino acid sequence. However, as was discussed in detail by Chase *et al.* (1991*b*), several of the peptide effects derived from the net charge rather than the specific amino acid sequence. An appropriate means to rule out uncertainties towards the specificity of peptide results is the application of scrambled peptides as a control. These peptides carry the same amino acids as the native peptides, but are arranged in a random order. Generally, the effects exerted by scrambled peptides are considerably weaker than those brought about by specific peptide action. If not, the specificity of the peptide results has to be questioned.

Covalent modifications of peptides

It is a common method to introduce into a peptide specific covalent modifications of amino acid side chains. This allows, for example, the monitoring of peptide–protein interactions by alterations in the ^{1}H-NMR signals specifically arising from spin labels or by intensity changes of fluorescence specifically arising from fluorescent labels that have been covalently bound to side chains of the peptide (e.g. Cachia *et al.* 1983; Chaussepied and Morales 1988). It has to be borne in mind that these covalent modifications might alter several properties of the peptide, such as hydrophobicity, PK_i values of neighbouring charged groups and thus the net charge depending on pH, and three-dimensional structure.

The user of peptides also has to be aware that some covalent modifications are essential for the specificity of peptide action; this was pointed out in detail by Talbot and Hodges (1981*a*): in an unmodified peptide, the C-terminal α-carboxyl group represents a negative charge which—in the parent protein—would have been involved in a peptide bond and, hence, be uncharged. This additional charge can dramatically alter the peptide's affinity for the interaction site it is targeted at. The same, of course, holds true for the positively charged free amino group at the N-terminal end of the peptide. Thus, acetylation of the N-terminus and amidation of the C-terminus are wide-spread methods to circumvent this problem. Talbot and Hodges (1981*a*) demonstrated that Lys^{105}, when present as the N-terminal residue of a synthetic TnI inhibitory peptide, had to be acetylated to evoke the petide's full inhibitory effect on actomyosin Mg^{2+}ATPase. The inhibitory potency of a N-terminally non-acetylated derivative was considerably reduced.

Cysteinyl residues present in a peptide might be subject to non-specific disulfide bonds. One means—and when working with proteolytic fragments rather than synthetic peptides the only means—to circumvent this situation is to add to the assay solution sufficient reducing capacity, e.g. dithiothreitole. When working with synthetic peptides, a feasible alternative is to replace cystein by, for example, alanine (cf. Van Eyk and Hodges 1991).

Application of peptides in different assay systems

Peptides as a tool for investigating the mechanisms that govern muscle contraction and regulation offer the great advantage that they are applicable in various different assay systems. They were first employed in solution studies aimed to elucidate the biochemical properties of the interactions between muscle proteins. However, they have also been used extensively in skinned fibre studies and, more recently, even in intact muscle preparations. The use of peptides in skinned muscle fibres is possible because there are no membrane barriers to impede diffusion into the myofilament lattice. An alternative approach widely employed in solution studies, namely to specifically interfere with protein–protein interactions immunologically using mono- or polyclonal antibodies, unfortunately is not feasible in skinned fibres. The enormous size of antibodies or even their F(ab) fragments ($\approx$ 50 000 Da) results in limitations to their diffusability into skinned fibres (Chase *et al.* 1991*a*).

Experiments with various peptide sequences carried out in different laboratories showed a valid tendency for a 10- to 100-fold higher peptide concentration being required to elicit specific effects in skinned fibres as compared to solution studies (Chase *et al.* 1991*b*). In solution, the frequency of protein–protein interactions depends on the concentrations of all proteins involved. However, in a muscle fibre (especially, but not only, in striated muscles) the regularity of the myofilament lattice creates geometrical factors that influence the likelihood of interaction (Goldman and Brenner 1987). Thus, the apparent concentration of two myofilament proteins interacting with each other may be considerably higher than the actual concentration per unit cell volume, which could account for the differences in the effective peptide concentrations in skinned fibres and in solution.

The use of peptides in intact cells sometimes may be limited by the lack of stability towards proteolysis by intracellular proteolytic enzymes. Fernandez *et al.* (1991) had to undertake considerable efforts to render satisfactory stability to their protein kinase A-inhibiting peptide when working with intact rat embryo fibroblast. Two-fold covalent modification and co-injection with a truncated peptide to inhibit proteases was required to allow final examination of the peptide's effects in intact cells.

5.4 Concluding remarks and future prospects

Peptide mimetics of proteins involved in signal transduction may exhibit a high specificity in mimicking or blocking signal transduction processes. In muscle research, the peptide approach—especially when combined with structural investigations, such as X-ray crystallography or NMR—has been found to be very useful in identifying segments of a specific protein recognition site. The recent advances in molecular genetics have offered a complementary approach to examine specific protein–protein interactions, since *in vitro* mutagenesis allows for the replacement of single amino acid residues within the polypeptide chain of a protein or the creation of deletion mutants for proteins lacking putative interaction sites. The concerted application of both approaches may be expected to considerably increase our knowledge about contractile and regulatory proteins in muscle cells within the next decade.

The effects exerted by peptides may be so specific that it may be possible to use them as a therapeutic principle in medicine. Nature itself makes use of peptides as regulators of biological activity. Think, for example, of the tripeptide glutathione, which plays a central role in hepatic metabolism, or of peptide hormones such as angiotensin II, vasopressin, and the hypothalamic releasing factors. If we learn to solve the major problems of how to apply peptides in order to make them systemically available in the organism and how to provide them with the stability necessary for prolonged action we will perhaps be able to design drugs with maximum specificity towards the pathologic mechanisms they are aimed at.

References

Atassi, M.Z. (1975) Antigenic structure of myoglobin: the complete immunochemical anatomy of a protein and conclusions relating to antigenic structure of proteins. *Immunochem.* **12**, 423–38.

Barth, Z., Strauss, J.D., Heyder, S., Van Eyk, J., Wiesner, R.J. and Rüegg, J.C. (1995) Ca^{2+} sensitizing effects of EMD 53998 after troponin replacement in skinned fibres from porcine atria and ventricles. *Pflügers Arch.* **430**, 220–9.

Brenner, B., Yu, L.C. and Chalovich, J.M. (1991) Parallel inhibition of active force and relaxed fiber stiffness in skeletal muscle by caldesmon: implications for the pathway to force generation *Proc. Natl. Acad. Sci. (USA)* **88**, 5739–43.

Cachia, P.J., Sykes, B.D. and Hodges, R.S. (1983) Calcium-dependent inhibitory region of troponin: a proton nuclear magnetic resonance study on the interaction between troponin C and the synthetic peptide N^{α}-acetyl[$FPhe^{106}$]TnI-(104–115) amide. *Biochemistry* **22**, 4145–52.

Cai, S., Ferguson, D.G., Martin, A.F. and Paul, R.J. (1995) Smooth muscle contractility is modulated by myosin tail-S2-LMM hinge region interation. *Am. J. Physiol.* **269**, C1126–32.

Cartoux, L., Chen, T., DasGupta, G., Chase, P.B., Kushmerick, M.J. and Reisler, E. (1992) Antibody and peptide probes of interactions between the SH_1–SH_2 region of myosin subfragment 1 and actin's N-terminus. *Biochemistry* **31**, 10 929–35.

Chase, P.B. and Kushmerick, M.J. (1989) Ca-dependence of a myosin heavy chain (HC) peptide binding to and dissociation from skinned fibers from rabbit psoas muscle. *Biophys. J.* **55**, 406a.

Chase, P.B., Bursell, J. and Kushmerick, M.J. (1990) Peptide inhibition of rabbit psoas muscle skinned fiber mechanics and myofibriller ATPase activity. *Biophys. J.* **57**, 538a.

Chase, P.B., Beck, T.W., Bursell, J. and Kushmerick, M.J. (1991*a*) Competitive inhibition of maximum Ca-activated force in skinned muscle fibers by cationic peptides from the

SH-1 region of myosin heavy chain. In *Peptides as probes in muscle research* (ed. J.C. Rüegg), pp. 49–55. Springer-Verlag, Berlin, Heidelberg and New York.

Chase, P.B., Beck, T.W., Bursell, J. and Kushmerick, M.J. (1991*b*) Molecular charge dominates the inhibition of actomyosin in skinned muscle fibers by SH_1 peptides. *Biophys. J.* **60**, 352–9.

Chase, P.B., Martyn, D.A. and Hannon, J.D. (1994) Activation dependence and kinetics of force and stiffness inhibition by aluminiofluoride, a slowly dissociating analogue of inorganic phophate, in chemically skinned fibres from rabbit psoas muscle. *J. Muscle Res. Cell Motil.* **15**, 119–29.

Chaussepied, P. and Morales, M.F. (1988) Modifying preselected sites on proteins: the stretch of residues 633–642 of the myosin heavy chain is part of the actin-binding site. *Proc. Natl. Acad. Sci. (USA)* **85**, 7471–5.

Chaussepied, P. and Kasprzak, A.A. (1989) Change in the actin-myosin subfragment 1 interaction during actin polymerization. *J. Biol. Chem.* **264**, 20 752–9.

DasGupta, G., White, J., Bulinski, J.C. and Reisler, E. (1988) An immunochemical study of the N-terminal segment of actin and its role in macromolecular interactions. *Biophys. J.* **53**, 574a.

DasGupta, G., White, J., Cheung, P. and Reisler, E. (1990) Interactions between G-actin and myosin subfragment 1: immunochemical probing of the NH_2-terminal segment on actin. *Biochemistry* **29**, 8503–8.

Duffy, P.S. and Cobbold, P.H. (1995) Protection of metabolically poisoned cardiomyocytes by a peptide. *J. Mol. Cell. Cardiol.* **27**, A175.

Fernandez, A., Mery, J., Vandromme, M., Basset, M., Cavadore, J-C. and Lamb, N.J.C. (1991) Effective intracellular inhibition of the cAMP-dependent protein kinase by microinjection of a modified form of the specific inhibitor peptide PKI in living fibroblasts. *Exp. Cell Res.* **195**, 468–77.

Gailly, P., Wu, X., Haystead, T.A.J., Somlyo, A.P., Cohen, P.T.W., Cohen, P. *et al.* (1996) Regions of the 110-kDa regulatory subunit M_{110} required for regulation of myosin-light-chain-phosphatase activity in smooth muscle. *Eur. J. Biochem.* **239**, 326–32.

Goldman, Y.E. and Brenner, B. (1987) Special topic: molecular mechanism of muscle contraction. *Ann. Rev. Physiol.* **49**, 629–36.

Grabarek, Z., Drabikowski, W., Leavis, P.C., Rosenfeld, S.S. and Gergely, J. (1981) Proteolytic fragments of troponin C. *J. Biol. Chem.* **256**, 13121–7.

Hamm, H.E. (1991) Defining sites and mechanisms of interaction between rhodopsin and transducin. In *Peptides as probes in muscle research* (ed. J.C. Rüegg), pp. 141–9. Springer-Verlag, Berlin, Heidelberg and New York.

Hardie, G. (1988) Pseudosubstrates turn off protein kinases. *Nature* **335**, 592–3.

Henry, G.D., Winstanley, M.A., Dalgarno, D.C., Scott, G.M.M., Levine, B.A. and Trayer, I.P. (1985) Characterization of the actin-binding site on the alkali light chain of myosin. *Biochem. Biophys. Acta* **830**, 233–43.

Horowitz, A., Clément-Chomienne, O., Walsh, M.P., Tao, T., Katsuyama, H. and Morgan, K.G. (1996) Effects of calponin on force generation by single smooth muscle cells. *Am. J. Physiol.* **270**, H1858–63.

Hunt, J.T., Floyd, D.M., Lee, V.G., Little, D.K. and Moreland S. (1989) Minimum requirements for inhibition of smooth-muscle myosin light-chain kinase by synthetic peptides. *Biochem. J.* **257**, 73–8.

Ikebe, M., Slepinska, M., Kemp, B.E., Means, A.R. and Hartshorne D.J. (1987) Proteolysis of smooth muscle myosin light chain kinase. *J. Biol. Chem.* **260**, 13 828–34.

Itoh, T., Ikebe M., Kargacin, G.J., Hartshorne D.J., Kemp, B.E. and Fay F.S. (1989) Effects of modulators of myosin light-chain kinase activity in single smooth muscle cells. *Nature* **338**, 164–7.

Itoh, T., Suzuki, A., Watanabe, Y., Mino, T., Naka, M. and Tanaka, T. (1995) A calponin peptide enhances Ca^{2+} sensitivity of smooth muscle contraction without affecting myosin light chain phosphorylation. *J. Biol. Chem.* **270**, 20 400–3.

Kabsch, W., Mannherz, H.G., Suck, D., Pai, E.F. and Holmes, K.C. (1990) Atomic structure of the actin : DNase I complex. *Nature* **347**, 37–44.

Kargacin, G.J., Ikebe, M. and Fay, F.S. (1990) Peptide modulators of myosin light chain kinase affect smooth muscle cell contraction. *Am. J. Physiol.* **259**, C315–24.

Katoh, T., Katoh, H. and Morita, F. (1985) Actin-binding peptide obtained by the cyanogen bromide cleavage of the 20-kDa fragment of myosin subfragment-1. *J. Biol. Chem.* **260**, 6723–7.

Katsuyama, H., Wang, C.-L.A. and Morgan K.G. (1992) Regulation of vascular smooth muscle tone by caldesmon. *J. Biol. Chem.* **267**, 14 555–8.

Kay, L.E., Forman-Kay, J.D., McCubbin, W.D. and Kay, C.M. (1991) Solution structure of a polypeptide dimer comprising the fourth Ca^{2+}-binding site of troponin C by nuclear magnetic resonance spectroscopy. *Biochemistry* **30**, 4323–33.

Keane, A.M., Trayer, I.P., Levine, B.A., Zeugner, C. and Rüegg, J.C. (1990) Peptide mimetics of an actin-binding site on myosin span two functional domains on actin. *Nature* **344**, 265–8.

Kemp, B.E., Pearson, R.B., Guerriero, V., Bagchi, I.C. and Means, A.R. (1987) The calmodulin binding domain of chicken smooth muscle myosin light chain kinase contains a pseudosubstrate sequence. *J. Biol. Chem.* **262**, 2542–8.

Kögler, H., Moir, A.J.G., Trayer, I.P. and Rüegg, J.C. (1991) Peptide competition of actin activation of myosin-subfragment 1 ATPase by an amino terminal actin fragment. *FEBS Lett.* **294**, 31–4.

Kögler, H., Trayer, I.P. and Rüegg, J.C. (1996) A synthetic myosin heavy chain peptide mimicks the calcium-sensitizing effect of cycling cross-bridges in cardiac skinned fibres. *Pflügers Arch.* **431**, R140.

Kraft, T., Rommel, U., Trayer, I.P. and Brenner, B. (1991) Interference of myosin peptides with weak and strong actin interaction of cross-bridges in skeletal muscle fibers. In *Peptides as probes in muscle research* (ed. J.C. Rüegg), pp. 69–80. Springer-Verlag, Berlin, Heidelberg and New York.

Kraft, T., Chalovich, J.M., Yu, L.C. and Brenner, B. (1995) Parallel inhibition of active force and relaxed fiber stiffness by caldesmon fragments at physiological ionic strength and temperature conditions: additional evidence that weak cross-bridge binding to actin is an essential intermediate for force generation. *Biophys. J.* **68**, 2404–18.

Levine, B.A. and Moir, A.J.G. (1985) Protein recognition sites of actin: ^{1}H-n.m.r. spectroscopic study. *J. Muscle Res. Cell Motil.* **6**, 76–7.

Levine, B.A., Moir, A.J.G. and Perry, S.V. (1988) The interaction of troponin-I with the N-terminal region of actin. *Eur. J. Biochem.* **172**, 389–97.

Levine, B.A., Moir, A.J.G., Audemard, E., Mornet, D., Patchell, V.B. and Perry, S.V. (1990) Structural study of gizzard caldesmon and its interaction with actin. *Eur. J. Biochem.* **193**, 687–96.

Lu, X., Xu, L. and Meissner, G. (1994) Activation of the skeletal muscle calcium release channel by a cytoplasmic loop of the dihydropyridine receptor. *J. Biol. Chem.* **269**, 6511–6.

Lu, X., Xu, L. and Meissner, G. (1995) Phosphorylation of dihydropyridine receptor II-III loop peptide regulates skeletal muscle calcium release channel function. *J. Biol. Chem.* **270**, 18459–64.

Lukas, T.J., Burgess, W.H., Prendergast, F.G., Lau, W. and Watterson, D.M. (1986) Calmodulin binding domains: characterization of a phosphorylation and calmodulin binding site from myosin light chain kinase. *Biochem.* **25**, 1458–64.

Marsden, B.J., Hodges, R.S. and Sykes, B.D. (1989) A ^{1}H-NMR determination of the solution conformation of a synthetic peptide analogue of calcium-binding site III of rabbit skeletal troponin C. *Biochemistry* **28**, 8839–47.

Marston, S.B. and Huber, P.A.J. (1995) Caldesmon. In *Biochemistry of smooth muscle contraction* (ed. M. Bárány), pp. 77–90. Academic Press, London.

Melzer, W., Hermann-Frank, A. and Lüttgau, H.Ch. (1995) The role of Ca^{2+}ions in excitation–contraction coupling of skeletal muscle fibres. *Biochim. Biophys. Acta* **1241**, 59–116.

Mezgueldi, M., Fattoum, A., Derancourt, J. and Kassab R. (1992) Mapping of the functional domains in the amino-terminal region of calponin. *J. Biol. Chem.* **267**, 15 943–51.

Mezgueldi, M., Derancourt, J., Calas, B., Kassab, R. and Fattoum, A. (1994) Precise identification of the regulatory F-actin and calmodulin-binding sequences in the 10-kDa carboxyl-terminal domain of caldesmon. *J. Biol. Chem.* **269**, 12 824–32.

Mezgueldi, M., Mendre, C., Calas, B., Kassab, R. and Fattoum, A. (1995) Characterization of the regulatory domain of gizzard calponin. *J. Biol. Chem.* **270**, 8867–76.

Mezgueldi, M., Strasser, P., Fattoum, A. and Gimona, M. (1996) Expressing functional domains of mouse calponin: Involvement of the region around alanine 145 in the actomyosin ATPase inhibitory activity of calponin. *Biochem.* **35**, 3654–61.

Miller, L., Kalnoski, M., Yunossi, Z., Bulinski, J.C. and Reisler, E. (1987) Antibodies directed against N-terminal residues on actin do not block acto-myosin binding. *Biochemistry* **26**, 6064–70.

Mittmann, K., Jaquet, K. and Heilmeyer Jr., L.M.G. (1990) A common motif of two adjacent phosphoserines in bovine, rabbit and human cardiac troponin I. *FEBS Lett.* **273**, 41–5.

Moir, A.J.G., Levine, B.A., Goodearl, A.J. and Trayer, I.P. (1987) The interaction of actin with myosin subfragment 1 and with pPDM-cross-linked S1: a ^{1}H-NMR investigation. *J. Muscle Res. Cell Motil.* **8**, 68–9.

Morano, I., Ritter, O., Bonz, A., Timek, T., Vahl, C.F. and Michel, G. (1995) Myosin light chain-actin interaction regulates cardiac contractility. *Circ. Res.* **76**, 720–5.

Morita, F., Katoh, T., Suzuki, R., Isonishi, K., Hori, K. and Eto, M. (1991) An actin-binding site on myosin. In *Peptides as probes in muscle research* (ed. J.C. Rüegg), pp. 39–48. Springer-Verlag, Berlin, Heidelberg and New York.

Mornet, D., Bertrand, R., Pantel, P., Audemard, E. and Kassab, R. (1981) Structure of the actin–myosin interface. *Nature* **292**, 301–6.

Mueller, H. and Perry, S.V. (1962) The degradation of heavy meromyosin by trypsin. *Biochem. J.* **85**, 431–9.

Ngai, S.-M. and Hodges, R.S. (1992) Biologically important interactions between synthetic peptides of the N-terminal region of troponin I and troponin C. *J. Biol. Chem.* **267**, 15 715–20.

Obara, K., Szymanski, P.T., Tao, T. and Paul, R.J. (1996) Effects of calponin on isometric force and shortening velocity in permeabilized taenia coli smooth muscle. *Am. J. Physiol.* **270**, C481–7.

Pearson, R.B., Misconi, L.Y. and Kemp, B.E. (1986) Smooth muscle myosin kinase requires residues on the COOH-terminal side of the phosphorylation site. *J. Biol. Chem.* **261**, 25–7.

Perets, T., Blumenstein, Y., Shistik, E., Lotan, I. and Dascal, N. (1996) A potential site of functional modulation by protein kinase A in the cardiac Ca^{2+} channel α_{1c} subunit. *FEBS Lett.* **384**, 189–92.

Pfitzer, G., Zeugner, C., Troschka, M. and Chalovich, J.M. (1993) Caldesmon and a 20 kDa actin binding fragment of caldesmon inhibit tension development in skinned gizzard muscle fiber bundles. *Proc. Natl. Acad. Sci. (USA)* **90**, 5904–8.

Rayment, I., Holden, H.M., Whittaker, M., Yohn, C.B., Lorenz, M., Holmes, K.C. *et al.* (1993*a*) Structure of the actin–myosin complex and its implications for muscle contraction. *Science* **261**, 58–65.

Rayment, I., Rypniewski, W.R., Schmidt-Bäse, K., Smith, R., Tomchick, D.R., Benning, M.M. *et al.* (1993*b*) Three-dimensional structure of myosin subfragment-1: a molecular motor. *Science* **261**, 50–8.

Reid, R.E., Clare, D.M. and Hodges, R.S. (1980) Synthetic analog of a high affinity calcium binding site in rabbit skeletal troponin C. *J. Biol. Chem.* **255**, 3642–6.

Reid, R.E., Gariépy, J., Saund, A.K. and Hodges, R.S. (1981) Calcium-induced protein folding. Structure-affinity relationships in synthetic analogs of the helix-loop-helix calcium binding unit. *J. Biol. Chem.* **256**, 2742–51.

Rüegg, J.C. (1990) Muscle protein interaction; competition by peptide mimetics. *J. Muscle Res. Cell Motil.* **11**, 189–90.

Rüegg, J.C. (1992) *Calcium in muscle contraction. Cellular and molecular physiology* (2nd edn.). Springer-Verlag, Berlin and Heidelberg.

Rüegg, J.C., Strauss, J.D., Zeugner, C. and Trayer, I.P. (1993) Effect of myosin heavy chain peptides on contractile activation of skinned cardiac muscle fibres. In *Mechanism of myofilament sliding in muscle contraction* (eds. H. Sugi and G.H. Pollack), pp. 173–81. Plenum Press, New York.

Rüegg, J.C., Zeugner, C., Strauss, J.D., Paul, R.J., Kemp, B., Chem, M. *et al.* (1989*a*). A calmodulin-binding peptide relaxes skinned muscle from guinea-pig *Taenia coli. Pflügers Arch.* **414**, 282–5.

Rüegg, J.C., Zeugner, C., Van Eyk, J., Kay, C.M. and Hodges, R.S. (1989*b*) Inhibition of TnI–TnC interaction and contraction of skinned muscle fibres by the synthetic peptide TnI [104–115]. *Pflügers Arch.* **414**, 430–6.

Rüegg, J.C., Zeugner, C., Van Eyk, J.E., Hodges, R.S. and Trayer, I.P. (1991) Myosin and troponin peptides affect calcium sensitivity of skinned muscle fibres. In *Peptides as probes in muscle research* (ed. J.C. Rüegg), pp. 95–109. Springer-Verlag, Berlin, Heidelberg and New York.

Shaw, G.S., Hodges, R.S. and Sykes, B.D. (1990) Calcium-induced peptide association to form an intact protein domain: ^{1}H-NMR structural evidence. *Science* **249**, 280–3.

Shaw, G.S., Hodges, R.S. and Sykes, B.D. (1992) Stoichiometry of calcium binding to a synthetic heterodimeric troponin-C domain. *Biopolymers* **32**, 391–7.

Solaro, R.J., Moir, A.J.G. and Perry, S.V. (1976) Phosphorylation of troponin I and the inotropic effect of adrenaline in the perfused rabbit heart. *Nature* **262**, 615–7.

Stapleton, M.T., Duffy, P.S., Cobbold, P.H. and Allshire, A.P. (1995) Troponin-I 137–48 peptide abolishes rigor in cardiomyocytes. *Circulation* **92**, 1–189.

Strauss, J.D., de Lanerolle, P. and Paul R.J. (1992*a*) Effects of myosin kinase inhibiting peptide on contractility and LC_{20} phosphorylation in skinned smooth muscle. *Am. J. Physiol.* **262**, C1437–45.

Strauss, J.D., Zeugner, C., Van Eyk, J.E., Bletz, C., Troschka, M. and Rüegg, J.C. (1992*b*) Troponin replacement in permeabilized cardiac muscle. Reversible extraction of troponin I by incubation with vanadate. *FEBS Lett.* **310**, 229–34.

Strauss, J.D., Barth, Z., Van Eyk, J.E. and Rüegg, J.C. (1993) Probing the regulation of contractility in cardiac and smooth muscle skinned fibers with synthetic peptides. *Methods: a Companion to Meth. Enzymol.* **5**, 281–90.

Strauss, J.D., Van Eyk, J.E., Barth, Z., Kluwe, L., Wiesner, R.J., Maeda, K. *et al.* (1996) Recombinant troponin I substitution and calcium responsiveness in skinned cardiac muscle. *Pflügers Arch.* **431**, 853–62.

Sutoh, K. (1982) Identification of myosin-binding sites on the actin sequence. *Biochemistry* **21**, 3654–61.

Sutoh, K. (1983) Mapping of actin-binding sites on the heavy chain of myosin subfragment 1. *Biochemistry* **22**, 1579–85.

Suzuki, R., Morita, F., Nishi, N. and Tokura, S. (1990) Inhibition of actomyosin subfragment 1 ATPase activity by analog peptides of the actin-binding site around the Cys(SH1) of myosin heavy chain. *J. Biol. Chem.* **265**, 4939–43.

Syska, H., Wilkinson, J.M., Grand, R.J.A. and Perry, S.V. (1976) The relationship between biological activity and primary structure of troponin I from white skeletal muscle of the rabbit. *Biochem. J.* **153**, 375–87.

Szent-Györgyi, A.G. (1953) Meromyosins, the subunits of myosin. *Arch. Biochem. Biophys.* **42**, 305–20.

Szpacenko, A., Wagner, J., Dabrowska, R. and Rüegg, J.C. (1985) Caldesmon-induced inhibition of ATPase activity of actomyosin and contraction of skinned fibers of chicken gizzard smooth muscle. *FEBS Lett.* **192**, 9–12.

Talbot, J.A. and Hodges, R.S. (1979) Synthesis and biological activity of an icosapeptide analog of the actomyosin ATPase inhibitory region of troponin I. *J. Biol. Chem.* **254**, 3720–3.

Talbot, J.A. and Hodges, R.S. (1981*a*) Synthetic studies on the inhibitory region of rabbit skeletal troponin I. *J. Biol. Chem.* **256**, 2798–802.

Talbot, J.A. and Hodges, R.S. (1981*b*) Comparative studies on the inhibitory region of selected species of troponin-I. *J. Biol. Chem.* **256**, 12 374–8.

Tanabe, T., Beam, K.G., Adams, B.A., Niidome, T. and Numa, S. (1990) Regions of the skeletal muscle dihydropyridine receptor critical for excitation–contraction coupling. *Nature* **346**, 567–9.

Trayer, I.P., Trayer, H.R. and Levine, B.A. (1987) Evidence that the N-terminal region of A1-light chain of myosin interacts directly with the C-terminal region of actin. *Eur. J. Biochem.* **164**, 259–66.

Trayer, I.P., Keane, A.M., Murad, Z., Rüegg, J.C. and Smith, K.J. (1991) The use of peptide mimetics to define the actin-binding sites on the head of the myosin molecule. In *Peptides as probes in muscle research.* (ed. J.C. Rüegg), pp. 57–68. Springer-Verlag, Berlin, Heidelberg and New York.

Van Eyk, J.E. and Hodges, R.S. (1988) The biological importance of each amino acid residue of the troponin I inhibitory sequence 104–115 in the interaction with troponin C and tropomyosin-actin. *J. Biol. Chem.* **263**, 1726–32.

Van Eyk, J.E. and Hodges, R.S. (1991) A synthetic peptide of the N-terminus of actin interacts with myosin. *Biochemistry* **30**, 11 676–82.

Van Eyk, J.E. and Hodges, R.S. (1993) The use of synthetic peptides to unravel the mechanism of muscle regulation. *Methods: a Companion to Meth. Enzymol.* **5**, 264–80.

Van Eyk, J.E., Sönnichsen, F.D., Sykes, B.D. and Hodges, R.S. (1991) Interaction of actin 1–28 with myosin and troponin I and the importance of these interactions to muscle regulation. In *Peptides as probes in muscle research* (ed. J.C. Rüegg), pp. 15–31. Springer-Verlag, Berlin, Heidelberg and New York.

Van Eyk, J.E., Strauss, J.D., Hodges, R.S. and Rüegg, J.C. (1993) A synthetic peptide mimics troponin I function in the calcium-dependent regulation of muscle contraction. *FEBS Lett.* **323**, 223–8.

Weeks, R.A. and Perry, S.V. (1978) Characterization of a region of the primary sequence of troponin C involved in calcium ion-dependent interaction with troponin I. *Biochem. J.* **173**, 449–57.

Winder, S.J. and Walsh, M.P. (1990) Smooth muscle calponin. *J. Biol. Chem.* **265**, 10 148–55.

Zhan, Q., Wong, S.S. and Wang, C.-L.A. (1991) A calmodulin-binding peptide of caldesmon. *J. Biol. Chem.* **266**, 21 810–14.

Zhang, R., Sheng, Z., Jones, M. and Potter, J.D. (1996) Studies on the inhibitory region of skeletal troponin I. *J. Muscle Res. Cell Motil.* **17**, 153–4.

6 The use of recombinant molecules

Kazuo Sutoh

6.1 Background

Construction of *in vitro* as well as *in vivo* actin–myosin systems using recombinant actin and recombinant myosin is a promising technique for understanding the molecular mechanism of muscle contraction. *Dictyostelium discoideum*, a lower eukaryote, is a good expression system for both recombinant actin (Sutoh *et al.* 1991; Johara *et al.* 1993) and recombinant myosin (Manstein *et al.* 1989*a*,*b*).

Although *Dictyostelium* cells contain more than ten actin genes, many genes produce the same or very similar proteins and a single actin is the major protein expressed in the cells. By using one of the *Dictyostelium* actin genes, actin 15 gene, we expressed recombinant actins in *Dictyostelium* cells. When a recombinant actin gene is introduced into the cells, the gene, as well as the endogenous actin genes, are expressed at the same time. By a unknown mechanism, expression level of total actin in a cell is maintained at a constant level. Therefore, forced expression of the recombinant actin results in reduced levels of expression of the endogenous actins. In some cases, we could replace as much as two-thirds of the total actin in a cell with a recombinant actin. Fortunately, many actin mutants were not toxic to *Dictyostelium* cells, even when they could not sustain the motility *in vitro*. For example, when a non-motile actin mutant with D1H/E3H/D4H mutations (triple replacement of D1, E3 and D4 with histidine residues) was expressed in *Dictyostelium* cells, two-thirds of the total actin in the cells was the mutant actin, while the total amount of actin was the same as that of the wild-type cell. The transformed cell grew at the same rate as the wild-type cell and its motile properties was also the same, even though the major fraction of actin molecules in the transformed cells was non-motile (Sutoh, unpublished results). Since a recombinant *Dictyostelium* actin gene was expressed in *Dictyostelium* cells, no aggregation or other abnormal processing of the recombinant actin was observed. Moreover, normal post-translational modifications of the mutant actin were expected to be carried out in the cells. Various types of mutant actins were separated from the endogenous actins by using a single ion exchange column. Lysate of the transformed cells was applied on a HPLC DEAE column and well-separated peaks corresponding to wild-type and mutant actins were collected. In this way, we could obtain several hundreds micrograms of purified actin mutants from 1 l suspension culture of the transformed *Dictyostelium* cells (Sutoh *et al.* 1991; Johara *et al.* 1993).

Recombinant myosins were also expressed in *Dictyostelium* cells. A recombinant *Dictyostelium* myosin II heavy chain gene was inserted into a shuttle vector, which can replicate extrachromosomaly as multicopy plasmids in *Dictyostelium* cells. The shuttle vector carrying the recombinant myosin gene was introduced into *Dictyostelium* mutant cells (myosin-null cells), which are incapable of producing myosin II heavy chain, since in these cells the myosin II heavy chain gene is knocked out by homologous recombination (Manstein *et al.* 1989*a*). By using this expression system, only the recombinant myosin II heavy chain is expressed in the cells without comcomitant expression of the endogenous heavy chain (Uyeda *et al.* 1994). Since two types of light chains of myosin II (essential and regulatory light chains) are expressed in the myosin-null cells, recombinant myosin comprising recombinant heavy chain and two endogenous light chains is formed *in vivo*. Usually the expression level of the recombinant myosin heavy chain in the transformed cells is similar to that of the wild type in the parental AX2 cells. About 1 mg of purified recombinant myosin is obtained from 1 l culture of the transformed *Dictyostelium* cells.

The first advantage of using *Dictyostelium* cells to express recombinant myosins is that a large quantity of recombinant proteins can be obtained from transformed *Dictyostelium* cells. The second advantage is that post-translation modifications, such as trimethylation of a lysine residue, proceed normally in *Dictyostelium* cells. The third advantage is that myosin functions can be very easily checked by observing the phenotypes of living transformed cells. *Dictyostelium* myosin-null cells show a phenotype distinct from that of wild-type cells. In suspension culture, the volume of a cell increases without cell cleavage until it ruptures. Thus myosin-null cells cannot grow in suspension. On a plate, however, they can divide by unusual traction-fission and can grow, though many multi-nuclei cells are generated by this unusual step. Upon starvation, the wild-type cells aggregate and form fruiting bodies. The myosin-null cells show deficiency in this process: upon starvation, they aggregate to form multicellular structures called 'slugs' but cannot form the fruiting bodies. Upon introduction of the functional myosin heavy chain gene into the myosin-null cells, the myosin-null phenotype is suppressed and the wild phenotype is rescued fully or partially, depending on the structures of the recombinant myosins. Therefore, by observing the phenotype of the transformed cells, we can qualitatively tell how functional the recombinant myosin is.

6.2 Examples for using recombinant actins and myosins to reveal the molecular mechanism of force generation by the actin–myosin system

The *Dictyostelium* expression system

Two types of transformation vectors are available for expressing foreign genes in *Dictyostelium* cells. One is an integration vector carrying three components: a selectable marker for *Dictyostelium* cells, such as a resistance gene for neomycine (Neo^r); a foreign gene driven by *Dictyostelium* promoter and terminator, such as the actin 15 promoter and the actin 8 terminator; and a vector sequence for multicopy replication in *Escherichia coli* (Fig. 6.1). A large number (~100 copies) of this type of vector are tandemly inserted into *Dictyostelium* chromosomal DNA when introduced into the cells by electroporation. The other type is an extrachromosomal vector, which carries a portion of the *Dictyostelium* plasmid Ddp1 or Ddp2, together with three components of the integration vector, as mentioned above (Fig. 6.2). The vectors can replicate in *Dictyostelium* cells as multicopy plasmids independent of replication of the genomic DNA. Once these transformation vectors are introduced into *Dictyostelium* cells by electroporation as multicopy DNA, a foreign gene (a mutant actin gene or a mutant myosin gene) in the vector is expressed in the cells (Fig. 6.3).

Recombinant actins

Using the recombinant *Dictyostelium* actins, we have identified amino acid residues on the actin surface interacting with myosin during ATP hydrolysis. All actins obtained from a variety of organisms from yeast, *Dictyostelium* to human has a negative charge cluster at the N-terminus. For example, many types of actins,

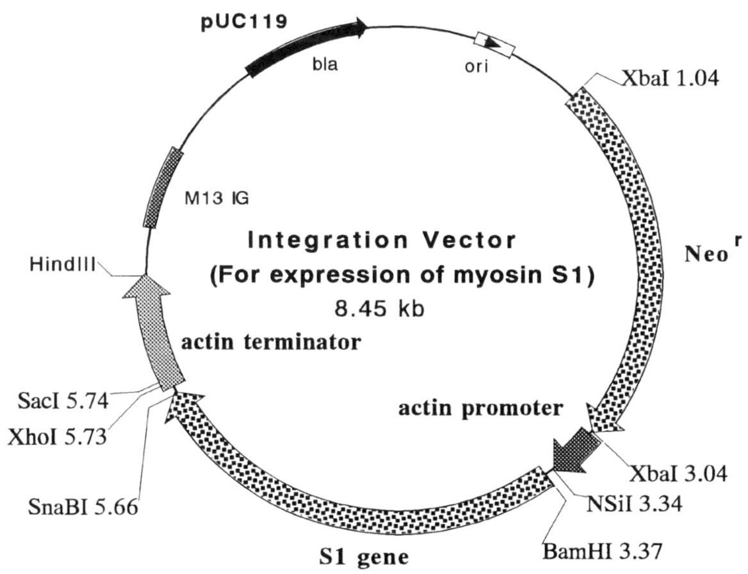

Fig. 6.1 Integration vector for transforming *Dictyostelium discoideum* cells. The vector is composed of three elements: the myosin subfragment 1 gene (or any gene to expressed) driven by the actin promoter and terminator; a selection marker (Neo^r); and pUC119 for replication in *E. coli* cells.

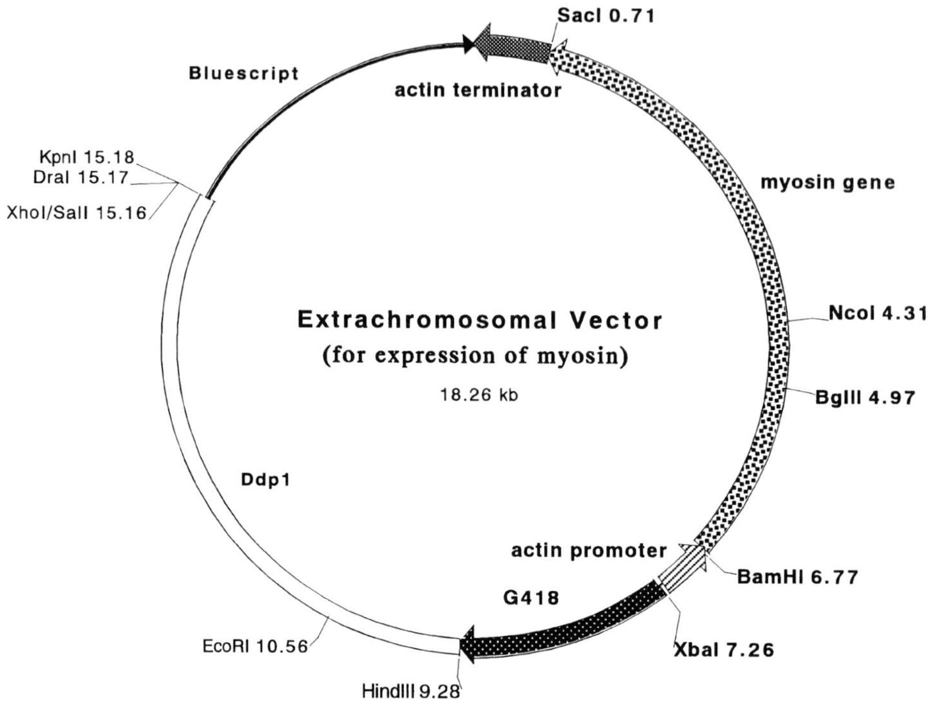

Fig. 6.2 Extrachromosomal vector for transformation of *Dictyostelium discoideum* cells. This vector is composed of four elements: the myosin gene (or any gene to be expressed) driven by the actin promoter and terminator; a selection marker (Neo^r); Bluescript for replication in *E. coli* cells; and a part of a *Dictyostelium discoideum* plasmid Ddp1. This vector exists in *Dictyostelium discoideum* cells as a multicopy plasmid.

including mammalian actin, has four consecutive aspartic acid and glutamic acid residues. The lower eukaryotes have a smaller number of acidic residues at the N-terminus, e.g. *Dictyostelium* actin has three acidic residues and yeast actin has two. From biochemical and structural studies, it has been suggested that these N-terminal acidic residues are involved in the actin–myosin interaction. By using the recombinant actin, we addressed the question as to whether or not these acidic residues are involved in the actin–myosin interaction *during force generation*. The most direct way to answer this question is to alter the acidic residues to other residues by site-directed mutagenesis and observe the consequences.

First the acidic residues of the N-terminal sequence of *Dictyostelium* actin (DGED-) were replaced by a His residue one by one (Sutoh *et al.* 1991). The three types of mutants constructed had the following N-terminal sequences: HGED- for the D1H mutant; HGHD- for the D1H/E3H mutant; HGHH- for the D1H/E3H/D4H mutant. These mutant actins were expressed in *Dictyostelium* cells using

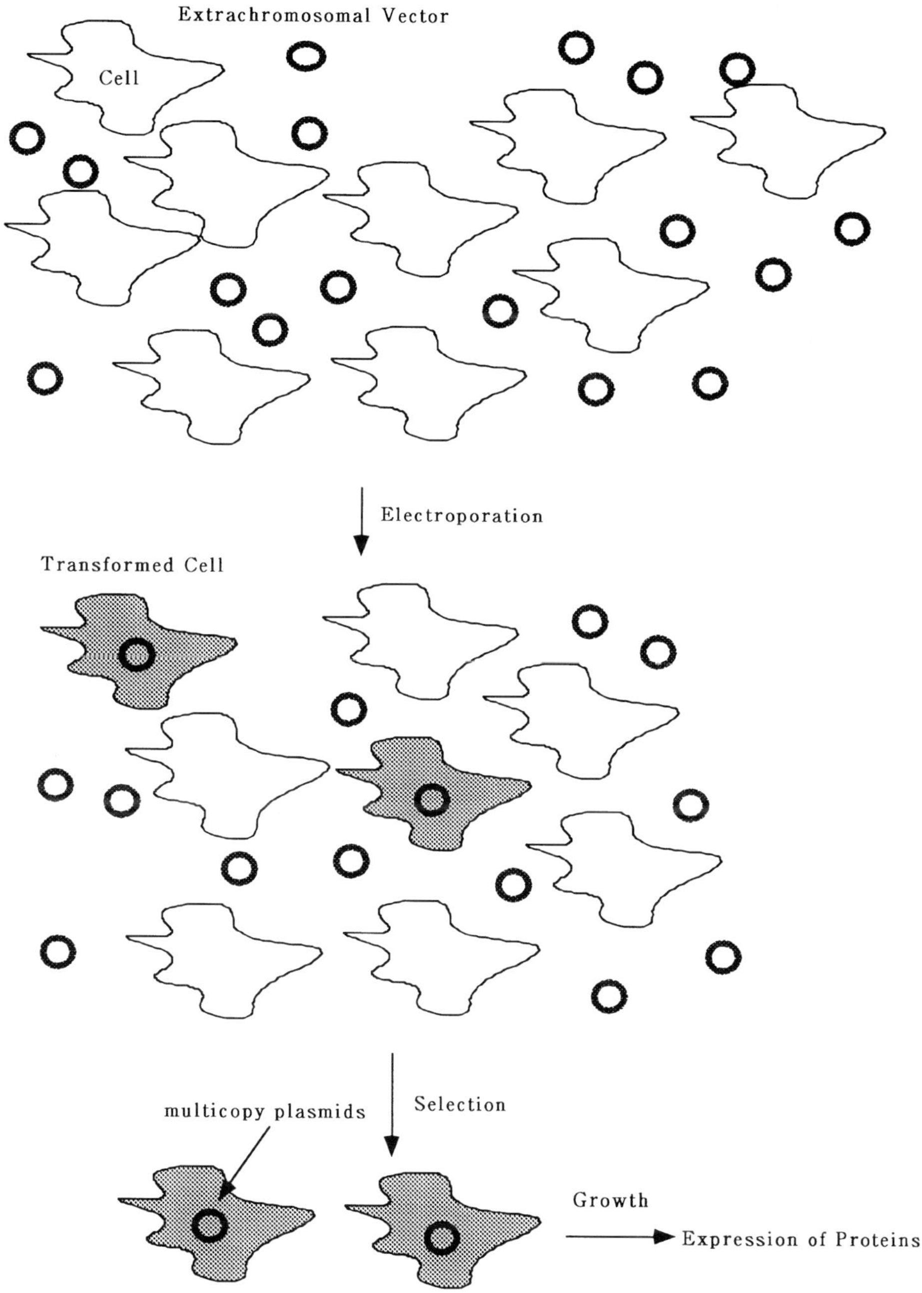

Fig. 6.3 Scheme for expressing a foreign gene in *Dictyostelium discoideum* cells by using the extrachromosomal vector. The transformation vector is introduced into the cells by electroporation, and the transformants are selected by using the selection marker.

an integration transformation vector. As mentioned above, expression of these mutant actins forces the cells to produce lower levels of the endogeneous actins, keeping the total amount of actin expressed in a cell constant. By using a single DEAE-column, these mutant actins and the endogeneous actins were separated and purified for use in biochemical studies.

By using these purified mutant actins, the activation of myosin ATPase activity and the velocity of ATP-dependent sliding on myosin were measured. By reducing the number of acidic residues at the N-terminus, velocity of the ATP-dependent sliding motion decreased. At the same time, V_{max} values of the actin-activated ATPase activity decreased while K_m values remained at the same level, consistent with the reduction of the sliding velocity. These results imply that the N-terminal acidic residues are actually involved in the actin–myosin interaction during the ATP hydrolysis in such a way that these residues directly activate myosin ATPase activity.

Immunological and structural studies have also shown that other negative charge clusters on loops, i.e. D24/D25 and E99/E100, may be involved in the actin– myosin interaction (for review see Milligan 1996). We expressed and purified single and double mutants, D24H, D24H/D25H, E99H, and E99H/E100H to assess the role of these clusters. Measurements of actin activation of myosin ATPase activity and ATP-dependent sliding velocity indicate that these acidic residues are involved in the actin–myosin association during the ATP hydrolysis, but not in the direct activation of myosin ATPase activity, since these replacements affected K_m values, not V_{max} values of the actin-activated ATPase activity of myosin (Johara *et al.* 1993; Sutoh unpublished data).

Experiments on these recombinant actins lead us to conclude that the N-terminal acidic residues contribute to direct activation of myosin ATPase activity, though its exact mechanism is unknown; while the clusters at the two loops are involved in the actin–myosin association through ionic interactions during the ATP hydrolysis.

Recombinant myosins

Recombinant *Dictyostelium* myosins are a good tool for understanding the molecular mechanism of onset of human familial hypertopic cardiomyopathy (FHC), an inherited heart disease resulting from mutations in various muscle proteins including myosin. To elucidate the impact of FHC mutations on myosin motor functions, we created myosin mutants equivalent to FHC mutations, using *Dictyostelium discoideum* myosin as a model system (Fujita *et al.* unpublished data). Although sequence homology between the motor domain of *Dictyostelium* myosin and that of human cardiac β myosin of 40%, their three-dimensional structure must be very similar to each other, given that the crystal structure of the *Dictyostelium* myosin motor domain is almost identical to that of chicken skeletal myosin, which share 47% sequence identity (Fisher *et al.* 1995). Thus, the

Dictyostelium myosin mutants can be good models for the FHC mutants as far as mutations in conserved sequences are concerned. By using the *Dictyostelium* myosin heavy chain gene, R397Q, F506C, G575R, A699R, K703Q and K703W mutants were constructed because these mutations are in the conserved stretch of the sequence and also because FHC mutations equivalent to them exhibit differences in clinical prognosis among the affected individuals. F506C is equivalent to F513C of FHC; this mutation exhibits a better prognosis among those examined here—the affected family members have a near normal life expectancy. A699R, K703Q and K703W are equivalent to G716R, R719Q and R719W of FHC, respectively; these FHC mutation exhibit a worse prognosis—the affected families have a high incidence of premature death and an average life expectancy in affected individuals of only 38 years. R397Q and G575R are equivalent to R403Q and G584R of FHC, respectively; the level of prognosis of individuals affected by these mutations is between that for F513C and the G716G/R719Q/R719W group.

Dictyostelium myosins equivalent to the FHC mutants were expressed in *Dictyostelium* myosin-null cells in which the myosin heavy chain gene was knocked out by homologous recombination (Manstein *et al.* 1989*a*). These mutant myosins were purified, and their motile and enzymatic activities were examined.

The first type of *Dictyostelium* mutants (A699R, K703Q and K703W) is characterized by a dramatically high K_m, only moderately reduced V_{max} of the actin-activated ATPase activity and low level of force. In the *Dictyostelium* motor domain, A699 and K703 are on an α helix in the C-terminal subdomain after the conservative SH1/SH2 helix. This C-terminal subdomain, rich in α helices and β sheets, is located at the distal end of the motor domain away from the actin-binding site. Therefore, it is very unlikely that A699R, K703Q and K703W mutations affect the actin–myosin interaction. Since the A699R, K703Q and K703W mutations would induce only minor distortion of the α helix, they would mainly change interactions between the C-terminal subdomain with another subdomain of myosin. It has been recently shown that the C-terminal subdomain can take two very different orientations against the remaining part of the motor domain, dependings on the states of the nucleotide at the ATPase pocket (Fisher *et al.* 1995). The C-terminal subdomain might swing from one orientation to the other during the cyclic ATP hydrolysis. This line of argument leads us to speculate that mutations at the C-terminal domain might reduce the affinity with actin filaments by affecting this swing motion, since the A699R, K703Q and K703W mutations would alter the ionic and hydrophobic properties of the surface of this subdomain. It must be mentioned here that the mutated residues are close to the essential light chain which binds to a long α helix after the C-terminal subdomain (Rayment *et al.* 1993). Dramatically weak interactions of this type of mutant to actin filaments would lead to the lowest level of force among those examined here, and to the worst prognosis of the corresponding FHC.

Moderately reduced V_{max} and K_m and nearly normal motile properties, are the characteristics of the second type of *Dictyostelium* mutant F506C. F506 is located on a long α helix away from the actin-binding site and the ATPase pocket. This helix is composed of the longest stretch of the conserved segment of myosin and is connected to the ATPase site through a loop containing the pivotal Gly457, which may play an essential role as a hinge to open and close of 50 K cleft. Replacement of F506 with another hydrophobic residue would not disrupt the helix, though sidechain rearrangements would take place. The notion is also consistent with our observation that the F506C mutation in *Dictyostelium* myosin showed nearly normal motor functions. Nearly normal motor functions, especially nearly normal force level, of the *Dictyostelium* F506C mutant is in concert with the better clinical prognosis of individuals affected by the corresponding FHC mutation.

The third type of *Dictyostelium* mutants (R397Q and G575) are characterized by low V_{max}, moderately increased K_m and moderately reduced force level. R397 resides at the base of a finger-like structure composed of two β strands, which is located at the tip of the motor domain and would be a part of the actin-binding site on myosin (Milligan 1996). Since the finger-like structure is stabilized by strong hydrophobic interactions, the R397Q mutation would rarely affect the structure. Thus it is likely that the R397Q mutation would affect the ionic interaction between actin and myosin during force generation. Since positive charges on the myosin surface interact with acidic residues at the N-terminus and on the loop containing D24 and D25 (Sutoh *et al.* 1991; Johara *et al.* 1993), loss of one positive charge at the actin-binding surface would weaken the actin–myosin interaction. The weak actin–myosin interaction of the corresponding FHC mutant would then result lower level of force, leading to the onset of FHC.

G575 is located close to the 50 K cleft and at a turn connecting two β strands. Replacement of the glycine residue with the bulky arginine residue can be accommodated into the original structure without much distortion of the peptide folding, though side-chain rearrangements may take place. The turn is in contact with the long α helix connected to the ATPase site through a loop where the pivotal G457 resides. Introduction of a positive charge to the contact site would affect the helix rearrangement during ATP hydrolysis. This might indirectly reduce the apparent affinity of the mutant myosin with actin filaments, and then its force level.

Comparison of the enzymatic and motile properties of these *Dictyostelium* mutants with clinical prognosis of individuals affected by the corresponding FHC mutants lead us to conclude that force level of mutant myosins depends on their affinity to actin filaments during ATP hydrolysis, and that the force level of these myosin mutants is a key factor in the prognosis of FHC.

6.3 Limitations of the use of recombinant *Dictyostelium* actin and myosin

We can construct and express recombinant actins and myosins by using *Dictyostelium* cells as an expression system. From large-scale culture of transformed cells

(1–101), we can purify easily milligram quantities of recombinant proteins for biochemical characterization. As far as *in vitro* experiments are concerned, the recombinant actins and myosins produced by using the *Dictyostelium* expression system are very useful. However, when we study the actin–myosin system in the organized state, e.g. muscle, isolated actins and myosins are no longer good models. Fruitfly flight muscle is the choice for *in vivo* studies. Since a single muscle-specific actin gene is expressed as a major component in the flight muscle, it is possible to make a mutant fly in which the flight muscle-specific actin gene is not active. The actin-null fly, which lacks the flight muscle actin gene, cannot fly but it is not lethal. By using the p-element, a recombinant actin gene can be introduced into the fly (Drummond *et al.* 1992). Thus, it is possible to make a fly with flight muscle in which the endogenous actin is replaced with the recombinant actin. Using the flight muscle from the mutant fly, physiological studies are now possible. Contrary to the *Dictyostelium* system, however, it is rather difficult to purify a large amount of recombinant protein from the fly. Thus, the *Dictyostelium* system and the fruitfly system should be regarded as complementary.

References

Drummond, D.R., Hennessey, E.S. and Sparrow, J.C. (1992) The binding of mutant actins to profilin, ATP and DNaseI. *Eur. J. Biochem.* **209**, 171–9.

Fisher, A.J., Smith, C.A., Thoden, J., Smith, R., Sutoh, K., Holden, H.M. *et al.* (1995) Structures of Beryllium and Aluminium metallofluoride-Mg-ADP complexes of the *Dictyostelium discoideum* myosin motor domain. *Biochemistry* **34**, 8960–72.

Johara, M., Toyoshima, Y.Y., Ishijima, A., Kojima, H., Yanagida, T. and Sutoh, K. (1993) Charge-reversion mutagenesis of *Dictyostelium* actin to map the surface recognized by myosin during ATP-driven sliding motion. *Proc. Natl. Acad. Sci.* (*USA*) **90**, 2127–31.

Manstein, D.J., Rupel, K.A. and Spudich, J.A. (1989*a*) Expression and characterization of a functional myosin head fragment in *Dictyostelium discoideum. Science* **246**, 656–8.

Manstein, D.J., Titus, M.A., DeLOzanne, A. and Spudich, J.A. (1989*b*) Gene replacement of *Dictyostelium*: generation of myosin null mutants. *EMBO J.* **81**, 923–32.

Milligan, R.A. (1996) Protein–protein interactions in the rigor actomyosin complex. *Proc. Natl. Acad. Sci.* (*USA*) **93**, 21–6.

Rayment, I., Rypniewski, W.R., Schmidt, B.K., Smith, R., Tomchick, D.R., Benning, M.M., Winkelmann, D.A., Wesenberg, G. and Holden, H.M. (1993*a*) *Science* **261**, 50–8.

Sutoh, K., Ando, M., Sutoh, K. and Toyoshima, Y.Y. (1991) Site-directed mutagenesis of *Dictyostelium* actin. *Proc. Natl. Acad. Sci.* (*USA*) **88**, 7711–4.

Uyeda, T.Q.P., Ruppel, K.A. and Spudich, J.A. (1994) Enzymatic activities correlate with chimaeric substitutions at the actin-binding face of myosin. *Nature* **368**, 567–9.

7 *In vitro* motility assay

Haruo Sugi

7.1 Introduction

From the standpoint of muscle physiology, isolated single muscle fibres are no doubt the best material for studying the mechanism of muscle contraction, since the ultimate purpose of muscle physiology is to fully understand the molecular events taking place in contracting muscle fibres. The results obtained from single muscle fibres are, however, not readily explained at the molecular level, for many reasons including the complex muscle fibre structures, the huge number of actin and myosin molecules involved, and the non-uniformity of contractile responses along the fibre length. This situation is essentially the same in demembranated muscle fibres, though these have advantages for changing conditions around the myofilaments. To investigate the molecular mechanism of muscle contraction without complications, it is desirable to construct simplified experimental systems, which consist of a small number of actin and myosin molecules but retain the basic characteristics of muscle contraction.

Although early attempts to construct experimental systems from extracted actin and myosin samples could produce directional movement *in vitro* (e.g. Oplatka and Tirosh 1973; Yano *et al.* 1982), these systems were inadequate for quantitative studies. The experimental systems, later developed for quantitative studies on the ATP-dependent actin–myosin sliding, are called the *in vitro* motility assay systems, and are roughly classified into two different types:

(1) 'the fluorescent actin filament *versus* myosin system', in which fluorescently labelled actin filaments are made to slide on myosin molecules;

(2) 'the actin cable *versus* myosin system', in which myosin-coated beads are made to slide on actin filament arrays (actin cables) of algal cells. Both assay systems have been modified and improved in various ways.

In this chapter, we will discuss the advantages, problems and limitations of these *in vitro* motility assay systems hitherto developed by various researchers. Because of space limitation, we shall only focus on assay systems for studying the ATP-dependent actin–myosin sliding responsible for muscle contraction, but not on the ATP-dependent microtubule–kinesin and microtubule–dynein sliding. For the same reason, this article aims at giving general readers an idea about the advantages, problems and limitations of *in vitro* motility assay techniques as described in selected papers, but it is not an exhaustive survey of the literature.

7.2 The fluorescent actin filament *versus* myosin system

This motility assay system, developed by Kron and Spudich (1986), consists of actin filaments fluorescently labelled with rhodamine-phalloidin (Yanagida *et al.* 1984) and myosin molecules fixed on a glass surface, and has the following advantages:

1. The ATP-dependent movement of actin filaments on myosin can be directly observed on a monitor screen with a video light microscope of appropriate sensitivity, and can be recorded on videotapes for analysis.
2. This system is composed of purified protein samples, so that they can be readily replaced by their proteolylic fragments or by recombinant molecules in order to study how these replacements affect the sliding movement. Reflecting the random orientation of myosin molecules on the substratum, however, actin filaments do not move in a straight line but with winding paths; nevertheless, the movement is unidirectional depending on the actin filament polarity. It is not clear how the leading end of a moving actin filament actually selects myosin heads, though some models of this have been presented (Uyeda *et al.* 1990).

Measurement of sliding velocity

Considering the extremely small size of actin filaments, inertial force is negligible compared to viscous drag, so that the filament stops moving at the moment when driving force is removed. This implies that, for a continuous filament movement, at least one myosin head must be attached in order to drive filament sliding at any time. The sliding velocity of individual filaments exhibits a fairly large scatter, so that it is necessary to use averaged values (each from more than 10 different filaments) to study factors influencing the sliding velocity. The filament sliding velocity on skeletal muscle myosin increases with increasing ATP concentration in the medium, reaching a maximum at 100 μM ATP (3–4 μm/s at 24 °C, Kron and Spudich 1986). Half-maximum velocity is attained at ~50 μM ATP. These results contrast with estimates of the half-maximum actin-activated myosin ATPase activity in solution, where the half-maximum is only 6 μM ATP. As the free energy derived from hydrolysis of one ATP molecule by a single myosin head (cross-bridge) is estimated to be much larger than that required to overcome the viscous drag on a moving actin filament, the high ATP concentration required to attain the half-maximum velocity of actin–myosin sliding indicates that the sliding velocity must be kinetically limited, probably by the rate of detachment of myosin head from actin after completion of its powerstroke to drive filament sliding, as has been thought to be the case in contracting muscle fibres.

Though no external load is applied to actin filaments, the maximum filament sliding velocity with saturating ATP concentrations is not necessarily the maximum sliding velocity in unloaded conditions. It has been pointed out that damaged myosin heads fixed on the substratum may bind actin in a rigor-like fashion, thus providing substantial internal resistance against filament sliding (Harada *et al.* 1990). Homsher *et al.* (1992) have shown that the filament sliding becomes more reproducible if the myosin-coated glass surface is previously exposed to a solution of unlabelled actin filaments—a procedure intended to reduce the number of myosin heads that form rigor-like linkages with actin. On the other hand, the actin filament sliding velocity exhibits a fairly wide variation (a factor of more than two) even at the same temperature and pH, depending not only on the myosin sample but also on the material with which the glass surface is coated (nitro-cellulose or silicone) (Huxley 1990; Oplatka 1994). These results indicate that the actin filament sliding velocity depends on many experimental conditions and, therefore, may not reliably inform about fundamental kinetic properties of actin–myosin sliding.

Homsher *et al.* (1992) carefully compared the actin filament sliding velocity in the assay system (V_f) with the unloaded shortening velocity of muscle fibres (V_u) in the published literature under a variety of conditions, including ATP, ADP and inorganic phosphate concentrations, pH, ionic strength and temperature. V_f and V_u were qualitatively similar with respect to their dependence on the substrate and product concentrations and temperatures above 20 °C, while V_f was more sensitive to low pH and to temperatures below 20 °C than V_u. Both V_f and V_u decreased with decreasing ionic strength from 150 to 50 mM, while V_f was markedly reduced with further decrease of ionic strength. Unexpectedly, although ATPγS is thought to increase the number of myosin heads weakly attached to actin, and thereby reduce the number of myosin heads to drive filament sliding, it has no serious effect on V_f. These results also indicate that V_f is not always a good analogue of V_u.

Despite the complicated factors influencing the filament sliding velocity, this assay system has been effective in deciding which part of a myosin molecule is essential for producing the actin filament sliding. Toyoshima *et al.* (1987) showed that actin filament sliding could be supported not only by myosin or heavy meromyosin (HMM) but also by myosin subfragment 1 (S-1), i.e. the globular head domain of a myosin molecule. The head domain is further divided into two smaller domains: 'ATPase domain', which consists only of the heavy chain and contains both the actin binding and the ATPase sites, and the 'neck domain', which consists of a part of the long α-helix of the heavy chain and two light chains (Rayment *et al.* 1993). Itakura *et al.* (1993) genetically truncated the S-1 head from a slime mould *Dictyostelium* to obtain a recombinant fragment composed only of the 'ATPase domain' (S-1dc), and examined whether this S-1dc domain alone could support actin filament sliding. Although *Dictyostelium* myosin S-1 randomly fixed on a glass surface supported actin filament sliding ($\sim$0.13 μm/s, 25 °C), S-1dc molecules randomly fixed on a glass surface produced no sliding. They could demonstrate, however, that S-1dc can support the sliding well (with velocities similar to those of S-1), if the S-1dc molecule, to which a C-terminal segment of *Dictyostelium* actin had been genetically fused, was held onto a glass surface at its C-terminus with an avidin–biotin system. Significantly, the velocity of actin filament sliding by S-1 (also with a genetically fused actin filament segment and fixed onto the glass surface with the avidin–biotin system) was similar to that observed with S-1dc. These results indicate that, in this type of motility assay, absence of actin filament sliding gives no definite conclusion about the motile ability of a molecule to support filament sliding, since the apparent absence of filament sliding can just as well result from unfavourable orientation, inadequate fixation, etc.

This assay system has also been used to study the effect of chemical modification of myosin and actin molecules. Warshaw *et al.* (1990) examined actin filament sliding on mixtures of phosphorylated and unphosphorylated mammalian smooth muscle myosin, and showed that the latter acted as a load on the filament sliding caused by the former. Lowey *et al.* (1993) demonstrated that the filament sliding velocity was reversibly reduced from 8.8 μm/s to 0.8 μm/s (30 °C) by removal of light chains from skeletal muscle myosin, even though the removal did not significantly affect actin-activated ATPase activity. In the foregoing experiments, the myosin samples were fixed either directly on a glass surface or indirectly via a monoclonal antibody to the C-terminus of the myosin rod in order to minimize unfavourable orientation of the myosin sample on the substratum; the filament velocity was not significantly affected by this procedure. The procedures for inducing the 'correct' orientation of the molecule do not always seem effective in facilitating the filament sliding, suggesting that ordered location of the molecule, similar to that in the muscle thick filament, is an important factor, as will be shown later.

That S-1 alone supports actin filament sliding does not logically exclude the possibility that myosin subfragment 2, connecting the myosin head to the thick filament backbone, plays a role in muscle contraction (Harrington *et al.* 1988; Sugi *et al.* 1992). To explore this possibility, Oishi *et al.* (1996) prepared two *Dictyostelium* mutant HMMs lacking 30 and 59 amino acid residues in the N-terminal sequence of S-2, respectively. They compared the actin filament sliding velocities on these as well as on the wild-type *Dictyostelium* HMM. The HMM molecules were uniformly attached by use of a monoclonal antibody. The ATP-dependent actin filament sliding velocity decreased progressively as the S-2 region was shortened, suggesting possible involvement of S-2 in muscle contraction.

To address the question of how the myosin heads recognize actin, Johara *et al.* (1993) studied sliding of filaments of mutant *Dictyostelium* actins on myosin, and showed that amino acid residues (D24/25 and E99/E100) are essential for the filament sliding. Oishi and Sugi (1994) showed that substitution or removal of actin-bound nucleotide had no effect on filament sliding velocity on myosin, suggesting no significant role of actin-bound nucleotide in the sliding process. Schwyter *et al.* (1990) examined the effect of cleaving actin with a protease (subtilisin), which cuts between Met-47 and Gly-48 without changing its ability to activate HMM ATPase or the appearance of the actin filament made from cleaved actin monomers; they found that subtilisin cleavage markedly impairs the actin filament sliding on HMM.

Nishizaka *et al.* (1993) fixed the front end of an actin filament fixed to a glass surface, and found that the sliding of its rear part past fixed HMM exhibited right-handed rotation, reflecting its helical structure.

Myosin step size

Myofilament sliding in contracting muscle results from cyclic actin–myosin interaction in the course of ATP hydrolysis. According to the contraction model of A.F. Huxley (1957) and Huxley and Simmons (1971), the distance of myofilament sliding caused by the rotation of the myosin head on actin (the 'step size') should be of the order of 10 nm per ATP molecule hydrolysed. Using the fluorescent actin filament *versus* myosin assay system, two research groups have attempted to determine the myosin step size, by estimating the total lengths of actin filament moving on myosin in the system, the average filament sliding velocity, and the total amount of ATP hydrolysed for the actin–myosin sliding in a given time (Harada *et al.* 1990; Toyoshima *et al.* 1990; Uyeda *et al.* 1990). The myosin step size d is calculated as:

$$d = t_s V_s, \tag{7.1}$$

where t_s is the time in one cycle of ATP hydrolysis that the myosin head is firmly attached to actin thus moving the filament by its powerstroke, and V_s is the velocity of filament sliding. The value of t_s is estimated from the rate of ATPase activity associated with the filament sliding, i.e. actin-activated ATPase activity per myosin head, and the fraction of the heads attached to actin at any one moment by:

$$t_s = f t_c, \tag{7.2}$$

where t_c is the ATPase cycle time (1/ATPase) and f is the fraction of time within one ATP hydrolysis cycle that the head is attached to actin and exerting its powerstroke to move the filament. An alternate estimation of t_s is made from the ATPase activity per unit filament length (U_f) and the average fraction of the heads firmly attached to actin at any one moment (n) as:

$$t_s = n/U_f. \tag{7.3}$$

Specifically, the sliding velocity V_s at a saturating ATP concentration is approximately the same irrespective of filament lengths, implying that the ATPase activity per unit filament length may be approximately the same. Although the number of myosin heads involved in filament sliding cannot be readily determined in long filaments, the minimum length filaments can provide data for estimating *d* by eqn 7.1 – eqn 7.3. In such a minimum length filament, at least two myosin heads should be involved in sliding; otherwise, the filament would be detached from the substratum during the time at which one head is detached altogether from the filament. Thus, assuming that only one head is attached at a time in order to drive the movement of the minimum length filament, the value of *d* can be calculated from the measured values of V_0 and the number of ATP molecules hydrolysed.

The myosin step sizes thus obtained are extremely divergent between the two groups, being of the order of 10 nm according to the Spudich group and more than 100 nm according to the Yanagida group. The latter group suggests the possibility that a myosin head repeats multiple powerstrokes per hydrolysis of one ATP molecule. These studies are, however, based on oversimplified assumptions, including the uniform ATPase activity per unit filament as well as indirect determination of the minimum filament lengths. The discrepancy has been discussed in detail by H.E. Huxley (1990) and Burton (1992). Especially, it has been pointed out that, with high myosin head densities, the leading end of an actin filament would only interact with the heads that can most rapidly release Pi to start their powerstroke on binding with actin. As a result, the ATPase activity per unit filament length would be much higher in short filaments than long ones (Huxley 1990). More recently, Uyeda *et al.* (1991) again reported that *d* was 5–10 nm, based on velocity distribution of actin filaments sliding on HMM at low densities, though care was not taken about uniform orientation of the molecules. The estimation of the myosin step size may best be approached in future using motility assay systems, in which actin filaments slide along regularly arranged myosin heads in native or synthetic myosin thick filaments.

Measurement of force

Modifications of the fluorescent actin filament *versus* myosin system have made it possible to measure force exerted on an actin filament by myosin molecules fixed on the substratum. Kishino and Yanagida (1988) attached one end of an actin filament to the tip of a compliant glass microneedle, and let the filament interact with myosin heads on a glass surface in the presence of ATP. They could measure the force exerted on the filament by myosin heads from the bending of the needle (Fig. 7.1A). Based on some assumptions, they suggest that the force generated by one myosin head may be of the order of 1 pN. Combining the 'microneedle' method with a photoelectric device to record small needle displacements, Ishijima *et al.* (1991) analysed the force fluctuations, which were most marked when the needle was stiff enough to prevent unidirectional filament sliding (nearly 'isometric' condition). Power spectra of the force fluctuations indicate that the heads generated force randomly and independently of one another, being consistent with the Huxley contraction model (Huxley 1957). If the filament was allowed to slide on the myosin-coated glass surface, the force fluctuation in the

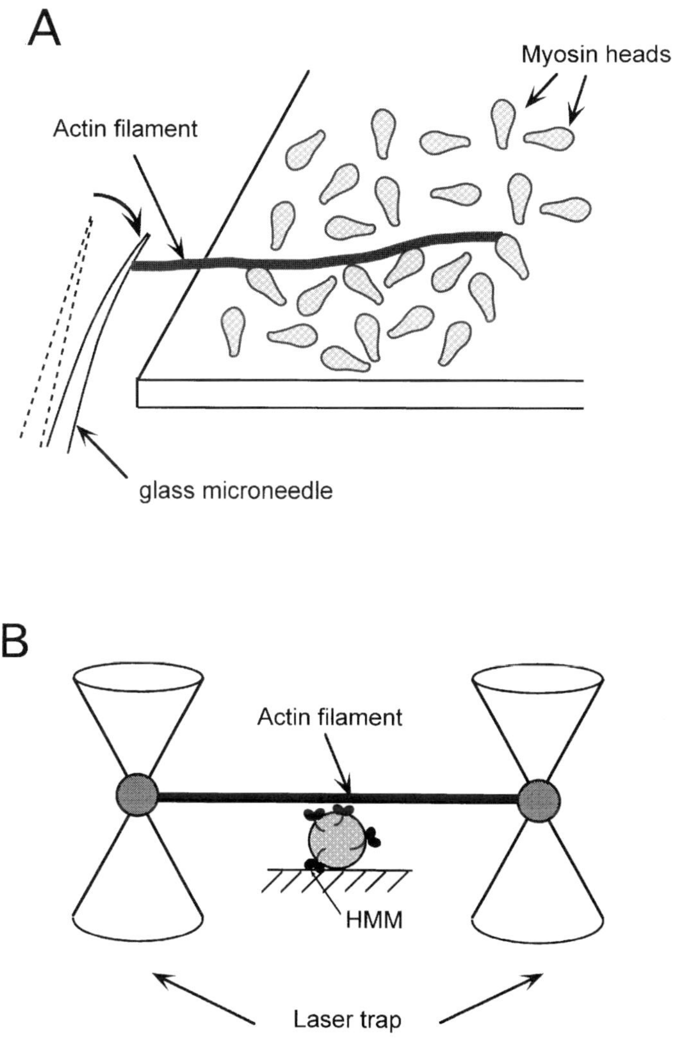

Fig. 7.1 Techniques to measure force and displacement produced by ATP-dependent actin–myosin interaction. (A) Measurement of force and displacement by bending of a glass microneedle connected to an actin filament interacting with myosin heads fixed on a glass surface. (B) Measurement of force and displacement by 'optical trap'. An actin filament, with beads attached to both ends, is held taut between two laser beams, and is made to interact with HMM molecules fixed on a bead surface. Force and displacement resulting from the interaction between the filament and the HMM molecules is measured by choosing appropriate trap stiffness.

needle became much less prominent. Comparing the force fluctuation data with the ATPase activity measurements, they suggest that each myosin head powerstroke is coupled with hydrolysis of one ATP molecule in the 'isometric' condition, while multiple powerstrokes occur per hydrolysis of one ATP molecule if the filament moves on myosin at high velocities. The 'microneedle' method was also used by Van Buren *et al.* (1994), who showed that the force exerted by skeletal muscle myosin was reduced by $\sim 50\%$ after removal of the essential light chain, while it was little affected after removal of the regulatory light chain.

A more sophisticated method of force measurement in motility assay has been developed by Finer *et al.* (1994). This utilizes an 'optical trap' (or 'optical tweezers') to manipulate a single filament under microscopic observation. This technique is based on the power of a focused laser beam to draw and trap a small object such as a latex bead within the beam, and allows moving the trapped object by moving the position of the laser beam (Ashkin and Dziedzic 1987). For detailed descriptions of the 'optical trap' method, see Block (1990) and Warrick *et al.* (1993). The advantages of the 'optical trap' method over the 'microneedle' method are:

(1) it does not require technical skill;

(2) the stiffness of the trap can be easily changed by changing the laser beam intensity;

(3) force measurement under nearly 'isometric' conditions is possible by using a feedback circuit.

Finer *et al.* (1994) held taut an actin filament with polystyrene beads attached at both ends by trapping the beads with two laser beams, and brought the filament towards an HMM-coated silica bead surface (Fig. 7.1B). Within a trap, the force on the bead is proportional to its displacement from the centre of the trap, so that the bead position change serves as a measure of both force and displacement. By choosing the stiffness of the trap large enough to reasonably reduce brownian motion, the authors succeeded in recording discrete bead displacement (average, 11 nm), which they believe to arise from a single or from very few HMM molecules interacting with the thin filament at very low HMM densities. They also recorded discrete force production by HMM (average, 3–4 pN) using a large trap stiffness attained with the feedback circuit. Similar discrete mechanical events resulting from the ATP-dependent actin–myosin interaction have also been reported by Miyata *et al.* (1994) and Molloy *et al.* (1995*b*) using the 'optical trap' technique.

In the experiments described above, discrete mechanical events above the level of thermal noise have been obtained by examining the records by eye. Molloy *et al.* (1995*a*), however, re-examined their previous results (Molloy *et al.* 1995*b*) based on the assumption that the magnitude of myosin head powerstroke is small and therefore cannot be readily distinguished from thermal noise by eye. They constructed histograms of amplitude distribution of the observed displacements from selected regions in the records, where the noise level was significantly smaller than in other regions reflecting attachment of myosin heads to the filament. The histograms were Gaussian with centres shifted by ~4 nm from the origin to the right, suggesting that the myosin head powerstroke amplitude is ~4 nm, being much smaller than that determined in previous studies already described. Thus, the present situation with respect to the distance of myosin head powerstroke seems to be somewhat puzzling, and much more experimental work is needed to settle the problem. The amplitude of myosin head powerstroke would be best studied using assay systems, in which mechanical events originating from a single myosin head can be recorded unambiguously.

Problems and limitations

As described above, the fluorescent actin filament *versus* myosin system has been widely used, because it enables direct observation of ATP-dependent actin–myosin sliding under a light microscope. Since this system consists of purified actin and myosin samples, it has been effectively used to study the ability of the fragments of myosin to support the actin filament sliding, and also to determine essential amino acid residues for actin–myosin sliding. This system has also been useful for studying regulatory mechanisms of contraction, though such studies are not mentioned in this article.

However, this motility assay system has not been so effective in studying the mechanism of muscle contraction, even though experiments with this system have raised interesting issues with respect to the myosin step size as well as with the amplitude of the myosin powerstroke. The difficulties mainly arise from random orientation and random location of myosin molecules fixed on the substratum. In some cases, incorrect orientation of the molecules results in the absence of ATP-dependent filament sliding, though the molecule has the potential ability to support filament movement. The orientation of the molecule can be made uniform by use of fusion proteins or of antibodies to a specific site in the molecule, but random location of the molecule is inherent in this assay system.

Careful comparison of the filament sliding velocity with the unloaded shortening velocity of muscle fibres has shown that the two phenomena are analogous to each other under some conditions but not under others, indicating limitations to deducing muscle contraction mechanisms based on the results obtained with this assay system.

7.3 Actin cable *versus* myosin system

This motility assay system was developed independently by Sheetz and Spudich (1983) and Shimmen and Yano (1984); it employs well-organized arrays of actin filaments occurring in giant internodal cells of green algae (*Nitella*, *Chara* or *Nitellopsis*) and small beads coated with myosin. The internodal cell is cut open to expose the actin filament arrays (actin cables), onto which the myosin-coated beads are applied. In the presence of ATP, the beads move in one direction only, as determined by the actin filament polarity. Although actin cables cannot be seen under a light microscope under ordinary optical conditions, they are known to be straight and parallel to the rows of chloroplasts, so that myosin-coated beads are observed to move parallel to the chloroplast rows. The direction of bead movement is reversed across the indifferent zone (where the chloroplasts are almost absent), reflecting the reversal of the actin filament polarity. The features of bead movement are similar to those of cytoplasmic streaming, which is produced as a result of ATP-dependent interaction between actin cables and cytoplasmic myosin.

The advantage of the actin cable *versus* myosin system over the fluorescent actin filament *versus* myosin system is that the myosin-coated bead movement is straight over a considerable distance (more than 100 μm), and can therefore provide experimental data comparable with those of contracting muscle if combined with appropriate devices for measuring forces arising in actin–myosin sliding. A disadvantage is that algal cell actin is not well characterized, although this is not a serious problem in most experiments.

Measurement of sliding velocity

As with actin filament sliding on myosin, the inertial force in this system is negligible compared to the viscous drag on a moving bead. The sliding velocity of the beads coated with filamentous skeletal muscle myosin increases with increasing ATP concentration, reaching a maximum at 100 μM ATP (2–6 μm/s at 22 °C, Sheetz and Spudich 1983). Half-maximum velocity is attained at 50 μM ATP. These features are similar to those of the actin filament sliding on myosin, again indicating that the bead sliding velocity must be limited by the kinetic properties of ATP-dependent actin–myosin interaction, rather than by the force generated by myosin molecules to drive the bead movement. The bead sliding velocities vary widely among myosin species used, and correlates well with their respective actin-activated ATPase activities (Sheetz and Spudich 1983; Sugiura *et al.* 1992). Hynes *et al.* (1987) attached HMM to the bead surface *via* antibodies, and showed that such beads slide on actin cables with velocities similar to those of beads coated with filamentous myosin. Bead aggregates move smoothly on actin cables with velocities similar to those of single beads without any sign of tumbling or reorientation, indicating that the bead–actin cable contact is fairly stable during the course of bead sliding.

This motility assay system has been used to investigate factors influencing the ATP-dependent actin–myosin sliding resulting from structural changes of myosin molecule. In accordance with the myosin-linked Ca^{2+} regulation of contraction in molluscan smooth muscles, Vale *et al.* (1984) showed that scallop myosin-coated beads started moving on actin cables when pCa^{2+} was lowered to < 5.84. Sellers *et al.* (1985) measured the sliding velocity of beads coated with a mixture of phosphorylated and unphosphorylated smooth muscle myosin, and reported that unphosphorylated myosin reduced the rate of sliding of phosphorylated myosin on the actin cables, a result qualitatively similar to that of Warshaw *et al.* (1990). Mammalian cardiac muscle is known to be composed of three myosin isozymes (V_1, V_2 and V_3), whose percentages in cardiac muscle vary depending on either physiological or pathological conditions. Yamashita *et al.* (1992) showed that both the actin-activated ATPase activity and the bead sliding velocity on actin cables were higher for V_1 than for V_3.

Measurement of force and work

The most basic characteristic of contracting muscle is the dependence of shortening velocity, or the rate of power output on the isotonic load (the force–velocity and force–energy relations) (Hill 1938; Huxley 1957, 1974). The *in vitro* motility assay systems hitherto described do not yield this characteristic, and do not allow comparison with intact muscle. From the standpoint of muscle physiology, it is essential to construct motility assay systems, which allow examination of the characteristic using a small number of actin and myosin molecules under simplified conditions.

Long and straight actin filament arrays (actin cables) are suitable for performing such experiments. Chaen *et al.* (1989) brought a skeletal muscle myosin-coated glass microneedle in contact with the actin cables. In the presence of ATP, the myosin-coated needle moved along actin cables in the direction expected from the actin filament polarity until it came to a complete stop due to the balance between forces generated by myosin heads and by the bent needle (total sliding distance, $< 150\,\mu m$) (see Fig. 7.2A). In such an experimental condition, the force (= load) against the actin–myosin sliding increases with the distance of sliding, and therefore corresponds with 'auxotonic' condition in which an active muscle shortens against a linear spring. The force–velocity relationship constructed from the above experiments is convex upwards, and was thus analogous in shape to the force–velocity relationship of single tetanized skeletal muscle fibres under auxotonic condition (Iwamoto *et al.* 1990, see Fig. 1.3B), indicating that this assay system retains the basic kinetic properties of myofilament sliding in muscle, despite the random orientation of myosin molecules on the needle.

It is well known that a contracting skeletal muscle varies its energy output (heat + work) depending on the load imposed on it (Fenn 1924), and this phenomenon is called the 'Fenn effect'. The molecular mechanism underlying the Fenn effect is not yet clarified, though the Huxley contraction model (Huxley 1957) suggests that the velocity of myofilament sliding determines the number of myosin heads (cross-bridges) involved in the work and heat production. Oiwa *et al.* (1991) have attempted to approach the problem using a method similar to that of Chaen *et al.* (1989), combined with the technique of iontophoretic ATP application, i.e. application of ATP by passing current through an ATP-containing microelectrode (see Fig. 7.2A). The myosin heads on the needle

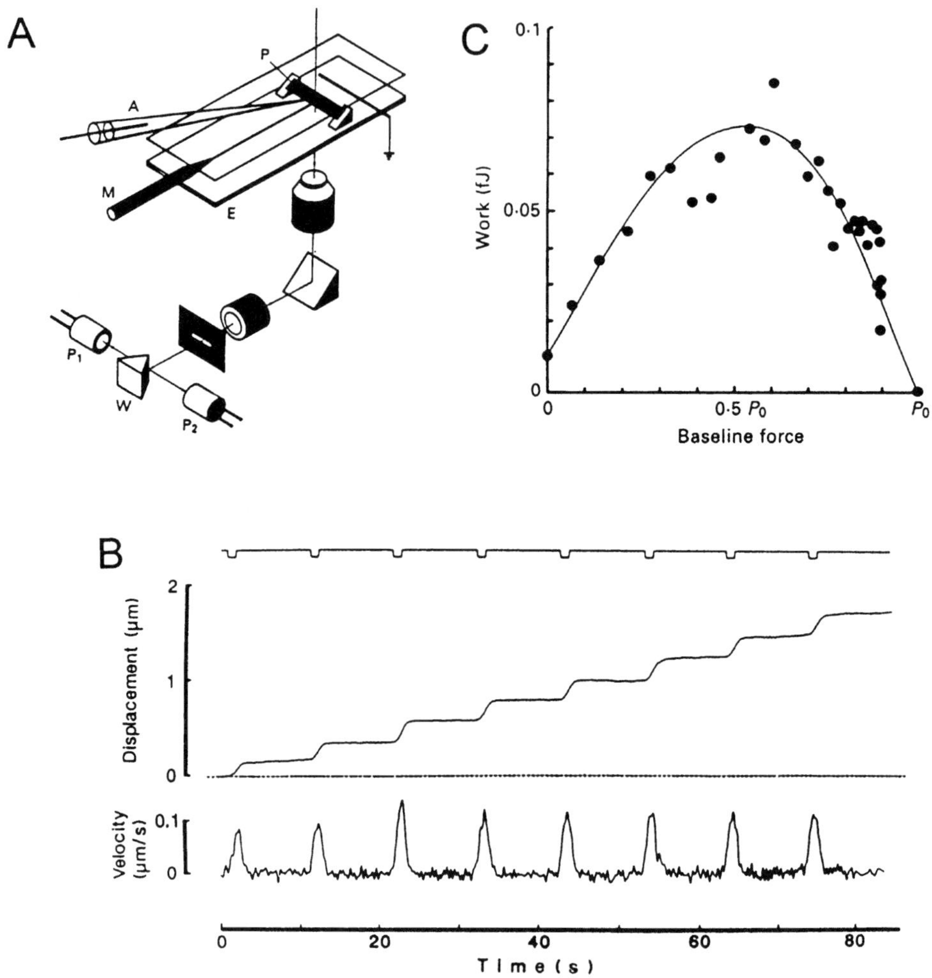

Fig. 7.2 Measurement of work done by the ATP-induced sliding between a myosin-coated glass microneedle and actin cables in algal cells. (A) Diagram of experimental arrangement showing the myosin-coated needle M in contact with actin cables on the surface of an algal cell strip P, and ATP-containing glass micro-electrode A for iontophoretic ATP application, and photo-electric device to record needle movement. The needle tip image is split with the wedge-shaped mirror W to be recorded with photodiodes P_1 and P_2. (B) Simultaneous recordings of ATP current pulse releasing a constant amount of ATP from the ATP-electrode (*upper trace*), ATP-induced movement of the myosin coated microneedle (*middle trace*), and velocity of microneedle movement (*lower trace*). Note that the magnitude and the velocity profile in each ATP-induced needle movement is nearly the same irrespective of the initial baseline force from which the needle starts moving on ATP application. (C) Relation between the amount of work done by the actin–myosin sliding induced by a constant amount of iontophoretically applied ATP and the initial baseline force. (From Oiwa *et al.* 1991.)

initially formed rigor linkages with actin cables in the absence of ATP, and on iontophoretic application of ATP they initiated needle sliding along actin cables until the applied ATP was used up by the myosin heads and the hexokinase–glucose system in the experimental solution; on complete exhaustion of ATP, the heads again formed rigor linkages with actin cables to terminate the needle sliding. Thus, by repeated application of constant amounts of ATP to the myosin molecules on the needle at appropriate intervals, a stepwise needle movement was constructed (Fig. 7.2B). A remarkable feature of the needle movement was that the magnitude of each step, i.e. the distance of the ATP-induced actin–myosin sliding for each ATP application, was nearly constant over many steps (Fig. 7.2B). The amount of work (W) done by the ATP-induced actin–myosin sliding when the needle moved from position x_1 to position x_2 is given by:

$$W = \int_{x_1}^{x_2} Kx\,dx = K(x_2^2 - x_1^2)/2, \tag{7.4}$$

where K is the elastic coefficient of the needle. From eqn 7.4, it is clear that the amount of work done by the ATP-induced actin–myosin sliding increases steeply as the baseline force, from which the needle starts moving, is increased. As shown in Fig. 7.2C, the work *versus* baseline force relationship is bell-shaped, and is similar to the work *versus* load relationship obtained from contracting skeletal muscle by Fenn (1924), indicating that the randomly oriented myosin molecules still show the work *versus* load relationship similar to that of contracting muscle. In both cases, the amount of work done increases to a maximum if the load is increased from zero to a certain value. Another interesting feature of the results shown in Fig. 7.2B is that the time course of the needle sliding velocity, i.e. the time derivative of the needle movement, is nearly the same for several consecutive steps. This suggests that the efficiency with which the myosin heads converts chemical energy derived from ATP hydrolysis into mechanical work is primarily determined by load but not by the velocity of actin–myosin sliding.

With the same assay system, Oiwa *et al.* (1993) attempted to approach the unitary distance of ATP-induced actin–myosin sliding (equivalent to the myosin head powerstroke) by reducing the density of myosin heads on the needle, as well as the amount of iontophoretically applied ATP, and obtained a minimum distance of the ATP-induced actin–myosin sliding of ~ 10 nm, a value similar to the minimum distance of myofilament sliding in glycerinated skeletal muscle fibres in response to ATP released by laser flash photolysis of caged-ATP (Yamada *et al.* 1993).

Steady-state force–velocity relationship

In contracting skeletal muscle or muscle fibres, there is a definite steady relationship between the velocity of shortening and the amount of isotonic load (= force) to give a hyperbolic force–velocity relationship (Hill 1938; Jewell and Wilkie

1958). Using a centrifuge microscope, with which constant centrifugal forces can be applied to a moving object, Oiwa *et al.* (1990) succeeded in obtaining the steady-state force–velocity relationship using the ATP-dependent sliding of myosin-coated beads on actin cables.

The centrifuge microscope is composed of a light microscope, a rotor to apply centrifugal forces to an object mounted on it, and a stroboscopic light source (Kamitsubo *et al.* 1989) (Fig. 7.3A). Polystyrene beads (diameter, 2.8 μm; specific gravity, 1.3) coated with skeletal muscle myosin were put into the algal cell segment together with an experimental solution containing ATP, and then the cell segment, ligated at both ends, was mounted on the rotor of the centrifuge microscope. The amount of centrifugal force (= load, F) on the bead is given by:

$$F = \Delta\rho \, V r \omega^2, \tag{7.5}$$

where $\Delta\rho$ is the difference in density between the bead and the surrounding medium, V is the bead volume, r is the effective radius of centrifugation, and ω is the angular velocity of the rotor. Under a given submaximal centrifugal force opposite to the direction of bead sliding (ordinary positive load), the bead continued to move on actin cables at a constant velocity, indicating the existence of a definite steady-state relationship between the load and the sliding velocity, as with contracting muscle. As actin cables are very long, the sliding velocities under different loads can be obtained from one and the same bead. The steady-state force–velocity relationship thus obtained is hyperbolic in shape except for the high load region, where the curve deviates from a hyperbola (Fig. 7.3B), being analogous to that in contracting skeletal muscle fibres (Edman 1988). The maximum isometric force P_0 generated by myosin molecules is <40 pN, suggesting that the number of myosin molecules driving the bead sliding may be <20, again indicating that the basic contraction characteristics are retained in the small number of randomly oriented myosin molecules.

Since the direction of bead sliding is reversed across the indifferent zone (Fig. 7.3A), it was possible to apply centrifugal forces in the same direction as that of the bead sliding, so that the bead was pushed forward by centrifugal forces (negative load). Unexpectedly, the bead sliding velocity was slowed down by a negative load (Fig. 7.3C). A similar slowdown of negative load has been reported for ATP-dependent microtubule-dynein sliding (Hall *et al.* 1993).

Sugiura *et al.* (1995) also used the centrifuge microscope to compare the force–velocity characteristics between the two myosin isozymes, V_1 and V_3, in mammalian cardiac muscle, and showed a marked difference in the shape of the force–velocity relationship between V_1 and V_3. They explained the difference as being due to a difference between V_1 and V_3 in regard to the fraction of time in one ATP cycle in which the myosin head is attached to actin to produce myofilament sliding, while the force-generating ability of the two isozymes seemed to be equal (Sugiura *et al.* 1996).

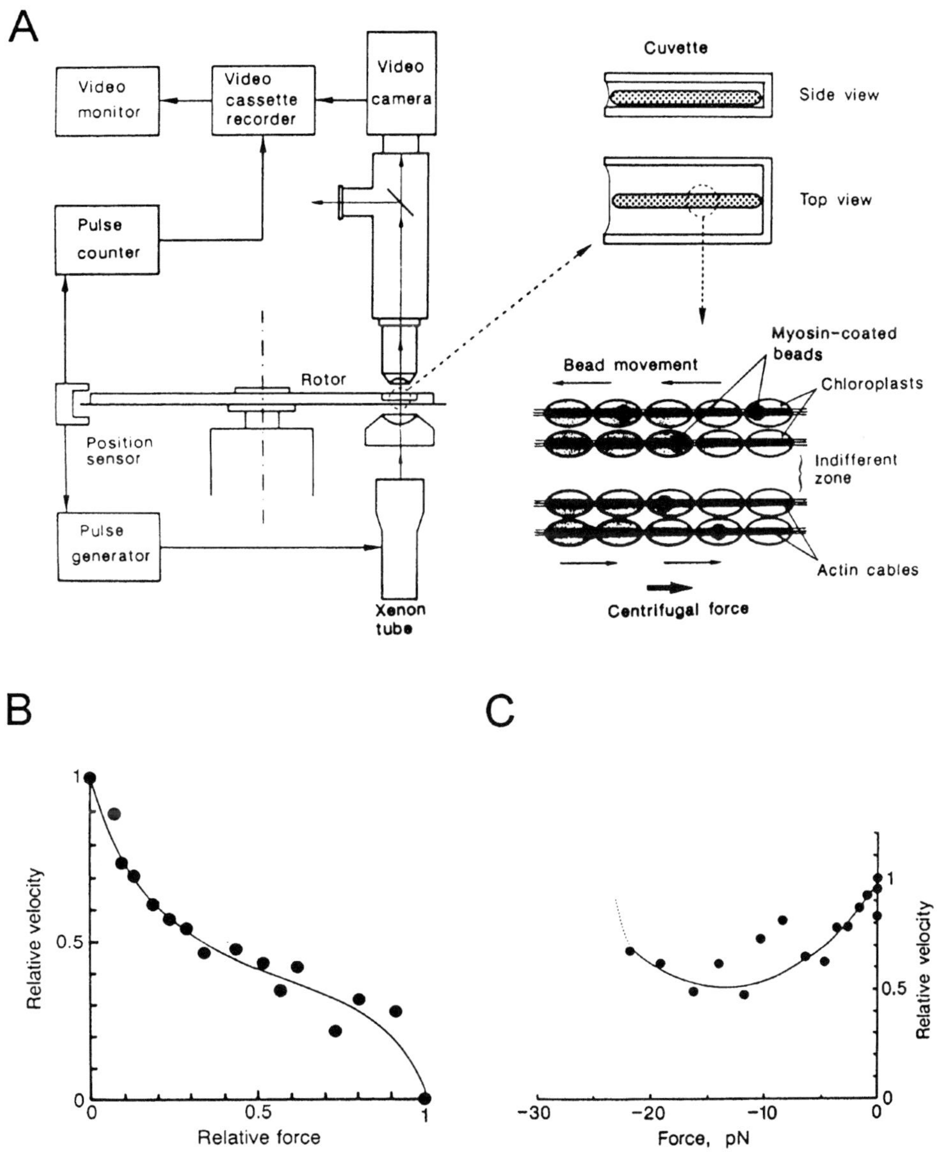

Fig. 7.3 An *in vitro* motility assay system combined with the centrifuge microscope to apply constant centrifugal forces to myosin-coated beads sliding along actin cables in an algal cell segment. (A) Diagram of experimental arrangement showing the centrifuge microscope system (*left*), and the sliding of myosin-coated beads along actin cables in the algal cell segment. (B) Steady-state force–velocity relationship for positive loads, produced by centrifugal forces in the direction opposite to the bead movement. (C) Steady-state force–velocity relationship for negative loads, produced by centrifugal forces in the same direction as that of the bead movement. (From Oiwa *et al.* 1990.)

On the other hand, Tregear *et al.* (1993) controlled the sliding of myosin-coated magnetizable beads on actin cables by the magnetic field of a strong magnet, and obtained a magnetic force *versus* bead velocity relationship analogous to that of Oiwa *et al.* (1990). A magnetic field could be used effectively to change quickly the load on the moving bead.

Problems and limitations

The actin cable *versus* myosin system has been used less frequently than the fluorescent actin filament *versus* myosin system, because the algal cells are not readily available to investigators. Both assay systems have the common problem of random orientation of the myosin molecules. The advantage of using actin cables is that the linear ATP-dependent actin–myosin sliding takes place for many seconds, so that it is possible to study its force–velocity characteristics and work production. The similarities in basic characteristics of ATP-dependent actin–myosin sliding and contracting muscle fibres (in both auxotonic and isotonic conditions) indicate that this assay system may be effectively used for studying the mechanisms underlying the characteristics with very small numbers of myosin molecules involved. On the other hand, this system seems to be unsuitable for exploring the mechanical events originating in a single myosin molecule, since the myosin molecules are attached to an object much larger than their size.

7.4 Use of myosin thick filament

Although myosin molecules are believed to have some degree of flexibility in order to drive actin filament movement at various orientations (Toyoshima *et al.* 1989), it has been shown that myosin molecules in the myosin thick filaments in muscle are under rather severe structural constraints when they drive myofilament sliding. Sellers and Kachar (1990) and Yamada *et al.* (1990) isolated native myosin thick filaments from molluscan smooth muscles, and observed the ATP-dependent sliding of a fluorescently labelled actin filament along a single thick filament, attention being focused on actin filament sliding across the centre (bare region) of the thick filament, where the polarity of myosin heads is reversed. Both groups found that:

(1) the actin filament slides not only toward the bare region, i.e. in the direction of myofilament sliding when a muscle shortens, but also away from the bare region, i.e. in the direction opposite to myofilament sliding in shortening muscle;

(2) the sliding velocity of the actin filament was much higher (by a factor of ~ 10) when it slides toward the bare region than when it slides away from the bare region.

These results demonstrate that the direction ATP-driven actin–myosin sliding is determined by the actin filament polarity, but its velocity is markedly dependent on the thick filament polarity, i.e. the orientation of myosin heads. Similar results have been obtained on actin filament sliding along the synthetic thick filaments from skeletal muscle myosin (Yamada and Wakabayashi 1993). Actin filament sliding on thick filaments therefore seems to have two different modes: the 'fast' mode toward the bare region driven by the 'correctly' oriented myosin heads, and the 'slow' mode away from the bare region driven by the 'incorrectly' oriented myosin heads.

Yamada and Takahashi (1992) made the further observation that an actin filament changed its mode of sliding from 'slow' to 'fast' by transferring from one thick filament to another across their crossover point; it started sliding in the 'fast' mode as soon as its leading end went into contact with another thick filament. This indicates that the 'incorrectly' oriented heads producing the 'slow' filament sliding do not impose a significant load on the 'fast' filament sliding by the 'correctly' oriented heads; in other words, the 'correctly' oriented heads produce much larger forces than the 'incorrectly' oriented ones.

On the basis of the findings described above, it seems clear that to study the molecular mechanism of muscle contraction most effectively with *in vitro* motility assay systems, myosin thick filaments should be used instead of randomly oriented myosin molecules. Some experiments along this line have already been made as described below.

Mechanical events and the force–velocity relationship

Ishijima *et al.* (1994) have attempted to explore the elementary mechanical events during the ATP-driven actin–myosin sliding without complications arising from the random orientation of the myosin molecules, using single myosin–myosin rod co-filaments, on which a single actin filament, attached to a microneedle at one end, was made to slide. By reducing the myosin : myosin rod ratio in the co-filament, they claimed that they could detect mechanical events originating from unitary actin–myosin interactions (peak force, 5–6 pN; distance of a power-stroke, $\leqslant 17$ nm), although their results are not free from ambiguities arising from uncertain myosin head density in the co-filament, as well as thermal noise. Conversely, Saito *et al.* (1994) held a single fluorescently labelled actin filament taut by a dual laser trap, and made single fluorescently labelled myosin filaments slide along the actin filament, and again estimated the myosin step size *d* to be $\geqslant 71$ nm, based on indirect measurement of minimum length myosin filaments that could slide on the actin filament, possible number of myosin head involved, and published value of the acto-HMM ATPase activity.

Using the centrifuge microscope (Oiwa *et al.* 1990), Ishii *et al.* (1997) have determined the steady-state force–velocity relationships in ATP-driven sliding between native myosin thick filaments isolated from a molluscan smooth muscle and actin cables, by attaching a polystyrene bead (diameter, 4 μm) to the thick filament in order to apply constant centrifugal forces to the correctly oriented myosin heads interacting with actin cables. The force–velocity relationship for positive loads $\leqslant P_0$ was strikingly similar to that in single tentanized skeletal muscle fibres (Edman 1988; Iwamoto *et al.* 1990), including the deviation from a hyperbola with loads $\geqslant 0.8\,P_0$ (Fig. 7.4A). Assuming that the time averaged isometric force per myosin head is 2 pN, the number of myosin heads driving the filament sliding may be $\leqslant 20$. If the amount of positive load is linearly increased with an appropriate rate, the thick filament-attached bead shows 'backward' sliding under loads $> P_0$ until it is detached from the actin cables (Fig. 7.4B), a phenomenon analogous to the lengthening of muscle fibres under loads $> P_0$

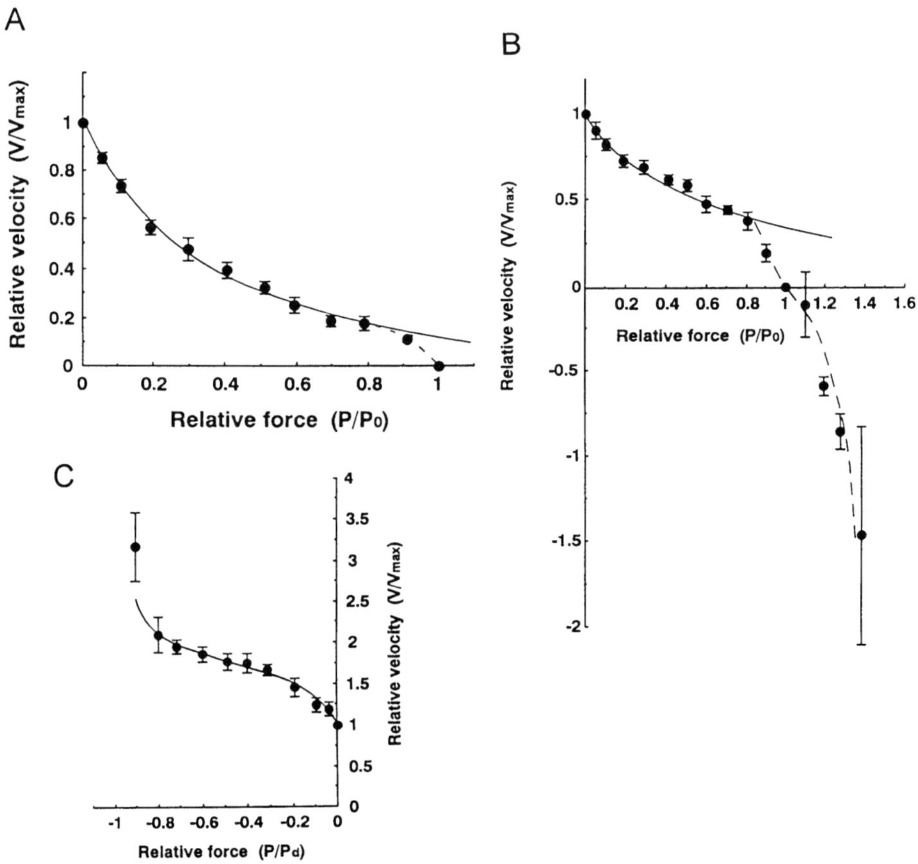

Fig. 7.4 Force–velocity relationships of the ATP-dependent sliding between a thick myosin filament-attached bead actin cables, determined with the centrifuge microscope system shown in Fig. 7.3A. (A) Steady-state force–velocity relationship for positive loads $\leqslant P_0$. (B) Quasi steady-state force–velocity relationship for positive loads $\leqslant P_0$ and for those $> P_0$. The load was linearly increased from zero to $> P_0$ in 100 s. (C) Steady-state force–velocity relationship for negative loads. (From Ishii *et al.* 1997.)

(Katz 1939; Sugi and Tsuchiya 1981). In contrast with the effect of negative load in slowing down myosin-coated bead sliding on actin cables (Oiwa *et al.* 1990), thick filament sliding on actin cables increased with negative loads (Fig. 7.4C).

7.5 Future prospects

The most straightforward approach to understanding muscle contraction mechanisms may be to examine the elementary processes with which a myosin head hydrolyses ATP and converts chemical energy derived from ATP hydrolysis into mechanical work. Recent development of techniques to visualize single fluorescent dye molecules in solution with very much reduced background fluorescence now enables detection of single fluorescently labelled actin, myosin and ATP molecules (Funatsu *et al.* 1995; Sase *et al.* 1995). It may be possible in the future to investigate the elementary mechanical events originating from a single myosin head associated with hydrolysis of one ATP molecule, though there would be a number of problems to overcome. The use of the atomic microscope (Ando *et al.* 1996), improved for recording single molecule events, would also be useful for investigating physico-chemical aspects of the actin–myosin interaction. Though not directly relevant to the topic of this chapter, Sugi *et al.* (1997) have succeeded in recording the ATP-induced movement of individual myosin heads on synthetic thick filaments that are kept in the 'living' state in the gas environmental (hydration) chamber mounted in an electron microscope. This new technique is promising for future studies of elementary molecular events, since it can record movement of individual myosin heads unambiguously. Recombinant actin and myosin molecules could also be effectively studied with the above techniques.

To fully understand the muscle performance at the molecular level, however, the behaviour of actin and myosin molecules as a group should also be clarified from the standpoint of physiology. In this connection, the actin filament *versus* thick filament system, combined with the centrifuge microscope (Fig. 7.4), may be suitable for future research work, since the molluscan muscle thick filament is a myosin–paramyosin core complex, in which the myosin molecules are regularly arranged around the paramyosin filament (Cohen *et al.* 1971); it is possible to replace the molluscan muscle myosin by the mammalian skeletal muscle myosin (Tameyasu *et al.* in preparation) or by the *Dictyostelium* mutant myosins (Akimoto *et al.* unpublished data). So, it is possible to study how the changes in myosin molecule structures affect characteristics of ATP-dependent filament sliding.

Acknowledgements

I wish to thank my colleagues who carried out research work using various *in vitro* motility assay techniques described in this article. My thanks are also due to Professor Manuel F. Morales for his valuable comments on the manuscript.

References

Ashkin, A. and Dziedzic, J.M. (1987) Optical trapping and manipulation of viruses and bacteria. *Science* **235**, 1517–20.

Ando, T., Kunioka, Y., Adachi, K. and Nakajima, H. (1996) Uni-molecular physiology of actomyosin motor II. *J. Muscle Res. Cell Motil.* **17**, 283.

Block, S.M. (1990) Optical tweezers: a new tool for biophysics. In *Noninvasive techniques in cell biology* (ed. J.K. Foskett and S. Grinstein), pp. 375–402. Wiley–Liss, New York.

Burton, K. (1992) Myosin step size: estimates from motility assays and shortening muscle. *J. Muscle Res. Cell Motil.* **13**, 590–607.

Chaen, S., Oiwa, K., Shimmen, T., Iwamoto, H. and Sugi, H. (1989) Simultaneous recordings of force and sliding movement between a myosin-coated glass microneedle and actin cables *in vitro*. *Proc. Natl. Acad. Sci.* (*USA*) **86**, 1510–4.

Cohen, C., Szent-Györgyi, A.G. and Kendrick-Jones, J. (1971) Paramyosin and the filaments of molluscan "catch" muscles I. Paramyosin: structure and assembly. *J. Mol. Biol.* **56**, 223–37.

Edman, K.A.P. (1988) Double-hyperbolic force–velocity relation in frog muscle fibres. *J. Physiol.* **404**, 301–21.

Fenn, O.W. (1924) The relation between the work performed and the energy liberated in muscular contraction. *J. Physiol.* **58**, 373–95.

Finer, J.T., Simmons, R.M. and Spudich, J.A. (1994) Single myosin molecule mechanics: piconewton forces and nanometre steps. *Nature* **368**, 113–9.

Funatsu, T., Harada, Y., Tokunaga, M., Saito, K. and Yanagida, T. (1995) Imaging of single fluorescent molecules and individual ATP turnovers by single myosin molecules in aqueous solution. *Nature* **374**, 555–9.

Hall, K., Cole, D.G., Yeh, Y., Scholey, J.M. and Baskin, R.J. (1993) Force–velocity relationships in kinesin-driven motility. *Nature* **364**, 457–9.

Harada, Y., Sakurada, K., Aoki, T., Thomas, D.D. and Yanagida, T. (1990) Mechanochemical coupling in actomyosin energy transduction studied by *in vitro* movement assay. *J. Mol. Biol.* **216**, 49–68.

Harrington, W.F., Ueno, H. and Davis, J.S. (1988) Helix-coil melting in rigor and activated cross-bridges of skeletal muscle. In *Molecular mechanism of muscle contraction* (ed. H. Sugi and G.H. Pollack), pp. 307–18. Plenum, New York.

Hill, A.V. (1938) The heat of shortening and the dynamic constants of muscle. *Proc. R. Soc. Lond. B.* **126**, 136–95.

Homsher, E., Wang, F. and Sellers, J.R. (1992) Factors affecting movement of F-actin filaments propelled by skeletal muscle heavy meromyosin. *Am. J. Physiol.* **31**, C714–23.

Huxley, A.F. (1957) Muscle structure and theories of contraction. *Prog. Biophys. Chem.* **7**, 255–318.

Huxley, A.F. (1974) Review lecture, muscular contraction. *J. Physiol.* **243**, 1–43.

Huxley, A.F. and Simmons, R.M. (1971) Proposed mechanism of force generation in striated muscle. *Nature* **233**, 533–8.

Huxley, H.E. (1990) Sliding filaments and molecular motile systems. *J. Biol. Chem.* **265**, 8347–50.

Hynes, T.R., Block, S.M., White, B.T. and Spudich, J.A. (1987) Movement of myosin fragments *in vitro*: domains involved in force production. *Cell* **48**, 953–63.

Ishijima, A., Doi, T., Sakurada, K. and Yanagida, T. (1991) Sub-piconewton force fluctuations of actomyosin *in vitro*. *Nature* **352**, 301–6.

Ishijima, A., Harada, Y., Kojima, H., Funatsu, T., Higuchi, H. and Yanagida, T. (1994) Single-molecule analysis of the actomyosin motor using nano-manipulation. *Biochem. Biophys. Res. Commun.* **199**, 1057–63.

Ishii, N., Tsuchiya, T. and Sugi, H. (in press) An *in vitro* assay system retaining the steady-state force–velocity characteristics of muscle fibres under positive and negative loads. *Biochim. Biophys. Acta* **1319**, 155–62.

Itakura, S., Yamakawa, H., Toyoshima, Y.Y., Ishijima, A., Kojima, T., Harada, Y. *et al.* (1993) Force-generating domain of myosin motor. *Biochem. Biophys. Res. Commun.* **196**, 1504–10.

Iwamoto, H., Sugaya, R. and Sugi, H. (1990) Force–velocity relation of frog skeletal muscle fibres shortening under continuously changing load. *J. Physiol.* **422**, 185–202.

Jewell, B.R. and Wilkie, D.R. (1958) An analysis of the mechanical components in frog's striated muscle. *J. Physiol.* **143**, 515–40.

Johara, M., Toyoshima, Y.Y., Ishijima, A., Kojima, H., Yanagida, T. and Sutoh, K. (1993) Charge-reversion mutagenesis of *Dictyostelium* actin to map the surface recognized by myosin during ATP-driven sliding motion. *Proc. Natl. Acad. Sci.* (*USA*) **90**, 2127–31.

Kamitsubo, E., Ohashi, Y. and Kikuyama, M. (1989) Cytoplasmic streaming in internodal cells of *Nitella* under centrifugal acceleration: a study done with a newly constructed centrifuge microscope. *Protoplasma* **152**, 148–55.

Katz, B. (1939) The relation between force and speed in muscular contraction. *J. Physiol.* **96**, 45–64.

Kishino, A., and Yanagida, T. (1988) Force measurements by micromanipulation of a single actin filament by glass needles. *Nature* **334**, 74–6.

Kron, S.J. and Spudich, J.A. (1986) Fluorescent actin filaments move on myosin fixed to a glass surface. *Proc. Natl. Acad. Sci.* (*USA*) **83**, 6272–6.

Lowey, S., Waller, G.S. and Trybus, K.M. (1993) Skeletal muscle myosin light chains are essential for physiological speeds of shortening. *Nature* **365**, 454–6.

Miyata, H., Hakozaki, H., Yoshikawa, H., Suzuki, N., Kinosita, K. Jr., Nishizaka, T. *et al.* (1994) Stepwise motion of an actin filament over a small number of heavy meromyosin molecules is revealed in an *in vitro* motility assay. *J. Biochem.* **115**, 644–7.

Molloy, J.E., Burns, J.E., Kendrick-Jones, J., Tregear. R.T. and White, D.C.S. (1995*a*) Movement and force produced by a single myosin head. *Nature* **378**, 209–12.

Molloy, J.E., Burns, J.E., Sparrow, J.C., Tregear, R.T., Kendrick-Jones, J. and White, D.C.S. (1995*b*) Single-molecule mechanics of heavy meromyosin and S1 interacting with rabbit or *Drosophila* actins using optical tweezers. *Biophys. J.* **68**, 298s–305s.

Nishizaka, T., Yagi, T., Tanaka, Y. and Ishiwata, S. (1993) Right-handed rotation of an actin filament in an *in vitro* motile system. *Nature* **361**, 269–71.

Oishi, N. and Sugi, H. (1994) *In vitro* ATP-dependent F-actin sliding on myosin is not influenced by substitution or removal of bound nucleotide. *Biochim. Biophys. Acta* **1185**, 346–9.

Oishi, N., Sutoh, K., Suzuki, S. and Sugi, H. (1996) Evidence for the functional role of myosin subfragment-2 in the *in vitro* actin–myosin sliding. *J. Muscle Res. Cell Motil.* **17**, 281.

Oiwa, K., Chaen, S. and Sugi, H. (1991) Measurement of work done by ATP-induced sliding between rabbit muscle myosin and algal cell actin cables *in vitro*. *J. Physiol.* **437**, 751–63.

Oiwa, K., Kawakami, T. and Sugi, H. (1993) Unitary distance of actin–myosin sliding studied using an *in vitro* force–movement assay system combined with ATP iontophoresis. *J. Biochem.* **114**, 28–32.

Oiwa, K., Chaen, S., Kamitsubo, E., Shimmen, T. and Sugi, H. (1990) Steady-state force–velocity relation in the ATP-dependent sliding movement of myosin-coated beads on actin cables *in vitro* studied with a centrifuge microscope. *Proc. Natl. Acad. Sci. (USA)* **87**, 7893–7.

Oplatka, A. (1994) Myosin-specific adaptations of the motility assay, factors affecting movement of F-actin filaments propelled by skeletal muscle heavy meromyosin. *Chemtracts Biochemistry and Molecular Biology* **5**, 120–8.

Oplatka, A. and Tirosh, R. (1973) Active streaming in actomyosin solution. *Biochim. Biophys. Acta* **305**, 684–5

Rayment, I., Rypniewski, W.R., Schmidt-Bäse, K., Smith, R., Tomchick, D.R., Benning, M.M. *et al.* (1993) Three-dimensional structure of myosin subfragment-1: a molecular motor. *Science* **261**, 50–8.

Saito, K., Aoki, T., Aoki, T. and Yanagida, T. (1994) Movement of single myosin filaments and myosin step size on an actin filament suspended in solution by a laser trap. *Biophys. J.* **66**, 769–77.

Sase, I., Miyata, H., Corrie, J.E.T., Craik, J.S. and Kinosita, K. Jr. (1995) Real time imaging of single fluorophores on moving actin with an epifluorescence microscope. *Biophys. J.* **69**, 323–8.

Schwyter, D.H., Kron. S.J., Toyoshima, Y.Y., Spudich, J.A. and Reisler, E. (1990) Subtilisin cleavage of actin inhibits *in vitro* sliding movement of actin filament over myosin. *J. Cell Biol.* **111**, 465–70.

Sellers, J.R. and Kachar, B. (1990) Polarity and velocity of sliding filaments: control of direction by actin and of speed by myosin. *Science* **249**, 406–8.

Sellers, J.R., Spudich, J.A. and Sheetz, M.P. (1985) Light chain phosphorylation regulates the movement of smooth muscle myosin on actin filaments. *J. Cell Biol.* **101**, 1897–902.

Sheetz, M.P. and Spudich, J.A. (1983) Movement of myosin-coated fluorescent beads on actin cables *in vitro. Nature* **303**, 31–5.

Shimmen, T. and Yano, M. (1984) Active sliding movement of latex beads coated with skeletal muscle myosin on *Chara* actin cables. *Protoplasma* **121**, 132–7.

Sugi, H., and Tsuchiya, T. (1981) Isotonic velocity transients in frog muscle fibres following quick changes in load. *J. Physiol.* **319**, 219–38.

Sugi, H., Akimoto, T., Sutoh, K., Chaen, S., Oishi, N. and Suzuki, S. (1997) Dynamic electron microscopy of ATP-induced myosin head movement in living muscle thick filaments. *Proc. Natl. Acad. Sci. (USA)* **94**, 4378–82.

Sugi, H., Kobayashi, T., Gross, T., Noguchi, K., Karr, T. and Harrington, W.F. (1992) Contraction characteristics and ATPase activity of skeletal muscle fibres in the presence of antibody to myosin subfragment 2. *Proc. Natl. Acad. Sci. (USA)* **89**, 6134–7.

Sugiura, S., Kobayakawa, N., Momomura, S., Chaen, S., Omata, M. and Sugi, H. (1996) Different cardiac myosin isoforms exhibit equal force-generating ability *in vitro. Biochim. Biophys. Acta* **1273**, 73–6.

Sugiura, S., Yamashita, H., Sata, M., Momomura, S., Serizawa, T., Oiwa, K. *et al.* (1995) Force–velocity relations of rat cardiac myosin isozymes sliding on algal cell actin cables *in vitro. Biochim. Biophys. Acta* **1231**, 69–75.

Sugiura, S., Yamashita, H., Serizawa, T., Iizuka, M., Shimmen, T. and Sugimoto, T. (1992) Active movement of cardiac myosin on Characeae actin cables. *Pflüg. Arch.* **421**, 32–6.

Tameyasu, T., Shibayama, R., Akimoto, T., Hirohata, Y. and Sugi, H. Force–velocity characteristics in the ATP-driven sliding between skeletal muscle myosin regularly arranged in the thick filament and actin cables *in vitro*. (In preparation.)

Toyoshima, Y.Y., Kron, S.J. and Spudich, J.A. (1990) The myosin step size: measurement of the unit displacement per ATP hydrolyzed in an *in vitro* assay. *Proc. Natl. Acad. Sci.* (*USA*) **87**, 7130–4.

Toyoshima, Y.Y., Toyoshima, C. and Spudich, J.A. (1989) Biodirectional movement of actin filaments along tracks of myosin heads. *Nature* **341**, 154–6.

Toyoshima, Y.Y., Kron, S.J., McNally, E.M., Niebling, K.R., Toyoshima, C. and Spudich, J.A. (1987) Myosin subfragment-1 is sufficient to move actin filaments *in vitro*. *Nature* **328**, 536–9.

Tregear, R.T., Oiwa, K., Chaen, S. and Sugi H. (1993) Relation between magnetically-applied force and velocity in beads coated with rabbit myosin, sliding on actin cables in *Nitellopsis* cells. *J. Muscle Res. Cell Motil.* **14**, 412–5.

Uyeda, T.Q.P., Kron, S.J. and Spudich, J.A. (1990) Myosin step size estimation from slow sliding movement of actin over low densities of heavy meromyosin. *J. Mol. Biol.* **214**, 699–710.

Uyeda, T.Q.P., Warrick, H.M., Kron, S.J. and Spudich, J.A. (1991) Quantized velocities at low myosin densities in an *in vitro* motility assay. *Nature* **352**, 307–11.

Vale, R.D., Szent-Györgyi, A.G. and Sheetz, M.P. (1984) Movement of scallop myosin on *Nitella* actin filaments: regulation by calcium. *Proc. Natl Acad. Sci.* (*USA*) **81**, 6775–8.

Van Buren, P., Waller, G.S., Harris, D.E., Trybus, K.M., Warshaw, D.M. and Lowey, S. (1994) The essential light chain is required for full force production by skeletal muscle myosin. *Proc. Natl. Acad. Sci.* (*USA*) **91**, 12403–7.

Warrick, H.M., Simmons, R.M., Finer, J.T., Uyeda, T.Q.P., Chu, S. and Spudich, J.A. (1993) *In vitro* methods for measuring force and velocity of the actin–myosin interaction using purified proteins. In *Methods in cell biology* Vol. 39, *Motility assays for motor proteins* (ed. J.M. Scholey), pp. 1–21. Academic Press, New York.

Warshaw, D.M., Desrosiers, J.M., Work, S.S. and Trybus, K.M. (1990) Smooth muscle myosin cross-bridge interactions modulate actin filament sliding velocity *in vitro*. *J. Cell Biol.* **111**, 453–63.

Yamada, A. and Takahashi, K. (1992) Movement of actin filaments along thick filaments isolated from the anterior byssus retractor muscle of *Mytilus edulis*. *J. Muscle Res. Cell Motil.* **12**, 308.

Yamada, A. and Wakabayashi, T. (1993) Movement of actin away from the center of reconstituted rabbit myosin filament is slower than in the opposite direction. *Biophys. J.* **64**, 565–9.

Yamada, A., Ishii, N. and Takahashi, K. (1990) Direction and speed of actin filaments moving along thick filaments isolated from molluscan smooth muscle. *J. Biochem.* **108**, 341–3.

Yamada, T., Abe, O., Kobayashi, T. and Sugi, H. (1993) Myofilament sliding per ATP molecule in rabbit muscle fibres studied using laser flash photolysis of caged ATP. *J. Physiol.* **466**, 229–43.

Yamashita, H., Sugiura, S., Serizawa, T., Sugimoto, T., Iizuka, M., Katayama, E. *et al.* (1992) Sliding velocity of isolated rabbit cardiac myosin correlates with isozyme disribution. *Am. J. Physiol.* **263**, H464–72.

Yanagida, T., Nakase, M., Nishiyama, K. and Oosawa, F. (1984) Direct observation of motion of single F-actin filaments in the presence of myosin. *Nature* **307**, 58–60.

Yano, M., Yamamoto, Y. and Shimizu, H. (1982) An actomyosin motor. *Nature* **299**, 557–9.

PART II

Combined structural and mechanical experiments

8 The use of fluorescent probes

Shin'ichi Ishiwata

8.1 Introduction

Muscle contraction is attributed to the microscopic (to be more precise, nanoscopic) structural changes in protein motors. However, we have not yet succeeded in observing these critical structural changes despite nearly half a century of anticipating this breakthrough.

At present, it is possible to directly observe single actin filaments under fluorescence microscopy by labelling actin protomers with fluorescent probes (Yanagida *et al.* 1984; Honda *et al.* 1986). Moreover, the sliding movement of actin filaments can be visualized on a myosin-coated glass surface in real time (*in vitro* motility assay system) (Kron and Spudich 1986) and the active tension developed in single actin filaments can then be measured (Kishino and Yanagida 1988; Finer *et al.* 1994; Miyata *et al.* 1994, 1995). The sliding movement of actin filaments clearly shows a mutual displacement of actin and myosin. However, this does not necessarily mean that the structural changes occur within the motor (myosin) molecules. Does the motor actually change its structure? The answer to this question cannot be obtained only by observing sliding movement (Yanagida *et al.* 1993; Spudich 1994).

Structural changes in the constituents may not even be a prerequisite for motor function. Were the sliding movement to occur according to a mechanism similar to that of a man-made linear motor, structural changes would not be required. There would be, instead, a flow of small molecules such as protons and ions. Thus, to study the molecular mechanism of biological motors, it is essential to detect critical structural changes.

Spectroscopies such as fluorescence (phosphorescence), absorption, circular dichroism (CD)/optical rotatory dispersion (ORD), light scattering, X-ray scattering, neutron scattering, electron paramagnetic resonance (EPR) and nuclear magnetic resonance (NMR), are very powerful tools for studying structural changes in proteins (Cantor and Schimmel 1980). Among these, the fluorescence method is especially useful because not only quantitative time-resolved analysis is possible, but also imaging of the organized structure under fluorescence microscopy; in addition, different information can be imaged at the same time by using dual- or multi-view microscopy (Kinosita *et al.* 1991). For monitoring the spatio-temporal changes in the structure of a particular protein in an organized system, the fluorescent probe method is superior to that of other spectroscopies. If an epi-fluorescence dye is incorporated into a specific site on a particular protein, it is possible to selectively monitor information on the state (and dynamic structure) of the labelled protein. Furthermore, various kinds of fluorescent probes are now available (Haugland 1996) and can monitor local and gross structural changes in proteins, hydrophobicity (polarity) and charge distributions around binding sites, the concentrations of ions such as Ca^{2+}, Mg^{2+} and inorganic phosphates and so on (see concise description on environmental sensitivity of fluorescence by Johnson 1996). Thus, the fluorescent probe method has various advantages which overcome the disadvantages and limitations characteristic of labelling methods.

In this section, I would like to introduce fluorescent (and phosphorescent) probe studies reported over the past decade. However, this is not a comprehensive review. Preliminary results obtained in my laboratory are also described. I recommend that readers refer to the reviews of Cooke for research up to about 1980 (Cooke 1982) and of Thomas for that up to about 1986 (Thomas 1987). The following sections deal with the spin probes of Arata (1997) and, subsequently, the nucleotide fluorescent analogues of Bagshaw (1997), concisely and comprehensively reviewing not only nucleotide fluorescent analogues but also general fluorescence techniques and principles, including fluorescence resonance energy transfer (FRET), fluorescence quenching and so on. As to the FRET method, which is widely used in muscle research, the reviews by dos Remedios and colleagues (dos Remedios *et al.* 1987; dos Remedios and Moens 1995) and Miki *et al.* (1992) also are useful.

8.2 Methods and results

The fluorescence labelling method is not complicated. The most important point is incorporating the label at the specific site (amino acid side chains for covalent bonds or the micro environment for non-covalent bonds) on the protein by selecting environmental conditions (pH, temperature, ionic strength, ion species and so on) suitable for the labelling reaction (for the actual labelling method for each probe, see references below and the next section, also Arata 1997).

The most difficult challenge is maintaining the physiological functions of the labelling proteins; functions are modified to some extent by labelling. It is thus important to characterize labelled proteins, for comparison with unlabelled proteins, and then determine the limitations of interpreting the data obtained. To incorporate probes at a specific site in order to monitor structural changes directly coupled to physiological functions without modification of functions entails two seemingly contradictory tasks. We must overcome the disadvantages inherent in probe methods by developing innovative experimental methods and systems.

Here, I will describe examples of fluorescent probe methods applied to the muscle system: actin, myosin and regulatory proteins. Further elaboration on applications to the muscle fibre (fibril) system is provided below.

Actin

Phallodin, a toxin extracted from the mushroom *Amanita phalloides*, is a useful substance for monitoring and visualizing the structure of actin, even in association with other organized elements, because it is selectively incorporated into actin filaments (Wieland and Faulstich 1978; Wulf *et al.* 1979). Observation of single actin filaments under fluorescence microscopy was made possible by labelling with fluorescent-dye conjugated phalloidin (Yanagida *et al.* 1984) or stabilization of the polymer structure of actin with phalloidin, upon labelling with fluorescein isothiocyanate (FITC) (Honda *et al.* 1986) and other SH-reactive fluorescent probes. Recently, the location of the binding site of phalloidin was determined to be the boundary region of actin protomers (Lorenz *et al.* 1993). Applying a FRET method to fluoresceinisothiocyanatophalloidin-labelled and tetramethylrhodamineisothiocyanatophalloidin-labelled actin filaments, Heidecker *et al.* (1995) reported that the distance between adjacent phalloidins measured about 3.7 nm with a radial co-ordinate of 1.45 nm, which was consistent with the atomic model.

The only problem with incorporating the fluorescent-phalloidin into actin filaments in muscle fibres is that the binding is disturbed by nebulin (Yasuda *et al.* 1994) and probably some other proteins as well. In practice, in skeletal muscle, where nebulin is one of the constituents of thin filaments, the binding site of phalloidin is limited to the ends of the thin filaments (Yasuda *et al.* 1994) (see also Fig. 8.2). Some investigators have reported that prolonged incubation of phalloidin can induce specific binding to a broader thin filament region (Szczesna and Lehrer 1993; Bukatina *et al.* 1996). On the other hand, in cardiac muscle phalloidin immediately attaches to the entire thin filament, probably because nebulin is absent (Yasuda *et al.* 1994).

This technique was first applied to muscle fibres by Prochniewitz-Nakayama *et al.* (1983). Although they did not observe the binding region of FITC-phalloidin in sarcomeres, they showed that the concentration of incorporated FITC-phalloidin coincided with that of actin. Based on the results of fluorescence polarization of the attached FITC under rigor, relaxing and contracting conditions, they suggested that the orientation of actin molecules changes with the activation of muscle. The results were consistent with the flexibility changes in reconstituted thin filaments previously obtained by solution experiments (Fujime and Ishiwata 1971; Ishiwata and Fujime 1971, 1972). Here and below, the fluorescence polarization, *P*, is defined as follows, unless stated otherwise:

$$P_{\parallel} = ({}_{\parallel}\mathrm{I}_{\parallel} - {}_{\parallel}\mathrm{I}_{\perp})/({}_{\parallel}\mathrm{I}_{\parallel} + {}_{\parallel}\mathrm{I}_{\perp}),$$

$$P_{\perp} = ({}_{\perp}\mathrm{I}_{\perp} - {}_{\perp}\mathrm{I}_{\parallel})/({}_{\perp}\mathrm{I}_{\perp} + {}_{\perp}\mathrm{I}_{\parallel}),$$

where ${}_{\parallel}\mathrm{I}_{\parallel}$, ${}_{\parallel}\mathrm{I}_{\perp}$, ${}_{\perp}\mathrm{I}_{\perp}$, ${}_{\perp}\mathrm{I}_{\parallel}$ are the intensities of four components of polarization. The subscripts on the left and right, respectively, show the directions of polarization of

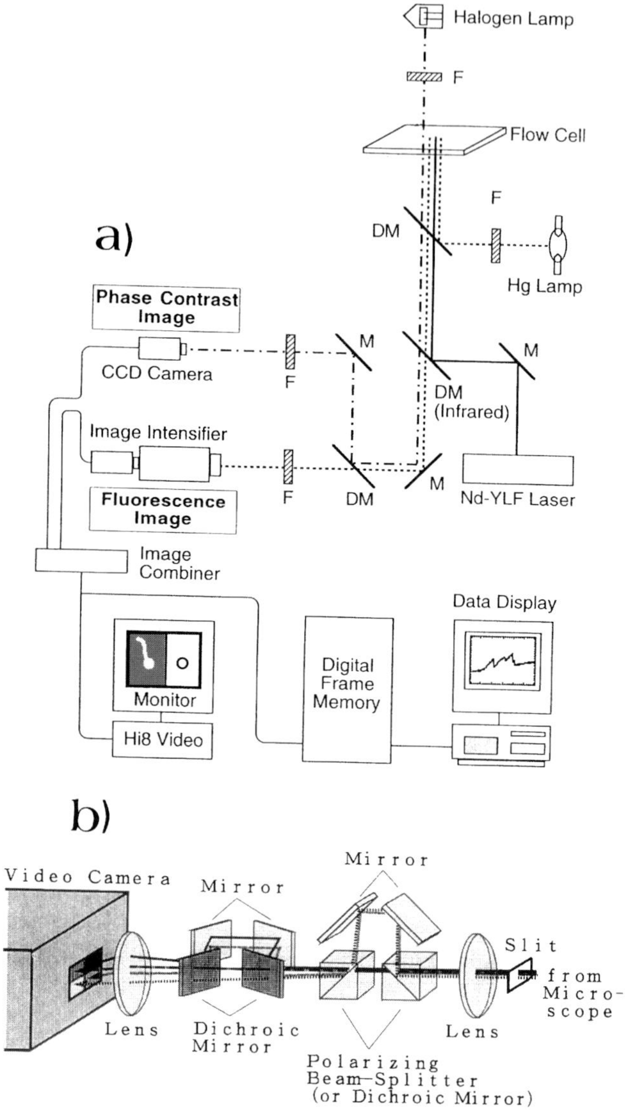

Fig. 8.1 Schematic illustration of a multi-view microscope equipped with optical tweezers. (a) Dual-view microscopic system (Kinosita *et al.* 1991) used for nanometer–piconewton measurements in an *in vitro* motility assay. Infrared laser light serves as the light source of the optical tweezers used to manipulate small particles. F, optical filter; DM, dichroic mirror; M, mirror; Flow Cell, sample stage. Taken from doctoral thesis of Nishizaka (1996). (b) Optical system for multi-view microscopy. The optical system shown at the *upper left* in (a) can be replaced by this system. Different images (four images in this illustration) can be observed simultaneously according to the direction of polarization or the wavelength. Taken from Suzuki *et al.* (1995).

incident and emitted light relative to the fibre axis; $\parallel$ and $\perp$, respectively, show the directions parallel and perpendicular to the fibre axis.

The same technique was used for detecting changes in the orientation of actin protomers in single actin filaments in an *in vitro* motility assay system (Kinosita *et al.* 1991). The fluorescence polarization of rhodamine-phalloidin attached to single actin filaments was measured with dual-view microscopy (Kinosita *et al.* 1991; Suzuki *et al.* 1995) (cf. Fig. 8.1). Here, the fluorescence polarization, P, was defined as $P = (V - H)/(V + H)$, where V and H, respectively, indicate the intensities of the vertical and horizontal polarization components of fluorescence obtained under isotropic excitation. No difference in the fluorescence polarization of rhodamine was detected between in rigor and sliding conditions. That is, the average orientation of rhodamine molecules attached to hundreds of actin protomers did not change upon activation.

Recently, Heidecker *et al.* (1995) used a FRET method to show that the distance between adjacent phalloidin binding sites does not change during sliding movement in and *in vitro* motility system, which supports our results (Kinosita *et al.* 1991).

Thus, there are two avenues of research which are inconsistent with each other; either the orientation of actin protomers changes or it does not change (or, more precisely, is undetectable) under different conditions. It is not clear at present whether this apparent discrepancy is attributable to differences among fluorescent dyes or experimental systems. This important point awaits clarification, as interest focuses on the role of structural changes in actin filaments in the molecular mechanism of contraction.

On the other hand, it recently became possible to image single fluorophores not only by fluorescence intensity (Funatsu *et al.* 1995; Sase *et al.* 1996) but also by fluorescence polarization (Sase 1996; Sase *et al.* 1997). Therefore, it is possible to image changes in the orientation of each actin protomer in single actin filaments. Sase *et al.* (1997) found that a single rhodamine-phalloidin rotates at a nearly constant rate accompanying sliding movement of an actin filament in an *in vitro* motility assay system (Sase 1996; Sase *et al.* 1997). This result supports the previous finding that a sliding actin filament, of which the front portion is fixed on a glass surface, forms a left-handed supercoil at the middle portion, suggesting that right-handed rotation around the filament axis occurs with sliding movement (Tanaka *et al.* 1992; Nishizaka *et al.* 1993).

The fluorescent dye-conjugated phalloidin could be used not only for studying structural changes in actin filaments as described above but also for monitoring changes in the local environment (such as pH, the concentrations of free ions, hydrophobicity, electrostatic potential and so on). Recently, we synthesized SNAFL-phalloidin for the purpose of monitoring and imaging the pH value around the phalloidin binding site (Kato *et al.* 1995). SNAFL (Haugland 1996) is a pH-sensitive dye, of which the excitation spectrum changes around pH 7.0. Figure 8.2 shows fluorescent images of skeletal myofibrils labelled with SNAFL-phalloidin in a rigor solution of different pH values (Kato *et al.* 1995; Kato 1996).

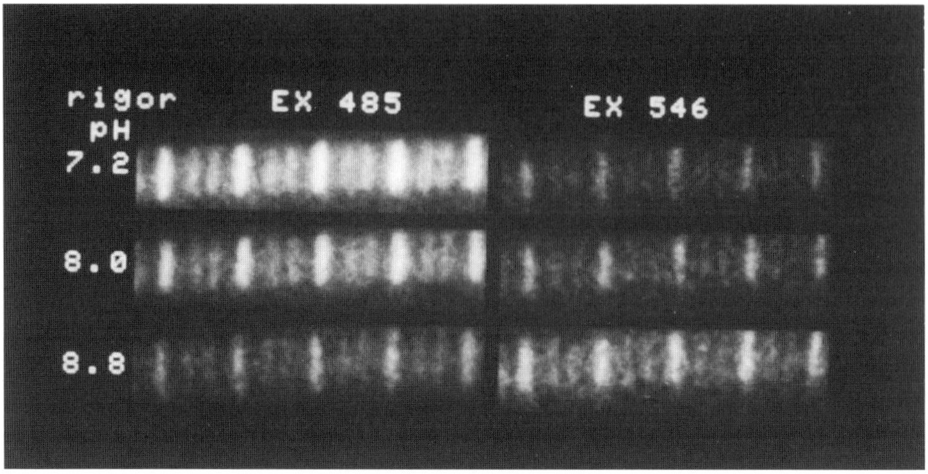

Fig. 8.2 Fluorescence images of a single skeletal myofibril obtained with SNAFL-phalloidin labelling (taken from Kato 1996). Only the ends of thin filaments are clearly labelled. The vertical lines with stronger and weaker fluorescence intensity, respectively, correspond to the Z line and the free ends of thin filaments. The images were taken at the excitation (Ex) wavelength of 485 nm (*left*) or 546 nm (*right*) and at an emission wavelength of about 590 nm. The pH value in the rigor solution was raised from 7.2 to 8.0 and finally to 8.8. As the pH value was raised, the fluorescence intensity at Ex. 485 nm decreased, whereas that at Ex. 546 nm increased. Thus, from the intensity ratio, the pH value around the phalloidin binding site in the myofibrils could be imaged and determined.

To apply dual-view microscopy and obtain real-time imaging of probe-labelled protein systems, it is convenient to use fluorescent probes of which the emission, but not excitation, spectrum is sensitive to the environment, allowing excitation at a single wavelength and imaging at two different wavelengths (Fig. 8.1a) or through vertical and horizontal polarization components (Fig. 8.1b). Recently, we custom-ordered Molecular Probes Inc. to synthesize SNARF-phalloidin of which the emission spectrum changes in response to pH changes. Experiments like those shown in Fig. 8.2 but using dual-view microscopy, are now possible.

Both association and dissociation rate constants were measured by monitoring changes in the fluorescence intensity of rhodamine phalloidin, which increases several-fold upon binding to actin filaments (De La Cruz and Pollard 1994). (Note: the binding constant of phalloidin with actin filaments is in the order of 10^7/M according to Haugland 1996). We found that both rate constants increased with increasing temperature, between 25 and 45 °C with an activation enthalpy in the order of 10^5 J/mol (Hotta and Ishiwata 1996). This approach, using environment-sensitive fluorescent dye-conjugated phalloidin, yields valuable information about the local environment of actin filaments.

Ando and associates recently examined the electrostatic potential around the phalloidin binding site using the 'diffusion-enhanced FRET' method (Yamamoto *et al.* 1994). For this purpose, they synthesized terbium (Tb)-diethylenetriamine-N,N,N′,N″,N″-Pentaacetic acid (DTPA)-phalloidin to serve as the fluorescent donor. They found that the electrical potential around the phalloidin binding site was negative and that this negative potential was slightly reduced by the binding of S1 (Ando *et al.* 1993).

The extent of modification of the structure and function of actin filaments by phalloidin labelling is not known; there is reportedly a tension–pCa curve shift to the left, i.e. the regulatory mechanism is modified by the attachment of phalloidin (Bukatina *et al.* 1996). We have confirmed that the maximum tension itself does not change appreciably with the binding of phalloidin to thin filament-reconstituted cardiac muscle fibres (Fujita *et al.* 1996).

As is well known, there is one reactive SH group in actin, i.e. Cys374, located near the actin filament surface (Holmes *et al.* 1990). Because SH-reactive reagents can be preferentially incorporated into Cys374, among the five cysteines in actin, the SH-reactive fluorescent probes have been widely used for specific labelling of actin molecules. Among these probes, N-(1-pyrene) iodoacetamide (PIA) appears to have been the most successful for monitoring structural changes in actin molecules, i.e. the fluorescence intensity is enhanced by more than 20-fold upon polymerization of actin (Kouyama and Mihashi 1981). Such an extraordinarily large change in fluorescence intensity is exceptional; this study is a typical demonstration of the remarkable potential uses of fluorescent probes.

In general, the fluorescence intensity of the probe incorporated into Cys374 is sensitive not only to the polymerization–depolymerization of actin (for tetramethylrhodamine-iodoacetamide (IATR) labelling, see Tait and Frieden 1982), but also to the interaction of actin filaments with myosin and regulatory proteins. For example, the fluorescence intensity of pyrene, incorporated into Cys374 of actin filaments, decreased to 1/3–1/5 with S1 binding (Geeves 1989). However, few studies taking this approach have been reported recently.

The labelling with PIA and N-((2-(iodoacetoxy)ethyl)-N-methyl)amino-7-nitrobenz-2-oxa-1, 3-diazole (IANBD) of actin (Ishiwata and Funatsu 1985) minimally affected polymerizability and the interaction with myosin. However, in general, the properties of actin filaments are modified to some extent with labelling at Cys374. Although rhodamine, for example, is a useful dye in the sense of slow photobleaching, large quantum yield and so on, tetramethylrhodamine-maleiamide (MTR) and IATR labelling reduces the polymerizability of actin. However, polymerizability is restored by co-polymerization with unlabelled actin (Ishiwata, unpublished observation). Needless to say, the attachment of phalloidin also stabilizes these fluorescent dye-labelled filaments, allowing application to an *in vitro* motility assay.

Other than Cys374, there are several sites of potential fluorescent probe labelling, such as Cys10, Gln41 (Kasprzak *et al.* 1988), and Lys61 and Tyr69 (see review, Miki *et al.* 1992). These sites have been used mainly as loci for FRET measurements. It has been shown that the polymerizability of actin was impaired by the labelling of, e.g. Lys61, but was restored with the addition of phalloidin (Miki 1987). The motion of subdomain 2 in actin filaments, in which Lys61 is located, was detected upon binding of S1, i.e. via changes in the distance between Lys61 and the adenine moiety (Miki and Kouyama 1994).

Thus, several studies have been done in solution but relatively few in an organized system that is relevant to the contraction mechanism. The main reason is that there has been no way to selectively incorporate the fluorescent dyes into a particular residue, such as Cys374, while maintaining the organized structure of muscle fibres or myofibrils. The only option, currently available, is to use a reconstitution system as described below in Section 8.4 on future prospects.

Myosin

Phallotoxin has no counterpart that specifically attaches to myosin molecules (or filaments) without modifying the structure and function of myosin. The most popular way to incorporate fluorescent dyes into myosin is the use of reactive cysteine residues called SH1 (Cys707) and SH2 (Cys697) present in the main chain (Reisler 1982) or SH groups in the light chain such as Cys177 on the alkali light chain (Trayer and Trayer 1988) or Cys127 and 156 (skeletal), and Cys108 (gizzard) on the regulatory light chain (LC2) of myosin.

There is an established method of labelling SH1 selectively (Reisler 1982). Figure 8.3 shows that both *N*-(iodoacetyl)-*N'*-(1-sulfo-5-napthyl)-ethylenediamine (5-(2-iodoacetylaminoethyl) aminonaphthalene-1-sulfonic acid (IAEDANS) and 5-(iodoacetamido) fluorescein (IAF) are incorporated into SH1 (Ishiwata, unpublished data). However, a serious problem exists in the labelling of SH1 and SH2, i.e. ATPase activities are to some extent modified (Root and Reisler 1992). As is well known, the K^+-ATPase activity of myosin is lost by labelling of SH1 (or SH2) with SH reagents such as N-ethylmaleimide (NEM) (Sekine and Kielley 1964; Seidel 1969). The actin-activated Mg^{2+} ATPase activity, which is essential for motor function, is also damaged and actin filaments remain attached to the labelled myosin even in the presence of ATP. Interestingly, these labelling effects are highly dependent on the label species. For most labels, however, the actin-activated Mg^{2+} ATPase is reduced so that the sliding movement of actin filaments is somewhat modified (Root and Reisler 1992). It would still be overly hasty to assert that SH1 (or SH2) labelling studies are useless. In fact, if an appropriate label is selected, important characteristics of intact myosin, for example, the dissociation from actin filaments on addition of ATP and binding affinity in the absence of ATP, are preserved. Using such labels, one can examine the dynamic structure of myosin in the attached and detached states.

In this respect, we selectively introduced eosin-maleimide into SH1 of myosin in myofibrils and studied the rotational motion of myosin heads in the presence

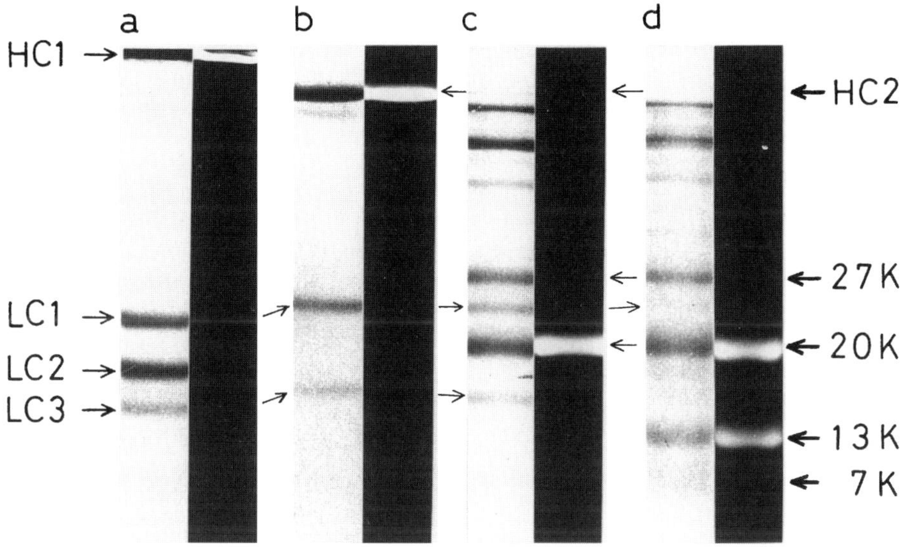

Fig. 8.3 An SDS PAGE pattern of SH1-labelled myosin and its derivatives separated on a 16% acrylamide gel. Approximately 10 μg of protein were loaded onto each lane. In each pair of photographs, the *left lane* shows proteins stained with Coomassie Brilliant Blue R, the *right line* shows fluorescence before staining. (a) Doubly-labelled HMM, where HC1, LC1, LC2 and LC3, respectively, indicate heavy chain and the three light chains of myosin (Ishiwata *et al.* 1996, 1997). The doubly-labelled HMM was prepared by labelling first with IAEDANS (fluorescence donor), then with IAF (fluorescence acceptor). (b) S1 separated from doubly-labelled myosin, where HC2 indicates heavy chain of S1. (c) Products of digestion of S1 (b) by trypsin at 25 °C. (d) Hydrolysis of S1 with hydroxylamine for 5.5 h at 45 °C (Sutoh 1981). 27 K, 20 K, 13 K and 7 K, respectively, indicate the 27, 20, 13 and 7 kDa fragments. The observation of fluorescence only at 13 kDa fragments indicates that the label was specifically incorporated into SH1 (Sutoh 1981). Similar results were reported by Miyanishi and Borejdo (1989) for the labelling of SH1 with IATR, or SH2 with MTR.

and absence of ATP using the phosphorescence anisotropic decay method (Ishiwata *et al.* 1987, 1988). Under rigor conditions, the anisotropy did not decay within milliseconds (at a minimum). This observation indicated that the rotational motion of myosin heads was suppressed. Under relaxing conditions, on the contrary, anisotropic decay was observed; the decay curve was decomposed into two exponentials having respective relaxation times of sub-microseconds and a few microseconds. We interpreted these results according to the double-cone model (Kinosita *et al.* 1977): the detached myosin heads rotate in sub-microseconds and, additionally, slow motion is superimposed because one end of the heads is connected to the movable neck portion. This result is not peculiar to a condensed system such as that of muscle fibres; similar results were obtained for myosin filaments in dilute solution (Eads *et al.* 1984; Kinosita *et al.* 1984; Ludescher *et al.* 1987; Mendelson and Wilson 1987; Ludescher and Thomas 1988).

To introduce eosin-maleimide specifically into SH1 in myofibrils, we adopted the method developed by Duke *et al.* (1976). After pre-treatment with NEM under rigor conditions, i.e. blocking reactive SH residues (such as Cys374 in actin) other than SH1 in advance, the label was introduced under relaxing conditions with pyrophosphate (Ishiwata *et al.* 1987); the same technique was previously used for spin-labelling (Ishiwata *et al.* 1986). We noticed the problem that ATPase activity under relaxing conditions tended to approach that under activating conditions, making relaxation difficult to achieve. The labelling method we used was the best available at that time but, at present, removable SH blocking reagents should be used instead of NEM. For example, SH-reactive reagents, such as thimerosal (Emoto *et al.* 1985), are recommended [recently, 5,5′-dithiobis-(2-nitrobenzoic acid) (DTNB) has also been used; see next section and Arata 1997]. The SH-reactive reagents can be removed after selective labelling using S–S bond reductants, such as dithiothreitol. The aforementioned side-effects will thus be avoided. Pre-treatment of S1 with thimerosal was recently adopted to block undesirable SH residues from reacting with the Au (gold) surface and to allow creation of a mono-layer of S1 molecules on the Au surface (Sasaki *et al.* 1997). Thus, the selective labelling method can be improved by choosing an appropriate labelling procedure and using different types of reagents.

On the other hand, contractile function is not damaged by LC2 labelling. Physiological functions may be essentially unchanged, despite the sliding velocities in an *in vitro* motility assay system being diminished by removal of, not only alkali light chains (LC1 and LC3), but also regulatory light chains (LC2) (Lowey *et al.* 1993). Caution should be exercised in applying the same technique to other muscle systems, such as smooth and molluscan muscle, in which regulatory light chains are especially important for regulation.

It is now believed that the tip, i.e. the SH1 (Thomas *et al.* 1995), and nucleotide binding regions (Yanagida 1981; Tanner *et al.* 1992) of the myosin head do not rotate during force generation; rather, it may be that the neck portion of the head changes its orientation. For these two reasons, research interest has focused on how to detect changes in the orientation of the neck portions of myosin heads.

The LC2 of chicken gizzard muscle contains only one SH residue (Cys108), although there are two SH residues (Cys127 and 156) in the LC2 of rabbit skeletal muscle. Thus, a new method has been developed, in which the intrinsic LC2 or rabbit skeletal muscle fibres (or fibrils) is substituted for the extrinsic chicken gizzard LC2 of which Cys108 has been labelled beforehand. Recently, Irving *et al.* (1995) expressed wild-type chicken gizzard LC2 in *Escherichia coli* and labelled its Cys108 with 6-IATR, thereby measuring the polarization (Q) of the label under contraction, relaxation and rigor conditions, to investigate changes in response to quick release and stretch of the fibre. They found that, during contraction, $Q_{\parallel}$ increased and $Q_{\perp}$ decreased simultaneously on quick release of the fibre, whereas $Q_{\parallel}$ decreased and $Q_{\perp}$ increased simultaneously on quick stretch of the fibre (definitions of $Q_{\parallel}$ and $Q_{\perp}$ differ slightly from those of $P_{\parallel}$ and $P_{\perp}$). On the other hand, with rigor, the direction of polarization change was opposite that of contraction. The orientation of the fluorescence dipole was estimated according to a model calculation (Irving 1996). They concluded that, 'the average probe orientations in the two states are on opposite sides of the plane at 90° to the filament axis' (see Fig. 4 in Irving *et al.* 1995). This appears to provide direct evidence for the tilting mechanism of force generation. The only problem is that the estimated degree of the tilting angle, i.e. less than 3° on average, is too small to account for the angle predicted in cross-bridge tilting models.

As pointed out by A.F. Huxley (1995), 'there are still many uncertainties in the interpretation', an ever-present feature of fluorescence polarization methods. First the orientation of the dipole relative to the long axis of myosin heads is unknown; second, myosin heads may have different orientations and so on.

More recently, the above study has been extended by combining X-ray diffraction with fluorescence microscopic observation (Ling *et al.* 1996) and flash photolysis of caged-ATP and ADP (Allen *et al.* 1995, 1996). The changes in LC2 orientation observed under different conditions could not be detected when sarcomeres were stretched up to 4.0 μm, where there was no overlap between thick and thin filaments (Ling *et al.* 1996).

Some of the difficulties recognized in the above studies might be overcome, if the orientation of a single fluorophore could be followed over time in the fibre system. This is not an unrealistic idea, because the first step in such an approach has already been taken in an *in vitro* motility assay system (Sase 1996; Sase *et al.* 1997).

Instead of labelling LC2, the conventional method of monitoring changes in the orientation of SH1 and SH2 regions in muscle fibres (or fibrils) has recently been re-examined by measuring fluorescence polarization (Cooke 1982; Thomas 1987). The time courses of changes in polarization and fluorescence intensities of labels incorporated into SH1 in skeletal muscle fibres were examined together with that of tension development following flash photolysis of caged nucleotides (Tanner *et al.* 1992; Berger *et al.* 1995, 1996). Tanner *et al.* found that the orientation of rhodamine fluorophores changed before force generation. However, the IATR they used was a mixture of 5 and 6 isomers, such that the estimated values of polarization contained some ambiguity. Furthermore, these isomers appear to have different reactivities to the labelling sites in muscle fibres (Ajtai *et al.* 1992). Accordingly, Berger *et al.* (1995, 1996) re-examined the same properties using either the 5 or the 6 isomer (Corrie and Craik 1994) and confirmed the previous results. They suggested that the observed changes may be related to nucleotide binding but not cross-bridge rotation because the fluorescence changes occurred before cross-bridge detachment.

Recently, Thomas *et al.* (1995) re-examined the time-resolved phosphorescence anisotropic decay technique to refine earlier results and have confirmed the microsecond rotational motions of the cross-bridges in SH1-labelled or LC2-labelled muscle fibres and fibrils. Erythrosin iodoacetamide and eosin maleimide were again used as the phosphorescent dyes. They recognized disordered rotational motions of both portions, instead of the ordered tilting motions hypothesized in the classical cross-bridge model, and have proposed that sliding force is generated in association with a disorder–order transition in the rotational motion of myosin heads.

Borejdo and associates used a fibre system to arrange the orientation of labels incorporated into either SH1 or SH2 of S1: the labelled S1 was diffused into glycerinated fibres and the orientation of labels was assessed based on the absorption dichroism (Borejdo and Burghardt 1987). The absorption dipole of IATR on SH1 was oriented at 65° with respect to the fibre axis under rigor conditions and at 51° in the presence of MgADP, in agreement with previous results, while that of MTR on SH2 was oriented at 42° regardless of MgADP, suggesting that the two SH sites behave differently (Miyanishi and Borejdo 1989).

The fluorescent dyes incorporated into SH1 and SH2 residues are sensitive to environmental changes. For example, Hiratsuka has shown that the binding of nucleotides increases the fluorescence intensity of 2-(4′-maleimidylanilino)-naphthalene-6-sulfonic acid (MIANS) attached to SH2 (Hiratsuka 1992*a*), while decreasing that of 4-fluoro-7-sulfamoylbenzofurazan (ABDF) attached to SH1 (Hiratsuka 1993). This suggests that the SH2 region moves toward the surface, the less hydrophobic region, whereas the SH1 region moves toward the interior of the protein. Fluorescence quenching studies with acrylamide and KI supported this conclusion (Hiratsuka 1992*a*, 1993). Measuring the fluorescence intensities of the IAF-labelled SH1 and IANBD-labelled SH2 in S1 showed that both sites behave differently depending on the species of bound nucleotides and phosphate analogues (Phan *et al.* 1996). It has thus been suggested that the SH1 and SH2 regions play essential roles in the communication pathway between nucleotide and actin binding sites (Botts *et al.* 1989; Berger *et al.* 1995, 1996).

Also, both SH1 and SH2 have been used extensively for distance measurements with FRET, especially under rigor conditions, because the capacity of rigor complex formation is maintained, allowing the assumption that the geometrical relationship is not greatly modified by the labelling. For an extensive review, see dos Remedios and Moens (1995) and Miki *et al.* (1992). Herein, I will briefly discuss our recent studies (Ishiwata *et al.* 1996, 1997).

We have used FRET to measure the distances between probes attached to SH1 on myosin (HMM), between those on adjacent S1s and between those on an HMM head and S1, and so on, when attached to F-actin under rigor conditions. The SH1 was specifically labelled with either IAEDANS employed as a fluorescent energy donor (D) or with IAF as an acceptor (A). We confirmed that, in free solution, the distance between SH1s on two HMM heads was too large to be measured by FRET. However, in the rigor complex with F-actin, the intermolecular distances between S1 molecules (S1(D) and S1(A)) and the heads of

HMM molecules (HMM(D) and HMM(A)) were determined to be 6.0–6.3 nm, whereas the intramolecular inter-head distance of HMM determined using doubly-labelled HMM (D and A) was larger than the values obtained above. We interpreted this to mean that the binding constant of the rigor bond is diminished for the second head of doubly-labelled HMM (D and A) molecules due to heterogeneous chemical modification of the SH1 groups. In any case, this is apparently the first report estimating the intramolecular distance between two HMM heads in complex with F-actin by spectroscopy.

Intrinsic fluorophores, such as the tryptophan residue, can be used as a fluorescence donor for FRET measurements. Hiratsuka (1992*b*) has shown that fluorescence energy is transferred from intrinsic tryptophan to MIANS attached to SH2, indicating a distance of 2.7 nm. He proposed that Trp510 is the most probable candidate for this energy transfer. Such an inspection, based on the results of fluorescence labelling techniques, must be reconciled with the three-dimensional structure of S1 (Rayment *et al.* 1993*a*) and the acto-S1 complex (Rayment *et al.* 1993*b*; Schröder *et al.* 1993).

Are there any other sites or techniques involving fluorescence labelling? As a possible means of overcoming limitations, site-directed mutagenesis may open a new approach to incorporating probes into specific sites of proteins. Saraswat and Lowey have succeeded in inserting cysteine residues into LC2, which are reactive to SH reagents (Saraswat and Lowey 1991). Very recently, Hopkins *et al.* (1996) attempted to detect tilting motions in the LC2 region by labelling five different SH sites introduced at positions 2, 73, 94, 126 or 155 by genetic engineering. Although the results were similar to those described above, such trials will ultimately yield valuable results.

Regulatory proteins

The conformational changes in TnC accompanying Ca^{2+} binding have been examined using labels incorporated into either Cys84 or Cys35 (Ingraham and Hodges 1988; Fuchs *et al.* 1989; Dong and Cheung 1996). It was suggested that Cys84 of cardiac TnC (cTnC), into which 2-(4′-(iodoacetamido)anilino)naphthalene-6-sulfonic acid (IAANS) or 6-acrylol-2-(dimethylamino)naphthalene (acrylodan) had specifically been incorporated, became externalized upon Ca^{2+} binding (Dong and Cheung 1996).

Miki measured the distance between Cys133 in troponin-I (TnI) and Lys61 in actin in reconstituted thin filaments, of which the filament structure had been stabilized with phalloidin (Miki 1990; Miki and Iio 1993). Either IAEDANS or 4-acetamido-4′-maleimidylstilbene-2,2′-disulfonic acid (AMS) was labelled at Cys133 in TnI, as a fluorescent donor, and FITC was labelled at Lys61 in actin, as a fluorescent acceptor. The kinetics of structural changes in the reconstituted thin filaments, which occurred co-operatively at a pCa of 6.25, were examined based on changes in the distance between TnI and actin by combining the FRET technique with a stopped-flow method (Miki and Iio 1993).

As to selective labelling of regulatory proteins in muscle fibres, the replacement of troponin and troponin-C (TnC) with the labelled counterparts is frequently used because selective removal and reconstitution of these proteins are relatively easy, with reliability similar to that of LC2 (see review of Moss 1992). Replacement is achieved by incubating the fibre or fibrils in rigor solutions without divalent cations. Accompanying this treatment, a portion of the LC2 molecules usually dissociates, necessitating reconstitution of LC2 together with troponin (or TnC).

For example, TnC labelled with 5-dimethylaminonaphthalene-1-sulfonyl aziridine was exchanged for TnC in muscle fibre and the structural changes upon Ca^{2+} binding were studied (Zot and Potter 1987). The same label was used for Met25 in TnC by Allen *et al.* (1992) to distinguish the bindings of Ca^{2+} to high and low affinity binding sites; they confirmed that muscle contraction was triggered by TnC structural changes that occured upon Ca^{2+} binding to the low affinity binding site. Hannon *et al.* (1992) introduced cTnC, in which Cys84 had been primarily labelled and Cys35 partly labelled with IAANS (sensitive to hydrophobicity), into cardiac or skeletal muscle fibres. The correlation between changes in fluorescence intensity and tension development (reflecting strong cross-bridge attachment) was examined. Liou and Fuchs (1992) examined cardiac myofibrils in which two Cys residues in TnC had been labelled with N-(1-pyrene)maleimide (PMA) or N-(1-pyrene)iodoacetamide (PIA).

As to tropomyosin, Cys190 is the usual labelling site (Ishii and Lehrer 1991). Either PIA (Ishii and Lehrer 1990), PMA (Ishii and Lehrer 1987, 1991) or 1,5-IAEDANS (Borovikov *et al.* 1993) can be labelled to detect local conformational changes around the SH residue of tropomyosin. Stopped-flow experiments have shown that structural changes occur in the tropomyosin–troponin complex within several tens of milliseconds after Ca^{2+} binding. The excimer fluorescence of PIA incorporated into Cys190 of two chains of skeletal tropomyosin molecules has allowed monitoring of intramolecular chain–chain interactions in reconstituted thin filaments (Ishii and Lehrer 1990).

Physico-chemical studies on regulatory proteins in which SH groups are genetically mutated started several years ago (for TnC, see Grabarek *et al.* 1990 and Fujimori *et al.* 1990 and for a review of TnC studies, see Grabarek *et al.* 1992). Genetically engineered regulatory proteins have since been used for fluorescence labelling of TnC and tropomyosin. For example, Zhao *et al.* (1995) have used a genetically introduced tryptophan residue of troponin-T (TnT) fragments as a fluorescent donor and dansyl aziridine or IAEDANS attached to TnC or TnI, in which a single Cys was genetically introduced at positions 9, 96 or 117, as a fluorescent acceptor site. They found that the distance between these sites in the complex of TnT–TnC or TnT–TnI changes upon binding of Ca^{2+} or Mg^{2+} to TnC. On the other hand, Dong and colleagues (Dong *et al.* 1996; Dong and Cheung 1996) prepared a mono-cysteine mutant of cardiac TnC, which contains only Cys84 (Cys35, one of two intrinsic cysteines, had been replaced with Ser), and labelled Cys84 with IAANS. They investigated the kinetics of the conformational changes in the regulatory domain of TnC, which accompany the association and dissociation of Ca^{2+} and Mg^{2+}.

8.3 Problems and limitations

The problems and limitations of the fluorescent probe method are the same, in many respects, to those of the spin probe method described in the next section (Arata 1997). The weakness of the fluorescent probe method is the difficulty of incorporating a bulky, hydrophobic probe into the functionally important site or, at least, a region close to that site, in order to monitor the structural changes tightly coupled to function. Chemical modification of the functional site inevitably affects the function being studied. Thus, the incorporation of probes into the functionally important site is, in principle, incompatible with the requirement that the physiological function not be modified.

In contrast to the spin probe method, the sensitivity of the fluorescence probe method is extremely high, such that the probe concentration necessary for measurements is several orders of magnitude smaller. In the fluorescence method, imaging and spectroscopy at the single fluorophore level are now possible (Funatsu *et al.* 1995, 1996; Sase 1996; Sase *et al.* 1995, 1997). The fluorescence intensity (spectrum) usually reflects only local conformational and/or environmental changes around the labelling site, i.e. secondary changes, such that the interpretation becomes sometime difficult. Thus, the fluorescence probe method needs many-sided control experiments.

To monitor the rotational motions (orientation changes) of the labelled protein using fluorescence (phosphorescence) (de)polarization measurements (either static or dynamic), rigid binding is essential. The orientation of the probes against the protein is also important. In such measurements, however, the labelling site can be a physiologically insignificant site, such as the SH residues in LC2. If the tilting motion of myosin heads in actually essential to the molecular mechanism of muscle contraction, one should be able to detect a sizeable rotational angle. It is ironic that the largest problem in the fluorescent probe approach to muscle research is that no probe has yet been found which reflects structural changes essential for motor function. This critical breakthrough in muscle research must be made.

One must be cautious in that probes usually contain hydrophobic residues, making dissolution in aqueous solution difficult (especially, at high probe concentrations) and forming non-specific complexes (especially, in muscle fibres and fibrils). The fluorophores frequently form dimers in solution or at the binding site (Ajtai *et al.* 1992), which may produce excimer fluorescence.

It is also possible for probes to be incorporated into binding sites non-randomly, i.e. the binding at one site may affect the affinity of subsequent binding to adjacent sites. In fact, there is a report suggesting that the binding of phalloidin to actin filaments is co-operative (Moens *et al.* 1994). Similarly, one must be careful when labelling two myosin heads, and in the exchange of LC2 and TnC. Whether or not co-operative binding occurs should be examined in each case. Imaging and spectroscopy of single fluorophores (Funatsu *et al.* 1995, 1996; Sase *et al.* 1995, 1997; Sase 1996) will overcome some of these difficulties.

8.4 Future prospects

As described above, various fluorescent probe techniques have been developed. How can we optimally apply these techniques to future muscle research? I consider the next advance in the application of probe methods to depend on the development of reconstitution techniques for the muscle (motor) contractile system. Needless to say, the reconstituted *in vitro* motility assay systems widely used nowadays are powerful investigative tools. On the other hand, in the contractile system, the incorporation of labelled proteins is limited to a few cases, such as TnC and LC2. In this sense, our thin(actin)-filament reconstitution system is attractive (Funatsu *et al.* 1994; Fujita *et al.* 1996). Labelled actin and regulatory proteins (even genetically engineered) can be incorporated into thin filament-free fibres (fibrils), allowing measurement of the contractile function and, at the same time, monitoring of the dynamics of labelled proteins and the local environment of labelling sites. Unfortunately, we have not yet devised an appropriate way to reconstitute thick (myosin) filaments in the contractile apparatus using purified proteins (Ishiwata 1980).

On the other hand, phalloidin is a unique substance: the binding to actin is highly specific, such that the aforementioned reconstitution of actin filaments is not required for incorporating into specific muscle fibre sites. Furthermore, many useful fluorescent-probe conjugates have been synthesized. Unfortunately, monitoring structural changes closely related to the molecular mechanism of contraction has not yet been achieved. I look forward to the development of a new fluorescent phallotoxins, which will reveal essential information on the structure of actin.

Finally, I would like to point out that the future of fluorescent probe methods is still bright. Chemical techniques allowing incorporation of fluorescent probes into specific sites have not yet reached an impasse. Genetic engineering has just been introduced to this field. Innovative fluorescent probes, e.g. temperature-sensitive (Kato *et al.* 1997), stress-sensitive and specific-ion sensitive probes, are anticipated. I also expect that new techniques will be devised that allow different protein states to be distinguished, monitoring of ion flow in organized structures, and so on. Moreover, we can expect technological advances in optics, microscopy, image-processing and high-speed cameras suitable for detecting the properties of individual fluorophores. The combination of these technologies will make it possible to detect the key event, which must occur in a single molecule, underlying muscle contraction and its regulation.

8.5 Conclusion

In this short review, I have focused on a limited range of topics pertaining to the application of fluorescent probe techniques. Although the use of fluorescent probes has inevitable problems, I believe that the potential of these techniques has not yet been fully developed. The future possibilities of innovative techniques and applications remain to be explored.

Acknowledgments

I acknowledge my colleagues, especially, Mr H. Kato, for permitting the use of unpublished data; unpublished data described herein were obtained mainly in my laboratory and partly in collaboration with other laboratories. A part of the unpublished data was obtained in the Advanced Research Laboratory of the Graduate School of Science and Engineering. Waseda University. I thank Dr M. Miki of Fukui University and Dr Y. Ishii of ERATO for their valuable discussions, and Dr H. Sugi (editor) of Teikyo University for his encouragement in completing this manuscript. I also thank Drs C.R. Bagshaw and T. Arata for having sent me their preprints, which were very useful in preparing this manuscript. Finally, I thank Dr K. Kinosita, Jr. for his continuous co-operation of fluorescence techniques. The unpublished studies described herein were partly supported by Grants-in-Aid for Scientific Research from the Ministry of Education, Science, Sports and Culture of Japan and CREST.

References

Ajtai, K.P., J.K. Ilich, A. Ringler, S.S. Sedarous, D.J. Toft and T.P. Burghardt (1992) Stereospecific reaction of muscle fiber proteins with the 5′ or 6′ isomer of (iodoacetamido) tetramethylrhodamine. *Biochemistry* **31**, 12 431–40.

Allen, T.S.C., L.D. Yates and A.M. Gordon (1992) Ca^{2+}-dependence of structural changes in troponin-C in demembranated fibers of rabbit psoas muscle. *Biophys. J.* **61**, 399–409.

Allen, T.S.C., C. Sabido-David, N. Ling, M. Irving and Y.E. Goldman (1995) Transients of fluorescence polarization in skeletal muscle fibers labeled with rhodamine on the regulatory light chain. *Biophys. J.* **68**, 81s–86s.

Allen, T.S.C., N. Ling, M. Irving and Y.E. Goldman (1996) Orientation changes in myosin regulatory light chains following photorelease of ATP in skinned muscle fibers. *Biophys. J.* **70**, 1847–62.

Ando, T., N. Kobayashi and E. Munekata (1993) Electrostatic potential around actin. *Adv. Exp. Med. Biol.* **332**, 361–76.

Arata, T. (1997) The use of spin probes. In this volume, Chapter 9, pp. 223–40.

Bagshaw, C.R. (1997) The use of fluorescent nucleotides. In this volume, Chapter 4, pp. 91–132.

Berger, C.L., J.S. Craik, D.R. Trentham, J.E.T. Corrie and Y.E. Goldman (1995) Fluorescence polarization from isomers of tetramethyl rhodamine at SH-1 in rabbit psoas muscle fibers. *Biophys. J.* **68**, 78s–80s.

Berger, C.L., J.S. Craik, D.R. Trentham, J.E.T. Corrie and Y.E. Goldman (1996) Fluorescence polarization of skeletal muscle fibers labeled with rhodamine isomers on the myosin heavy chain. *Biophys. J.* **71**, 3330–43.

Borejdo, J. and T.P. Burghardt (1987) Cross-bridge order and orientation in resting and active muscle fibers studied by the linear dichroism of fluorescence. In *Optical studies of muscle cross-bridges* (ed. R.J. Baskin and Y. Yeh), pp. 99–121. CRC Press, Boca Raton, FL.

Borovikov, Y.S., E. Nowak, M.I. Khoroshev and R. Dabrowska (1993) The effect of Ca^{2+} on the conformation of tropomyosin and actin in regulated actin filaments with or without bound myosin subfragment 1. *Biochim. Biophys. Acta* **1163**, 280–6.

Botts, J., J.F. Thomason and M.F. Morales (1989) On the origin and transmission of force in actomyosin subfragment 1. *Proc. Natl. Acad. Sci.* (*USA*) **86**, 2204–8.

Bukatina, A.E., F. Fuchs and S.C. Watkins (1996) A study on the mechanism of phalloidin-induced tension changes in skinned rabbit psoas muscle fibres. *J. Musc. Res. Cell Motil.* **17**, 365–71.

Cantor, C.R. and P.R. Schimmel (1980) *Biophysical chemistry*. W.H. Freeman, San Francisco.

Cooke, R. (1982) Fluorescence as a probe of contractile system. *Meth. Enzymol.* **85**, 574–93.

Corrie, J.E.T. and J.S. Craik (1994) Synthesis and characterisation of pure isomers of iodoacetamidotetramethyl rhodamine. *J. Chem. Soc. Perkin Trans.* **1**, 2967–73.

De La Cruz, E.M. and T.D. Pollard (1994) Transient kinetic analysis of rhodamine phalloidin binding to actin filaments. *Biochemistry* **33**, 14 387–92.

Dong, W.-J. and H.C. Cheung (1996) Calcium-induced conformational change in cardiac troponin C studied by fluorescence probes attached to Cys-84. *Biochim. Biophys. Acta* **1295**, 139–46.

Dong, W.-J., S.S. Rosenfeld, C.-K. Wang, A.M. Gordon and H.C. Cheung (1996) Kinetic studies of calcium binding to the regulatory site of troponin C from cardiac muscle. *J. Biol. Chem.* **271**, 688–94.

dos Remedios, C.G. and P.D.J. Moens (1995) Actin and the actomyosin interface: a review. *Biochim. Biophys. Acta* **1228**, 99–124.

dos Remedios, C.G., M. Miki and J.A. Barden (1987) Fluorescence resonance energy transfer measurements of distances in actin and myosin. A critical evaluation. *J. Musc. Res. Cell Motil.* **8**, 97–117.

Duke, J., R. Takashi, K. Ue and M.F. Morales (1976) Reciprocal reactivities of specific thiols when actin binds to myosin. *Proc. Natl. Acad. Sci.* (*USA*) **73**, 302–6.

Eads, T.M., D.D. Thomas and R.H. Austin (1984) Microsecond rotational motions of eosin-labeled myosin measured by time-resolved anisotropy of absorption and phosphorescence. *J. Mol. Biol.* **179**, 55–81.

Emoto, Y., T. Kawamura and K. Tawada (1985) Characterization of the ATPase active site in myosin subfragment-1 with the use of vanadate plus ADP as a reversible 'affinity-labeling' reagent: evidence for heterogeneity in the active sites. *J. Biochem.* **98**, 351–77.

Finer, J.T., R.M. Simmons and J.A. Spudich (1994) Single myosin mechanics: piconewton forces and nanometer steps. *Nature* **368**, 113–19.

Fuchs, F., Y.M. Liou and Z. Grabarek (1989) The reactivity of sulfhydryl groups of bovine cardiac troponin-C. *J. Biol. Chem.* **264**, 20 344–9.

Fujime, S. and S. Ishiwata (1971) Dynamic study of F-actin by quasielastic scattering of laser light. *J. Mol. Biol.* **62**, 251–65.

Fujimori, K., M. Sorenson, O. Herzberg, J. Moult and F.C. Reinach (1990). Probing the calcium-induced conformational transition of troponin C with site-directed mutants. *Nature* **345**, 182–4.

Fujita, H., K. Yasuda, S. Niitsu, T. Funatsu and S. Ishiwata (1996) Structural and functional reconstitution of thin filaments in the contractile system of cardiac muscle. *Biophys. J.* **71**, 2307–18.

Funatsu, T., T. Anazawa and S. Ishiwata (1994) Structural and functional reconstitution of thin filaments in skeletal muscle. *J. Musc. Res. Cell Motil.* **15**, 158–71.

Funatsu, T., Y. Harada, M. Tokunaga, K. Saito and T. Yanagida (1995) Imaging of single fluorescent molecules and individual ATP turnovers by single myosin molecules in aqueous solution. *Nature* **374**, 555–9.

Funatsu, T., Y. Ishii, Y. Harada, A. Iwane, T. Wazawa, J. Watai *et al.* (1996) Imaging and spectroscopy of single bio-molecules. *Biophysics* (*Japanese*) **36** (Supplement), S216 (Abstract).

Geeves, M.A. (1989) Dynamic interaction between actin and myosin subfragment 1 in the presence of ADP. *Biochemistry* **28**, 5864–71.

Grabarek, Z., R.-Y. Tan, J. Wang, T. Tao and J. Gergely (1990) Inhibition of mutant troponin C activity by an intra-domain disulphide bond. *Nature* **345**, 132–5.

Grabarek, Z., T. Tao and J. Gergely (1992) Molecular mechanism of troponin-C function. *J. Musc. Res. Cell Motil.* **13**, 383–93.

Hannon, J.D., D.A. Martyn and A.M. Gordon (1992) Effects of cycling and rigor crossbridges on the conformation of cardiac troponin C. *Circ. Res.* **71**, 984–91.

Haugland, R.P. (1996) *Handbook of fluorescent probes and research chemicals*. Molecular Probes Inc., Eugene, OR.

Heidecker, M., Y. Yan-Marriott and G. Marriott (1995) Proximity relationships and structural dynamics of the phalloidin binding site of actin filaments in solution and on single actin filaments on heavy meromyosin. *Biochemistry* **34**, 11 017–25.

Hiratsuka, T. (1992*a*) Movement of Cys-697 in myosin ATPase associated with ATP hydrolysis. *J. Biol. Chem.* **267**, 14 941–8.

Hiratsuka, T. (1992*b*) Spatial proximity of ATP-sensitive tryptophanyl residue(s) and Cys-697 in myosin ATPase. *J. Biol. Chem.* **267**, 14 949–54.

Hiratsuka, T. (1993) Behavior of Cys-707 (SH1) in myosin associated with ATP hydrolysis revealed with a fluorescent probe linked directly to the sulfur atom. *J. Biol. Chem.* **268**, 24 742–50.

Holmes, K.C., D. Popp, W. Gebhard and W. Kabsch (1990) Atomic model of the actin filament. *Nature* **347**, 44–9.

Honda, H., H. Nagashima and S. Asakura (1986) Directional movement of F-actin *in vitro*. *J. Mol. Biol.* **191**, 131–3.

Hopkins, S.C., C. Sabido-David, L.D. Saraswat, Y.E. Goldman, M. Irving and S. Lowey (1996) Tilting motions of rhodamine probes at five different sites on the myosin regulatory light chain (RLC) following length steps in contracting muscle fibers. *Biophys. J.* **70**, A17.

Hotta, M. and S. Ishiwata (1996) Temperature dependence of association and dissociation kinetics of rhodamine phalloidin with actin filaments. *Biophysics* (*Japanese*) **36**(Supplement), S12.

Huxley, A.F. (1995) Crossbridge tilting confirmed. *Nature* **375**, 631–2.

Ingraham, R.H. and R.S. Hodges (1988). Effects of Ca^{2+} and subunit interactions on surface accessibility of cysteine residues in cardiac troponin. *Biochemistry* **27**, 5891–8.

Irving, M. (1996) Steady-state polarization from cylindrically-symmetric fluorophores undergoing rapid restricted motion. *Biophys. J.* **70**, 1830–5.

Irving, M., T.S.C. Allen, C. Sabido-David, J.S. Craik, B. Brandmeier, J. Kendrick-Jones *et al.* (1995) Tilting of the light-chain region of myosin during step length changes and active force generation in skeletal muscle. *Nature* **375**, 688–91.

Ishii, Y. and S.S. Lehrer (1987) Fluorescence probe studies of the state of tropomyosin in reconstituted muscle thin filaments. *Biochemistry* **26**, 4922–5.

Ishii, Y. and S.S. Lehrer (1990) Excimer fluorescence of pyrenyliodoacetamide-labeled tropomyosin: a probe of the state of tropomyosin in reconstituted muscle thin filaments. *Biochemistry* **29**, 1160–6.

Ishii, Y. and S.S. Lehrer (1991) Two-site attachment of troponin to pyrene-labeled tropomyosin. *J. Biol. Chem.* **266**, 6894–903.

Ishiwata, S. (1980) *Partial destruction and reconstruction of an ordered structure in striated muscle.* Sixth Taniguchi International Symposium on Biophysics, Santa City, The Taniguchi Foundation, pp. 262–93.

Ishiwata, S. and S. Fujime (1971) Effect of Ca^{2+} on dynamic properties of muscle proteins studied by quasielastic light scattering. *J. Phys. Soc. Japan* **31**, 1601.

Ishiwata, S. and S. Fujime (1972) Effect of calcium ions on the flexibility of reconstituted thin filaments of muscle studied by quasielastic scattering of laser light. *J. Mol. Biol.* **68**, 511–22.

Ishiwata, S. and T. Funatsu (1985) Does actin bind to the ends of thin filaments in skeletal muscle? *J. Cell Biol.* **100**, 282–91.

Ishiwata, S., B.A. Manuck, J.C. Seidel and J. Gergely (1986) Saturation transfer electron paramagnetic resonance study of the mobility of myosin heads in myofibrils under conditions of partial dissociation. *Biophys. J.* **49**, 821–8.

Ishiwata, S., K. Kinosita, Jr., H. Yoshimura and A. Ikegami (1987) Rotational motions of myosin heads in myofibril studied by phosphorescence anisotropy decay measurements. *J. Biol. Chem.* **262**, 8314–7.

Ishiwata, S., K. Kinosita, Jr., H. Yoshimura and A. Ikegami (1988) Optical anisotropy decay studies of the dynamic structure of myosin filaments. *Adv. Exp. Med. Biol.* **226**, 267–76.

Ishiwata, S., M. Miki, I. Shin, T. Funatsu, K. Yasuda and C.G. dos Remedios (1996) Distance between myosin heads in rigor complex measured by fluorescence resonance energy transfer. *Biophys. J.* **70**, A14.

Ishiwata, S., M. Miki, I. Shin, T. Funatsu, K. Yasuda and C.G. dos Remedios (1997) Inter-head distances in myosin attached to F-actin estimated by fluorescence energy transfer spectroscopy. *Biophys. J.* **73**, No. 2, in press.

Johnson, I.D. (1996) Introduction to fluorescence techniques. In *Handbook of fluorescent probes and research chemicals*, (ed. R.P. Haugland), pp. 1–6. Molecular Probes Inc., Eugene, Oregon.

Kasprzak, A.A., R. Takashi and M.F. Morales (1988) Orientation of the actin monomer in the F-actin filament: radial coordinate of glutamine-41 and effect of myosin subfragment-1 binding on the monomer orientation. *Biochemistry* **27**, 4512–22.

Kato, H. (1996) *Imaging of mechano-chemical coupling in the contractile system of muscle by using pH-sensitive fluorescent dyes.* Master thesis. Department of Physics, School of Science and Engineering. Tokyo, Waseda University.

Kato, H., M. Kawase, T. Nishizaka, K. Someya, G. Marriott, K. Kinosita, Jr. *et al.* (1995) Imaging of mechanochemical coupling of muscle contractile system by pH-sensitive fluorescent dyes. *J. Musc. Res. Cell Motil.* **16**, 337–8 (Abstract).

Kato, H., T. Nishizaka, T. Iga, K. Kinosita, Jr. and S. Ishiwata (1997) Thermal activation of actomyosin motors with temperature-pulse microscopy. *Biophys. J.* **72**, A183 (Abstract).

Kinosita, K. Jr., S. Kawato and A. Ikegami (1977) A theory of fluorescence polarization decay in membranes. *Biophys. J.* **20**, 289–305.

Kinosita, K. Jr., S. Ishiwata, H. Yoshimura, H. Asai and A. Ikegami (1984) Submicrosecond and microsecond rotational motions of myosin head in solution and in myosin synthetic filaments as revealed by time-resolved optical anisotropy decay measurements. *Biochemistry* **23**, 5963–75.

Kinosita, K. Jr., H. Itoh, S. Ishiwata, K. Hirano, T. Nishizaka and T. Hayakawa (1991) Dual-view microscopy with a single camera: real-time imaging of molecular orientations and calcium. *J. Cell Biol.* **115**, 67–73.

Kishino, A. and T. Yanagida (1988) Force measurements by micromanipulation of a single actin filament by glass needles. *Nature* **334**, 74–6.

Kouyama, T. and K. Mihashi (1981) Fluorimetry study of N-(1-pyrenyl) iodoacetamide-labeled F-actin. Local structural change of actin protomer both on polymerization and on binding of heavy meromyosin. *Eur. J. Biochem.* **114**, 33–8.

Kron, S.J. and J.A. Spudich (1986) Fluorescent actin filaments move on myosin fixed to glass surfaces. *Proc. Natl. Acad. Sci.* (*USA*) **83**, 6272–6.

Ling, N., C. Shrimpton, J. Sleep, J. Kendrick-Jones and M. Irving (1996) Fluorescent probes of the orientation of myosin regulatory light chains in relaxed, rigor and contracting muscle. *Biophys. J.* **70**, 1836–46.

Liou, Y.-M. and F. Fuchs (1992) Pyrene-labeled cardiac troponin C. Effect of Ca^{2+} on monomer and excimer fluorescence in solution and in myofibrils. *Biophys. J.* **61**, 892–901.

Lorenz, M., D. Popp and K.C. Holmes (1993) Refinement of the F-actin structure against X-ray fiber diffraction data by the use of a direct mutation algorithm. *J. Mol. Biol.* **234**, 826–36.

Lowey, S., G.S. Waller and K.M. Trybus (1993) Skeletal muscle myosin light chains are essential for physiological speeds of shortening. *Nature* **365**, 454–6.

Ludescher, R.D. and D.D. Thomas (1988) Microsecond rotational dynamics of phosphorescent-labeled muscle cross-bridges. *Biochemistry* **27**, 3343–51.

Ludescher, R.D., T.M. Eads and D.D. Thomas (1987) Triplet anisotropy studies of restricted rotational motion in myosin monomers and filaments. In *Optical studies of muscle cross-bridges*, (ed. R.J. Baskin and Y. Yeh), pp. 33–65. CRC Press, Boca Raton, FL.

Mendelson, R. and M.G.A. Wilson (1987) Fluorescence polarization studies of myosin and muscle cross-bridges. In *Optical studies of muscle cross-bridges*. (ed. R.J. Baskin and Y. Yeh) pp. 67–98. CRC Press, Boca Raton, FL.

Miki, M. (1987) The recovery of the polymerization of Lys-61-labeled actin by the addition of phalloidin. *Eur. J. Biochem.* **164**, 229–35.

Miki, M. (1990) Resonance energy transfer between points in a reconstituted skeletal muscle thin filament. *Eur. J. Biochem.* **187**, 155–62.

Miki, M. and T. Iio (1993) Kinetics of structural changes of reconstituted skeletal muscle thin filaments observed by fluorescence resonance energy transfer. *J. Biol. Chem.* **268**, 7101–6.

Miki, M. and T. Kouyama (1994) Domain motion in actin observed by fluorescence resonance energy transfer. *Biochemistry* **33**, 10 171–7.

Miki, M., S.I. O'Donoghue and C.G. dos Remedios (1992) Structure of actin observed by fluorescence resonance energy transfer spectroscopy. *J. Musc. Res. Cell Motil.* **13**, 132–45.

Miyanishi, T. and J. Borejdo (1989) Differential behavior of two cysteine residues on myosin head in muscle fibers. *Biochemistry* **28**, 1287–94.

Miyata, H., H. Hakozaki, H. Yoshikawa, N. Suzuki, K. Kinosita Jr., T. Nishizaka *et al.* (1994) Stepwise motion of an actin filament over a small number of heavy meromyosin molecules is revealed in an *in vitro* motility assay. *J. Biochem.* **115**, 644–7.

Miyata, H., H. Yoshikawa, H. Hakozaki, N. Suzuki, T. Furuno, A. Ikegami, K. Kinosita, Jr., T. Nishizaka and S. Ishiwata (1995). Mechanical measurements of single actomyosin motor force. *Biophys. J.* **68**, 286s–90s.

Moens, P.D.J., D.J. Yee and C.G. dos Remedios (1994) Determination of the radial coordinate of Cys-374 in F-actin using fluorescence resonance energy transfer spectroscopy: effect of phalloidin on polymer assembly. *Biochemistry* **33**, 13 102–8.

Moss, R. (1992) Ca^{2+} regulation of mechanical properties of striated muscle. Mechanistic studies using extraction and replacement of regulatory proteins. *Circ. Res.* **70**, 865–84.

Nishizaka, T. (1996) *Microscopic analysis of function and mechanical properties of single actomyosin motors with optical tweezers*. Doctoral thesis, p. 160. Department of Physics, School of Science and Engineering. Waseda University, Tokyo.

Nishizaka, T., T. Yagi, Y. Tanaka and S. Ishiwata (1993) Right-handed rotation of an actin filament in an *in vitro* motile system. *Nature* **361**, 269–71.

Phan, B.C., P. Cheung, W.F. Stafford and E. Reisler (1996) Complexes of myosin subfragment-1 with adenosine diphosphate and phosphate analogs: probes of active site and protein conformation. *Biophys. J.* **59**, 341–9.

Prochniewicz-Nakayama, E., T. Yanagida and F. Oosawa (1983) Studies on conformation of F-actin in muscle fibers in the relaxed states, rigor, and during contraction using fluorescent phalloidin. *J. Cell Biol.* **97**, 1663–7.

Rayment, I., W.R. Rypniewski, K. Schmidt-Base, R. Smith, D.R. Tomchick, M.M. Benning *et al.* (1993*a*) Three-dimensional structure of myosin subfragment-1: a molecular motor. *Science* **261**, 50–8.

Rayment, I., H.M. Holden, M. Whittaker, C.B. Yohn, M. Lorenz, K.C. Holmes *et al.* (1993*b*) Structure of the actin-myosin complex and its implications for muscle contraction. *Science* **261**, 58–65.

Reisler, E. (1982) Sulfhydryl modification and labeling of myosin. *Meth. Enzymol.* **85B**, 84–93.

Root, D.D. and E. Reisler (1992) Cooperativity of thiol-modified myosin filaments. ATPase and motility assays of myosin function. *Biophys. J.* **63**, 730–40.

Saraswat, L.D. and S. Lowey (1991) Engineered cysteine mutants of myosin light chain 2. Fluorescent analogues for structural studies. *J. Biol. Chem.* **266**, 19 777–85.

Sasaki, Y.C., K. Yasuda, Y. Suzuki, T. Ishibashi, I. Satoh, Y. Fujiki *et al.* (1997) Two-dimensional arrangement of a functional protein by cysteine-gold substrate interaction: Enzyme activity and characterization of a protein monolayer on a gold substrate. *Biophys. J.* **72**, 1842–8.

Sase, I. (1996) *Real time imaging of single fluorophores and single small cells with high-sensitive epifluorescence microscope*. Doctoral thesis, p. 156. Department of Physics, Faculty of Science and Technology. Keio University, Yokohama.

Sase, I., H. Miyata, J.E.T. Corrie, J. Craik and K. Kinosita, Jr. (1995) Real time imaging of single fluorophores on moving actin with an epifluorescence microscope. *Biophys. J.* **69**, 323–8.

Sase, I., H. Miyata, S. Ishiwata and K. Kinosita, Jr. (1997) Axial rotation of sliding actin filaments revealed by single-fluorophore imaging. *Proc. Natl. Acad. Sci.* (*USA*) **94**, 5646–50.

Schröder, R.R., D.J. Manstein, W. Jahn, H. Holden, I. Rayment, K.C. Holmes *et al.* (1993) Three-dimensional atomic model of F-actin decorated with *Dictyostelium* myosin S1. *Nature* **364**, 171–4.

Seidel, J.C. (1969) Similar effects on enzymic activity due to chemical modification of either of two sulfhydryl groups of myosin. *Biochim. Biophys. Acta* **180**, 216–9.

Sekine, T. and W.W. Kielley (1964) Enzymic properties of N-ethylmaleimide modified myosin. *Biochim. Biophys. Acta* **81**, 336–45.

Spudich, J.A. (1994) How molecular motors work. *Nature* **372**, 515–8.

Sutoh, K. (1981) Location of SH1 and SH2 along a heavy chain of myosin subfragment 1. *Biochemistry* **20**, 3281–5.

Suzuki, K., Y. Tanaka, Y. Nakajima, K. Hirano, H. Itoh, H. Miyata *et al.* (1995) Spatiotemporal relationships among early events of fertilization in sea urchin eggs revealed by multiview microscopy. *Biophys. J.* **68**, 739–48.

Szczesna, D. and S.S. Lehrer (1993) The binding of fluorescent phallotoxins to actin in myofibrils. *J. Musc. Res. Cell Motil.* **14**, 594–7.

Tait, J.F. and C. Frieden (1982) Polymerization-induced changes in the fluorescence of actin labeled with iodoacetamidotetramethylrhodamine. *Arch. Biochem. Biophys.* **216**, 133–41.

Tanaka, Y., A. Ishijima and S. Ishiwata (1992) Super helix formation of actin filaments in an *in vitro* motile system. *Biochim. Biophys. Acta* **1159**, 94–8.

Tanner, J.W., D.D. Thomas and Y.E. Goldman (1992) Transients in orientation of a fluorescent cross-bridge probe following photolysis of caged nucleotides in skeletal muscle fibers. *J. Mol. Biol.* **223**, 185–203.

Thomas, D.D. (1987) Spectroscopic probes of muscle crossbridge rotation. *Ann. Rev. Physiol.* **49**, 691–709.

Thomas, D.D., S. Ramachandran, O. Roopnarine, D.W. Hayden and E.M. Ostap (1995) The mechanism of force generation in myosin: a disorder-to-order transition, coupled to internal structural changes. *Biophys. J.* **68**, 135s–41s.

Trayer, H.R. and I.P. Trayer (1988) Fluorescence resonance energy transfer within the complex formed by actin and myosin subfragment 1. Comparison between weakly and strongly attached states. *Biochemistry* **27**, 5718–27.

Wieland, T. and T. Faulstich (1978) Amatoxins, phallotoxins, phallolysin, and antamanide: the biologically active components of poisonous Amanita mushrooms. *CRC Crit. Rev. Biochem.* **5**, 185–260.

Wulf, E., A. Deboben, F.A. Bautz, H. Faulstich and T. Wieland (1979) Fluorescent phallotoxin, a tool for the visualization of cellular actin. *Proc. Natl. Acad. Sci.* (*USA*) **76**, 4498–502.

Yamamoto, T., S. Nakayama, N. Kobayashi, E. Munekata and T. Ando (1994) Determination of electrostatic potential around specific locations on the surface of actin by diffusion-enhanced fluorescence resonance energy transfer. *J. Mol. Biol.* **241**, 714–31.

Yanagida, T. (1981) Angles of nucleotides bound to cross-bridges in glycerinated muscle fiber at various concentrations of e-ATP, e-ADP and e-AMP.PNP detected by polarized fluorescence. *J. Mol. Biol.* **146**, 539–60.

Yanagida, T., Y. Harada and A. Ishijima (1993) Nano-manipulation of actomyosin molecular motors *in vitro*: a new working principle. *TIBS* **18**, 319–24.

Yanagida, T., M. Nakase, K. Nishiyama and F. Oosawa (1984) Direct observation of motion of single F-actin filaments in the presence of myosin. *Nature* **307**, 58–60.

Yasuda, K., H. Fujita, Y. Fujiki and S. Ishiwata (1994) Length regulation of thin filaments without nebulin. *Proc. Jap. Acad.* **70**(Ser. B), 151–6.

Zhao, X., T. Kobayashi, H. Malak, I. Gryczynski, J. Lakowicz, R. Wade *et al.* (1995) Calcium-induced troponin flexibility revealed by distance distribution measurements between engineered sites. *J. Biol. Chem.* **270**, 15 507–14.

Zot, A.S. and J.D. Potter (1987) The effect of [Mg^{2+}] on the Ca^{2+} dependence of ATPase and tension development of fast skeletal muscle. The role of the Ca^{2+}-specific sites of troponin C. *J. Biol. Chem.* **262**, 1966–9.

9 The use of spin probes

Toshiaki Arata

9.1 Introduction

As already explained in Section 8.1, various spectroscopic probes have been developed to obtain information about changes in orientation and rotational motion of the contractile proteins, actin and myosin, during their production of force and motion associated with the actin-activated ATPase cycle of myosin, and have proved to be useful in constructing the contraction models (Morales *et al.* 1982; Cooke 1986; Thomas 1987). These probes are externally applied to demembranated muscle fibres, so that they can diffuse into the myofilament-lattice and attach to specific sites on the contractile proteins. This section aims at describing the recent development of electron paramagnetic resonance (EPR) techniques applied to demembranated muscle fibres. Some EPR studies on purified proteins are also mentioned in case the techniques can be applied to fibres in the future. Attention is focused on the comparison of the techniques with respect to their advantages, problems and limitations, but not on the details of each technique.

9.2 Electron paramagnetic resonance

Magnetic resonance spectroscopy has high spectral resolution, which detects different orientations or motions of probes in the steady-state spectrum (Berliner 1976; Jost and Griffith 1978). Nuclear magnetic resonance (NMR) has the advantage of no requirement of extrinsic probes, but its low sensitivity and low specificity have limited most NMR studies to qualitative studies on purified myosin or bound nucleotides. When the problems of specific and rigid spin-labelling and its system perturbation (dysfunction) are solved, electron magnetic resonance (EPR) provides the sensitivity and resolution to detect orientations and rotational motions in purified proteins or muscle fibre bundles.

Nitroxyl-type spin labels can sense orientation with respect to the applied magnetic field, $\mathfrak{H}$ ($\mathfrak{H}$ can be usually made parallel to the fibre axis). Figure 9.1 shows a series of spectra obtained at different molecular orientations. The interactions giving the EPR spectral lines are described in the spin-Hamiltonian operator,

$$\mathfrak{H} = \beta\mathfrak{S}\mathfrak{g}\mathfrak{H} + \mathfrak{S}\mathfrak{A}\mathfrak{I}. \tag{9.1}$$

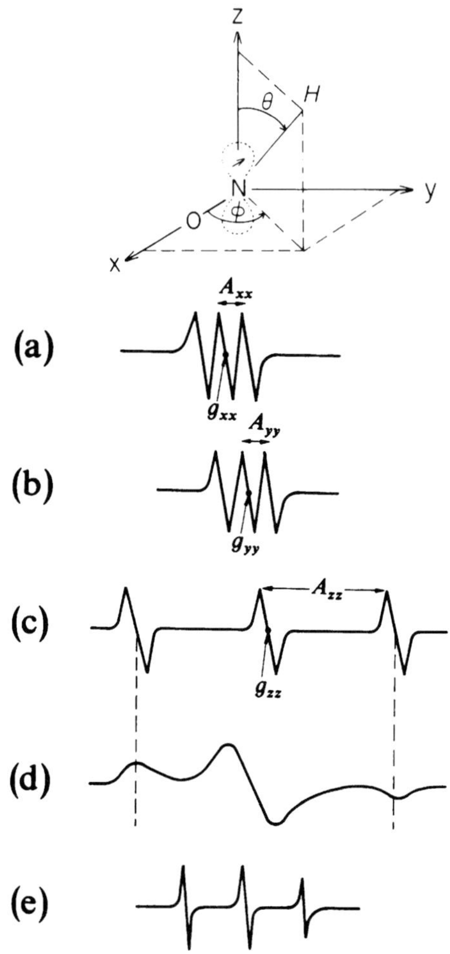

Fig. 9.1 Electron paramagnetic resonance spectra at 9.5 GHz (X band) of nitroxide free radicals. The molecular co-ordinate system is shown at the top. (a)–(c) Crystal spectra with the external magnetic field along the principal axes x, y, and z of the g and A tensors, respectively. (d) Spectrum of glassy powder. (e) Spectrum in aqueous solution.

The interaction between the electron spin, $\mathfrak{S}$, and $\mathfrak{H}$ is described by the tensor, $\mathfrak{g}$, and that between the nuclear spin, $\mathfrak{I}$, and $\mathfrak{S}$, by the tensor $\mathfrak{A}$. β is the Bohr magneton. These interactions give three energy levels corresponding to $I_z = -1$, 0, $+1$ for ^{14}N nucleus. The density of the unpaired electron is approximately cylindrically symmetric about an axis perpendicular to the plane of the label ring. We thus assume that the interaction can be described by cylindrically symmetric tensors, A_{xx} ($= A_{yy}$) and A_{zz} and g_{xx} ($= g_{yy}$) and g_{zz}. The magnetic field strength where resonance occurs is a function of I_z and the angle θ between H and the principal axis z of spin label:

$$H(\theta, I_z) = g_{zz} H(0,0)/g - I_z A, \tag{9.2}$$

where

$$g = (g_{xx}^2 \sin^2\theta + g_{zz}^2 \cos^2\theta)^{1/2}, \tag{9.3}$$

and

$$A = (A_{xx}^2 \sin^2\theta + A_{zz}^2 \cos^2\theta)^{1/2}. \tag{9.4}$$

The corresponding three peaks in the EPR spectrum are determined by θ. Because the assumption of cylindrical symmetry is not precise, in practice it is easier to fit experimental spectra by comparing with positions generated by substituting proposed θ-distributions into eqn 9.2. In general, we can simulate the whole spectrum by adding line shapes (Lorentzian function determined by the spin–spin relaxation time T_2) positioned at $H(\theta, I_z)$ over a wide range of θ (Barnett *et al.* 1986). Figure 9.2 shows examples of EPR spectra simulated by assuming a superimpose of two angular distributions with various spreads. In principle, if we know experimentally the whole EPR spectrum, it can be inverted to find the θ-distributions. We usually measure and analyse the spectra from parallel-oriented fibres. Perpendicular spectra are also measured to ensure a prefered orientation of probes. Random-oriented fibres are used to determine how the probes are immobilized on protein surface by comparing their spectra to a rigid limit spectrum obtained in the absence of motion (Berliner 1976). The rigid limit spectrum is obtained by treating samples (random-oriented) in viscous solution at low temperature or precipitated with ammonium sulphate, etc; its example is shown at the bottom in Fig. 9.2. The maximum splitting (between first maximum and last minimum) and its centre approximate $2A_{zz}$ and g_{zz} values, and the minimum splitting (between first minimum and last maximum) and its centre approximate $2A_{xx}(=2A_{yy})$ and $g_{xx}(=g_{yy})$ values. Precise computer determination of asymmetric tensors $\mathfrak{g}$ and $\mathfrak{A}$ from a spectrum of randomly oriented samples (rigid limit) enables us to find distributions of both tilt θ and twist ϕ angles without any assumption (see Ajtai *et al.* 1989; Fajer *et al.* 1990).

The fast isotropic molecular motion almost completely averages out in anisotropic terms, $g = (1/3)(g_{xx} + g_{yy} + g_{zz})$ and $A = (1/3)(A_{xx} + A_{yy} + A_{zz})$. When the viscosity of the solution is increased, the EPR lines gradually broaden. If the sample has either no or less macroscopic order, then the spectrum is determined almost extensively by nanosecond rotational motion, which narrows the lines. In a steady-state EPR experiment, a motionally narrowed spectrum is easily resolved from a broad spectrum that corresponds to no motion (slower than 0.1 μs motion) (Fig. 9.1d,e). Thus the conventional EPR experiment is useful for measuring rapid motions, such as intramolecular flexibility of polypeptide chain attached to spin

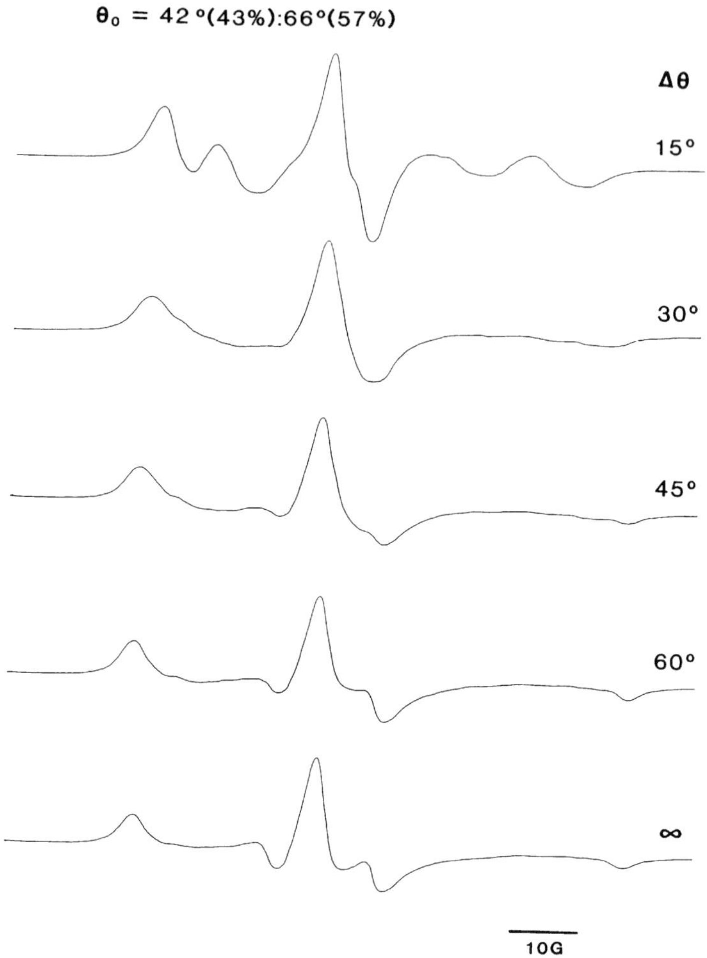

Fig. 9.2 Examples of composite EPR spectra simulated for two Gaussian angular distributions with various full widths of half maximum ($\Delta\theta$): θ is the angle between the direction of the applied magnetic field and the molecular z axis. Altogether, 43% of nitroxide spin labels have mean angle, $\theta_0 = 42°$ and 57% have $\theta_0 = 66°$. Spectra are first derivatives of a Lorentzian line with peak-to-peak line-width of 2.71 G using $g_{xx} = g_{yy} = 2.0074$, $g_{zz} = 2.0024$, $A_{xx} = A_{yy} = 7$ G, and $A_{zz} = 35$ G.

label. It can resolve conformational states of muscle proteins, such as the chemical intermediate state of myosin during ATPase reaction (Stone 1973; Seidel 1982; Barnett and Thomas 1987).

The range of rotational motion detected by EPR measurements has been greatly expanded by the development of saturation transfer spectroscopy. Many excellent reports have described the theory and applications in detail (Thomas *et al.* 1976; Hyde 1978; Hyde and Dalton 1979; Seidel 1982). This spectroscopy uses much more intense microwave radiation than conventional EPR and a different detection protocol (out-of-phase detection with larger field modulation). Saturation transfer depends on rotational diffusion of saturated spins on a time scale comparable to the spin-lattice relaxation time T_1; for nitroxides approximately 10 μs. Thus the saturation transfer EPR is useful for measuring the rotational motion of contractile protein in muscle fibres, while conventional EPR of oriented sample determines the amplitude of this motion. Low sensitivity and complicated detection protocol may limit its applications.

9.3 Spin labels

The probes used in EPR studies of muscle proteins are usually nitroxyl-type spin labels, stable free radicals that absorb microwaves when they are placed in a magnetic field. Most myosin EPR studies have been done on maleimide (MSL, 4-maleimido-2,2,6,6-tetramethyl-1-piperidinyloxy or 4-maleimido-TEMPO) or iodoacetamide derivatives (IASL or 4-(2-iodoacetamido)-TEMPO), which can be reacted selectively to SH1 thiol (Cys707) in myosin (Seidel 1982) or glycerinated muscle fibres (Thomas and Cooke 1980; Arata and Shimizu 1981). Until now, many other spin probes have been synthesized and successfully used to attach the SH group (Ajtai *et al.* 1990; Hambly *et al.* 1991; Roopnarine and Thomas 1994; Raucher *et al.* 1995): IPSL (N-(iodoacetyl)amino)-2,2,5,5-tetramethylpyrolidine-1-oxyl or iodoacetamido-proxyl), IKSL(3-(2-iodopropyonyl)-1-oxyl-2,5-dihydro-2,2,5,5-tetramethyl-1H-pyrrol), FDNA-SL (3-(5-fluoro-2,4-dinitroanilino)-2,2,5,5-tetramethyl-1-pyrrolidinyloxy) and InVSL (2-(oxyl-2,2,5,5-tetramethylpyrrolin-3-yl)methenyl)iodane-1,3-dione). These thiol-reacted spin labels have also been used for the selective labelling of SH2 (Cys697) in myosin heavy chain and a cysteine in myosin regulatory light chain. Spin trapping techniques may be useful but have not been used yet; labelled (xanthene) dyes are reduced photochemically to the radical form, which shows a strong EPR signal (Ajtai and Burghardt 1992). Spin-labelled nucleotide analogues (Crowder and Cooke 1987) and spin-labelled phalloidin (Naber *et al.* 1993) have also been synthesized and used to bind active site or specific site non-covalently in skinned fibres. Paramagnetic metal ions, such as Mn^{2+}, may also be useful (Bagshaw and Reed 1976; Bagshaw and Kendrick-Jones 1979; Hambly *et al.* 1991). Spin labels are more compact than optical probes, and they tend to bind more rigidly to proteins. Another advantage of spin labels is their susceptibility to the destruction of the free radical following treatment with a mild reducing or oxidizing reagent, potassium ferricyanide or cysteine. The latter advantage has proved to selectively eliminate signals from sites other than SH1. Detailed procedures of spin labelling will be discussed later (Section 9.6).

9.4 Applications of EPR to muscle physiology

If the sample is macroscopically oriented by placing labelled fibres parallel to the magnetic field, the spectrum is sensitive to the orientational distribution of probes relative to the system symmetry axis (the fibre axis). Owing to the shape of samples (long muscle body) and dielectric loss or heating by aqueous samples in microwave cavity, many special sample cells have been constructed. The most optimal aqueous sample cell design have been a quartz flat cell of dimensions $15 \times 10 \times 0.3$ mm for TE_{102} mode rectangular resonant cavity. The flat cell with a cover lid made of quartz (quartz flat tissue cell) or dielectric materials (Rexolite, Teflon, Kel-F, etc.) can be designed for mounting several muscle fibre bundles (0.5 and 4–7 mm in diameter and length respectively) horizontally in a ladder-like configuration (Thomas and Cooke 1980; Ajtai *et al.* 1989; Arata 1990). For easier experiments, we have used a sealed-off capillary (1 mm i.d. and 7–10 mm long) held horizontally on a holder as a sample cell (Arata and Shimizu 1981). In these sample cells, sample rotation is easy but cell positioning reproducibility is variable. For perpendicular or solution experiments, we used a sealed-off disposable capillary (1 mm i.d. and 5–10 cm long) inserted vertically in to a quartz tube (Arata and Shimizu 1981). They can be used only for stable rigor and relaxed states of muscle fibre.

For active contraction, the bathing solution should flow continuously to ensure sufficient ATP supply (Fig. 9.3). Two ends of fibre bundles in a capillary

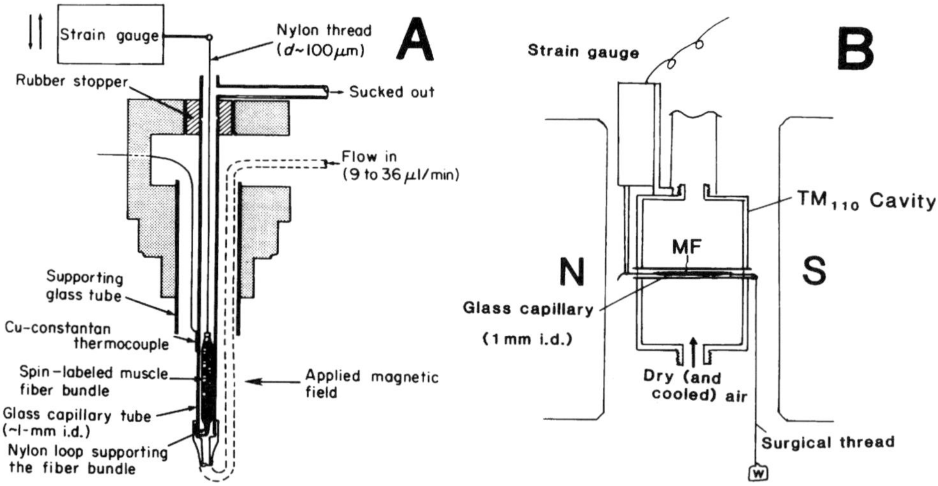

Fig. 9.3 EPR cell designs for mounting muscle fibres perpendicular (A) and parallel (B) to the magnetic field. The capillary is attached to a flow system that allow solutions to flow through and rapidly diffuse into the fibres. Tension can be continuously monitored when one end of the fibres is secured by a thread to a tension transducer mounted on the top or side face of the cavity. (A) (From Arata and Shimizu 1981.)

(0.8–1 mm i.d.) are secured with silk or nylon thread to prevent length changes. The capillary is attached to a flow system that allow solutions to flow through and rapidly diffuse into the fibres. The solution should contain adequate concentrations of creatine kinase (1–5 mg/ml) and creatine phosphate for rephosphorylation of ADP to ATP in the centre of a fibre bundle. The creatine phosphate level (usually 20–30 mM) can be determined by ensuring that the EPR spectrum is unchanged when its concentration is decreased or increased by two-fold (Arata and Shimizu 1981; Cooke *et al.* 1982; Zhao *et al.* 1995*a*). For perpendicular experiments, the capillary (5–10 cm long) is inserted into the centre of a quartz tube, which is inserted into the TE_{011} or TE_{102} cavity (Arata and Shimizu 1981; Barnett and Thomas 1989). For experiments with parallel orientation, the cavity can be modified by drilling holes (1–2 mm diameter) into the side faces of the cavity. The capillary is then inserted through the cavity, so that the fibre axis is aligned along the magnetic field of the spectrometer (TM_{110} or TE_{011} cavity gives higher sensitivity (higher Q) than TE_{011} or TE_{102}) (Cooke 1981; Cooke *et al.* 1982; Arata 1995). Tension can be monitored continuously when one end of the fibres is secured by a thread to a tension transducer mounted on top or on the side face of the cavity. A force can be exerted on the other end of the rigor fibres using weights that are attached by a length of thread (Cooke 1981; Arata 1995). For active contraction, the data acquisition time is usually 1–4 min and reliability is checked by ensuring the identical rigor spectra obtained before and after active contraction. Caged-ATP can be photolysed during EPR experiments through the radiation slits of a cavity (Fajer *et al.* 1990).

9.5 Main results

In the absence of ATP, actin and myosin form a tight bond known as the rigor complex. Conventional EPR on oriented fibres, labelled with thiol-reacted spin-label (IASL, MSL, InVSL, IPSL, IKSL, etc.) at SH1 (Cys707) on a main domain (catalytic domain) of myosin heads, was used to characterize the orientational distribution, and saturation transfer EPR on minced fibres, or fibres oriented perpendicularly (minimizing orientational effects), was used to determine whether the disorder was static or dynamic on the microsecond time-scale. The results showed that probes on the catalytic domain of myosin heads are uniformly and rigidly oriented in rigor; they have the same narrow (full-width 10–15°) orientational distribution as observed for isolated heads (S1) bound to actin in fibres (Thomas and Cooke 1980; Roopnarine and Thomas 1994). Similar results were obtained for insect flight muscle, although more orientational disorder was observed in rigor, which is consistent with predictions that steric constraints in insect muscle may prevent uniform binding of cross-bridges (Thomas *et al.* 1983).

Extending from the main domain (catalytic domain) toward the thick filament is the neck region (regulatory domain) formed by a long α-helix wrapped with essential and regulatory light chains. The regulatory light chain (LC2) was labelled with MSL at Cys156 and successfully exchanged into isolated proteins (S1 and HMM) and endogenous myosin in skinned fibres (Arata 1990). The probes attached to the light chain are not uniformly ordered, as at the other sites of catalytic domain, and they have provided more complicated and flexible motions. Although the probes on S1 bound to actin in fibres is uniformly ordered (mean angle 66° with full width 23°), those on HMM heads have two mean angles (42° and 66°) with narrow distribution (full width 15°) as shown in Fig. 9.4. The probes on endogenous myosin in skinned fibres reported greater degrees of disorder (the same two mean angles with full width 30–40°) than those of HMM (Arata 1990, 1995). The spectrum is very much like the second ($\Delta\theta = 30°$) or third one ($\Delta\theta = 45°$) in Fig. 9.2. Thus the neck can be distorted significantly by the interaction of the two heads of myosin or by the mismatch in the filament helix geometry when bound to actin. Similar results were obtained with different probes FDNA-SL at Cys127, although they were more mobile and could not resolve this distortion (Hambly *et al.* 1991). The IASL placed at Cys108 of gizzard light chain detected this distortion by a few degrees (Zhao *et al.* 1996).

In contrast to rigor, relaxation produced a considerable orientational disorder of the probes mentioned above, which is dynamic on the microsecond time-scale

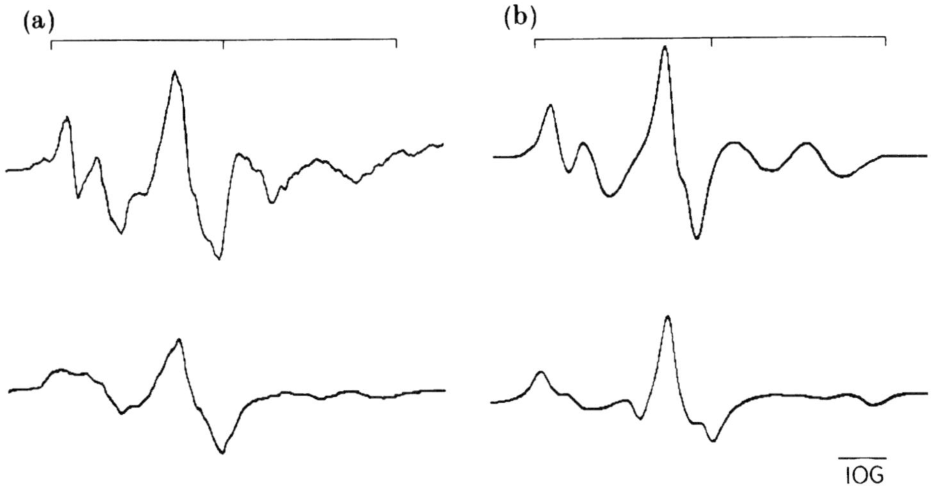

Fig. 9.4 (a) Experimental and (b) simulated EPR spectra for unlabelled fibres decorated with HMM containing spin-labelled light chain (LC2) in rigor solution. The fibres were parallel (*top*) and perpendicular (*bottom*) to the magnetic field. The simulated spectra were obtained by adding approximately equal proportions of two components: 43% of the spin labels (maleimido-TEMPO or MSL) have a Gaussian distribution with mean angle (θ_0) = 42° and full width of half maximum ($\Delta\theta$) = 15°, while 57% have $\theta_0 = 66°$ and $\Delta\theta = 15°$. The upper scale indicates centre and the extrema defined by A_{zz}. (From Arata 1990.)

(Thomas and Cooke 1980; Barnett and Thomas 1989; Arata 1990, 1995; Hambly *et al.* 1991). The spectrum is like that presented at the bottom of Fig. 9.2. Some of this axial disorder of the probes in relaxation could come from torsional disorder of the heads (Mendelson and Wilson 1982). Similar results were obtained for a spin-labelled ATP analogue bound to the catalytic site of myosin heads in skinned fibres (Crowder and Cooke 1987). The observed spectrum was decomposed into a component due to bound spin-labelled ATP and an unbound component.

Many ATP analogues have been used for studying an intermediate state. The probes mentioned above showed a small change of orientation and almost no change of motion upon addition of ADP to skinned fibres (Ajtai *et al.* 1989; Fajer *et al.* 1990*b*), which is consistent with a conformational change propagated from an active site when ADP binds. Almost no spectral change was also observed for probes at light chains (Arata 1990; Hambly *et al.* 1991). As expected from a previous saturation transfer EPR study using isolated actin–myosin in solution (Furukawa and Tonomura 1982), AMPPNP and pyrophosphate produced a spectrum of probes at SH1 resolved as a composite of two orientational distributions indistinguishable from rigor (actin-bound rigid heads) and relaxation (unbound mobile heads) (Fajer *et al.* 1988; Pate and Cooke 1988). The probes at light chains also showed that in the presence of AMPPNP a fraction of probes are ordered the same as those of HMM attached to actin in rigor fibres (Arata 1990; Hambly *et al.* 1991). In contrast, ATPγS produced relaxed-like disorder and microsecond motion of SH1 probes when a substantial fraction of heads were attached weakly to actin (Fajer *et al.* 1995). Vanadate and aluminum fluoride are phosphate transition-state analogues. They trap an intermediate state (actomyosin-ADP-vanadate or actomyosin-ADP-aluminum fluoride) in the presence of ATP and Ca^{2+} and produced a relaxed-like disorder of SH1 probes and spin-labelled ATP, but their rotational dynamics were highly restricted on a submillisecond time-scale (Raucher and Fajer 1994). 2,3-Butanedione stabilizes a myosin-ADP-phosphate intermediate state as an uncompetitive inhibitor of active tension and produces a relaxed-like disorder of SH1 probe in the presence of ATP and Ca^{2+} (Zhao *et al.* 1995*a*). In contrast, elevated concentrations of inorganic phosphate do not shift the SH1-probe orientation while reducing 70% of active tension (Zhao *et al.* 1995*b*). A force applied to a muscle fibre in rigor should produce a mechanically intermediate state. When a force equivalent to active contraction was applied to rigor fibres, there was no detectable rotation of probes attached to SH1 or light chain nor of spin-labelled ADP (Cooke 1981; Crowder and Cooke 1987; Hambly *et al.* 1991; Arata 1995). These observations show that chemical or mechanical means of reversing the power stroke produces no new fixed angles but chemical means tends to increase a population of disorder.

The conventional EPR spectrum of fibres labelled at SH1 during active isometric contraction shows two resolved components (Cooke *et al.* 1982; Fajer *et al.* 1990*a*; Roopnarine and Thomas 1995). One, which represents about 80% of myosin heads, is indistinguishable from that in relaxation and thus corresponds to considerable orientational disorder. The other component, which represents about 20% of the heads, has a narrow line shape that is indistinguishable from the spectrum in rigor. Saturation transfer EPR studies of contracting MSL-myofibrils or perpendicular-oriented MSL-fibres show almost as much microsecond rotational mobility as in relaxation (Barnett and Thomas 1989; Berger and Thomas 1993; Roopnarine and Thomas 1995). A small rigor-like population was also estimated from our earlier EPR experiments detecting a nanosecond motion of SH1-bound IASL as described later (Arata and Shimizu 1981). In contrast, probes (MSL and FDNA-SL) placed on the light chains have reported great degrees of orientational disorder, suggesting this end of the molecule is undergoing motions that are different from the catalytic domain (Hambly *et al.* 1992; Arata 1995). The straightforward interpretation is that more than 20% of heads are attached to actin, but at least some attached heads are dynamically disordered and the remainder have the same orientation at the catalytic domain as in rigor with their neck regulatory domain having a variety of orientations.

Actin and regulatory proteins were first spin-labelled by Tonomura *et al.* (1969) showing conformational changes in Ca control. Actin was then selectively labelled at Cys374 with thiol-reacted probes by many workers, and at the nucleotide binding site with a spin-labelled ATP and a paramagnetic ion Mn^{2+} (Loscalso and Reed 1976). Probe orientation can be measured while actin filaments and myosin heads are well oriented by flow into capillary (Ostap *et al.* 1992; Naber *et al.* 1994; Naber and Cooke 1994). Specific labelling of actin in fibres has been done using a spin-labelled phalloidin (Naber *et al.* 1993). No appreciable changes have been observed during activity. Tropomyosin can be labelled at Cys197 and Cys36 with MSL and reconstituted into ghost fibres (Szczesna and Fajer 1995). Troponin-C can also be labelled with MSL at Cys97 and reconstituted into muscle fibres in which troponin-C has been removed in the presence of EDTA (Li and Fajer 1994). Large changes were observed in orientation of troponin-C-bound probes when Ca^{2+} binds to it and myosin heads bind to thin filaments, suggesting a large domain movement of troponin-C. The binding of paramagnetic ions Mn^{2+} to a divalent cation binding site can be measured directly and a dipolar interaction between Mn^{2+} and spin labels decreases the amplitude of spin-label spectrum as a result of the proximity (<2.5 nm) between the two probes (Burley *et al.* 1972).

Using IASL bound to SH1, a correlation of local nucleotide-induced nanosecond motions of purified myosin with chemical intermediate states has been applied to fibres perpendicular-oriented (minimizing orientational effects) in order to detect force-generating intermediate states (Arata and Shimizu 1981; Ostap *et al.* 1995). Figure 9.5 shows spectra of IASL-fibres in rigor, contraction, and relaxation (A) and estimation of the mole fraction of rigor-like component as a force-generating chemical state at various ATP concentrations (B). During isometric contraction at millimolar ATP concentrations in the presence of Ca, the EPR spectrum appears closer to relaxation than rigor and resolves the same mobile components observed in relaxation (corresponding to (actin-)myosin-ATP and

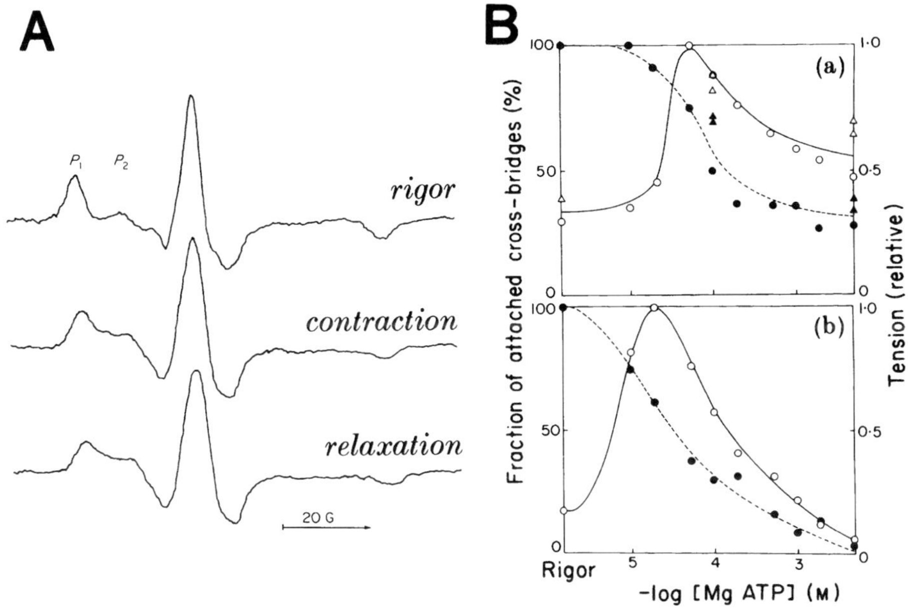

Fig. 9.5 (A) EPR spectra of iodoacetamidoTEMPO(IASL)-labelled muscle fibres oriented perpendicular to the magnetic field, in rigor, contraction, and relaxation. Spectral changes are caused by nanosecond rotational motions of labels at SH1 (Cys707) within the myosin head, which can be correlated with the conformational states of the specific biochemical intermediates during force generation. (B) Cross-bridge attachment (*closed symbols*) and tension development (*open symbols*) are shown as a function of ATP concentrations in the presence (a) and absence (b) of Ca. The fraction of attached cross-bridges (rigor-like component) can be calculated as $1-(P_2/P_1-0.03)/(0.43-0.03)$, where P_1 and P_2 are the 1st and 2nd peaks in the lower magnetic field and P_2/P_1 is 0.03 in rigor and 0.43 in relaxation. In the absence of Ca, reduced tension is developed at low ATP levels. During contraction at millimolar ATP concentrations in the presence of Ca, 30–40% have rigor-like component that may represent force-generating chemical intermediates. From Arata and Shimizu 1981.

(actin-)myosin-ADP-phosphate), and a small fraction (20–35%) of immobile component that is observed in rigor or rigor + ADP. The rigor-like component may represent intermediate states holding a high force (corresponding to actomyosin-ADP).

9.6 Problems and limitations

The success of using spin labels to study protein dynamics in muscle depends on how and where the label can be attached. The label must be attached rigidly to a single site at which its spectrum will reflect the molecular changes, and its binding to the protein should produce minimal changes in physiological activity.

Based on techniques developed for fluorescence labelling (Borejdo and Putnum 1977), selective spin-labelling of SH1 on purified myosin with thiol-reacted probes (MSL, IASL, IPSL and InVSL) have been applied successfully to skinned fibres (glycerinated fibres) by some modifications (Thomas and Cooke 1980; Arata and Shimizu 1981). One modification is to reduce the EPR signal from non-specific labelling with potassium ferricyanide (Thomas and Cooke 1980), which impairs muscle function seriously and has not been used recently. Another recent modification is pre-blocking other SH groups with a reversible SH reagent such as DTNB (Roopnarine and Thomas 1994; Zhao *et al.* 1995*a*), which is removed by washing with DTT after spin labelling of SH1 in the presence of 5–10 mM pyrophosphate. Iodoacetamido spin labels (IASL and IPSL) are very specific to SH1 and thus may not require any pre- or post-treatments (Arata and Shimizu 1981; Ajtai *et al.* 1990; Ostap *et al.* 1995). The fraction of heads labelled at SH1 can be determined from the fractional inhibition of K-EDTA ATPase activity, while the amount of total spin-labels can be estimated from EPR signals (line heights of alkaline-denatured fibres or double integration). For most preparations, about 50% of heads are labelled specifically at SH1. Active tension is usually reduced to 70–80% by spin labelling. Thus spin-labelled heads produce about a half of active tension. However, the co-operative behaviour of inhibition suggests that labelled heads produce tension much less than a half that produced by unlabelled heads (Root *et al.* 1991). Extensively (>95% SH1) labelled fibres can still generate 20–30% of active tension (Bell *et al.* 1995). Actin-activated ATPase rate of spin-labelled S1 is about 20% of unlabelled S1 (Svensson and Thomas 1986). In summary, heads spin-labelled at SH1 can still cycle through the entire contractile cycle, but the results obtained with spin-labelled fibres require some inspection. Recently, IKSL was reported to spin-label SH1 without imparing active tension and ATPase rate of fibres (Raucher *et al.* 1995).

Regulatory light chains (LC2) have been spin-labelled and exchanged into isolated myosin or myofibrils (Okamoto and Yagi 1976; Chantler and Szent-Gyorgyi 1978). LC2-bound spin probes (MSL or FDNA-SL) can be used to determine the angular distribution of the neck region of myosin in muscle fibres (Arata 1990; Hambly *et al.* 1991). Skeletal LC2 has two cysteines (Cys127 and Cys156). Specific labelling at one of two SH residues can be achieved successfully using differential labelling rates (MSL-Cys156 and FDNA-SL-Cys127). Gizzard regulatory light chains have a single cysteine (Cys108) that can be labelled easily (Zhao *et al.* 1996). Chemically skinned (glycerinated) fibre preparations are usually obtained from rabbit skeletal psoas muscle. Labelled LC2 can be exchanged into isolated myosin subfragments or endogenous myosin in fibres at high temperature (30 °C) in the presence of EDTA (Arata 1990; Hambly *et al.* 1991). The fraction of spin-labelled light chains is usually 20–50% of the heads. In scallop striated muscle, all the regulatory light chains are exchanged for spin-labelled ones (Roopnarine *et al.* 1995). Labelled fibres can generate the same active tension as unlabelled fibres, although X-ray diffraction shows that its structure is not exactly the same as that of unlabelled fibres (Ling *et al.* 1996). However, rigid labelling at the light chains is not easy, as the mechanism by which immobilization occurs is unclear. Cys156 of skeletal LC2 has been rigidly labelled with MSL (Arata 1990). The other probes at different cysteines undergo restricted nanosecond motions (Hambly *et al.* 1991) or microsecond motions with great degrees of disorder (Roopnarine *et al.* 1995).

Spin-labelled ATP analogues can induce active tension as high as ATP can, but cannot relax the fibre at low concentrations that minimize free nucleotides for experimental requirement (Crowder and Cooke 1987). When the ATP regenerating system (creatine or pyruvate kinase) cannot work, an alternative system should be used. Recently, a new spin-labelled ATP was synthesized and overcame these disadvantages by using 3-phosphoglycerate kinase as the ATP regenerating system (Alessi *et al.* 1992).

Another disadvantage is low sensitivity of EPR measurements. The measurement needs large bundles of muscle fibres and a long acquisition time. Use of large bundles may cause inhomogeneous activation between each single fibre, although labelled fibres are desected into small bundles and accumulated into larger bundles (0.5 mm). Prolonged activation (2–4 min) may produce deterioration of the sarcomere structure, which will recover to the original state after activation. This problem can be checked with rapid scan in narrow field range, and was recently overcome by a use of flash photolysis of caged-ATP inside the fibres (Fajer *et al.* 1990*a*).

Probes sense molecular change at their attached points. If the principal z-axis is perpendicular to molecular rotation, EPR can give very small spectral changes. This problem can be solved by labelling with many different probes and by using deuterized spin labels narrowing the line width, which provide high angular sensitivity (determining tilt and twist angles). It is very difficult to tell whether the observed change is local or global. Probe mobility relative to proteins may be checked by conventional or saturation transfer EPR on labelled proteins attached to glass beads (Thomas and Cooke 1980; Hambly *et al.* 1991; Li and Fajer 1994).

9.7 Future prospects

One of most crucial problems is low sensitivity of EPR, which prevents time-resolved experiments such as mechanical transients or rapid shortening using a single fibre preparation. A loop-gap resonator may solve this problem although its sample space is not dimensionally suitable for mounting a fibre (Froncisz and Hyde 1982).

Another problem is that rigid labelling is not easy. A bifunctional spin-label (cross-linker) will be synthesized like a fluorescent cross-linker (Mornet *et al.* 1985). Drug design techniques using computer molecular simulations may be useful for synthesizing new spin labels rigidly attached to protein surfaces. Protein-engineering can be used for inserting cysteines at desired points, e.g. especially in light chains. Spin-labelled amino acids may also be useful (Berliner 1976).

The spin-trapping technique can be viewed as a modified spin-label method. The technique requires the use of a non-paramagnetic molecule that can react rapidly with labile free radicals, leading in turn to a stable paramagnetic product or free radical. The family of molecules that has been used as spin-traps are nitrone or nitroso compounds. They can produce stable nitroxides upon reaction with labile free radicals in an irradiated solution of amino acids (e.g. Makino 1980). EPR is able to detect signals from spin-trapped radicals at the specific sites in muscle fibres or protein solution irradiated by ultraviolet light, etc., in the presence of spin-traps.

Much work is aimed at extending probe studies to other sites (essential light chains, S-2, loops, etc.). There are some discrepancies between conclusions from spin probe studies and other studies; optical probe studies, muscle X-ray diffraction, electron microscopy, and X-ray crystallography. Fluorescent spin labels have been produced photochemically to solve the apparent discrepancy between optical and spin probe studies (Ajtai and Burghardt 1992). These studies using the same preparation and the same probe will provide us with unique conclusion.

References

Ajtai, K. and Burghardt, T.P. (1992) Luminescent/paramagnetic probes for detecting order in biological assemblies: transformation of luminescent probes into π-radicals by chemical reduction. *Biochemistry* **31**, 4275–82.

Ajtai, K., French, A.R. and Burghardt, T.P. (1989) Myosin cross-bridge orientation in rigor and in the presence of nucleotide studied by electron spin resonance. *Biophy. J.* **56**, 535–41.

Ajtai, K., Poto, L. and Burghardt, T.P. (1990) Specificity and orientation of (iodoacetamido)proxyl spin-labeled myosin subfragment 1 decorating muscle fibers: localization of protein-bound spin labels using SDS-PAGE. *Biochemistry* **29**, 7733–41.

Alessi, D.R., Corrie, J.E.T., Fajer, P.G., Ferenczi, M.A., Thomas, D.D., Trayer, I.P. *et al.* (1992) Synthesis and properties of a conformationally restricted spin-labeled analog of ATP and its interaction with myosin and skeletal muscle. *Biochemistry* **31**, 8043–54.

Arata, T. (1990) Orientation of spin-labeled light chain 2 of myosin heads in muscle fibers. *J. Mol. Biol.* **214**, 471–8.

Arata, T. (1995) Orientation of spin-labeled light chain 2 of myosin heads in force-generating muscle fibres. *Jap. J. Physiol.* **45** (Supplement 2), S96.

Arata, T. and Shimizu, H. (1981) Spin-label study of actin–myosin–nucleotide interactions in contracting glycerinated muscle fibers. *J. Mol. Biol.* **151**, 411–37.

Bagshaw, C.R. and Kendrick-Jones, J. (1979) Characterization of homologous divalent metal ion binding sites of vertebrate and molluscan myosins using electron paramagnetic resonance spectroscopy. *J. Mol. Biol.* **130**, 317–36.

Bagshaw, C.R. and Reed, G.H. (1976) Investigations of equilibrium complexes of myosin subfragment 1 with the manganous ion and adenosine diphosphate using magnetic resonance techniques. *J. Biol. Chem.* **251**, 1975–83.

Barnett, V.A. and Thomas, D.D. (1987) Resolution of conformational states of spin-labeled myosin during steady-state ATP hydrolysis. *Biochemistry* **26**, 314–23.

Barnett, V.A. and Thomas, D.D. (1989) Microsecond rotational motion of spin-labeled myosin heads during isometric muscle contraction. *Biophys. J.* **56**, 517–23.
Barnett, V.A., Fajer, P. and Thomas, D.D. (1986) High-resolution detection of muscle crossbridge orientation by electron paramagnetic resonance. *Biophys. J.* **49**, 144–6.
Bell, M.G., Matta, J.J., Thomas, D.D. and Goldman, Y.E. (1995) Changes in cross-bridge kinetics induced by SH-1 modification in rabbit psoas fibres. *Biophys. J.* **68**, 360s.
Berger, C.L. and Thomas, D.D. (1993) Rotational dynamics of actin-bound myosin heads in active myofibrils. *Biochemistry* **32**, 3812–21.
Berliner, L.J. (ed.) (1976) *Spin labeling: theory and applications.* Academic Press, New York.
Borejdo, J. and Putnum, S. (1977) Polarization of fluorescence from single skinned glycerinated rabbit psoas fibers in rigor and relaxation. *Biochim. Biophys. Acta* **459**, 578–95.
Burley, R.W., Seidel, J.C. and Gergely, J. (1972) The effect of divalent metal binding on the ESR spectra of spin-labeled actin. *Arch. Biochem. Biophys.* **150**, 792–6.
Chantler, P.D. and Szent-Gyorgyi, A.G. (1978) Spectroscopic studies on invertebrate myosins and light chains. *Biochemistry* **17**, 5440–8.
Cooke, R. (1981) Stress does not alter the conformation of a domain of the myosin cross-bridge in rigor muscle fibres. *Nature* **294**, 570–1.
Cooke, R. (1986) The mechanism of muscle contraction. *Crit. Rev. Biochem.* **21**, 53–118.
Cooke, R., Crowder, M.S. and Thomas, D.D. (1982) Orientation of spin-labels attached to cross-bridges in contracting muscle fibres. *Nature* **300**, 776–8.
Crowder, M.S. and Cooke, R. (1987) Orientation of spin-labeled nucleotides bound to myosin in glycerinated muscle fibers. *Biophys. J.* **51**, 323–33.
Fajer, P.G., Fajer, E.A., Brunsvold, N.J. and Thomas, D.D. (1988) Effects of AMPPNP on the orientation and rotational dynamics of spin-labeled muscle cross-bridges. *Biophys. J.* **53**, 513–24.
Fajer, P.G., Fajer, E.A. and Thomas, D.D. (1990*a*) Myosin heads has a broad orientational distribution during isometric contraction: time-resolved EPR studies using caged ATP. *Proc. Natl. Acad. Sci.* (*USA*) **87**, 5538–42.
Fajer, P.G., Fajer, E.A., Matta, J.J. and Thomas, D.D. (1990*b*) Effect of ADP on the orientation of spin-labeled myosin heads in muscle fibres: A high-resolution study with deuterized spin labels. *Biochemistry* **29**, 5865–71.
Fajer, E.A., Ostap, E.M., Thomas, D.D. and Fajer, P. (1995) Orientation and dynamics of myosin heads in ATPγS and Ca^{2+}. *Biophys. J.* **68**, 322s.
Froncisz, W. and Hyde, J.S. (1982) The Loop-gap resonator: a new microwave circuit ESR sample structure. *J. Magn. Reson.* **47**, 515–21.
Furukawa, K.-I. and Tonomura, Y. (1982) The conventional and saturation transfer electron paramagnetic resonance of spin-labeled myosin subfragment-1 in the presence of F-actin and nucleotides. *J. Biochem.* (*Tokyo*) **92**, 1219–25.
Hambly, B., Franks, K. and Cooke, R. (1991) Orientation of spin-labeled light chain-2 exchanged into myosin cross-bridges in glycerinated muscle fibers. *Biophys. J.* **59**, 127–38.
Hambly, B., Franks, K. and Cooke, R. (1992) Paramagnetic probes attached to a light chain on the myosin heads are highly disordered in active muscle fibers. *Biophys. J.* **63**, 1306–13.
Hyde, J.S. (1978) Saturation-transfer spectroscopy. *Methods Enzymol.* **49**, 480–511.
Hyde, J.S. and Dalton, L.R. (1979) Saturation-transfer spectroscopy. In *Spin labeling II* (ed. L.J. Berliner), pp. 1–70. Academic Press, New York.

Jost, P.C. and Griffith, O.H. (1978) The spin-labeling technique. *Methods Enzymol.* **49**, 369–418.

Li, H.-C. and Fajer, P.G. (1994) Orientational changes of troponin C associated with thin filament activation. *Biochemistry* **33**, 14 324–32.

Ling, N., Shimpton, C., Sleep, J., Kendrick-Jones, J. and Irving, M. (1996) Fluorescent probes of the orientation of myosin regulatory light chains in relaxed, rigor, and contracting muscle. *Biophys. J.* **70**, 1836–46.

Loscalso, J. and Reed, G.H. (1976) Spectroscopic studies of actin-metal-nucleotide complexes. *Biochemistry* **15**, 5407–17.

Makino, K. (1980) Studies on spin-trapped radicals in γ-irradiated aqueous solutions of L-isoleucine and L-leucine by high performance liquid chromatography and ESR spectroscopy. *J. Phys. Chem.* **84**, 1968–74.

Mendelson, R.A. and Wilson, M.G. (1982) Three-dimensional disorder of dipolar probes in a helical array. Application to muscle cross-bridges. *Biophys. J.* **39**, 221–7.

Morales, M.F., Borejdo, J., Botts, J., Cooke, R., Mendelson, R.A. and R. Takashi (1982) Some physical studies of the contractile mechanism in muscle. *Ann. Rev. Phys. Chem.* **33**, 319–51.

Mornet, D, Ue, K. and Morales, M.F. (1985) Stabilization of a primary loop in myosin subfragment 1 with a fluorescent crosslinker. *Proc. Natl. Acad. Sci.* (*USA*) **82**, 1658–62.

Naber, N. and Cooke, R. (1994) Mobility and orientation of spin probes attached to nucleotides incorporated into actin. *Biochemistry* **33**, 3855–61.

Naber, N., Ostap, E.M., Thomas, D.D. and Cooke, R. (1993) Orientation and dynamics of spin-labeled phalloidin bound to actin in muscle fibers. *Proteins: Struct. Funct. Genet.* **17**, 347–54.

Naber, N., Lorentz, M. and Cooke, R. (1994) The orientation of spin probes attached to Cys374 on actin in oriented gels. *J. Mol. Biol.* **236**, 703–9.

Okamoto, Y. and Yagi, K. (1976) Ca^{2+}-induced conformational changes of spin-labeled g2 chain bound to myosin and the effect of phosphorylation. *J. Biochem.* (*Tokyo*) **80**, 111–20.

Ostap, E.M., Yanagida, T. and Thomas, D.D. (1992) Orientational distribution of spin-labeled actin oriented by flow. *Biophys. J.* **63**, 966–75.

Ostap, E.M., Barnett, V.A. and Thomas, D.D. (1995) Resolution of three structural states of spin-labeled myosin in contracting muscle. *Biophys. J.* **69**, 177–88.

Pate, E. and Cooke, R. (1988) The energetics of the actomyosin bond in the filament array of muscle fibers. *Biophys. J.* **53**, 561–73.

Raucher, D. and Fajer, P.G. (1994) Orientation and dynamics of myosin heads in alminum fluoride induced pre-power stroke states: an EPR study. *Biochemistry* **33**, 11 993–9.

Raucher, D., Fajer, E.A., Sar, C., Hideg, K., Zhao, Y., Kawai, M. *et al.* (1995) A novel electron paramagnetic resonance spin label and its application to study the cross-bridge cycle. *Biophys. J.* **68**, 128s–34s.

Roopnarine, O. and Thomas, D.D. (1994) A spin label that binds to myosin heads in muscle fibers with its principal axis parallel to the fiber axis. *Biophys. J.* **67**, 1634–45.

Roopnarine, O. and Thomas, D.D. (1995) Orientational dynamics of iodane dione spin-labeled myosin heads in relaxed and contracting skeletal muscle fibers. *Biophys. J.* **68**, 1461–71.

Roopnarine, O., Szent-Gyorgyi, A.G. and Thomas, D.D. (1995) Saturation transfer electron paramagnetic resonance of spin-labeled myosin regulatory light chains in contracting muscle fibers. *Biophys. J.* **68**, 337s.

Root, D.D., Cheung, P. and Reisler, E. (1991) Catalytic cooperativity induced by SH1 labeling of myosin filament. *Biochemistry* **30**, 286–94.

Seidel, J.C. (1982) EPR of contractile systems. *Methods Enzymol.* **85**, 594–624.

Stone, D.B. (1973) Interaction of actin with spin-labeled heavy meromyosin in the presence of nucleotides. *Biochemistry* **12**, 3672–9.

Svensson, E.C. and Thomas, D.D. (1986) ATP induces microsecond rotational motions of myosin heads crosslinked to actin. *Biophys. J.* **50**, 999–1002.

Szczensna, D. and Fajer, P.G. (1995) The tropomyosin domain is flexible and disordered in reconstituted thin filaments. *Biochemistry* **34**, 3614–20.

Thomas, D.D. (1987) Spectroscopic probes of muscle cross-bridge rotation. *Ann. Rev. Physiol.* **49**, 691–709.

Thomas, D.D. and Cooke, R. (1980) Orientation of spin-labeled myosin heads in glycerinated muscle fibers. *Biophys. J.* **32**, 891–906.

Thomas, D.D., Dalton, L.R. and Hyde, J.S. (1976) Rotational diffusion studied by passage saturation transfer electron paramagnetic resonance. *J. Chem. Phys.* **65**, 3006–24.

Thomas, D.D., Cooke, R. and Barnett, V.A. (1983) Orientation and rotational mobility of spin-labeled myosin heads in insect flight muscle in rigor. *J. Muscle Res. Cell Motil.* **4**, 367–78.

Tonomura, Y., Watanabe, S. and Morales, M. (1969) Conformational changes in the molecular control of muscle contraction. *Biochemistry* **8**, 2171–6.

Zhao, L., Naber, N. and Cooke, R. (1995*a*) Muscle cross-bridges bound to actin are disordered in the presence of 2,3-butanedione monoxime. *Biophys. J.* **68**, 1980–90.

Zhao, L., Pate, E., Baker, A.J. and Cooke, R. (1995*b*) The myosin catalytic domain does not rotate during the working power stroke. *Biophys. J.* **69**, 994–9.

Zhao, L., Gollub, J. and Cooke, R. (1996) Orientation of paramagnetic probes attached to gizzard regulatory light chain bound to myosin heads in rabbit skeletal muscle. *Biochemistry* **35**, 10158–65.

10 Time-resolved X-ray diffraction

John M. Squire

10.1 Introduction

In order to follow in detail the structural changes involved in dynamic molecular events in muscle that are central to the production and regulation of force and movement it is necessary to use a fast structural technique that can be applied to living intact tissue. There is currently only one technique that can achieve this, namely that of X-ray diffraction (Huxley and Brown 1967; Squire 1981). X-ray diffraction patterns from live, intact muscle, in close to its normal physiological state and undergoing normal mechanical activity, can be recorded with time-resolutions better than 1 ms and with spatial resolutions ranging from 500 Å down to 5 Å (e.g. Podolsky *et al.* 1976; Sugi *et al.* 1978; Amemiya *et al.* 1980; Huxley *et al.* 1980, 1982, 1983; Lowy and Poulsen 1982, 1987*a*,*b*; Matsubara *et al.* 1984; Yu *et al.* 1984, 1985; Wakabayashi *et al.* 1985, 1988, 1991, 1993; Kress *et al.* 1986; Maeda *et al.* 1988, 1994; Yagi and Matsubara 1988, 1989; Cecchi *et al.* 1990, 1991; Harford and Squire 1990, 1992; Bordas *et al.* 1991, 1993; Yagi 1991; Irving *et al.* 1992; Griffiths *et al.* 1993; Squire and Harford 1993; Bagni *et al.* 1994; Martin-Fernandez *et al.* 1994; Mitsui *et al.* 1994; Lombardi *et al.* 1995; Chew *et al.*, in press; there are many more). In principle, then, one can solve these diffraction patterns to produce a series of 'images' of the muscle in each time slice, and these images can be put together to produce a movie of molecular activity in the muscle (Squire *et al.* 1993, 1994*a*,*b*). However, time-resolved X-ray diffraction studies have been on the go since the early 1960s; why is the movie not already available? The main problem is that X-ray diffraction does not give direct images of the diffracting object. Diffraction patterns either have to be solved or they have to be interpreted and in many cases this is by no means an easy task.

A useful analogy in thinking about this is a light microscope. Conventional microscopes (Fig. 10.1a) have an incoming light beam (possibly from a condenser system) that impinges on the specimen. The specimen scatters the light into the objective lens of the microscope. If one looks at the back focal plane of the objective lens then what is seen is actually the diffraction pattern (related to the Fourier transform) of the specimen. If the specimen is fairly regular with periodicities in this case of about 0.2 μm or more, then the diffraction pattern seen in the back focal plane of the objective is a set of separate spots of light. What the eyepiece (or projector lens) then does is to collect the light from the objective and

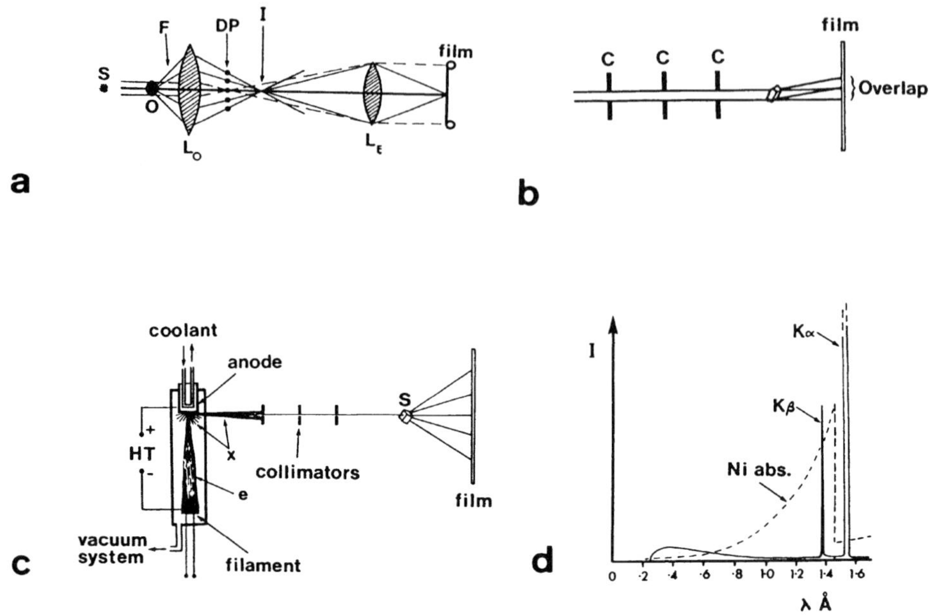

Fig. 10.1 (a) Schematic ray diagram of a light microscope: Source, S; Image, I; Lenses, L_O and L_E; diffraction plane, DP; F, film position for recording unfocused diffraction pattern. (b) Illustration of the size of a collimated X-ray beam and the problem of overlapping low-angle diffraction peaks. (c) Schematic diagram of a sealed X-ray 'tube': HT, high voltage (tension); e, electron beam; X, emitted X-rays; S, specimen. (d) *Full line*: part of the wavelength spectrum of emitted X-ray from, say, a copper target as in (c). *Dashed line*: the absorption edge of Ni, which is often used to remove the CuKβ line from the Kα—see p. 244. (From Squire 1981.)

use it to reconstruct an image of the specimen. The way the optical system achieves this is to maintain the optical path length through the system for each of the different diffracted beams. Each beam is associated with an amplitude and a relative phase. If both the amplitudes and the phases of the beams are maintained throughout the optical system, then, when the beams are recombined in the eyepiece, they add and interfere in such a way that a faithful image of the specimen is reconstructed. The latter process is known as Fourier synthesis.

The light microscope can achieve both diffraction (the Fourier transform) and image formation (Fourier synthesis) because visible light can be focused by the lenses (or mirrors) in the system. Without these focusing elements there would be no light microscopes. X-rays have a very much shorter wavelength ($\lambda \approx 1$ to 2 Å) than visible light ($\lambda \approx 5000$ to 8000 Å) and a rule of physics is that radiation (e.g. visible light or X-rays) can only reveal structural features of objects greater in size than about half of its wavelength. The structures of molecules typically 2 to 1000 Å in size cannot therefore be studied by light microscopy, whereas X-ray wavelengths are very suitable for this. Unfortunately, the refractive index of glass or any other potential refractive material is about 1.5 for visible light, but about 1.0 for X-rays; there is virtually no deviation of the X-rays from their original path. This means that, although X-ray diffraction patterns can be produced and recorded, equivalent to putting a film or other detector in front of the objective lens in Fig. 10.1a, there are currently no lenses that will allow an X-ray microscope to be produced with the kind of resolution needed to follow molecular events. Although X-ray mirrors are used for various purposes, the angles involved are so small that it is not possible to produce X-ray mirror microscopes. Note that X-ray microscopes *are* being produced that make use of what are known as zone plates (Michette *et al.* 1992; Kirz *et al.* 1995); however current resolutions are around 30 nm at best, and the technique cannot readily be applied to living biological tissue because of the problems of radiation damage (Shinohara and Ito 1991; Foster *et al.* 1992; Bennett *et al.* 1993; Fujisaki *et al.* 1996; see also a very useful discussion of imaging and radiation damage with different radiations by Henderson 1995).

To summarize, then, X-rays, with a wavelength short enough to resolve structural features of atomic or molecular size, can produce very rich diffraction patterns, but they cannot yet be recombined optically to produce good direct images of the diffracting molecules. What has to be done is to record the X-ray diffraction pattern with a film or other detector at a position equivalent to F in Fig. 10.1a and then to try to carry out the Fourier synthesis part of the operation on a computer. But clearly, in order to do this, one needs both the amplitude ($|F|$) and phase information for each diffraction peak. Unfortunately, the recorded diffraction pattern only provides half of the necessary information. The recorded intensities (I) of the diffraction peaks provide information about the amplitudes, since $I \propto |F|^2$, but all information about the relative phases is normally lost when the pattern is recorded. Protein crystallographers have got over this 'phase problem' by, for example, adding heavy atoms to their proteins. Compared with diffraction patterns from the original molecule, and assuming that the heavy atom derivative is still isomorphous with the parent structure (i.e. the unit cell is the same size and shape and the rest of the molecule has not moved), then the changes in the observed diffraction pattern produced by the heavy atom can be used to find the phases of the X-ray reflections. Fourier synthesis can then be carried out using computers to give a 3-D electron density map of the protein. In general, muscles do not have the order of single crystals and the spacings involved are often very much larger than in protein crystals. Direct methods of solving the phase problem for muscle are therefore not readily available. Muscle diffraction patterns require interpretation and modelling and the methods of doing this have taken a long time to work out. As a result, even though the first muscle X-ray diffraction patterns were recorded in the 1950s and 1960s, extracting unambiguous structural information from them has not been easy, and is still not trivial. Nevertheless, as detailed above, since X-ray diffraction is probably the ONLY structural method with the capability of revealing molecular structural details on a millisecond time-scale in intact muscle, finding ways to solve muscle diffraction patterns is well worth the effort.

This chapter is designed to provide practical help in understanding the design of X-ray diffraction cameras and detectors, in carrying out time-resolved X-ray diffraction experiments on muscle and also in interpreting and modelling the diffraction patterns that are produced. The limitations of the technique are also described. The reader is referred to other sources for details of the results that have been obtained using this technique and the ideas formulated from those results (e.g. Faruqi and Huxley 1981; Squire 1981; Huxley and Faruqi 1983; Huxley 1984, 1994; Huxley and Kress 1985; Lowy and Poulsen 1987*a*,*b*; Wakabayashi and Amemiya 1991; Harford and Squire 1990, 1992, in press).

10.2 Developments in X-ray production

Laboratory X-ray sources

The earliest practical method of producing X-rays for diffraction experiments was to bombard a cooled target with medium-energy electrons accelerated from a heated filament (Cullity 1978). The whole assembly (a so-called X-ray *tube*) was located within an evacuated glass chamber (Fig. 10.1c) and accelerating voltages of about 30 000 to 50 000 volts (30 to 50 kV) were commonly used. By addition of a small bias cup placed around the heated filament and kept slightly negative compared to the filament, focusing of the electron beam could be achieved. In this way a small electron beam focus on the target (anode) could be produced, and hence a relatively small source of X-rays. The mechanism of X-ray production is of two kinds; there is either *bremmstrahlung*, giving a continuous spectrum over a range of wavelengths, or there is a *characteristic* spectrum, specific to the target material, consisting of high intensity lines at specific wavelengths (Fig. 10.1d). One of the characteristics of charged particles such as electrons is that if they are deflected (i.e. accelerated) away from their original path then they emit radiation. Thus, when an incoming electron experiences the influence of the positive and negative charges in the anode material (often copper or molybdenum) and is deflected by these charges, *bremmstrahlung* or 'braking radiation' is emitted. This occurs over a range of wavelengths giving the observed continuous or 'white' spectrum. On the other hand, if the incoming electrons excite electrons in the anode atoms to higher energy levels, then, when the remaining electrons drop down between energy levels to fill the vacancies, radiation is emitted that is characteristic of the energy differences between the various electron levels. This radiation forms the *line spectrum*; in the case of copper the most prominent line is the CuKα line at $\lambda = 1.5418$ Å (Fig. 10.1d). This wavelength proved to be very useful for X-ray diffraction; it is about the same size as atomic dimensions, and it is a wavelength where the absorption in many materials is not great, but where their ability to diffract (their diffracting power) is reasonable.

When it is desired to carry out time-resolved X-ray diffraction studies, it is evident that the stronger the X-ray beam is, the shorter is the exposure of the detector that will be needed to achieve a usefully strong diffraction pattern. Efforts were therefore made to increase the X-ray flux from X-ray tubes. However, the processes of X-ray production discussed above are very inefficient. Only about 0.5% of the energy of the incident electrons appears as X-rays; the remaining 99.5% appears as heat that needs to be removed by cooling of the anode material (Fig. 10.1c). If the electron beam carries too much power then the anode will melt. As we shall see in Section 10.3, a necessity for muscle diffraction is that the source is also very small. Thus the electron beam focus in the X-ray tube needs to be as small as possible, thereby increasing the local heating of the anode, and the power in the electron beam needs to be as great as possible, both without actually melting the anode. Two tricks have been used to alleviate the problem. The first was to use a focus on the anode that was much bigger in one direction than was finally wanted, and then to foreshorten it by looking at the X-ray source from the side, usually at an angle of about 4° to the anode surface. In this way a gain of about 10 could be achieved, reducing a typical focus of 1×0.1 mm on the anode to 0.1×0.1 mm in its foreshortened view. The second method was to dispense with the sealed X-ray tube (Fig. 10.1c) and instead to use a continuously pumped system in which the anode was actually a rotating, copper-coated, drum which thus presented a continuously changing face to the incident electron beam, effectively spreading the heating effect over the circumference of the drum. Such rotating-anode X-ray generators (commercially available from, for example, Enraf-Nonius or Rigaku-Denki) represent the most powerful laboratory X-ray sources that are currently available. Such sources were used successfully in early time-resolved diffraction studies of muscle (Huxley and Brown 1967) and, indeed, it was largely because of the need for higher intensity in such muscle studies that fine focus rotating-anode X-ray generators were originally developed. Such generators are now used for many other purposes, including protein crystallography.

Synchrotron X-ray sources

In the 1950s and 1960s physicists started to study high-energy particle interactions using particle accelerators. To achieve maximal energies it was found to be convenient to constrain the particles (often electrons or positrons both of which carry charge) in a large circular orbit. This was achieved by passing the particle beam through a set of bending magnets (vertical field) equally spaced around the circumference of the horizontal orbit plane (Fig. 10.2a,b). The orbit size was clearly fixed, so acceleration of the particles to higher energies, using transitory (radio frequency) electric fields in synchrony with the orbiting particles, required gradually increasing magnetic fields in the bending magnets. The ultimate energy of the accelerator was therefore fixed by the maximum field of these magnets. The effect of the vertical magnetic fields on particles moving along a horizontal path was a lateral force which deflected the particles towards the centre of the orbit. An inevitable consequence of this force on the particles was the production of *bremmstrahlung* radiation. To the particle physicists this unwanted radiation was more of a nuisance than anything else, but in the late 1960s it was realized by biophysicists such as Hugh Huxley and Ken Holmes wishing to carry out fast time-resolved X-ray diffraction studies of muscle, that the enormous X-ray fluxes

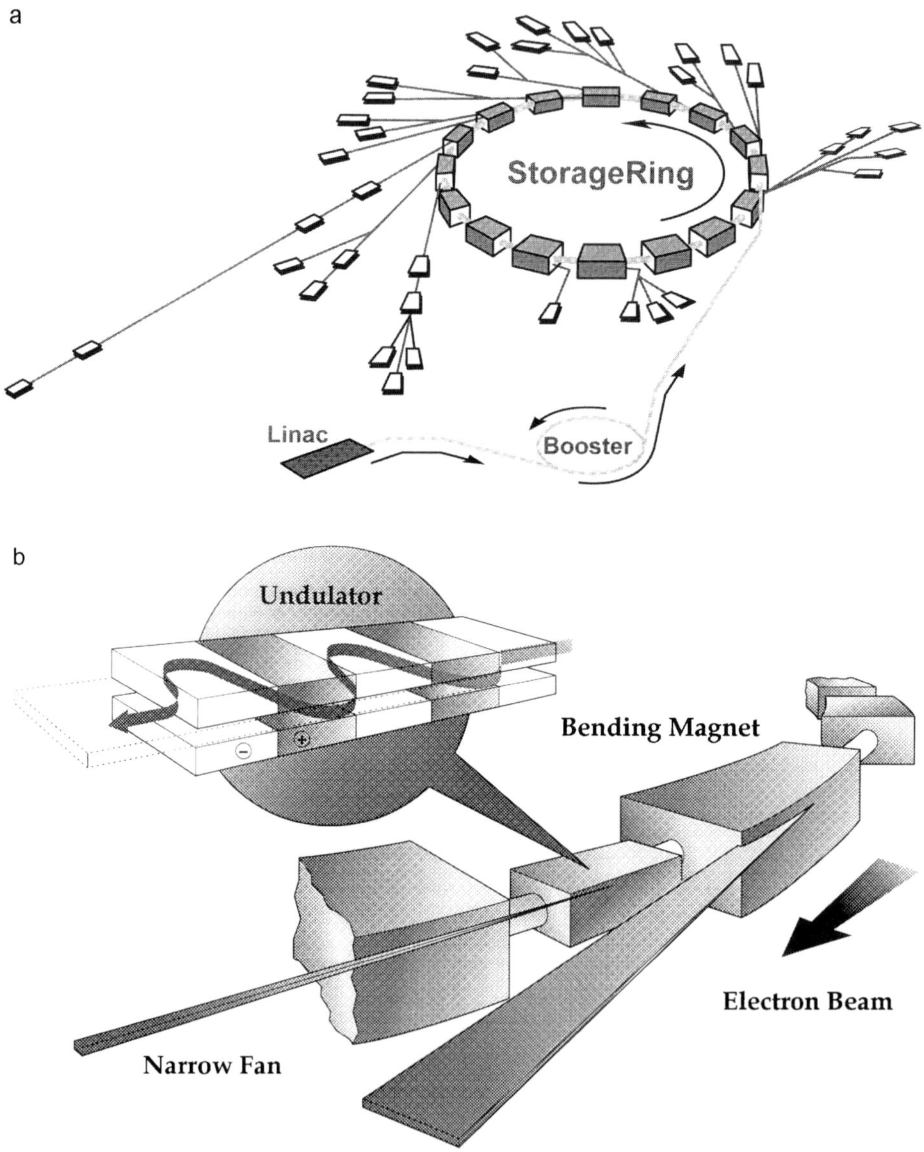

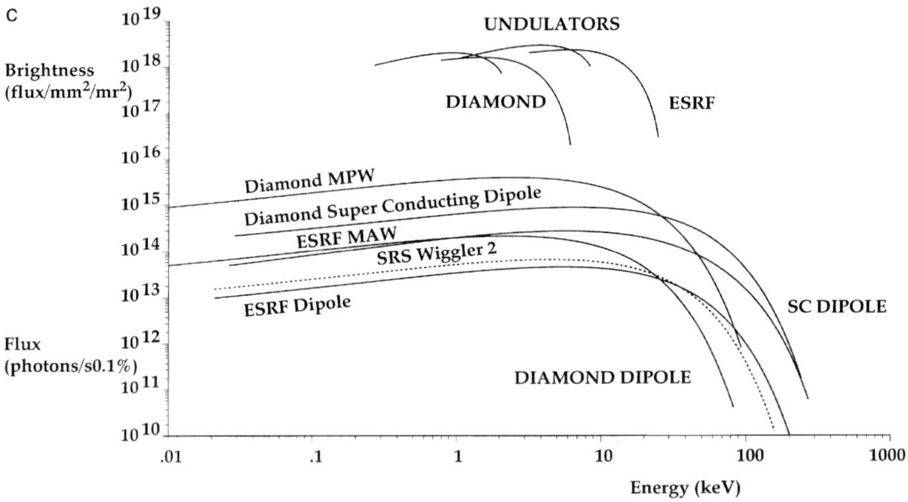

Fig. 10.2 (a) Layout of the storage ring and beam lines at the CCLRC Daresbury Synchrotron Radiation Source (SRS). (b) Schematic of the beam shape arising from a bending magnet and an undulator for the proposed storage ring DIAMOND at Daresbury, where the broad and narrow fans, respectively, show the shape of the emanating X-ray (or other) beam. (c) Comparison of the spectral outputs of various sources at the SRS, ESRF and DIAMOND. Dipole is a bending magnet, MPW is a multipole wiggler. (All illustrations courtesy of CCLRC Daresbury Laboratory, see pp. 246–8 for details.)

produced by synchrotrons could be just what they needed (Rosenbaum *et al.* 1971). It was these pioneers who started to use synchrotron radiation for biological work and the rewards were immediately obvious. Unfortunately, the synchrotrons were used *by* the particle physicists *for* the particle physicists and those using the parasitic X-rays had to take second place. However, others, including protein crystallographers in the field of biology and also those working on materials (e.g. synthetic polymers), soon realized the great advantages of synchrotron radiation. With this increased 'user pressure', a second generation of synchrotrons was developed with the primary aim of providing high radiation flux over a wide wavelength range including the range important for X-ray diffraction, namely 0.5 to 2 Å. Such synchrotrons were termed '*storage rings*' (Fig. 10.2a) and among the first were the Synchrotron Radiation Source at Daresbury in the United Kingdom and the Photon Factory at Tsukuba in Japan. Such storage rings have more than proved themselves and new rings are springing up in many countries around the world.

Since the first use of X-ray beams from bending magnets it has been found that the use of so-called *insertion devices* could provide even higher X-ray fluxes. In essence, the tighter the bend that a particle of given energy is constrained to move in, the greater is the flux of emitted radiation. The first such insertion device was a '*wiggler*'. By the use of carefully shaped high strength magnetic fields, these devices essentially 'wiggled' the particle beam through tight bends and then returned the beams back to their original track. At the major wiggle, great increases of X-ray flux were obtained (gains of about 100 ×). Since then the insertion devices of choice are the *multipole wiggler* and the *undulator*. By deflection of the beam in a series of *N* closely spaced wiggles, the combined effect of a multipole wiggler in the forward direction is flux that is *N* times that from one wiggle. Undulators are somewhat similar to wigglers in that the beam is subjected to a series of bends (Fig. 10.2b) but this time the amplitude of the bends is sufficiently small that interference occurs between the beams emitted from successive wiggles. These effects can be optimized to give very much higher fluxes at particular, chosen, wavelengths, which can be adjusted over a small range if the gap between the magnetic poles can be varied. Figure 10.2c compares the achievable fluxes from the Daresbury SRS, the European Synchrotron Radiation Facility (ESRF) at Grenoble and the different kinds of device planned for the new synchrotron, Diamond, at CCLRC Daresbury Laboratory.

With new synchrotrons, then, the bending magnets are essentially there to constrain the particles into a (segmented) circular path and the straight sections between these bends are where the insertion devices go that provide optimized flux at different wavelengths. At the time that the early storage rings such as the Daresbury SRS were built, the importance of the straight sections was not realized; in those days the bending magnets were everything. As a result, the straight sections at the SRS, for example, are relatively short and are too short for very long insertion devices to be included. The other thing that has happened is that bending magnet technology has greatly improved, especially with liquid Helium-cooled superconducting magnets, so that by using higher magnetic fields the volume occupied by a magnet to achieve a given beam deflection has been reduced.

In summary, synchrotron radiation sources using insertion devices are currently the best available sources of high flux X-ray beams. Sources optimized for the inclusion of long insertion devices, such as the European Synchrotron Radiation Facility at Grenoble in France, are generally known as 'third generation' sources. One aspect of the new designs is that the particle beams themselves are much more compact in cross-section and this means that as well as increased flux, the flux is concentrated into very small beams of high 'brilliance' (essentially flux per unit beam cross-section at a given distance from the source) and this is important for many kinds of muscle diffraction experiment (e.g. study of single muscle fibres) as discussed in later sections.

Other possible X-ray sources

Before leaving the topic of X-ray production, it is as well to note that, in the future, powerful X-ray sources may be totally different. There would be great experimental advantages in being able to have very powerful X-ray beams from laboratory-based X-ray generators, rather than from synchrotrons. Synchrotrons have their advantages and are the best that can be achieved at the moment, but there are usually only one or two in an advanced country and access is inevitably restricted. How might a small laboratory generator produce very high flux? Two approaches may prove to be useful. One is to try to collect more of the X-rays produced by target bombardment, where at present the vast majority of the X-rays produced are lost. The second is to use a totally different kind of source, namely a plasma X-ray source. Here a very high voltage is passed for a very short time through a small stream of gas under vacuum and a very high X-ray flux is produced (Cullmann *et al.* 1987). The immediate practical problem is in defining precisely in the gas where the source will be, and being able to repeat the voltage pulse at a rapid enough rate to be useful. Another method is to use X-rays from plasmas that have been generated by illumination of a suitable target by a laser beam (Fujisaki *et al.* 1996; Norreys *et al.* 1996). In principle, the development of a short wavelength X-ray laser would immediately allow imaging with molecular detail. However, developments of X-ray lasers, although rapid, have so far produced wavelengths only of about 35 Å; too long to be of immediate use in structural biology (Da Silva *et al.* 1992; Eder and Matthews 1994). It remains to be seen whether these approaches will prove to be useful for X-ray diffraction experiments.

10.3 X-ray camera geometry and detectors for fibre diffraction

Camera design

In the case of muscle, the main periodicities of interest are those of the actin and myosin filaments. Myosin filaments in vertebrate striated muscles have axial repeats of about 429 Å, the axial separation of *crowns* of myosin heads is about 143 Å, and the peak characteristic of the pitch of the α-helical coiled-coil myosin rods is about 5.1 Å (Huxley and Brown 1967; Squire 1975, 1981; Chew and Squire 1995). Actin filaments have a long repeat of about 720 to 760 Å, with a pseudo-halving to between 360 and 380 Å giving predominant 1st order reflections at spacings of about 360 to 380 Å, and further orders of this, particularly the 2nd (180 Å), 3rd (120 Å), 6th (59 Å), 7th (51 Å) and 13th (27.5 Å) orders (Huxley and Brown 1967; Squire, 1981). As in conventional X-ray diffraction theory, one can use Bragg's Law to define the angles of diffraction at which reflections corresponding to these various spacings will be observed with radiation of wavelength (λ) about 1.5 Å. The angle of diffraction between the incident and diffracted beams in Fig. 10.3a is 2θ, where θ is the angle of incidence of the beam on the diffracting planes (Fig. 10.3a), and is related to the spacing d between these planes by:

$$n\lambda = 2d\sin(\theta), \tag{10.1}$$

where n is the order of diffraction (for our present purposes assumed to be 1).

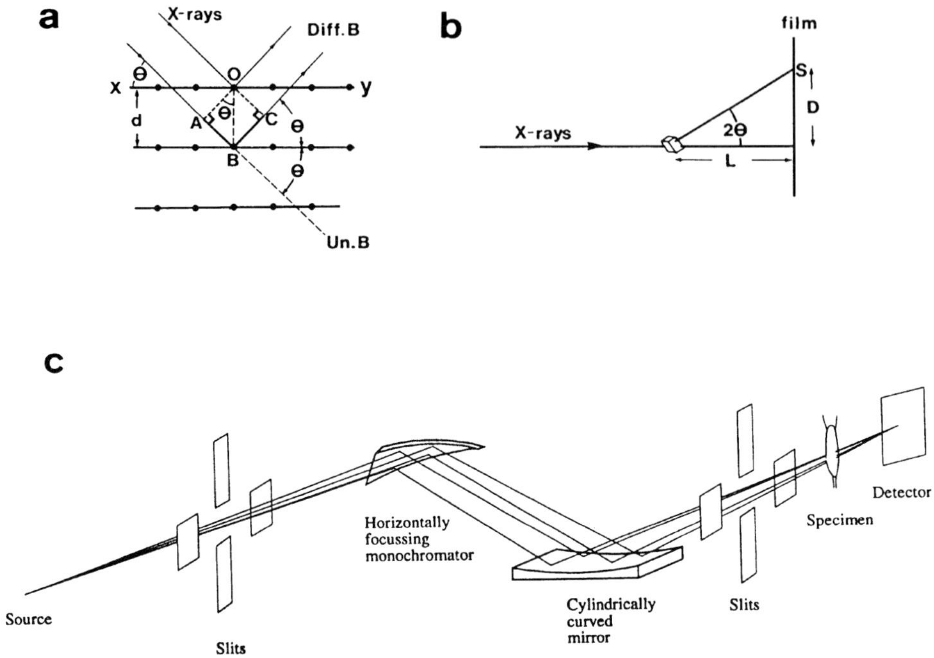

Fig. 10.3 (a) Illustration of diffraction geometry used to derive Bragg's Law. Diffracted beam, Diff. B; undiffracted beam, Un.B. Successive planes such as xy are separated by d, θ is the angle of incidence of the X-rays and 2θ is the angle of diffraction. This angle is seen again in (b) which illustrates the geometry required to determine θ from the separation D of a spot S from undiffracted beam in a camera of length L. (c) Schematic of the layout of a typical low-angle X-ray diffraction camera on a synchrotron. This represents line 2.1 at the Daresbury SRS; only two of the several slit systems are shown. ((a) and (b) from Squire 1981; (c) courtesy of Dr J. Barry.)

We have values of d ranging from several hundred Å to about 5 Å when considering muscle diffraction patterns. Taking $d = 500$ Å and $\lambda = 1.5$ Å, gives $2\theta = 0.17^\circ$. Taking $d = 5$ Å, gives $2\theta = 17.25^\circ$. The diffraction angles involved in studying myosin cross-bridge spacings or the actin filament helix are therefore very small (less than 1°), whereas angles involved in studying, for example, the α-helical backbone structure of the myosin filaments are moderately large. These regions of diffraction are often distinguished, because different camera geometries are often needed to record the appropriate diffraction peaks, into low-angle or small-angle diffraction (spacings 500 to say 30 Å), medium angle diffraction (say 30 to 5 Å) and high-angle diffraction (spacings 10 to 1 Å) although the boundaries are somewhat arbitrary. To a first approximation, high-angle diffraction resolves atomic positions in proteins, medium-angle diffraction resolves secondary structure and locates protein subdomains, and low-angle diffraction defines the structures of molecular assemblies. However, as shown later, low-angle diffraction can be used to provide high precision information if other high-resolution knowledge is made use of.

With these various diffraction angles in mind, most of the past time-resolved X-ray diffraction work on muscle has been in the low-angle region using specially designed low-angle diffraction cameras. The problem is simply this: the size of the X-ray beam coming from synchrotrons is defined initially by the size of the particle beam and then by the divergence of the X-ray beam as it leaves the synchrotrons. Synchrotron radiation is naturally very well collimated in the vertical direction, but in the horizontal direction the radiation is produced as a continuous fan (Fig. 10.2b). In order to have as much flux going through the muscle as possible, a large horizontal fan of X-rays is required, but this needs to be collected in some way so that, at the detector end of the X-ray camera, diffraction peaks at very small diffraction angles can be separated. This is usually achieved by using a curved single crystal (Fig. 10.3c), usually of Silicon or Germanium, which diffracts the fan of X-rays by Bragg reflection from particular crystal planes. By bending the crystal in an appropriate way, the diffracted beams can be made to converge to a focus at the detector plane. The wide fan of X-rays collected by the crystal contains a very broad spectrum of wavelengths. An intrinsic advantage of using such crystals to collect this fan is that they also primarily diffract just a single wavelength (with a small spread) as demanded by Bragg's Law; they act as *monochromators*. In addition to this, focusing in the vertical plane is often achieved by reflection of the X-ray beam at glancing angles from highly polished but slightly bent X-ray mirrors (Fig. 10.3c). By these means a typical low-angle diffraction camera, such as beam-line 16.1 at the Daresbury SRS, has a beam size of about 5 mm horizontally × 1 mm vertically at the specimen and about 3 mm horizontally and 0.5 mm vertically at the detector. Unfortunately, the optical elements, crystal and mirror, in the system do not diffract or reflect perfectly and these and other factors generate so-called '*parasitic scatter*'. This is usually minimized by the use of various sets of collimating slits along the optical path of the system up to the specimen.

The next factor to consider, with this size of X-ray beam at the detector, is what the specimen to detector distance needs to be to resolve various diffraction features. Say we wish to study the 429 Å reflection from vertebrate striated muscle myosin filaments, with a beam size that is 0.5 mm high at the detector. Let us say that, for comfort, this peak should be 10 mm away from the undiffracted beam since the parasitic scatter around the undiffracted beam is appreciable and lead backstops about 1 cm in size are needed to absorb this region. It is an easy matter to calculate the camera length required. As shown in Fig. 10.3b, for a given diffraction angle 2θ the separation D of peaks at the detector is a simple function of the specimen to detector distance L, usually just called the *camera length*:

$$\tan(2\theta) = D/L. \tag{10.2}$$

Putting eqn 10.1 and eqn 10.2 together relates D to the spacing of interest d:

$$L = D/\tan[2\{\sin^{-1}(\lambda/2d)\}]. \tag{10.3}$$

Taking D as 10 mm, gives L as 286 cm. A value close to 3 m is, in fact, a typical camera length for muscle diffraction experiments, although low-angle diffraction beam lines such as those at the Daresbury SRS have camera lengths adjustable from about 10 cm to 9 m, depending on which details of the diffraction pattern are of interest. The volume between the specimen and detector is usually occupied by an evacuated flight tube since air absorbs and scatters X-rays a great deal (about 10% of the beam is lost in 10 cm) and X-ray scattering by air very close to the detector needs to avoided. The entry window into the flight tube can be made quite small in diameter (a few mm), but the exit window needs not only to hold the vacuum, but also to be X-ray transparent and quite large (usually about 30 cm in diameter), so as not to cut too much into the higher angle regions of the diffraction pattern. Appropriate exit windows are often made from a thin polymer film such as Mylar. Various practical low-angle camera geometries for synchrotrons are described in the literature (e.g. Rosenbaum *et al.* 1971; Haselgrove *et al.* 1977; Hendrix *et al.* 1979; Bordas *et al.* 1980; Rosenbaum and Holmes 1980; Wakabayashi *et al.* 1986; Towns-Andrews *et al.* 1989; Bliss *et al.* 1995).

Detectors

In the early days of X-ray diffraction from muscle the only X-ray detector available was X-ray film. However, because of the high penetrating power of X-rays, such films were rather inefficient. In order to provide as much emulsion thickness as possible, while not making the emulsions too thick for the processing chemicals to soak into rapidly, most X-ray films were based on a cellulose acetate (or similar) sheet with emulsion on both faces. The image of the diffraction pattern was therefore double, which was acceptable for reflections at small diffraction angles incident more or less normally to the plane of the film. However, for precision spatial work, it was often necessary to measure reflections in each emulsion layer separately. Since the grain size in X-ray films is relatively small (say 25 μm), the spatial precision achievable in measuring diffraction peak positions is very high indeed and has still not been bettered. Intensity values were originally determined by eye relative to a calibrated scale and later by the application of 1-D microdensitometers, but automatic 2-D microdensitometers are now used routinely to convert the analogue data in the X-ray film into 2-D digital data.

Early work at synchrotron sources also used film for many purposes, but very soon both 1-D and 2-D X-ray detectors based on the principles of proportional counters were produced (Charpak *et al.* 1968; Gabriel 1977; Faruqi and Bond 1980; Lewis *et al.* 1988, 1991, 1992; Worgan *et al.* 1990, 1994; Lewis 1994). Here an incoming X-ray photon ionizes the gas in the chamber across which a high voltage is applied. The positive ions of the gas are attracted towards the cathode, and the emitted electrons travel towards the anode. However, the latter are accelerated sufficiently to cause further ionization; a localized avalanche of further ions and electrons is produced, giving something like 10^3–10^5 electrons arriving at the anode for each incoming X-ray photon. When the electron avalanche reaches the electrodes, a pulse (event) is recorded. In the case of a single wire electrode, the position of the discharge along the wire can be determined by timing the arrival of the event pulses at each end of the wire. Such 1-D detectors were very useful in early synchrotron studies, but these have mostly been replaced nowadays by so-called multiwire detectors (multiwire proportional counters or MWPCs). Modern MWPCs (Fig. 10.4a) typically consist of a central anode plane of fine wires, each perhaps 10 μm in diameter and 1 mm apart, with cathode planes of 30 to 50 μm wires about 3 to 10 mm from the anode and aligned orthogonally to each other. By using this *area detector* system the location of the event can be determined in 2-dimensions. The arrival of each photon is detected; MWPCs are photon counting devices. In early detectors, the photon position was determined by linking the wires in one plane with so-called delay-line electronics and matching the arrival times of the pulses in the two orthogonal cathode planes. However, this poses a problem in that there is a limit to the speed at which the detectors can operate without saturating the electronics. At Daresbury, using an active area of about 20×20 cm, the current saturation limit of such detectors is about 10^6 counts/sec. The spatial resolution achieved is about 300–400 μm. A disadvantage of this system is the substantial separation between the two cathode planes, which leads to the problem of parallax. If the incoming photons are not incident more or less normally to the detector face, unwanted peak spreading and distortion can arise. Such detectors have therefore been used successfully in low-angle diffraction studies where the 2θ angles are very small, but they are not appropriate for high-angle diffraction studies unless special geometries are adopted (Fig. 10.4c). More recent area detectors, for example, those using the RAPID readout system developed by the Detector Group at the Daresbury Laboratory (Lewis *et al.* 1991; Worgan *et al.* 1994), have enabled the signal from each wire to be read out separately, thus greatly increasing the saturation limit of the electronics (to 2×10^7 counts/sec or better). Other new developments on the way include the use of microstrip, microgap and high pressure detectors, which will improve the spatial resolution of the detectors and reduce the problems of parallax (Oed 1988; Lewis *et al.* 1992; Lewis 1994; Worgan *et al.* 1994).

Recently a third type of area detector has become very popular, namely the so-called image plate (Amemiya *et al.* 1987, 1988, 1989; Amemiya and Wakabayashi 1991). This is a flat plastic sheet coated on one side with very small crystals (about 5 μm) of a photo-stimulable phosphor held in place with an organic binding material. When an incident X-ray photon reaches the plate, electrons in the valence band are stimulated to higher energy levels (the conduction band) where they are trapped in so-called colour centres. The number of electrons trapped in this way is proportional to the number of photons absorbed. When the exposed image

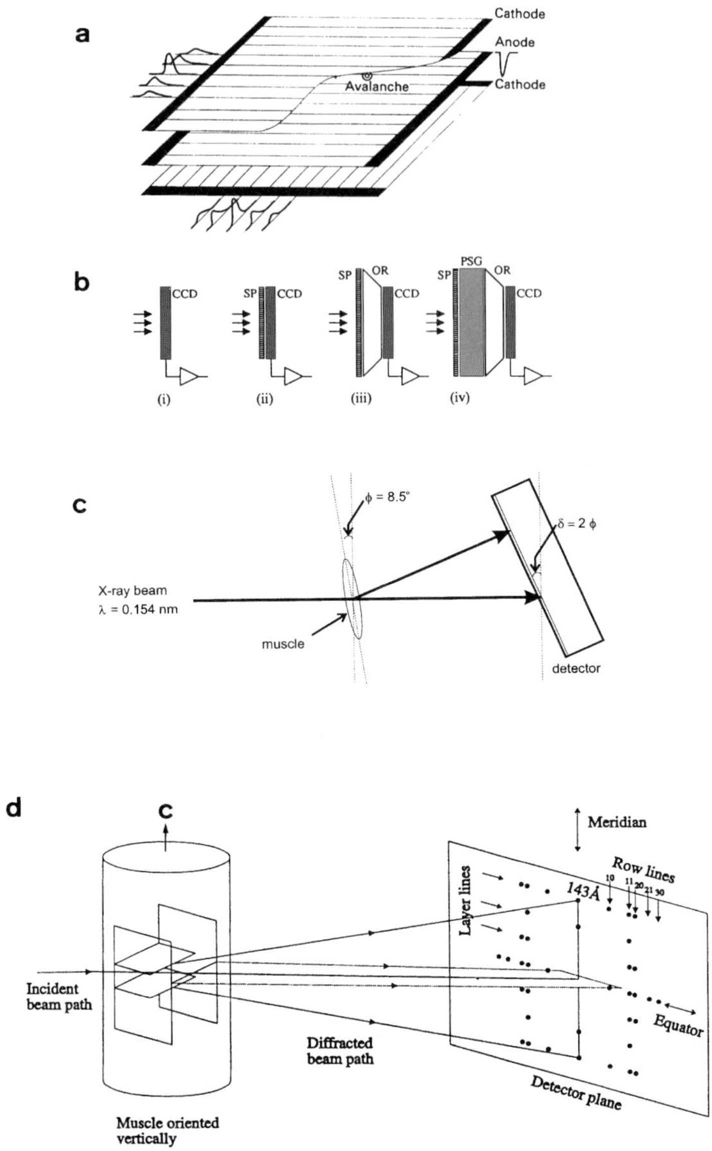

Fig. 10.4 (a) Schematic layout of the anode and cathode wires in a 2-D proportional counter showing the creation of an avalanche when an incoming photon ionizes an atom in the detector gas (from Lewis 1994). (b) Illustration of various forms of CCD detector (*arrows* indicate incoming X-rays; SP is a scintillator or phosphor, OR is an optical relay, PSG is prestorage gain, e.g. image intensifier) (from Allinson 1994). (c) Geometry needed to record high-angle diffraction features using a present-day 2-D multiwire proportional counter, which suffers from parallax problems. Note the need to tilt the specimen for this kind of study. (Courtesy of Dr M. Chew.) (d) Geometry of a rotating-crystal or fibre diffraction pattern showing the production of layer-lines and row-lines. Locations of the meridian and equator are indicated. (Courtesy of Dr J. Barry.)

plate is then scanned with a laser beam of appropriate wavelength (e.g. 6328 Å), the trapped electrons are given enough energy to allow them to fall back to the valence band, at the same time emitting visible photons by luminescence. The wavelength of this emitted radiation is sufficiently different (3900 Å) from the incident laser light that it can be detected separately by a high quantum efficiency photomultiplier tube.

Another detector which may be useful for some muscle studies but which has not been used a great deal for this is the CCD-based detector system (Fig. 10.4b; Allinson 1994; Amemiya *et al.* 1994; Arndt 1994). In the most useful devices, X-ray photons are incident on a phosphor, which produces visible photons that are collected by a bundle of tapering optical fibres and delivered to the CCD array. In this way a high resolution CCD (say 1024 pixels square) of rather small physical dimensions (2 to 3 cm) can be matched to a useful detection area, say 10 to 12 cm across.

Finally, it may be possible to design useful *silicon pixel detectors* (Hall 1995). These are based on a single silicon wafer on the back of which is bonded a second wafer carrying a 2-D array of amplifiers on a pitch of about 200 μm. In this way photons absorbed by the silicon in a given pixel are read out individually. Both the efficiency and speed of such detectors could be excellent, but a great deal of development needs to be done to make practical detectors for fast X-ray diffraction studies.

Comparison of detectors

In designing X-ray diffraction experiments on muscle it is essential to take into account the different characteristics of the various detectors that are available. In doing this a number of features are important. These include: useful detection area, spatial resolution, dynamic range (ratio of weakest to strongest peaks that can be recorded on the same device), efficiency (how much of the incident beam is actually detected), readout time (crucial for time-resolved experiments), background level, intensity linearity, spatial distortions (do the positions of the recorded peaks correspond exactly to the relative positions of the arriving X-rays?), etc.. In summary, X-ray film has the highest spatial resolution, effectively has no count-rate limit, has a moderate dynamic range and is relatively distortion free. But it is very slow to read out, has a high background (i.e. fog level) and is of little use in dynamic experiments. Film response can be fairly linear over a moderate range, but this requires very precisely controlled chemical development. Image plates are very convenient, have quite high resolution (limited by the read-out system, not by the plate itself), have high dynamic range, do not have a count-rate limit and have spatial distortions that can mostly be calibrated out. They have a lower background level than film and have now become the detector of choice for protein crystallographers, especially using systems where the plates can be exposed and read-out on-line (e.g. the Mar Research and Rigaku systems). However, they have slow readout times and it should be noted that they are only linear above a certain threshold level; they are not good detectors for relatively week intensities. Image plates have limited applications in time-resolved studies, where they or a blanking plate can either be moved laterally, say between resting and active parts of a contraction, so that different parts of the same plate are exposed in the two states (Huxley *et al.* 1994), or they can be wrapped around a rotating cylindrical drum with a horizontal slit just before the plate so that continuous time-resolved exposure in 1-D can be achieved (Amemiya *et al.* 1987, 1989). Another development for image plates is the use of a rapid plate exchanger that can present successive plates to the beam many times a second. Multiwire detectors have moderate resolution, correctable spatial distortion, large dynamic range, very fast readout, very low background and are very linear over a wide range, including very weak signals. New developments mean that their count rate is very high indeed. CCD systems have quite good resolution, a relatively small useful area, are relatively linear, but suffer from a high background level and relatively slow readout time. It is clear that the MWPCs are in many ways the optimal detectors for muscle diffraction experiments and it is clear that they are essential for dynamic experiments with very fast (μs or ms) time-resolutions.

Note that, whatever the detector, certain calibration checks or corrections are essential. Film may require an intensity linearity correction even after carefully controlled processing. Both image plate and MWPC data require corrections for deviations from spatial linearity. This is usually checked by placing a grid of wires of known geometry immediately in front of the detectors and then exposing them to a uniform field of radiation (often using a radioactive source placed 50 to 100 cm from the detector). The MWPC detectors also need correcting for variations in response associated with consistent local effects to do with individual wires. Once again, exposure with a uniform source can calibrate these response variations so that they can be divided out. Response variations also occur with CCD detectors, especially in relation to the fine structure of the tapering fibre optic array, but this also can be calibrated out.

10.4 Design of time-resolved experiments

The nature of fibre diffraction patterns

Before embarking on a description of common experimental designs in muscle diffraction experiments, it is appropriate to consider briefly the normal form of muscle diffraction patterns. This is being done here without recourse to mathematics. There is not space here to provide the full theory of fibre and helical diffraction and the reader is referred to other sources (Holmes and Blow 1965; Fraser and MacRae 1973; Hukins 1981; Harford and Squire, in press) for mathematical details of what is presented below. A more qualitative treatment of helical diffraction theory is given in Squire (1981).

A natural consequence of Bragg's Law (eqn 10.1) is that for diffraction of X-rays of wavelength (λ) from crystal planes of spacing d to occur, those planes must be oriented at the correct angle (θ) to the incident X-ray beam. In general, with a single crystal, diffraction from particular planes in the crystal will not be seen; only if the crystal is rotated to bring those planes to the correct tilt will diffraction occur. In practice, diffraction patterns from single crystals are either recorded by rocking the crystal through a small angle to allow several planes to satisfy the Bragg condition (at different times), or a full rotation diagram is produced by rotating the crystal through a full 360° around a vertical axis. This is usually carried out with one of the principle axes (say the c-axis) of the crystal unit cell parallel to the rotation axis. When this is done, it is found that the rotation X-ray pattern shows reflections along a series of horizontal lines, known as *layer-lines* (Fig. 10.4d). The layer-line which contains the undiffracted beam is known as the *equator* of the pattern, and reflections lying along the equator are known as *equatorial reflections* or simply *equatorials*.

These reflections, which arise from planes in the structure parallel to the rotation axis, carry information about the unit cell of the crystal when viewed down the c-axis. The positions of the peaks provide information on the size and shape of the unit cell when viewed down c (Fig. 10.4d); in other words the lengths of a and b and the angle γ between them (Fig. 10.5a). The intensities of the equatorials provide information about the distribution of the unit cell contents when projected down the c-axis. Because of the reciprocal nature of diffraction as evidenced by Bragg's Law (i.e. the larger d is, the smaller is the value of θ for a given wavelength λ), it follows that reflections seen along the equator at gradually increasing radii from the middle of the pattern correspond to gradually reducing d-spacings. As discussed in the Introduction, in principle, the diffraction peaks can be used in a computer-calculated Fourier synthesis to generate an image of the unit cell contents. Successive reflections along the equator correspond to successive terms in the Fourier series that describes the periodic function of which the crystallographic unit cell is one repeat. The outermost equatorial reflections therefore provide information of gradually increasing *resolution* (d_{min}).

Reflections which lie along the line in the diffraction pattern parallel to the rotation axis and also containing the undiffracted beam are said to be *meridional reflections* and the axis is known as the *meridian* (Fig. 10.4d). With a single crystal diffraction pattern each of the layer-lines will consist of reflections which also lie along lines parallel to the meridian. These vertical lines are known as *row-lines* (Fig. 10.4d). Just as the equator carries information about the unit cell in a view down the c-axis, so these other layer-line reflections carry information about views through the structure in other directions. In particular, the meridian carries information about the whole 3-dimensional structure projected onto the rotation axis from directions perpendicular to this axis. Spacings along the meridian provide information about repeats in the direction of the c-axis, and intensities of meridional reflections provide information about the unit cell contents projected onto c.

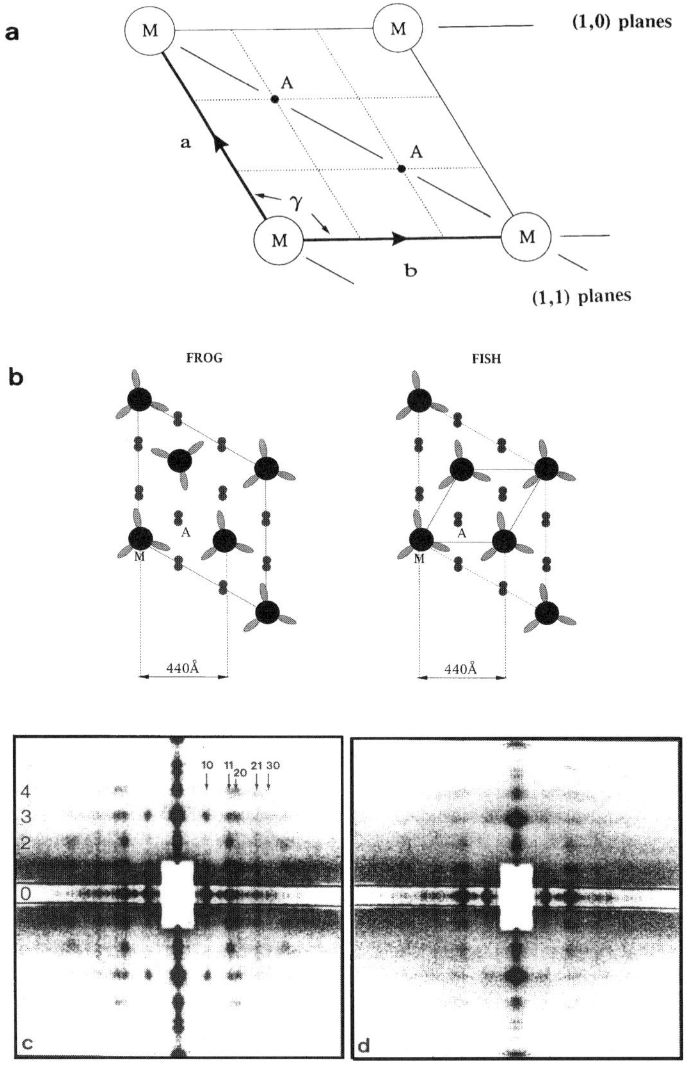

Fig. 10.5 (a) Definition of the parameters *a*, *b* and the angle γ in a hexagonal lattice. The (1, 0) and (1, 1) planes are illustrated. (b) Positions of the myosin and actin filaments in the unit cell in (a) for (*left*) frog and higher order vertebrate muscles and (*right*) bony fish and a few other fish muscles. The three-fold symmetry of the myosin filaments is seen, where each shaded ellipse represents a *pair* of myosin heads. (M, myosin filaments; A, actin filaments.) (c,d) Low-angle X-ray diffraction patterns from a plaice fin muscle in the relaxed state (c) and at the peak of a tetanus (d). Both patterns represent a total accumulated exposure of 6 s and both sets of intensities have been multiplied by the radial distance from the centre to aid visualization of outer, weaker reflections. Axial orders 0 to 4 of the myosin repeat of 429 Å are indicated. Note the horizontal layer-lines and vertical row-lines (some of which are indexed) indicating the existence of good 3-D order in the A-band of fish muscle. (a) From Dr J. Barry, (b) from L. Hudson; (c,d) from Harford and Squire (1992).

Muscle is not like a single crystal. However, muscle diffraction patterns have many similarities to crystal rotation patterns. In the case of vertebrate muscles, the myosin and actin filaments are positioned in a hexagonal array with appropriate *a* and *b* lattice dimensions (Fig. 10.5b). The filaments themselves have repeats along the *c*-axis that complete a kind a 3-D unit cell. However, a single muscle will contain a large number of fibres, each containing myofibrils with different orientations around the fibre (*c*) axis. There is therefore no need to rotate a muscle in the X-ray beam; different myofibrils already provide different crystallite orientations around *c*. For this reason a low-angle X-ray diffraction pattern from the highly ordered fish muscle (Fig. 10.5d; Harford and Squire 1986, 1992) contains both layer-lines and row-lines, just like a single crystal rotation pattern. In the case of fish muscle, the myosin and actin filaments are organized into a very well-defined, almost crystalline, 3-D unit cell (Fig. 10.5b), albeit with very large *a*, *b* and *c*-axis repeats. Diffraction patterns from the highly ordered invertebrate muscle, insect asynchronous flight muscle (Barrington-Leigh *et al.* 1977), are also very much like crystal rotation patterns.

Bony fish muscle and insect flight muscle represent the most ordered muscle samples available in vertebrates and invertebrates, respectively. Other muscles have much less order. These include the muscles of frog, rabbit, chicken and humans, in fact all higher vertebrate skeletal muscles (mammals, birds, amphibians, reptiles). Here the myosin filaments in the A-band are arranged into a statistical superlattice (Fig. 10.5b; Luther and Squire 1980; Luther *et al.* 1996), which means that it is not possible from knowledge of the orientation of any particular myosin filament in a given A-band to be able to predict unambiguously the orientations of most other myosin filaments in the same A-band. On the other hand, this superlattice is still based on a good hexagonal array. In general, the effects of disorder on diffraction patterns is to reduce the intensity of the Bragg reflections and to put intensity *between* these peaks (Vainshstein 1966; Hukins 1981). In the case of these higher vertebrate muscles this means that the equator is still sampled at the positions expected from a hexagonal unit cell, because in projection down the *c*-axis the superlattice has little effect. However, on other layer-lines the superlattice disorder is very evident and the layer-lines have slightly broadened Bragg peaks of reduced intensity with increased background intensity between them. In the limit of total disorder between myosin filaments, the diffraction pattern becomes such that the layer-lines still exist, assuming that each filament has good order within itself, but they are not sampled by Bragg peaks at all; the layer-lines are said to be continuous. This continuous layer-line pattern is essentially the diffraction pattern of a single filament, in this case one with approximately helical order.

The rigorous theory of diffraction from helical objects (Klug *et al.* 1958; see also simplified explanation in Squire 1981) shows that the diffraction pattern consists of a series of continuous layer-lines where the *c*-axis repeat is the axial repeat (R) of the helix (in the simplest case this is the same as the pitch P of the helix). However, the layer-line intensity actually *on* the meridian is zero on all layer-lines except those of spacing corresponding to orders of the subunit axial translation (h) along the helix. This distance (h) can be thought of as being the vertical rise per step of a spiral staircase. The pitch (P) corresponds to the height of the step immediately over the starting point on the staircase. In the case of vertebrate striated muscle myosin filaments, the myosin heads are organized approximately on a helix of pitch (P) of 3 × 429 Å and the subunit axial translation (h) is about 429/3 = 143 Å (Huxley and Brown 1967; Squire 1975, 1981, 1992; Harford and Squire 1986). However, there are three coaxial strands of myosin heads, giving the whole filament three-fold rotational symmetry, and this reduces the *repeat* (R) of the helix from 3 × 429 Å to just 429 Å. Thus the layer-lines in frog or fish muscle diffraction patterns (Figs. 10.4d, 10.5c) correspond to axial d-spacings in Bragg's Law which are successive orders of 429 Å. In addition, the 3rd layer-line at 143 Å (and layer-lines that are further multiples of three), all have marked intensity *on* the meridian. The myosin filament diffraction pattern therefore has the hallmarks of a classical helical diffraction pattern. However, some intensity is also seen on the meridian on layer-lines 1, 2, 4, 5 and so on, of the 429 Å repeat. The 1st order is known to have a contribution from the extra myosin filament protein C-protein, which has an approximately 430 to 435 Å repeat along the thick filaments (Rome *et al.* 1973*a*). There is also a meridional reflection at 385 Å due to troponin (Rome *et al.* 1973*b*). However, it is evident that meridional reflections on the 2nd, 4th, 5th (and so on) layer-lines of the myosin repeat are to do with the fact that the myosin head array on the thick filaments is not *exactly* helical; there are so-called perturbations that give an extra 429 Å periodic character to the myosin filament (Huxley and Brown 1967; Squire *et al.* 1982; Harford and Squire 1986).

The myosin filaments in insect flight muscle are different from those in vertebrates although the subunit axial translation between *crowns* of myosin heads is still about 143 Å (actually it is 145 Å in this case; Reedy *et al.* 1965; Reedy 1968; Barrington-Leigh *et al.* 1977). In this case the filaments have four-fold rotational symmetry (Morris *et al.* 1991); there are four coaxial strands of myosin head helices, and the true repeat is after eight crowns (i.e. 8 × 145 Å = 1160 Å). In addition, the array appears to be perfectly helical; there is little evidence of so-called 'forbidden' meridional reflections. Because of the particular geometry of this myosin head array, the prominent layer-lines are the 3rd order of 1160 Å at 385 Å, the 5th at 232 Å and the 8th (the meridional) at 145 Å. The pattern of strong layer-lines then repeats: 11, 13, 16, 19, 21, 24, etc., with multiples of eight particularly strong on the meridian.

Actin filaments are characterized by a subunit axial translation (h) of 27.5 Å and a pitch (P) of 59 Å. However, since the ratio P/h is close to, but not exactly 2, the filament appears as two slowly twisting strands of actin monomers. This gives the filament a variable 'cross-over' repeat of about 360 to 370 Å in vertebrate muscles increasing to 385 Å in insect flight muscles. In the case of vertebrate muscles, since the myosin repeat of 429 Å is incommensurate with the actin repeat of, say, 365 Å, the diffraction pattern consists of two different sets of layer-lines; a so-called myosin pattern and a so-called actin pattern. In many ways these two diffraction patterns can be treated separately when analysis is carried out. This is only partially true in the case of insect flight muscle where the 3rd myosin layer-line and the 1st strong actin layer-line both have axial spacings of 385 Å. In all cases, both sets of filaments contribute to the equator of the diffraction patterns.

Whole-muscle experiments

The strategy to be adopted in carrying out X-ray diffraction studies of whole muscle depends firstly on the particular muscle involved and the features of the pattern that are of most interest. Most synchrotron X-ray beams are horizontal lines of intensity at the specimen and they converge slightly to smaller horizontal lines at the focus. Much better spatial resolution is therefore achieved in the vertical direction than the horizontal. If the cameras manage to achieve a point-like focus then this is clearly not a factor. But in many cameras a focus approximating to a point can only be achieved by stopping down the beam using the various sets of slits, at the expense of the total flux in the beam. If very high time-resolution experiments are being attempted, such a loss in flux is very undesirable. The most highly ordered vertebrate muscle is that in bony fish (Luther *et al.* 1981, 1995; Harford and Squire 1986) and as a result the layer-lines due to the myosin head arrays on the myosin filaments are sampled by row-lines (Fig. 10.5c). In order to achieve optimal resolution along the equator of the diffraction pattern, the muscle must be oriented with the fibre axis horizontal. This can be advantageous for the layer-line pattern in studies of highly ordered muscles, such as those in bony fish muscle or insect flight muscle, where the row-lines can be clearly resolved. However, the penalty to pay is that closely spaced layer-lines, such as the 1st myosin layer-line at 429 Å and the 1st actin layer-line at 365 Å, are then relatively hard to separate. Much better definition of the layer-lines along the fibre axis direction is obtained if the muscle long axis is vertical. In the absence of a perfect point focus, a careful decision has to be made about the orientation of the muscle, or whether it is worth sacrificing flux to achieve good resolution in both directions.

Because of the relatively large amount of diffracting material in a whole muscle and the relatively large size of the beam at the specimen (say 3 mm by 1 mm), very successful time-resolved experiments can be carried out on whole muscle. Time-resolved studies of isometric and isotonic contractions have been achieved (e.g. Huxley *et al.* 1980, 1982; Kress *et al.* 1986; Lowy and Poulsen 1982, 1987*a,b*; Harford and Squire 1992; and see more complete list in the Introduction), as have studies of whole muscles subjected to rapid mechanical interventions (e.g. step length changes) during the plateau of a tetanus (e.g. Huxley *et al.* 1983). The large quantity of diffracting material permits the whole of a 2-D diffraction pattern, out to a *d*-spacing of about 65 Å, to be recorded with a time resolution of, say, 5 ms throughout a tetanic contraction (Harford and Squire 1992). This is achieved by cycling the muscle through a succession of contractions and reading out the diffraction intensities from each cycle into 5-ms time bins. This use of repeated cycles and the addition of counts into the appropriate time bins, permits sufficient *total* counts to be accumulated to provide reliable count statistics. A problem is that muscles inevitably fatigue over the sequence of contractions and the tension maximum reduces. In addition, there may be a real effect of radiation damage on the muscle. This is usually negligible for a total accumulated exposure of seconds to a minute, but may become very noticeable after several minutes. This also depends on the state of the muscle; for example, the effect is less bad on skinned muscle preparations, but then electrical stimulation is not possible.

Single-fibre experiments

There are several disadvantages to using whole muscle. One is that there is a natural variation is sarcomere length between different fibres in the muscle, even with very carefully dissected muscles, and as a consequence there are slight variations in lattice spacing (i.e. *a* and *b*). In addition, there may be slight misalignment of different fibres. As a result, the observed diffraction patterns may be slightly less well-defined than patterns from single fibres would be if they could be recorded easily. Another potential problem is that different fibre types may occur within a single muscle and these may have different contractile properties (e.g. speed). In experiments designed to probe the structural effects of precisely controlled mechanical experiments it is also desirable to control the muscle sarcomere length. This is readily achieved with single fibres using laser diffraction from the sarcomere array in the X-ray beam and monitoring the first order diffraction peak using an appropriate detector and length feedback circuit. Alternatively one can use a more sophisticated striation follower device (Huxley *et al.* 1981). Unfortunately, laser diffraction is rather unsatisfactory with whole muscles, which tend to be too thick, and the striation follower is only of use with single fibres. Another problem is that it is sometimes desirable to skin the muscle to allow rapid changes of the bathing medium or to exchange specific proteins. The efficacy of this is limited by the diffusion rate of the solution (Yu and Podolsky 1990). Diffusion into the middle of a large muscle will be very slow and a uniform concentration of, say, ATP may never, in fact, be reached. The use of single fibres greatly reduces all of these problems; by definition there is only one fibre type, it has a single orientation, sarcomere length control is relatively easy and diffusion is not a major problem. Single fibres have therefore been used very successfully in time-resolved diffraction experiments. But the paucity of diffracting material, together with the fact that the beam is often substantially bigger than the specimen, means that strong diffraction patterns of high quality are not usually obtained. At best, so far, only the strongest features of the low-angle diffraction pattern can be studied with high time resolution. However, the 145 Å meridional reflection from frog muscle fibres has been recorded with 100 μs time resolution during the imposition of rapid stretches and releases (Irving *et al.* 1992; Lombardi *et al.* 1995). In this way it was shown that the intensity of the 145 Å reflection, perhaps monitoring the state of the myosin head, changes in synchrony with some parts of the tension transient resulting from the rapid length step.

To carry out these single fibre experiments it was found to be useful to focus the synchrotron beam at the specimen rather than at the detector (sometimes between the two). This allowed more flux to pass through the specimen, which of necessity needed to be horizontal, giving optimal resolution in the equatorial direction, but it resulted in slightly defocused diffraction patterns. Facilities to achieve both high flux through the specimen and a good focus at the detector are coming on-line with the new generation of storage rings (e.g. the ESRF at Grenoble), since these are intrinsically more brilliant sources. They promise to provide very high quality single fibre diffraction data, although it is clear that radiation damage may become a serious problem.

Experiment control and data acquisition

In order to carry out time-resolved experiments on whole muscle, the specimen needs to be mounted in a temperature-controlled chamber that can hold the muscle firmly onto a transducer to monitor tension and possibly with a motor to control sarcomere length. The chamber needs to be supplied with electrodes (e.g. platinum) for electrical stimulation, with input port and output vent for oxygen circulation, and with thin, X-ray transparent windows through which the incident and diffracted beams pass. In the case of recording high-angle diffraction patterns (Fig. 10.4c), the outlet window in particular needs to have a relatively high angular aperture so as not to block the higher-angle diffraction peaks. In some cases it is also desirable to monitor the muscle appearance by light microscopy and a suitable light source and optics for this can be incorporated. A laser beam to monitor sarcomere length can also be included, sometimes by deflecting the laser beam into a path coincident with the X-ray beam. This also has the advantage of aiding specimen alignment.

A typical sequence of events in a time-resolved diffraction experiment on whole muscle is illustrated in Fig. 10.6c (Faruqi *et al.* 1978). Here, the duration of the exposure of the muscle to the X-ray beam is kept to a minimum to reduce radiation damage by means of a very fast-operating local shutter. An electrical stimulus (e.g. 50 Hz pulses) is then applied to generate a tetanic contraction of the muscle, its tension response is monitored and the diffraction pattern is recorded on a MWPC in time bins of chosen length (e.g. 5 ms) over the duration of the full contractile cycle. This whole procedure is then repeated after a suitable rest period (two or three minutes is good for fish muscle) and further counts are accumulated. This whole sequence of events is controlled by a time-frame generator; on line 16.1 at Daresbury

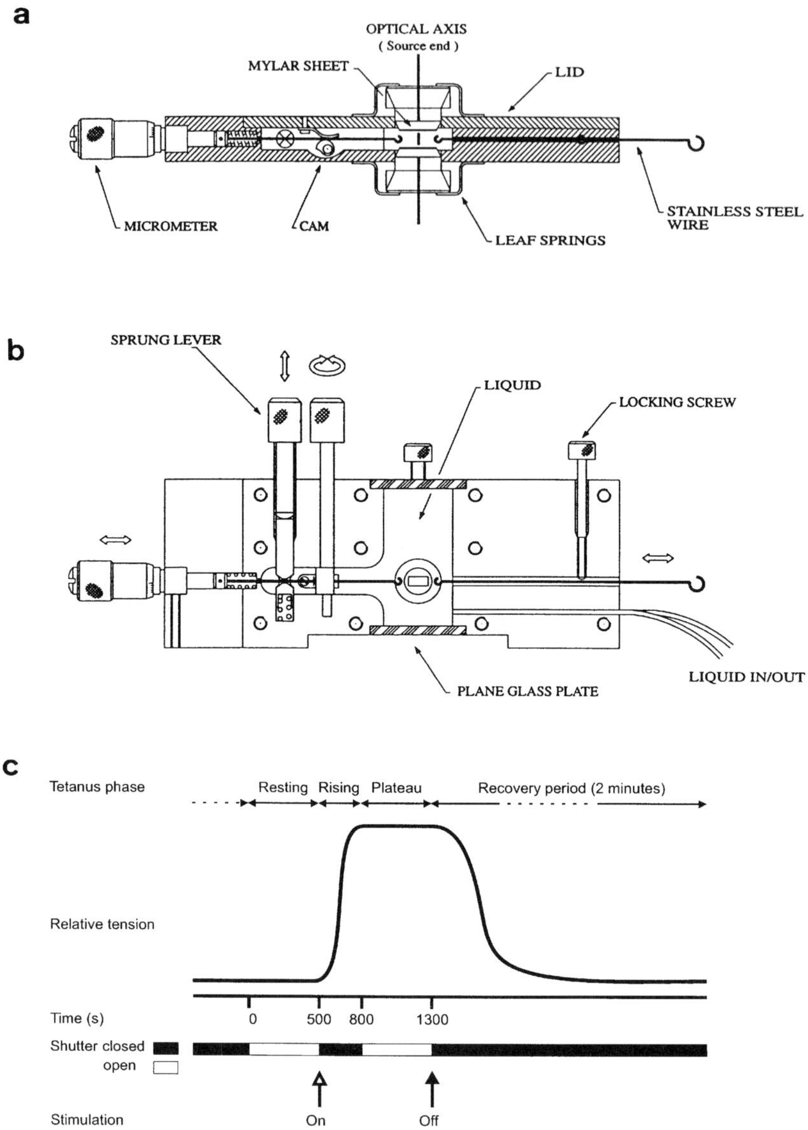

Fig. 10.6 (a,b) Views from the top and side of a perspex cell used to hold fish muscle (either whole or fibres), to stimulate the muscle (electrodes not shown), to record its tension, to monitor the sarcomere length by laser diffraction and to view the fibre by a suitably placed microscope viewing the muscle upwards through the lower plane glass plate in (b). The position of the muscle is adjustable in three dimensions, as is the muscle length. The windows containing mylar sheets next to the specimen are adjustable and can be closed in towards the specimen. (Courtesy of Dr J. Barry.) (c) A typical (simple) time sequence involved in carrying out time-resolved diffraction experiments on a muscle through an isometric contraction. (Courtesy of Dr M. Chew.)

this is based on a VME (Versa Module Europa—see http://www.leroy.com/lrs/introume/introume.htm) system which controls the whole experimental control and data acquisition process. If more sophisticated experiments are required, perhaps involving isotonic shortening or rapid mechanical interventions, it is common to link home-based control electronics into the VME system.

10.5 Data processing

Correcting and remapping the data

Typical muscle diffraction patterns consist of:

(1) some sharp (Bragg) peaks;

(2) a diffuse background from the disordered parts of the specimen;

(3) diffuse scatter from the muscle bathing solution in the specimen;

(4) a mixture of diffuse scattering and anomalous sharp intensities from the beamline itself.

The useful intensity is in (1) and (2) and as much as possible of (3) and (4) needs to be removed in data processing. Much of this unwanted scatter can be reduced firstly by making the X-ray path between the two windows of the specimen chamber as short as possible (very thin windows with adjustable separation really helps with this) and secondly by taking great care in setting the adjustable slits on the beamline. In particular, the second last set of slits should be used to define the required beam size and the final slits should be closed in so as to reduce as much as possible the parasitic scatter without actually touching the main beam. A third important factor is to keep as short as practicable the path of the X-ray beam in air. However, having minimized these factors, which generally improve the quality and aesthetics of the recorded diffraction patterns, it is then sensible to carry out an equivalent exposure of the muscle chamber containing solution but no specimen. This background pattern, appropriately weighted, can then be subtracted from the observed diffraction pattern with the specimen present to give patterns that are reasonably close to the sum of (1) and (2) only.

In principle, the background diffuse scatter contains important information about the disordered parts of the muscle, including any disordered myosin heads that may be present. Analysis of this diffuse scatter is by no means easy, but the approach is discussed in detail in Lowy and Poulsen (1987*a*,*b*).

Either before or after the background has been subtracted from the Bragg peaks, the various calibration corrections should be applied to yield patterns that are spatially accurate and with linear intensities. The resulting pattern, a combination of (1) and (2), then needs to be processed to provide usable data. The first process that should be applied, especially to 2-D diffraction patterns, is to remap the data so that the spacings in the resulting 2-D image are linearly related to dimensions in reciprocal space. For example, Bragg's Law shows that it is the angular distribution of intensity that relates to the *d* spacings, but patterns recorded on a flat detector do not represent angular distributions. It is also true that meridional reflections from muscles accurately oriented perpendicular to the incident X-ray beam do not satisfy the Bragg condition; the appropriate crystal planes are parallel to the incident beam direction. This problem is to some extent alleviated by the slight convergence of the X-ray beam and the lack of perfect fibre alignment in the muscle, but the net effect is that what is seen on the meridian of the diffraction pattern actually corresponds to intensity in the diffraction pattern that is slightly *off* the meridian. (This point is more easily dealt with in reciprocal space using the Ewald sphere construction and the reciprocal lattice, but this is not the place in which to teach basic crystallography. The reader is referred to other sources for this, e.g. Holmes and Blow 1965; Blundell and Johnson 1976; Cullity 1978; Harford and Squire, in press.)

1-D diffraction data

The simplest kind of analysis, either from the output of 1-D detectors or from 1-D data stripped from a 2-D detector pattern, is to fit both the remaining background and the superimposed Bragg peaks using already available software (Fig. 10.7a,b). Either commercial programs like Peakfit (Jandel Scientific) or programs available from CCP13 (Denny 1994; http://www.dl.ac.uk/CCP/CCP13) can be used. In either case, it is usual to represent the background by a sum of an exponential term, to fit the steeply varying intensity close to the origin of the diffraction pattern, and a polynomial (say four orders) to represent the more slowly varying background. The Bragg peaks are often modelled as Gaussian or Lorentzian peaks, to some extent determined by the geometry of the camera that is being used and the precise intensity profile of the incident X-ray beam. Once an initial fit has been obtained, it is desirable to refine the background and peak parameters together to optimize the fit. In the case of the muscle equatorial reflections, it is usually the first few orders, e.g. $(1,0)$, $(1,1)$, $(2,0)$, $(2,1)$ and $(3,0)$, from the hexagonal A-band lattice ($a = b = 470$ Å) that are recorded, together with the superimposed peaks from the pseudo-tetragonal I-band region ($a' = b' = 290$ Å) close to the Z-line (Yu *et al.* 1977; Harford *et al.* 1994). The Z-line $(1,0)$ reflection [Z(1, 0)] lies between the A-band $(1,0)$ and $(1,1)$ reflections [the A(1, 0) and A(1, 1) peaks]. However the Z-line $(1,1)$ peak [Z(1, 1)] is coincident with the A-band $(2,0)$ reflection [A(2, 0)]. In addition, the A(1, 1), which is often relatively strong, is very close to the Z(1, 0) and Z(1, 1) + A(2, 0) peaks, and these sometimes appear just as shoulders on the A(1, 1). These various points provide problems in the fitting of equatorial diffraction data. If the positions of the Z(1, 0) and Z(1, 1) + A(2, 0) model peaks are not constrained in the fitting they often get 'sucked into' the A(1, 1) profile and estimates of the intensity and the positions of these peaks become unreliable. It is sometimes necessary to use prior knowledge of the relative spacings expected, for example, for the A-band peaks and to constrain the A(2, 0)

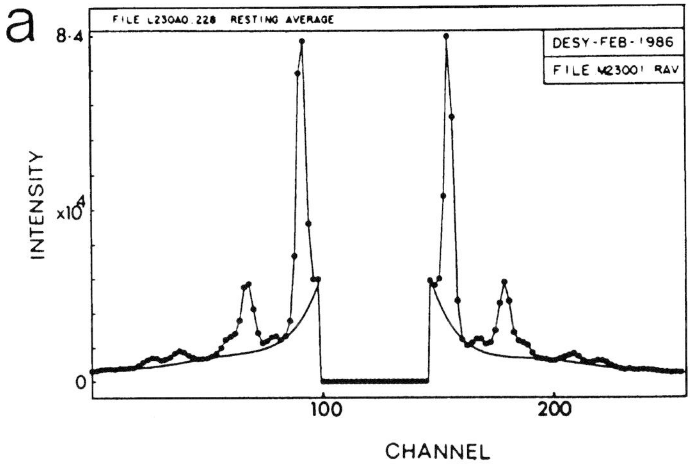

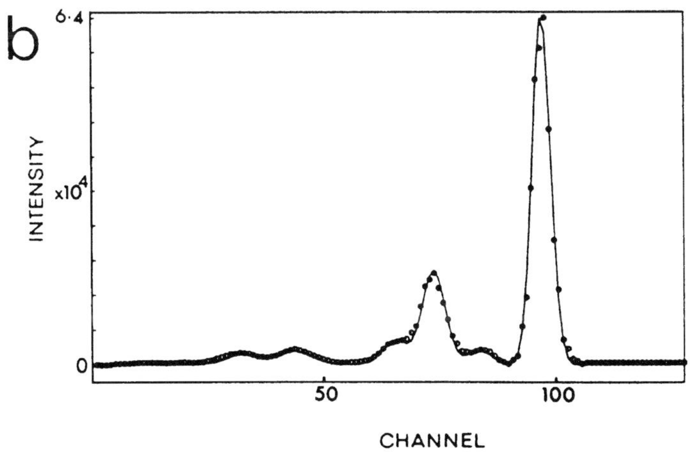

Fig. 10.7 Method of analysis of equatorial X-ray diffraction data from a vertebrate striated muscle. (a) Shows the result of fitting Lagrangian functions (*smooth line*) to the backgrounds on each side of the diffraction pattern under the observed data (*filled circles*). With these backgrounds removed the patterns can be folded across the middle and added to give curves such as (b, *circles*). (b) Can then be fitted with overlapping model functions (e.g. Gaussians, Lorentzians) representing in this case the A(1, 0), A(1, 1), A(2, 0), A(2, 1) and A(3, 0) reflections and the Z(1, 0) reflection. Present day analysis then refines both the background and the model peaks to give a best fit to the data. In this case the smooth line is the modelled intensity profile calculated to fit the data (*circles*). (From Harford and Squire 1990.)

peak, say, to be spatially consistent with the position predicted from the other A-band reflections. Another problem is the separation of the A(2, 0) and the Z(1, 1) reflections. When carrying out Fourier synthesis from the equatorial intensity data such separation is obviously essential. A method has been devised to do this (Harford *et al.* 1994), making use of the peak broadening due to the paracrystallinity of the A-band lattice (Yu *et al.* 1985), but this is rather laborious. It showed that in the case of resting fish muscle the intensity of the Z(1, 1) may well be more than that of the A(2, 0), although this apparently reverses in other muscle states, where the Z(1, 1) appears to become very weak, presumably due to disorder in the Z-line.

2-D diffraction data

Analysis of 2-D diffraction data requires the fitting of some sort of 2-D background function, whether the intensities to be stripped are continuous layer-lines or discrete Bragg reflections. Various methods for this have been tried and useful programs are available, such as the CCP13 program LSQINT (Denny 1994) or FIT-2D (Hammersley *et al.* 1994). Some methods work by identifying regions of the observed diffraction pattern that do not contain the signal to be analysed. Smooth 2-D analytical functions are then fitted to these points to give an estimated 2-D background, which is subtracted from the diffraction pattern. Layer-line intensities are then collapsed onto the layer-line positions and the intensities collected as 1-D intensity profiles. Bragg peaks are fitted in LSQINT using a two-step process. The positions and shapes of the peaks, but not their intensities, are modelled in terms of, say, 2-D Gaussian functions with allowance made for the effects of disorientation in the sample (Fig. 10.8). The background is then estimated from the positions outside these so-called 'no-fit' peaks and is subtracted. The parameters of the remaining peaks at the 'no-fit' positions are then refined by least-squares analysis or simulated annealing to give reliable intensities for the Bragg reflections, together with standard deviations estimating the accuracy of fit of each peak. The indexed layer-line profiles or indexed Bragg peak intensities can then be used to provide information about the muscle sarcomere contents as detailed in the next section.

10.6 Interpretation and modelling of diffraction data

Intensity time-courses

In general, apart from provision of accurate periodicities, muscle diffraction patterns are used to provide two kinds of information. The observed intensities (Fig. 10.5c,d) can either be used as indicators that structural changes are taking place and to show the speed at which these changes might occur (assuming that the nature of the structural change can be identified) or they can be used for full structure determination, analogous to what protein crystallographers do, but

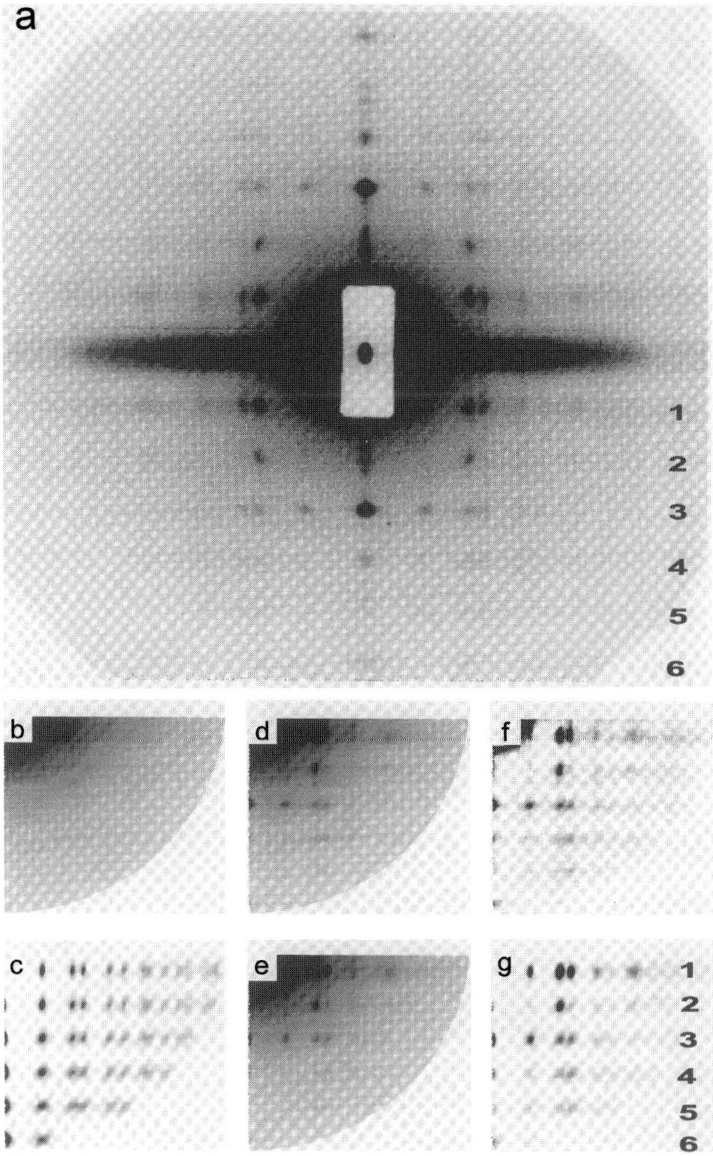

Fig. 10.8 (a) A resting diffraction pattern from plaice fin muscle showing the myosin layer lines 1 to 6 and various steps (b) to (g) in fitting this diffraction pattern. (c) A NOFIT representation of the shapes and positions of the Bragg peaks in (a) but not their intensities. (b) The background obtained by fitting the intensities between the NOFIT peaks. (f) The original data with the background in (b) removed and (g) the Bragg peaks modelled as in (c) but with the intensities fitted. Comparison can be made between the original data (d) and the modelled data (e), which is the sum of (b) and (g). Note that data in (a) were quadrant-folded to give (d), and (b) to (g) all represent the *bottom right quadrant* of (a); the undiffracted beam is at the *top left corner* of each figure in (b) to (g). (From Hudson *et al.*, in press.)

using a different approach. In the past it has been common practice to compare the intensity time-courses of, say, the A(1,0) and A(1,1) equatorial reflections and also the 143 Å meridional reflection with the tension produced by the muscle. This kind of approach was pioneered by Hugh Huxley and his colleagues (e.g. Huxley *et al.* 1980, 1982) and is exemplified by the analogous results in Fig. 10.9a for active bony fish muscle (Harford and Squire, 1992). Here it can be seen that the A(1,0) and A(1,1) intensities not only have time-courses different from each other, but also that only the A(1,0) time-course is similar to the tension record. The A(1,1) time-course is quite different; it may reflect approximately the *total* number of attached heads (weak plus strong binding heads), whereas the A(1,0) may reflect approximately just the strong-binding population (Harford and Squire 1992). It was shown many years ago (Elliott *et al.* 1963; Huxley 1968; Haselgrove and Huxley 1973) that, compared with diffraction patterns from relaxed muscle, those from active or rigor muscles showed A(1,0) reflections of reduced intensity and A(1,1) reflections of increased intensity. This was shown to be consistent with the movement of mass from around the myosin filament backbone in resting muscle, to being close to or on the actin filaments in active or rigor muscle.

This idea of the reciprocal changes in the intensities [I(1,0) and I(1,1)] of the A(1,0) and A(1,1) reflections came to be regarded as a well-established fact and the time-courses of these reflections were used almost universally as measures of the *number* of myosin heads attached to actin. However, it is very easy to show that there is no theoretical reason why the I(1,0) or I(1,1) values should be directly related to attachment number; these intensities also depend on the precise azimuthal position of the labelled actin-binding site around the actin filaments and on the exact configuration of the heads on their binding-sites on actin (Lymn and Cohen 1975; Squire 1981; Yu 1989; Squire *et al.* 1994*b*; Malinchik and Yu 1995; Harford and Squire, in press). It is because of these factors that the I(1,0) and I(1,1) time-courses in Fig. 10.9 are not the same; the observed changes in fact show that interesting structural transitions are taking place in the myosin head population. Unfortunately, there is no way of revealing the nature of these structural transitions just from inspection of the intensity changes. The only information that the equatorial reflections carry is information about the electron density in the A-band unit cell in projection down the *c* (fibre) axis. If one could reconstruct this density distribution then the equatorial reflections would have yielded all the information that they carry.

As discussed earlier, in order to produce images from X-ray diffraction data it is necessary to know both the amplitudes (*F*) of the diffraction peaks and the relative phases of the peaks. To a first approximation the vertebrate skeletal muscle A-band structure is centrosymmetric in projection down the fibre axis and this has the pleasing result that the phase angles associated with the reflections, at least out to the resolution at which centrosymmetry is still a valid assumption, are restricted to 0° or 180°, rather than being anything between 0 and 360°. In terms of the Fourier synthesis expression this means that the various terms

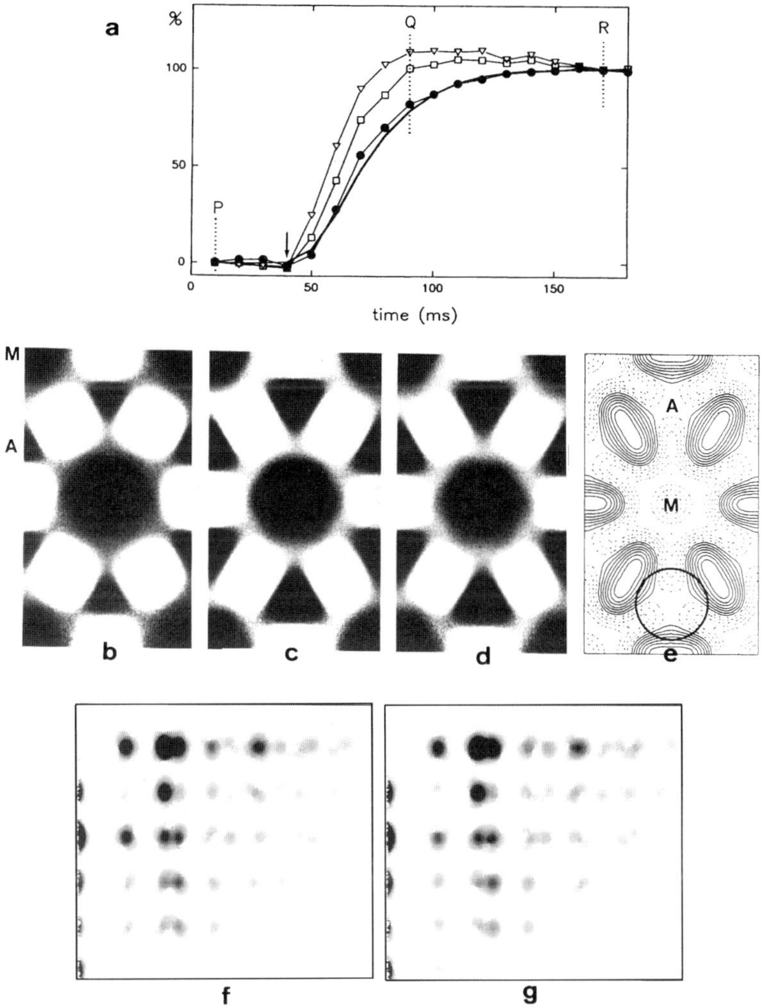

Fig. 10.9 (a) Time-courses of the intensity changes of the A(1, 0) and A(1, 1) equatorial reflections in the diffraction sequence, part of which is shown as Fig. 10.5c and d, relative to the recorded tension (*bold line*). Stimulus applied at *arrow*; all traces normalized to 0 and 100% for the resting a tension plateau phases. *Filled circles* represent the time sequence of mass arrival at actin (i.e. the mass within the *circle* in (e)); data were obtained from maps computed for 10-ms time intervals during the contraction cycle. (b) and (d) Electron density maps computed using phase factors 1, 1, −1, 1, 1 from the equatorial intensities in Fig. 10.5c and d, respectively and also (c) at an intermediate stage between. (e) Difference density (× 50) between (d) and (c) (i.e. d–c) shown as an contour plot; *solid lines* show increases, *broken lines* show decreases. The *circle* shows the region within which actin associated mass was counted for the mass time-course in (a). The density maps in (b) to (d) correspond to the time positions P, Q and R in (a). (From Harford and Squire 1992.) (f) and (g) The observed (f) and modelled (g) intensities from the optimized myosin filament structure in Fig. 10.10a. Each pattern corresponds to the same parts of the diffraction pattern as in Fig. 10.8b–g.

involving F require either to be added or subtracted to yield the required electron density (i.e. the *phase factor* is either $+1$ or -1). Using equatorial reflections out to the A(3, 0), there are five reflections in the Fourier synthesis summation [A(1, 0), A(1, 1), A(2, 0), A(2, 1) and A(3, 0)]. The phase factor for the first term can be assumed to be $+1$, leaving a combination of $2^4 = 16$ combinations of phase factor for the other reflections. Application of these different combinations in a trial and error process yields putative electron density maps that are usually accepted or rejected as plausible in comparison both with knowledge from electron microscopy and from the known mass ratios of the different A-band components (Harford and Squire 1990). On this basis there appear to be two main contenders for the correct phase combination: $+1$, $+1$, -1, $+1$, $+1$ and $+1$, $+1$, -1, -1 and $+1$. The former is preferred for fish muscle (Harford and Squire 1992; Harford *et al.* 1994), yields very satisfactory estimates of the filament masses, and is also consistent with the 3-D analysis discussed in the next section. The other combination has been preferred for frog muscle (Yu *et al.* 1985). In fact the A-band structure in frog is probably not identical to that in fish (Harford and Squire 1986); it is possible that the correct phase combination may differ for the two muscle types. Taking the fish muscle results, it is possible to calculate a series of electron density maps corresponding to the intensity time-courses shown in Fig. 10.9a and from these maps (Fig. 10.9b–d) one can extract an estimate of the amount of mass close to the actin filament (Fig. 10.9c) at any time during the contraction cycle. The result from this is shown in Fig. 10.9a. The time-course of this mass movement coincides neither with the intensity time-courses of the A(1, 0) and A(1, 1) peaks, nor with the tension. In particular, the mass time-course leads the tension time-course, suggesting that myosin heads initially attach to actin in a non-force producing state that is structurally different from the later state(s) in which force is generated.

This kind of analysis, in itself, is a major advance over attempts to use intensities on their own to provide structural information. However, it should be noted that the generation of electron density maps, as described here, not only depends on the assumption of centrosymmetry, but also assumes that the amplitudes of the various equivalent reflections at the same radial positions in reciprocal space, and which contribute to, say, the single observed A(1, 0) peak, are all equal. For the A(1, 0) there are six such components; the reflection is said to have a *multiplicity* of six. It is known that the myosin filament itself has three-fold rotational symmetry and, if it is assumed that on average at low resolution the whole A-band retains this symmetry (not unreasonable since there are six actin filaments around each myosin filament), then each of the six multiplets of the A(1, 0) will have the same intensity. Others of the first few equatorial reflections also have a multiplicity of six, except for the A(2, 1) peak which has a multiplicity of 12. In this case it is questionable whether the extra assumption of mirror symmetry needed to equate all 12 reflections in this multiplet is valid. Indeed, in the case of analysis of electron micrographs of rigor muscle, evidence against this kind of assumption is presented

(Hawkins and Bennett 1995). The main hope for this kind of analysis is to make use of the rest of the 2-D diffraction pattern, as detailed in the next section, because this allows the various symmetry assumptions to be tested rigorously.

In summary, the use of intensities (or even amplitudes) on their own as indicators of the timing of structural events has no theoretical justification (except in one or two special cases; Yu *et al.* 1979) and is not safe. The use of electron density maps from Fourier synthesis computations is very much to be preferred, but the assumptions on which these are based should be carefully scrutinized.

Note finally that in calculating electron density maps account should be taken of the problem of *series termination*. This is related to the process of Fourier synthesis from a limited dataset. In principle, in order to calculate a perfect equatorial electron density map for the muscle, the amplitudes and phases of every equatorial reflection out to very high resolution (very small d) need to be included in the Fourier summation. In fact the data used are usually limited to the $(3, 0)$ reflection at a spacing of about 120 Å. However, it is known that there are observable equatorial reflections beyond this. The fact that these have been cut out of the Fourier summation can lead to spurious detail within the computed electron density map. A practical way to avoid this is to multiply the observed amplitudes by an exponential function similar to a crystallographic temperature factor (Blundell and Johnson 1976; Cullity 1978). Thermal disorder in a system usually cuts down the observed intensities by an exponential factor so that beyond a certain resolution the reflections merge with the background. By imposing an artificial temperature factor so that any reflections beyond the outermost reflection included in the summation become negligibly small, then series termination errors are avoided (Franks *et al.* 1982; Harford and Squire 1992). The effect, in practice, is to make the computed images slightly blurred. However, what is then seen within the maps is more to be trusted than apparent details in maps obtained without a temperature factor correction.

Structure determination by modelling

The objective of X-ray diffraction studies of muscle is clearly to obtain unambiguous structural information about the distribution of molecules in the muscle and the way these molecules behave during contraction. In X-ray diffraction analysis so far, the positions of the various reflections, equatorials and layer-lines, provide unambiguous information about the *spacings* in the muscle, e.g. the unit cell size, the myosin and actin filament repeats and so on. But, until recently, they have not, for example, provided unambiguous data about the molecular arrangements in the A-band unit cell. However, enormous progress has been made by applying the technique of protein crystallography to various muscle components, particularly actin and the myosin S-1 head. The crystal structure of G-actin was determined from co-crystals of actin with DNase I, by Kabsch *et al.* in 1990. Ken Holmes then showed that a good fit could be obtained to high-angle X-ray (fibre) diffraction data from oriented gels if this actin monomer was placed in a particular way within the F-actin filament (Holmes *et al.* 1990). Since then, other crystals of actin-containing complexes have been solved (McLaughlin *et al.* 1993; Schutt *et al.* 1993); the actin monomer structures in each case were rather similar. A further milestone in muscle research was reached when the structure of myosin S-1 was solved using X-ray crystallography by Rayment *et al.* (1993). With these structures to hand, it is now possible to model the whole A-band unit cell in a realistic way and to refine the model against the observed low-angle X-ray diffraction data. In this way the structure of the A-band unit cell in bony fish muscle has been solved to quite high precision (Hudson *et al.*, in press) to reveal the myosin head positions and shape at the start of a contraction.

This approach is relatively easy to carry through with X-ray diffraction data from fish muscle, because the muscle has such good order that the myosin layer-lines are well-sampled to give Bragg peaks. The 2-D diffraction pattern can be carefully fitted using the CCP13 suite of programs (Fig. 10.8; Denny 1994; http://www.dl.ac.uk/CCP/CCP13). In addition, by parameterizing the A-band structure, computing the expected diffraction pattern as a function of these parameters and refining the parameter values against the observed diffraction intensities using a simulated annealing process followed by local refinement (Press *et al.* 1992), optimized structures, such as that shown in Fig. 10.10 can be obtained (Hudson *et al.*, in press). The observed and calculated intensity patterns from the myosin filaments in Fig. 10.10a are compared directly in Fig. 10.9f,g; they give a computed R-factor (measuring goodness of fit) of about 3%. Another advantage of using fish muscle is that the diffraction pattern is sensitive to the orientation of the myosin filaments within the A-band unit cell. It is therefore possible to find the positions of the myosin heads in resting muscle relative to the neighbouring actin filaments (Fig. 10.10b).

The full unit cell in Fig. 10.10b can be used not only as a starting point in modelling the time-series of X-ray diffraction patterns through, for example, a tetanic contraction (Squire *et al.* 1993, 1994*a*), but it can also be used to phase the diffraction pattern, including the equator, to test the assumptions on which the 2-D equatorial Fourier synthesis is based.

10.7 Problems and limitations: future prospects

Before summarizing what is likely to happen with the time-resolved X-ray diffraction technique over the next few years, it is as well to pause and reflect on the question, is it possible that the mechanism of force generation does not involve large scale structural changes of myosin and/or actin but that all the important events occur at an atomic level that would not be detectable by low-angle X-ray diffraction? Two comments on this spring to mind. The first is that it has been shown already that regulation of thin filaments does involve large scale

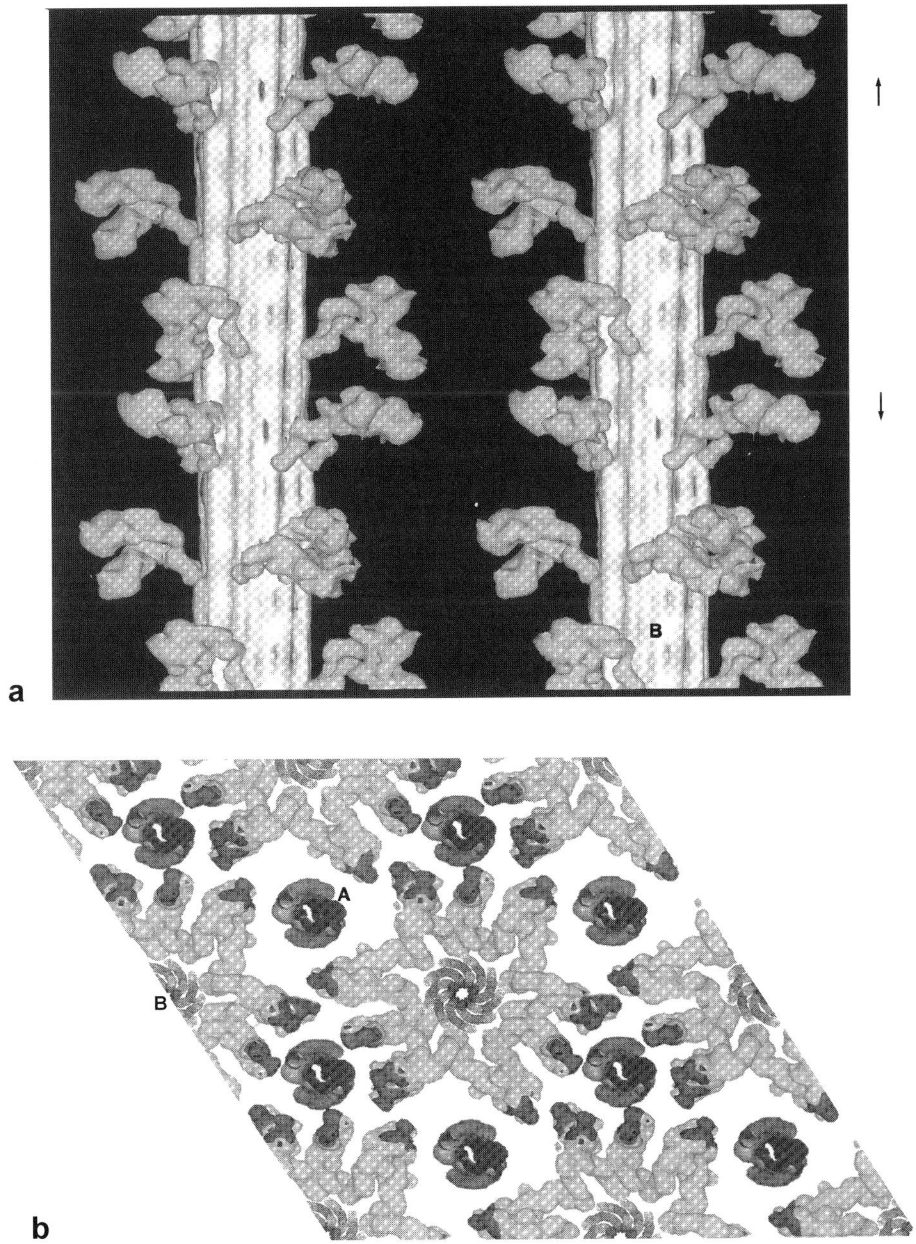

Fig. 10.10 (a) The 'best' model of the myosin head array in fish muscle obtained by simulated annealing and further local refinement of a parameterised filament structure. The filament backbone (B) is from Chew and Squire (1995). (b) The myosin filament in (a) has been put into the unit cell with its refined rotational position around its long axis and is shown together with actin filaments as modelled by AL-Khayat *et al.* (1995). A, actin filaments; B, myosin filament backbone; the unit cell side is about 440 Å and four unit cells are shown. (From Hudson *et al.*, in press.)

movements both of tropomyosin and of the actin monomer sub-domains (AL-Khayat *et al.* 1995). Of course this does not preclude the possibility that some aspects of regulation involve very subtle changes at the atomic level in actin, tropomyosin and the myosin head. Indeed it would be remarkable if such changes did not take place, but if they occur such changes will only be detectable by protein crystallography or other atomic resolution methods and that is not going to be easy. On the other hand there remain a number of central questions about regulation that low-angle diffraction can provide the answers to. The second comment is that the contractile event itself appears to involve a step length of about 55 to 120 Å (Huxley 1969; Finer *et al.* 1994; Molloy *et al.* 1995); just the sort of spacing that lies within the range of interest of low-angle X-ray diffraction. This either means that there is no change in the myosin head configuration or attitude on actin, but that there is some kind of head sliding action on the thin filament surface, or it means that there is a large scale change in configuration of the myosin heads attached to actin. In either case, results from low-angle X-ray diffraction can, in principle, distinguish between the two possibilities. In the end it is likely that details of the mechanism at atomic resolution will come from crystallography of actin/myosin co-crystals in different states, but such details will need to be related to events in whole muscle as defined by lower resolution techniques. Time-resolved low-angle X-ray diffraction still has a great deal to offer.

Because of its inherent advantages in providing reliable structural information about living, contracting muscle, it is inevitable that ways will be sought, not only to allow more rapid and direct interpretation of time-resolved X-ray diffraction data, but also to improve the quality and nature of the experimental methods. The development of new synchrotron sources, and possibly other kinds of X-ray source, will provide collimated X-ray beams of smaller cross-section and higher brilliance than at present. New generations of detectors, such as RAPID and its successors, will become available and more sophisticated data stripping software packages will gradually evolve. However, the future is not completely without its problems. In particular, the use of beams of ever higher photon density will inevitably lead to problems of radiation damage which at present are not well defined or understood. In addition, the acquisition of data at both higher spatial resolution and higher time resolution poses problems in the shear volume of data that needs to be stored and manipulated. Finally there are some basic aspects of diffraction that become more critical as specimens become smaller and X-ray beams are more brilliant and more highly parallel. An indication of possible problems to come was obtained in a diffraction experiment on line 16.1 at the Daresbury SRS using a stopped down beam to provide a nearly point focus and using an excellent *whole muscle* fish specimen (Harford and Squire, unpublished observations). Imagine the surprise at obtaining a beautifully ordered 2-D diffraction pattern (as in Fig. 10.8a) but with one major anomaly; the meridian was completely missing. This is easily explained if the muscle fibres are all parallel and the beam is highly collimated. As mentioned in the section on Camera design, this means that, if the muscle fibre axis is at 90° to the incident beam, Bragg's Law cannot be satisfied for the meridional reflections. One would need to tilt the muscle carefully away from perpendicular in order to record individual meridional reflections accurately. This instance occurred with a whole muscle where one might expect a small spread of fibre orientations, but even here the order proved to be too good to escape the problem. It is even more evident that, with the increased use in the future of single fibre preparations, such geometrical constraints will become increasingly important; muscle diffraction will get closer and closer to being like a form of crystallography of crystalline specimens, albeit with very large spacings. The future of time-resolved X-ray diffraction studies of muscle looks very promising, but such experiments will soon move into a different regime where aspects of the diffraction method itself will become more critical.

Acknowledgements

I am greatly indebted to my colleagues in the Muscle Group at Imperial College for their constant support and the pleasure of working with them, and also many people at the CCLRC Daresbury Laboratory who have helped, not only with setting up beam lines (i.e. Dr Liz Towns-Andrews, Mrs Sue Slawson), but also with provision and maintenance of detectors (i.e. Drs Rob Lewis and Chris Hall). It is also a pleasure to acknowledge the foresight of Professor Joan Bordas in pushing for improved low-angle diffraction facilities at the Daresbury SRS and for helping to create exceptionally fine beam-lines. My particular thanks are due to my colleague Dr Jeff Harford, who has become the main experimentalist in the Group working on time-resolved X-ray diffraction studies of fish muscle, and Dr Richard Denny, who, under the umbrella of CCP13, has developed much of the software used for stripping of data from our diffraction patterns. Finally, Liam Hudson has been involved in modelling the observed fish muscle diffraction data and has developed a program for this that we hope will eventually be suitable for release to other muscle diffractionists. Some of the figures included here are from my colleagues, John Barry and Liam Hudson. The research in the Muscle Group has been funded over the years by the SERC (now EPSRC), AFRC (now BBSRC), MRC, EMBO, the Wellcome Trust, the British Heart Foundation, the Leverhulme Trust, the Muscular Dystrophy Association of America and the Arthritis and Rheumatism Council. The support of all of these bodies is gratefully acknowledged.

References

Allinson, N.M. (1994) Development of non-intensified charge-coupled device area X-ray detectors. *J. Synchrotron Rad.* **1**, 54–62.

AL-Khayat, H.A., Yagi, N. and Squire, J.M. (1995) Structural changes in actin–tropomyosin during muscle regulation: computer modelling of low-angle X-ray diffraction data. *J. Mol. Biol.* **252**, 611–32.

Amemiya, Y. and Wakabayashi, K. (1991) Imaging plate and its application to X-ray diffraction from muscle. *Adv. Biophys.* **27**, 115–28.

Amemiya, Y., Tameyasu, T., Tanaka, H., Hashizume, H. and Sugi, H. (1980) Time-resolved X-ray diffraction from frog skeletal muscle during shortening against an inertial load and a quick release. *Proc. Japan Acad.* **56**, 235–40.

Amemiya, Y., Wakabayashi, K., Tanaka, H., Ueno, Y. and Miyahara, J. (1987) Laser-stimulated luminescence used to measure X-ray diffraction of a contracting striated muscle, *Science* **237**, 164–8.

Amemiya, Y., Matsushita, T., Nakgawa, A., Satow, Y., Miyahara, J. and Chikawa, J. (1988) Design and performance of an imaging plate system for X-ray diffraction study. *Nuc. Instr. Methods* **A266**, 645–53.

Amemiya, Y., Kishimoto, S., Matsishita, T., Satow, Y. and Ando, M. (1989) Imaging plate for time-resolved X-ray measurements. *Rev. Sci. Instr.* **60**, 1552–6.

Amemiya, Y., Yagi, N., Kishimoto, S., Oguchi, T., Suzuki, M., Ohnishi, Y. *et al.* (1994) The development of X-ray television detectors at the Photon Factory. In *Synchrotron radiation in the biosciences* (ed. B. Chance *et al.*) pp. 395–406. Oxford Science Publications.

Arndt, U.W. (1994) X-ray television detectors. In *Synchrotron radiation in the biosciences* (ed. B. Chance *et al.*) pp. 388–94. Oxford Science Publications.

Bagni, M.A., Cecchi, G., Griffiths, P.J., Maeda, Y., Rapp, G. and Ashley, C.C. (1994) Lattice spacing changes accompanying isometric tension development in intact single muscle fibers. *Biophys. J.* **67**, 1965–75.

Barrington-Leigh, J., Goody, R.S., Hofmann, W., Holmes, K.C., Mannherz, H.G., Rosenbaum, G. *et al.* (1977) The interpretation of X-ray diffraction from glycerinated flight muscle fibre bundles: new theoretical and experimental approaches. In *Insect flight muscle* (ed. R.T. Tregear) pp. 137–46. North-Holland, Amsterdam, New York and Oxford.

Bennett, P.M., Foster, G.F., Buckley, C.J. and Burge, R.E. (1993) The effect of soft X-radiation on myofibrils. *J. Microsc.* **172**, 109–19.

Bliss, N., Bordas, J., Fell, B.D., Harris, N.W., Helsby, W.I., Mant G.R. *et al.* (1995) W16.1: a new fixed wavelength diffraction station at the SRS Daresbury. *Rev. Sci. Instrum.* **66**, 1311–13.

Blundell, T.L. and Johnson, L.N. (1976) *Protein crystallography*. Academic Press, London and Orlando.

Bordas, J., Koch, M.H.J., Clout, P.N., Dorrington, E., Boulin, C. and Gabriel, A. (1980) A synchrotron radiation camera and data acquisition system for time-resolved X-ray scattering studies. *J. Phys. E Sci. Instrum.* **13**, 938–44.

Bordas, J., Diakun, G.P., Harris, J.E., Lewis, R.A., Mant, G.R., Martin-Fernandez, M.L. *et al.* (1991) Two-dimensional time resolved X-ray diffraction of muscle: recent results. *Adv. Biophys.* **27**, 15–33.

Bordas, J., Diakun, G.P., Diaz, F.G., Harries, J.E., Lewis, R.A., Lowy, J. *et al.* (1993) Two-dimensional time-resolved X-ray diffraction studies of live isometrically contracting frog sartorius muscle. *J. Musc. Res. Cell Mot.* **14**, 311–24.

Cecchi, G., Bagni, M.A., Griffiths, P.J., Ashley, C.C. and Maeda, Y. (1990) Detection of radial crossbridge force by lattice spacing changes in intact single muscle fibers. *Science* **250**, 1409–11.

Cecchi, G., Griffiths, P.J., Bagni, M.A., Ashley, C.C. and Maeda, Y. (1991) Time-resolved changes in equatorial X-ray diffraction and stiffness during rise of tetanic tension in intact length-clamped single muscle fibers. *Biophys. J.* **59**, 1273–83.

Charpak, G., Bouclier, R., Bressani, T., Favier, J. and Zupancic, C. (1968) The use of multiwire proportional counters to select and localise charged particles. *Nucl. Inst. Methods* **62**, 262–8.

Chew, M.W.K. and Squire, J.M. (1995) Packing of α-helical coiled-coil myosin rods in vertebrate muscle thick filaments. *J. Struct. Biol.* **115**, 233–49.

Cullity, B.D. (1978) *Elements of X-ray diffraction* (2nd edn) Addison-Wesley, Menlo Park, California.

Cullmann, E., Kunneth, T., Neff, W. and Stephan, K.H. (1987) Comparison of different X-ray sources using the same printing process parameters *J. Vac. Sci Technol.* **B5**, 638–40.

De Silva, L.B., Moreno, J.C., Barbee, T.W. Jr, Trebes, J.E., Cauble, R., Wan, A.S., *et al.* (1992) Imaging biological objects with X-ray lasers. *ICF Quarterly Report* (Lawrence Livermore National Laboratory, California) **3**, 10–4.

Denny, R. (1994) *CCP13 Newsletters 1994, 1995* (http://www.dl.ac.uk/CCP/CCP13).

Eder, D.C. and Matthews, D.L. (ed.) (1994) *X-ray lasers* Vol. 332 American Inst. Physics, Williamsburg, VA.

Elliott, G.F., Lowy, J. and Worthington, C.R. (1963) An X-ray and light diffraction study of the filament lattice of striated muscle in the living state and in rigor. *J. Mol. Biol.* **6**, 295–305.

Faruqi, A.R. and Bond, C.C. (1980) Parallel readout multiwire proportional chambers for time-resolved X-ray diffraction experiments. *Nucl. Inst. Methods* **176**, 71–7.

Faruqi, A.R. and Huxley, H.E. (1978) Time-resolved studies on contracting muscle using small-angle X-ray diffraction. 1: Design of Data Collection System. *J. Appl. Cryst.* **11**, 449–54.

Faruqi, A.R. and Huxley, H.E. (1981) A review of techniques for time-resolved X-ray studies on muscle. In *Scattering techniques applied to supramolecular and non-equilibrium systems* (ed. S.-H. Chen, B. Chu and R. Nossal) Plenum Publishing Corp. New York.

Finer, J., Simmons, R. and Spudich, J. (1994) Single myosin moleculae mechanics: piconewton forces and nanometre staps. *Nature* **368**, 113–9.

Foster, G.F., Buckley, C.J., Bennett, P.M. and Burge, R.E. (1992) Investigation of radiation damage to biological specimens at water window wavelengths. *Rev. Sci. Instrum.* **63**, 599–600.

Franks, N.P., Melchior, V., Kirschner, D.A. and Caspar, D.L.D. (1982) Structure of myelin lipid bilayers. Changes during maturation. *J. Mol. Biol.* **155**, 133–53.

Fraser, R.D.B. and MacRae, T.P. (1973) *Conformation in fibrous proteins and related synthetic polypeptides.* Academic Press, London and Orlando.

Fujisaki, H., Takahashi, S., Ohzeki, H., Sugisaki, K., Kondo, H., Nagata, H. *et al.* (1996) Soft X-ray damage to biological samples *J. Microscopy* **182**, 79–83.

Gabriel, A. (1977) Position sensitive X-ray detector. *Rev. Sci. Instrum.* **48**, 1303–5.

Griffiths, P.J., Ashley, C.C., Bagni, M.A., Maeda, Y. and Cecchi, G. (1993) Crossbridge attachment and stiffness during isotonic shortening of intact single muscle fibres. *Biophys. J.* **64**, 1150–60.

Hall, G. (1995) Silicon pixel detectors for X-ray diffraction studies at synchrotron sources. *Q. Rev. Biophys.* **28**, 1–32.

Hammersley, A.P., Svennson, S.O. and Thompson, A. (1994) Calibration and correction of spatial distortions in 2D detector systems. *Nucl. Inst. Methods* **A346**, 312–21.

Harford, J.J. and Squire, J.M. (1986) "Crystalline" myosin cross-bridge array in relaxed bony fish muscle: low-angle X-ray diffraction from plaice fin muscle and its interpretation. *Biophys. J.* **50**, 145–55.

Harford, J.J. and Squire, J.M. (1990) Static and time-resolved X-ray diffraction studies of fish muscle. In *Molecular mechanisms in muscular contraction* (ed. J.M. Squire), Vol. 13, pp. 287–300. Macmillan Press.

Harford, J.J. and Squire, J.M. (1992) Evidence for structurally different attached states of myosin cross-bridges on actin during contraction of fish muscle. *Biophys. J.* **63**, 387–96.

Harford, J.J. and Squire, J.M. (in press) Time-resolved diffraction studies of muscle using synchrotron radiation. *Reports on Progress in Physics.*

Harford, J.J., Luther, P.K. and Squire, J.M. (1994) Equatorial A-band and I-band X-ray diffraction from relaxed and active fish muscle: further details of myosin crossbridge behaviour. *J. Mol. Biol.* **239**, 500–12.

Haselgrove, J.C. and Huxley, H.E. (1973) X-ray evidence for radial cross-bridge movement and for the sliding filament model in actively contracting skeletal muscle. *J. Mol. Biol.* **77**, 549–68.

Haselgrove, J.C., Faruqi, A.R., Huxley, H.E. and Arndt, U.W. (1977) The design and use of a camera for low-angle X-ray diffraction experiments with synchrotron radiation. *J. Phys. E. Sci. Inst.* **10**, 1035–44.

Hawkins C.J. and Bennett, P.M. (1995) Evaluation of freeze substitution in rabbit skeletal muscle. Comparison of electron microscopy X-ray diffraction. *J. Musc. Res. Cell Mot.* **16**, 303–18.

Henderson, R. (1995) The potential and limitations of neutrons, electrons and X-rays for atomic resolution microscopy of unstained biological molecules. *Q. Rev. Biophys.* **28**, 171–93.

Hendrix, J., Koch, M.H.J. and Bordas, J. (1979) A double focusing X-ray camera for use with synchrotron radiation. *J. Appl. Cryst.* **12**, 467–72.

Holmes, K.C. and Blow, D.M. (1965) *The use of X-ray diffraction in the study of protein and nucleic acid structure.* Wiley, New York.

Holmes, K.C., Popp, D., Gebhard, W. and Kabsch, W. (1990) Atomic model of the actin filament. *Nature, London* **347**, 44–9.

Hudson, L., Harford, J.J., Denny, R. and Squire, J.M. (in press) Myosin head configuration in relaxed fish muscle; resting state myosin heads must swing axially by upto 150 Å or turn upside down to reach rigor. *J. Mol. Biol.*

Hukins, D.W.L. (1981) *X-ray diffraction by disordered and ordered systems.* Pergamon Press, Oxford and New York.

Huxley, A.F., Lombardi, V. and Peachey, L.D. (1981) A system for fast recording of longitudinal displacement of a striated muscle fibre. *J. Physiol. Lond.* **317**, 12–3P.

Huxley, H.E. (1968) Structural difference between resting and rigor muscle; evidence from Intensity changes in the low-angle equational X-ray diagram. *J. Mol. Biol.* **37**, 507–20.

Huxley, H.E. (1969) The mechanism of muscular contraction—recent structural studies suggest a revealing model for cross-bridge action at variable filament spacing. *Science* **164**, 1356–66.

Huxley, H.E. (1984) Time-resolved X-ray diffraction studies of cross-bridge movement and their interpretation. In *Contractile mechanisms in muscle* (ed. G.H. Pollack and H. Sugi) pp. 161–75. Plenum Press, New York.

Huxley, H.E. (1994) What X-rays tell us about muscle contraction—then and now. In *Synchrotron radiation in the biosciences* (ed. B. Chance *et al.*) pp. 409–17. Oxford Science Publications.

Huxley, H.E. and Brown, W. (1967) The low-angle X-ray diagram of vertebrate striated muscle and its behaviour during contraction and rigor. *J. Mol. Biol.* **30**, 383–434.

Huxley, H.E. and Faruqi, A.R. (1983) Time-resolved X-ray diffraction studies on vertebrate striated muscle. *Ann. Rev. Biophys. Bioeng.* **12**, 381–417.

Huxley, H.E. and Kress, M. (1985) Crossbridge behaviour during muscle contraction. *J. Musc. Res. Cell Motil.* **6**, 153–61.

Huxley, H.E., Stewart A. and Irving, T. (1994) X-ray diffraction measurements of the extensibility of actin and myosin filaments in contracting muscle. *Biophys. J.* **67**, 2411–21.

Huxley, H.E., Faruqi, A.R., Bordas, J., Koch, M.H.J. and Milch, J.R. (1980) The use of synchrotron radiation in time-resolved X-ray diffraction studies of myosin layer-line reflections during muscle contraction. *Nature* **284**, 140–43.

Huxley, H.E., Faruqi, A.R., Kress, M., Bordas, J. and Koch, M.H.J. (1982) Time-resolved X-ray diffraction studies of the myosin layer-line reflections during muscle contraction. *J. Mol. Biol.* **158**, 637–84.

Huxley, H.E., Simmons, R.M., Faruqi, A.R., Kress, M., Bordas, J. and Koch, M.H.J. (1983) Changes in the X-ray reflections from contracting muscle during rapid mechanical transients and their structural implications. *J. Mol. Biol.* **169**, 469–506.

Irving, M., Lombardi, V., Piazzesi, G. and Ferenczi, M.A. (1992) Myosin head movements are synchronous with the elementary force-generating process in muscle. *Nature* **357**, 156–8.

Kabsch, W., Mannherz, H.G., Suck, D., Pai, E.F. and Holmes, K.C. (1990) Atomic structure of the actin: DNaseI complex. *Nature* **347**, 37–44.

Kirz, J., Jacobsen, C. and Howells, M. (1995) Soft X-ray microscopes and their biological applications. *Q. Rev. Biophys.* **28**, 33–130.

Klug, A., Crick, F.H.C. and Wyckoff, H.W. (1958) Diffraction by helical structures. *Acta Crystallogr.* **11**, 199–213.

Kress, M., Huxley, H.E., Faruqi, A.R. and Hendrix, J. (1986) Structural changes during activation of frog muscle studied by time-resolved X-ray diffraction. *J. Mol. Biol.* **188**, 325–42.

Lewis, R. (1994) Multiwire gas proportional counters: decrepit antiques or classic performers? *J. Synchrotron Rad.* **1**, 43–53.

Lewis, R., Sumner, I., Berry, A., Bordas, J., Gabriel, A., Mant, G. *et al.* (1988) Multiwire X-ray detector systems at the Daresbury S.R.S. *Nuc. Instr. Methods* **A273**, 773–7.

Lewis, R., Worgan, J., Fore, N., d'Annunzio, F., Hall, C. and Parker, B. (1991) High count rate 2D proportional counter for synchrotron radiation. *Nuc. Instr. Methods* **A310**, 70–3.

Lewis, R.A., Helsby, W., Jones, A., d'Annunzio, F., Hall, C., Parker, B. *et al.* (1992) *Proceedings of the European workshop on X-ray detectors for synchrotron radiation sources.* (ed. A.H. Walenta) pp. 61, 257. University of Siegen, Germany.

Lombardi, V., Piazzesi, G., Ferenczi, M.A., Thirlwell, H., Dobbie, I. and Irving, M. (1995) Elastic distortion of myosin heads and repriming of the working stroke in muscle. *Nature* **374**, 553–5.

Lowy, J. and Poulsen, F.R. (1982) Time-resolved X-ray diffraction studies of the structural behaviour of myosin heads in a living contracting unstriated muscle. *Nature* **299**, 308–12.

Lowy, J. and Poulsen, F.R. (1987*a*) Fast X-ray diffraction studies of muscle; In *Fibrous protein structure* (ed. J.M. Squire and P.J. Vibert) pp. 451–94. Academic Press.

Lowy, J. and Poulsen, F.R. (1987*b*) X-ray study of myosin heads in contracting frog skeletal muscle. *J. Mol. Biol.* **194**, 595–600.

Luther, P.K. and Squire, J.M. (1980) Three-dimensional structure of the vertebrate muscle A-band II. The myosin filament superlattice. *J. Mol. Biol.* **141**, 409–39.

Luther, P.K., Munro, P.M.G. and Squire, J.M. (1981) Three-dimensional structure of the vertebrate muscle A-band III. M-region structure and myosin filament symmetry. *J. Mol. Biol.* **151**, 703–30.

Luther, P.K., Munro, P.M.G. and Squire, J.M. (1995) Muscle ultrastructure in the teleost fish. *Micron* **26**, 431–59.

Luther, P.K., Squire, J.M. and Forey, P.L. (1996) Evolution of myosin filament arrangements in vertebrate skeletal muscle. *J. Morph.* **229**, 325–35.

Lymn, R.W. and Cohen, G.H. (1975) Equatorial X-ray reflections and cross arm movement in skeletal muscle. *Nature* **258**, 770–2.

Maeda, Y., Popp, D. and McLaughlin, S.M. (1988) Cause of change in the thin filament-associated reflections on activation of frog muscle—myosin binding or conformational change of actin. In *Molecular mechanism of muscle contraction* (ed. H. Sugi and G.H. Pollack) pp. 381–90. Plenum Press, New York and London.

Maeda, Y., Cecchi, G., Ashley, C.C., Griffiths, P.J. and Bagni, M.A. (1994) Application of synchrotron radiation to studies of the contractile apparatus of single intact skeletal muscle fibres. In *synchrotron radiation in the biosciences* (ed. B. Chance *et al.*) pp. 478–85. Oxford Science Publications.

Malinchik, S. and Yu, L.C. (1995) Analysis of equatorial X-ray diffraction patterns from muscle fibers: factors that affect the intensities. *Biophys. J.* **68**, 2023–31.

Martin-Fernandez, M.L., Bordas, J., Diakun, G., Harries, J., Lowy, J., Mant, G.R. *et al.* (1994) Time-resolved X-ray diffraction studies of myosin head movements in live frog sartorius muscle during isometric and isotonic contractions. *J. Mus. Res. Cell Motil.* **15**, 319–48.

Matsubara, I., Yagi, N., Miura, H., Ozeki, M. and Izumi, T. (1984) Intensification of the 5.9-nm actin layer line in contracting muscle. *Nature* **312**, 471–3.

McLaughlin, P.J., Gooch, J.T., Mannherz, H.G. and Weeds, A.G. (1993) Structure of gelsolin segment 1—actin complex and the mechanism of filament severing. *Nature* **364**, 685–92.

Michette, A.G., Morrison, G.R. and Buckley, C.J. (ed.) (1992) *X-ray microscopy III*, Vol. 67. Springer series in optical sciences. Springer-Verlag, Berlin.

Mitsui, T., Wakabayashi, K., Wang, En-Z., Iwamoto, H., Tanaka, H., Kobayashi, T. *et al.* (1994) Frequency dependence of the variation of the X-ray diffraction pattern from tetanized frog skeletal muscle during sinusoidal length changes. In *Synchrotron radiation in the biosciences* (ed. B. Chance *et al.*) pp. 460–72. Oxford Science Publications.

Molloy, J., Burns, J., Kendrick-Jones, J., Tregear, R.T. and White, D. (1995) Movement and force produced by a single myosin head. *Nature* **378**, 209–12.

Morris, E.P., Squire, J.M. and Fuller, G.W. (1991) The 4-stranded helical arrangement of myosin heads on insect (*Lethocerus*) flight muscle thick filaments. *J. Str. Biol.* **107,** 237–49.

Norreys, P.A., Zept, M., Moustaizis, S., Fews, A.P., Zhang, J., Lee, P., *et al.* (1996) Efficient extreme UV harmonics generated from picosecond laser pulse interactions with solid targets. *Phys. Rev. Letters* **76**, 1832–5.

Oed, A. (1988) Position sensitive detector with microstrip anode for electron multiplication with gases. *Nucl. Inst. Methods* **A263**, 351–9.

Podolsky, R.J., Onge, R.St., Yu, L. and Lymn, R.W. (1976) X-ray diffraction of actively shortening muscle. *Proc. Nat. Acad. Sci.* (*USA*) **73**, 813–7.

Press, W.H., Teukolsky, S.A., Vetterling, W.T. and Flannery, B.P. (1992) *Numerical recipes in C*. Cambridge University Press, Cambridge, UK and New York.

Rayment, I., Rypniewski, W.R., Schmidt-Base, K., Smith, R., Tomchick, D.R., Benning, M.M *et al.* (1993) Three-dimensional structure of myosin subfragment-1: a molecular motor. *Science* **261**, 50–8.

Reedy, M.K. (1968) Ultrastructure of insect flight muscle I: screw sense and structural grouping in the rigor crossbridge lattice. *J. Mol. Biol.* **31**, 155–76.

Reedy, M.K., Holmes, K.C. and Tregear, R.T. (1965) Induced changes in orientation of the crossbridges of glycerinated insect flight muscle. *Nature* **207**, 1276–80.

Rome, E., Offer, G. and Pepe, F.A. (1973*a*) X-ray diffraction of muscle labelled with antibody to C-protein. *Nature* **244**, 152–4.

Rome, E.M., Hirabayashi, T. and Perry, S.V. (1973*b*) X-ray diffraction of muscle labelled with antibody to troponin-C. *Nature* **244**, 154–5.

Rosenbaum, G. and Holmes, K.C. (1980) Small-angle diffraction of X-rays and the study of biological structures. In *Synchrotron radiation research* (ed. H. Winick and S. Doniach) pp. 533–64. Plenum Press, New York.

Rosenbaum, G., Holmes, K.C. and Witz, J. (1971) Synchrotron radiation as a source of X-ray diffraction. *Nature* **230**, 434–7.

Schutt, C.E., Myslik, J.C., Rozycki, M.D., Goonesekere, N.C.W. and Lindberg, U. (1993) The structure of crystalline profilin-β-actin. *Nature* **365**, 810–6.

Shinohara, K. and Ito, A. (1991) Radiation damage in soft X-ray microscopy of live mammalian cells. *J. Microsc.* **161**, 463–72.

Squire, J.M. (1975) Muscle filament structure and muscle contraction. *Ann. Rev. Biophysics Bioeng.* **4**, 137–63.

Squire, J.M. (1981) *The structural basis of muscular contraction*. Plenum Press, New York and London.

Squire, J.M. (1992) The structures of striated and smooth muscles related to their function. In *Molecular and cellular aspects of muscle contraction and cell motility. Advances in comparative & environmental physiology*, Vol. 12 (ed. H. Sugi) pp. 87–131. Springer-Verlag.

Squire, J.M. and Harford, J.J. (1993) Time-resolved studies of crossbridge movement: why use X-rays? Why use fish muscle? In *Mechanism of myofilament sliding in muscle contraction* (ed. H. Sugi and G.H. Pollack) pp. 435–50. Plenum Press, New York.

Squire, J.M., Harford, J.J., Edman, A.C. and Sjostrom, M. (1982) Fine structure of the A-Band in Croysections. III: crossbridge distribution and the axial structure of the human C-zone. *J. Mol. Biol.* **155**, 467–94.

Squire, J.M., Harford, J.J. and Morris, E.P. (1993) Muscle—the movie. *Image Processing* **5** (issue 1, Spring), 22–3.

Squire, J.M., Harford, J.J. and AL-Khayat, H.A. (1994*a*) Molecular movements in contracting muscle: towards muscle—the movie. *Biophys. Chem.* **50**, 87–96.

Squire, J.M., Harford, J.J. and AL-Khayat, H.A. (1994*b*) Synchrotron radiation in time-resolved X-ray diffraction studies of molecular movements in muscle. In *Synchrotron radiation in the biosciences* (ed. B. Chance *et al.*) pp. 451–9. Oxford University Press.

Sugi, H., Amemiya, Y. and Hashizume, H. (1978) Time-resolved X-ray diffraction from frog skeletal muscle during an isotonic twitch under a small load. *Proc. Japan Acad.* **54**, 559–64.

Towns-Andrews, E., Berry, A., Bordas, J., Mant, G.R., Murray, P.K., Roberts, K. *et al.* (1989) *A time-resolved X-ray diffraction station: X-ray optics, detectors and data acquisition system*, 3rd International Conference on Synchrotron Radiation Instrumentation (SRI-88), Tsukuba, Japan, Aug. 28th to Sept. 2nd.

Vainshstein, B.K. (1966) *Diffraction of X-rays by chain molecules*. Elsevier, Amsterdam.

Wakabayashi, K. and Amemiya, Y. (1991) Progress in X-ray synchrotron diffraction studies of muscle contraction. In *Handbook on synchrotron radiation* 4 (ed. S. Ebashi, M. Koch and E. Rubenstein) pp. 597–678. Elsevier Science Publishers B.V., Tokyo.

Wakabayashi, K., Tanaka, H., Amemiya, Y., Fujishima, A., Kobayashi, T., Hamanaka, T. *et al.* (1985) Time-resolved X-ray diffraction studies on the intensity changes of the 5.9 and 5.1 nm actin layer lines from frog skeletal muscle during an isometric tetanus using synchrotron radiation. *Biophys. J.* **47**, 847–50.

Wakabayashi, K., Hamanaka, T., Hashizume, H., Wakabayashi, T., Amemiya, Y. and Matsushita, T. (1986) X-ray optics for a muscle diffractometer using synchrotron radiation. In *X-ray instrumentation for the photon factory* (ed. S. Hosoya, Y. Iitaka and H. Hashizume) pp. 61–74. KTK Scientific Publishers, Tokyo.

Wakabayashi, K., Ueno, Y., Amemiya, Y. and Tanaka, H. (1988) Intensity changes of actin-based layer lines from frog skeletal muscles during an isometric contraction. In *Molecular mechanisms of muscle contraction* (ed. H. Sugi and G.H. Pollack) pp. 353–67. Plenum Press, New York and London.

Wakabayashi, K., Tanaka, H., Saito, H., Moriwaki, N., Ueno, Y. and Amemiya, Y. (1991) Dynamic X-ray diffraction of skeletal muscle contraction: structural change of actin filaments. *Adv. Biophys.* **27**, 3–13.

Wakabayashi, K., Saito, H., Moriwaki, N., Kobayashi, T. and Tanaka, H. (1993) The first thin filament layer-line decreases in intensity during an isometric contraction of frog skeletal muscle. In *Mechanism of myofilament sliding in muscle contraction* (ed. H. Sugi and G.H. Pollack) pp. 451–60. Plenum Press, New York.

Worgan, J.S., Lewis, R., Fore, N.S., Summer, I.L., Berry, A., Parker, B. *et al.* (1990) The application of multiwire X-ray detectors to experiments using synchrotron radiation. *Nuc. Inst. Meth. Phys. Res.* **A291**, 447–454.

Worgan, J.S., Fore, N., Lewis, R.A., Hall, C., Helsby, W., Jones, A. *et al.* (1994) Developments in gas detectors for biophysics at the Daresbury SRS. In *Synchrotron radiation in the biosciences* (ed. B. Chance *et al.*) pp. 379–87. Oxford Science Publications.

Yagi, N. (1991) Intensification of the first actin layer-line during contraction of frog skeletal muscle. *Adv. Biophys.* **27**, 35–43.

Yagi, N. and Matsubara, I. (1988) Changes in the 5.9 nm actin layer-line on activation of frog skeletal muscles. In *Molecular mechanisms of muscle contraction* (ed. H. Sugi and G.H. Pollack) pp. 369–80. Plenum Press, New York and London.

Yagi, N. and Matsubara, I. (1989) Structural changes in the thin filament during activation studied by X-ray diffraction of highly stretched skeletal muscle. *J. Mol. Biol.* **208**, 359–63.

Yu, L.C. (1989) Analysis of equatorial X-ray diffraction patterns from skeletal muscle. *Biophys. J.* **55**, 433–40.

Yu, L.C. and Podolsky, R.J. (1990) Equatorial X-ray diffraction studies of single skinned muscle fibres. In *Molecular mechanisms in muscular contraction* (ed. J.M. Squire). pp. 265–86. Macmillan Press.

Yu, L.C., Lymn, R.W. and Podolsky, R.J. (1977) Characterization of a non-indexible equatorial X-ray reflection from frog sartorius muscle. *J. Mol. Biol.* **115**, 455–64.

Yu, L.C., Hartt, J.E. and Podolsky, J. (1979) Equatorial X-ray intensities and isometric force levels in frog sartorius muscle. *J. Mol. Biol.* **132**, 1–15.

Yu, L.C., Arata, T., Steven, A.C., Naylor, G.R.S., Gamble, R.C. and Podolsky, R.J. (1984). Structural studies of muscle during force development in various states. In *Contractile mechanism in muscle* (ed. G.H. Pollack and H. Sugi) pp. 207–20. Plenum Publishing Corporation, New York.

Yu, L.C., Steven, A.C., Naylor, G.R.S., Gamble, R.C. and Podolsky, R.J. (1985) Distribution of mass in relaxed frog skeletal muscle and its redistribution upon activation. *Biophys. J.* **47**, 311–21.

11 Electron microscopy coupled with quick freezing

Eisaku Katayama

11.1 Introduction

Transmission electron microscopy is a unique means of providing high resolution real-space images of biological ultrastructure and thus has greatly contributed to the detailed structural studies of muscle at various levels, from the morphology of tissues or cells to the submolecular structure of each contractile protein. However, the fixation and/or the staining procedures, which are required to visualize the fine structural details under the microscope, include time-consuming chemical processes in an aqueous phase, raising the possibility that the final images might include certain changes from the original native and thus physiological structures. Since it is difficult to evaluate the degree of such artifactual structural change from the real native one, the need for some kind of rapid fixation is apparent. Heuser *et al.* (1979) completed the practical experimental system of liquid (liq.). He-operated quick-freeze electron microscopy and captured the transient structure of a neuromuscular junction at the moment of synaptic transmission as vital images, showing for the first time the feasibility of such a technique as a powerful tool for coupling physiology with anatomy/cell biology. Since the surface of the specimen freezes in less than 1 ms (Heuser *et al.* 1979), this method is especially suitable for analysing the dynamic processes accompanying fast structural changes. The contraction of muscle involves quick and dynamic molecular events at cross-bridges coupled with cyclic enzymatic reactions, hence it is clear that this remarkable biological phenomenon is one of the most suitable subjects to be studied by such methodology.

The primary aim of this article is to give a brief explanation and critical evaluation of each technique used in quick-freeze electron microscopy of muscle and contractile proteins. Though there is a large literature on studies made with such techniques, the results of each study will not be discussed here, except for a brief summary on those studies related to the most intriguing structural change of myosin cross-bridges during muscle contraction.

11.2 Quick-freeze electron microscopy categorized by processing methods

There are too many studies to be listed in this section. Readers should refer to the many useful articles in the books dedicated to the description of principles, merits and problems for each methodology, e.g. Steinbrecht and Zierold (1987).

The whole procedure of quick-freeze electron microscopy consists of the initial freezing of biological materials and followed by processes to visualize them under an electron beam with reasonable contrast. The first freezing step is usually done in one of three ways:

(1) metal-contact freezing, where specimens are instantly slammed onto the polished surface of the copper block, pre-cooled to ultralow temperatures, preferably with liq. He;

(2) propane-jet freezing; where small and thin specimens are sandwiched in-between two pieces of thin copper plate and the total assembly is exposed to a massive stream of liq. N_2-cooled propane;

(3) plunge freezing, where thin specimens (usually dilute protein solutions put to the holey carbon film on the specimen grids) are dropped into liq. propane/ethane, which is cooled by vigorous mixing in liq. N_2.

Though some other methods have been devised for quick-freezing (e.g. Lang and Bronk 1978), they are not discussed here.

The second step, for contrast formation, is usually carried out in one of three ways:

(1) freeze-substitution (e.g. Bridgeman and Reese 1984);

(2) freeze-fracture deep-etch replication (e.g. Heuser and Salpeter 1979);

(3) direct observation of ice-embedded frozen-hydrated specimens (e.g. Stewart and Vigers 1986).

In (1), frozen specimens (materials should be solid tissues, otherwise they must be embedded in agar or collected by centrifugation at each step) are put into an excess volume of chilled organic solvents, such as acetone (solvents must remain liquid at $-80\,^{\circ}\mathrm{C}$) and kept there for a few days to completely remove water molecules. In most cases, chemical fixative such as osmium, tannate or aldehyde is added to the solvent so that it fixes the material as the temperature rises. These specimens are then embedded in plastic and the thin sections are examined under the electron microscope in the conventional manner.

In (2), the very surface of the frozen specimen (usually tissues or cells, but solution can be treated as will be described later) is lightly scraped off with a cold knife and exposed to high vacuum for the sublimation of surrounding water, leaving the cytoskeletal elements or membranous structures in the cell to protrude from the top level. The specimens are then rotary-shadowed and backed by carbon, followed by dissolution of the underlying biological materials, to give a stable replica film. With combined use of mica-flakes as adsorbents (Heuser 1983), it is also possible to examine the fine structure of soluble protein systems, essentially in the same way as above but with higher spatial resolution.

In (3), quickly frozen samples are set in a cold-stage and transferred into the electron microscope column. Unfixed and unstained proteins are directly examined under minimal dose electron-beam conditions. Since the image contrast here comes solely from the interference of a very small difference in density between protein and the surrounding water, it is usually necessary to take micrographs with large defocusing to enhance the visibility. The procedures listed above can be adequately coupled according to the aims and requirements of each experimental system.

11.3 Compatibility of processing methods with experimental materials used

In principle, it should be possible to select any combination of the freezing and processing methods listed above. In practice, however, the combination is somewhat restricted according to the experimental system setup or the materials used.

Muscle fibre

Whole muscle fibre (usually skinned or glycerinated) is naturally the material of choice when structural information is required, coupled with physiology experiments such as tension measurement. Freezing can be done by any of the methods above, though metal-contact freezing is the most practical and the most frequently used. Propane-jet freezing may be sometimes used, especially when the material is small, such as single or fine bundles of fibre (Suzuki and Pollack 1986; Suzuki *et al.* 1993). In some rare cases, plunge-freezing may be used, whereby a very small piece of muscle fibre is quickly dipped into liq. propane or N_2-thrush prepared instantly by the evacuation of liq. N_2 with rotary-pump. The performance of freezing, i.e. depth of reasonably frozen area, as well as the cost, decreases according to the order of the above three methods.

The selection of post-freezing process depends largely on the kind of information needed and the available equipment. Freeze-substitution can be carried out without any special equipment but is much more easily done if a $-80\,^{\circ}C$ deep-freezer is available. Since this procedure gives plastic-embedded thin sections of muscle tissues, a large accumulation of knowledge by conventional electron microscopy is a great help in interpreting the images. Furthermore, there are few restrictions in the experimental conditions of the materials. If necessary, intact muscle can be observed by the same protocol, though the image contrast might be less. Here, chemical fixation by dissolved fixative is very slow at the substitution temperature—it probably proceeds gradually while the solvent temperature rises, eventually to ambient. Plastic embedding is usually performed in a conventional manner including high temperature polymerization steps, except when special care is taken, for example, to preserve the antigenicity (e.g. Nicolas and Bassot 1993). Though protein molecules in the tissue are exposed to liquid (substitution solvent, dissolved fixative, plastic monomer), throughout the processes after the initial ice is replaced, morphological preservation seems good, judged by the diffraction pattern of control protein crystals processed by a standard protocol of freeze-substitution (Hirose *et al.* 1993). Based on such a test, this method is presumed to represent the fine structure of biological materials at the moment of freezing, if the initial quick-freezing is done properly to avoid ice crystal formation. Such a view would hold true, practically as long as the structural detail discernible with the resolution of conventional plastic-embedded section is concerned and if the morphological analyses are made on some statistical basis. However, it must be noted that the protein crystal used in the control experiment is tightly packed, leaving very little water space, while the cross-bridge lattice in muscle fibre has a large inter-protein space, retaining the possibility that the original cross-bridge structure might be subject to some delicate changes on a submicroscopic scale (Lenart *et al.* 1996). Thus, one must be very cautious not to over-interpret the very detailed and subtle shape of each cross-bridge visualized by this method.

The other option for examining quick-frozen muscle fibre is freeze-fracture deep-etch replication (Heuser 1983; Heuser and Cooke 1983; Ip and Heuser 1983). There are several factors that could potentially affect the fidelity of final images. The largest problem among them might be the distortion due to mandatory mechanical shear-force of knife-fracturing (Heuser 1983; Heuser and Cooke 1983). Skeletal muscle consists of a three-dimensional network of thick and thin filaments which are cross-connected by myosin cross-bridges. Though the situation might change according to the fundamental lattice organization pattern (vertebrate, skeletal, fast, slow, cardiac, invertebrate, flight, etc.) or chemical state (rigor, active or relaxed), knife-fracturing of frozen muscle tissue must cut certain intrinsic connections between proteins, especially under rigor conditions. The knife edge sometimes travels through and splits off the core of thick filaments, but when it scratches the cross-bridges, they are either cut off or sometimes pulled to distort the surrounding lattice through the connecting network (Heuser and Cooke 1983). In order to avoid the visible change of regular lattice, the direction of knife movement is usually set parallel to the long axis of the fibre. Though such distortion may become less conspicuous under lower cutting temperatures, it is very difficult either to eliminate completely or to evaluate correctly the effect of such manipulation, which is intrinsic to the methodology. It might be possible theoretically to observe the unfixed and unstained muscle fibre, directly in a cold stage (Ménétret and Craig 1994). However, there is no such application because of too much technical difficulty.

Solution of purified protein systems

Other than muscle fibres conveniently coupled with physiology experiments, a solution mixture of actin and myosin subfragment-1 (S1) is a useful material for examination by quick-freeze electron microscopy; especially when coupled with various biochemical experiments. Freeze-fracture deep-etch replication and cryo-electron microscopy of frozen-hydrated specimens are the processing methods of choice in this case. Since the replica represents the surface profile of the sample proteins, it is advantageous in that one can deal with samples of reasonably high protein concentration, in contrast with frozen-hydrated cryo-electron microscopy where high concentration is not compatible because this method gives the projected images. However, deep-etch replicas of quickly-frozen protein solution may not be easy to obtain, because individual particles of protein molecules or the complexes have been floating freely in solution before freezing; this means that the sample particles (especially, the small and globular ones) can easily escape from the surface, once exposed to the vacuum by the sublimation of surrounding water. Thus, one must devise some means of preventing the particles of interest from escaping before they are actually shadowed and replicated (Pollard *et al.* 1990, 1993; Funatsu *et al.* 1993). Usually this can be done by etching for a very short period; but unlike muscle fibre in which all the filaments run parallel, actin filaments are wavy and oriented randomly in solution, making it very difficult to obtain a longitudinal view of reasonable length. The low probability of observing adequately oriented particles will affect the final yield of good images showing well-preserved structural details. Another way to solve the problem is to use a relatively high concentration of actin filaments: they form a three-dimensional network and so might not escape, even under deep-etching conditions. However, the final images obtained by these methods might not be regarded as satisfactory, partly because the shadowed metal particles accumulate, not only on the wanted top-face but also on the lower-side. The overlapping of both side images of the filaments often makes the surface details obscure and very difficult to interpret by visual examination. Heuser (1983) has devised a very good alternative method by employing mica-flakes to avoid all of these difficulties: protein samples are pre-adsorbed onto the surface of finely-pulverized mica, and the whole slurry is processed exactly like usual tissue. Every step of the whole procedure (quick-freezing, knife-fracturing, deep-etching, rotary-shadowing and carbon-backing) is easily carried out and the dissolution of mica-flakes can be done simply by floating overnight on the surface of full-strength hydrofluoric acid. Though this method has not prevailed widely, it provides excellent images of single or complexed protein molecules, probably because the adsorption of the sample onto the very flat mica-surface prevents the lower half of the sample from metal accumulation as mentioned above. The only concern of this method is how the adsorption of protein samples affects their structure, which will be discussed shortly in the next section. Returning to the elementary steps of replica preparation, the effects of deep-etching and shadowing on protein molecules are the important issues to be considered. There is a report (Gross 1987) in which the behaviour of water molecules in a biological specimen was checked by using D_2O as a tracer, under the conditions of freeze-drying, i.e. slow temperature rise from $-130\,^{\circ}C$ to ambient temperature. According to this study, the sublimation of D_2O occurred in two separate phases. In the first phase (temperature goes up to $-80\,^{\circ}C$), water evaporation gradually occurred according to the rise in temperature but then declined and eventually stopped. In the second phase, up to room temperature, remaining D_2O started to sublime once more and abruptly depleted at the range from

-50 to -30 °C. Such a remarkable phenomenon was interpreted as showing the presence of two kinds of molecular species of water: so-called 'free' and 'bound' water, respectively. Such an interpretation means that under deep-etching conditions, only free water evaporates and protein-bound water, which would play an important role in maintaining the structure and function of proteins, remains unremoved. This is a very important and superior feature of freeze-drying compared to air-drying, which supports the good preservation of protein architecture (Morris *et al.* 1994) by deep-etching, i.e. partial freeze-drying. Rotary-shadowing is the final step for giving contrast to the sample object. Though the problems and solutions have been a matter of debate, it is clear that metal-shadowing at low temperature gives better resolution by far than that made at room temperature, probably because of less coalescence of metal particles after deposition (Zingsheim 1972). In summary it is feasible to reconstruct the three-dimensional surface profile of biological architecture from the projected images obtained by this methodology (Katayama *et al.* 1996).

The other means of visualizing quickly frozen protein solution is to observe the sample directly under an electron beam. Theoretically, it is the ideal and final method of choice for obtaining the least perturbed structural information. Here, the electron beam is scattered differentially by protein moiety and water: the small phase difference between them is converted to an intensity difference, producing the image contrast. Since the difference in the density between protein and water is very small, the final image contrast is very low, so the micrographs are usually taken under conditions of large defocus. Protein appears darker than the background water, a reversal of negative staining; this does not present much of a problem as long as low-resolution features are concerned. In order to get higher resolution, however, careful consideration of the relationship between defocus value and contrast transfer-function is required (e.g. Milligan and Flicker 1987; Milligan *et al.* 1990), since the contrast could be reversed in some features of intermediate resolution. Another problem encountered by this methodology is common to negative staining: the restriction in protein concentration of the sample solution. In order to observe individual molecules by transmission electron microscopy, it is necessary to prepare the specimen with lower concentration than is used routinely in various biochemical experiments. This is because the overlapping of the images under relatively high protein concentration makes the structural details hard to recognize by visual examination. On favourable occasions, the sample has some regularly repeated features, which are subject to various types of Fourier analysis. If this is not the case, one must devise the experimental system so that it is compatible with the low protein concentration. For example, low affinity active complex of acto-S1 in the presence of ATP is one of the most difficult materials in this sense. To observe such a complex in solution by this method, or by negative staining, two different approaches have been used: one is to chemically cross-link S1 with actin using an inert bifunctional reagent (Applegate and Flicker 1987) and the other is to use extremely low ionic conditions to enhance the interaction between them (Walker *et al.* 1994, 1995). It is of course necessary to check carefully the effect of such alterations made to the experimental system.

In vitro actin motility assay system

In vitro actin motility assay system is an experimental setup for observing directly the active sliding movement of fluorescently labelled actin filaments supported by myosin or its subfragments adsorbed onto the glass surface, under the fluorescence microscope. The total experimental protocol can be easily modified according to the purpose and either purified or crude protein samples can be used. Because of such flexibility, its application has rapidly expanded from the assay of motile activity of minute amounts of myosin in cell extract to the measurement of piconewton force produced by a single myosin head (Kishino and Yanagida 1988), despite the fact that this system was introduced only about 10 years ago (Kron and Spudich 1986; Harada *et al.* 1987). A number of physiological, biophysical and biochemical studies have been carried out using this system but, strangely, none of the structural studies compatible with this concept have been reported so far. In order to observe the structure of myosin cross-bridges supporting the sliding movement of actin filaments, the author has devised a protocol to substitute the glass surface with mica. Since mica is chemically analogous to glass, it was naturally expected that actin filaments would slide past myosin subfragments adsorbed onto the surface of mica. Though observation through mica sheets was a bit difficult, a similar sliding movement of actin certainly occurred and its velocity was somewhat slower than on more adequately-treated glass, as already reported by Harada *et al.* (1990). The reason for the slower sliding velocity has not yet been identified but possibly is due to some difference in its surface property. Whatever the reason, it is apparent that mica does work as a good substrate of *in vitro* actin motility assay, which means that myosin and its subfragments remain functional even after adsorption onto mica. Thus, the matching of quick-freeze deep-etch replication with an *in vitro* motility assay system is completely feasible. Since the protein components in the sample are supported by mica-surface and not by the protein network as in muscle fibre, shear-force of knife cutting applied to the frozen surface does not affect at all the deeply-etched layer of the sample, which is the field to be actually observed. Various features of these combined experimental systems, concomitantly, would provide new structural information which is probably difficult to obtain by other techniques. It is expected in future that some other techniques for providing structural information about the surface properties of the sample (e.g. atomic force microscopy) could be combined effectively with this superior *in vitro* functional assay system.

11.4 Summary of the results obtained by each technique

Since this article is not intended to provide an extensive review of the literature, only the main issues obtained by each technique on the structure of functioning cross-bridges will be described.

Freeze-substitution of muscle fibres

Tsukita and Yano (1984) made the first attempt to capture the structure of tension-developing glycerinated muscle fibre. They compared the diffraction pattern of a freeze-substituted plastic-embedded section of contracting muscle with that of rigor or relaxed muscle. They concluded that active cross-bridges follow the actin-based helix and their tilt-angle is closer to 90° than in the rigor state. Hirose and Wakabayashi (1993) used fresh muscle fibre, which was chemically skinned, expecting better structural preservation. They analysed the shape of each cross-bridge in longitudinal and cross-section and reported that active cross-bridges have a variety of shapes, different from rigor ones. Next, Hirose *et al.* (1993*a*) performed a 'flash and smash' experiment to combine the high time-resolution of quick-freeze electron microscopy with laser activated ATP generation system (caged-ATP). They followed the time sequence of the structural changes in cross-bridges and demonstrated that the gross structural features observed in the longitudinal sections are already mature by 50 ms after the release of ATP. It was also suggested that rigor cross-bridges are triangular-shaped, while the active ones look uniformly thin. Next, they examined the cross-sections of the same material by computer-aided correspondence analysis to classify the cross-bridge shapes into several discrete types (Hirose *et al.* 1993*b*). More recently, similar analyses were made with caged-Ca^{2+} as an initiator (Lenart *et al.* 1996).

Quick-freeze deep-etch replication of acto-S1 mixture

The first attempt to locate the structural changes of actin-associated myosin-heads (i.e. the protein counterpart of cross-bridges) at molecular resolution was made by the author (Katayama 1989) using a quick-freeze deep-etch technique combined with mica-flake methodology. The acto-S1 rigor-complex in the absence of ATP was easily visualized in a conventional way, but this was not the case for the active complex. So, the surface property of mica was modified and the morphology of actin-attached myosin head was successfully captured in its active form (weakly-bound myosin head). The major finding was that elongated myosin S1 is almost straight and bound to actin filament with about 45° tilt-angle under rigor condition, whereas its shape drastically changes to a short and rounded form when ATP is added. Such a change in the gross configuration of myosin heads was also demonstrated in the negatively stained images of acto-S1; they were covalently cross-linked not to dissociate under low protein concentration. Later, Pollard *et al.* (1990, 1993) and Funatsu *et al.* (1993) independently challenged the same issue by quick-freeze deep-etch replication but now coupled with quick-mixing or the photolysis of caged-ATP to trace the time-course of ATP-induced structural change. They obtained the images of globular particles bound to actin filaments, the quality or the resolution of which was similar to each other. However, since neither of them used mica-flake methodology, they could not define the clear elongated shape of the myosin heads and reached a different conclusion: that the tilt angle of the cross-bridges does not change upon binding of ATP (Pollard *et al.* 1993), and that it does (Funatsu *et al.* 1993); though the measured angle for each particle substantially differed in each care. The latter authors also analysed and denied the presence of any co-operativity in the association and dissociation of S1 and actin.

Cryo-electron microscopy of frozen-hydrated acto-S1

The first demonstration in frozen-hydrated specimen, of the conformational difference of ATP-hydrolyzing active S1 from rigor was made by Applegate and Flicker (1987) using chemically cross-linked acto-S1 as a sample. Their micrographs of acto-S1 in each state showed very similar morphology to that obtained by negative staining of the same sample (Craig *et al.* 1985; Katayama 1989), though there was still the argument, at the time, that such cross-bridges might represent merely tethered molecules by newly introduced artificial links—such a possibility was virtually ruled out later by the presence of a similar structural change found in uncross-linked acto-S1 observed by deep-etch replica by Katayama (1989); (see the previous section). Direct observation of uncross-linked frozen-hydrated acto-S1 in the presence of ATP was performed much later by Walker *et al.* (1994). They utilized the enhanced association of nucleotide-bound S1 with actin under extremely low salt and could actually visualize actin-attached S1 molecules in the presence of ATP. The defocused images they obtained under such condition exhibited a variable configuration of S1 particles. The structural features were very similar to those with covalently cross-linked complex in the presence of ATP as visualized in cross-linked materials by negative staining and frozen-hydrated specimens, or in uncross-linked materials by deep-etch replication with mica-flakes. Later, they modified the experimental system so that ATP could be rapidly mixed just before quick-freezing (Walker *et al.* 1995).

As regards the structural changes of myosin heads induced by nucleotides, Whittaker *et al.* (1995) and Jontes *et al.* (1995) examined the effect of ADP on the structure of actin-associated smooth muscle S1 and brush border myosin I, respectively. The most distal portion of each cross-bridge was displaced by about 3.5 nm to 5 nm by that treatment, from the original position under rigor conditions. They suggested the possibility that such movement could be part of an effective power-stroke postulated in a swinging cross-bridge mechanism; the prevailing explanation of tension-development.

Quick-freeze deep-etch replication of heavy meromyosin (HMM) cross-bridges supporting sliding actin filaments under *in vitro* motility assay conditions

The author has devised an experimental system for observing the configuration of myosin heads, which are actually in operation to translocate actin-filaments while hydrolyzing ATP (Katayama 1993*a*). Since there is no other published

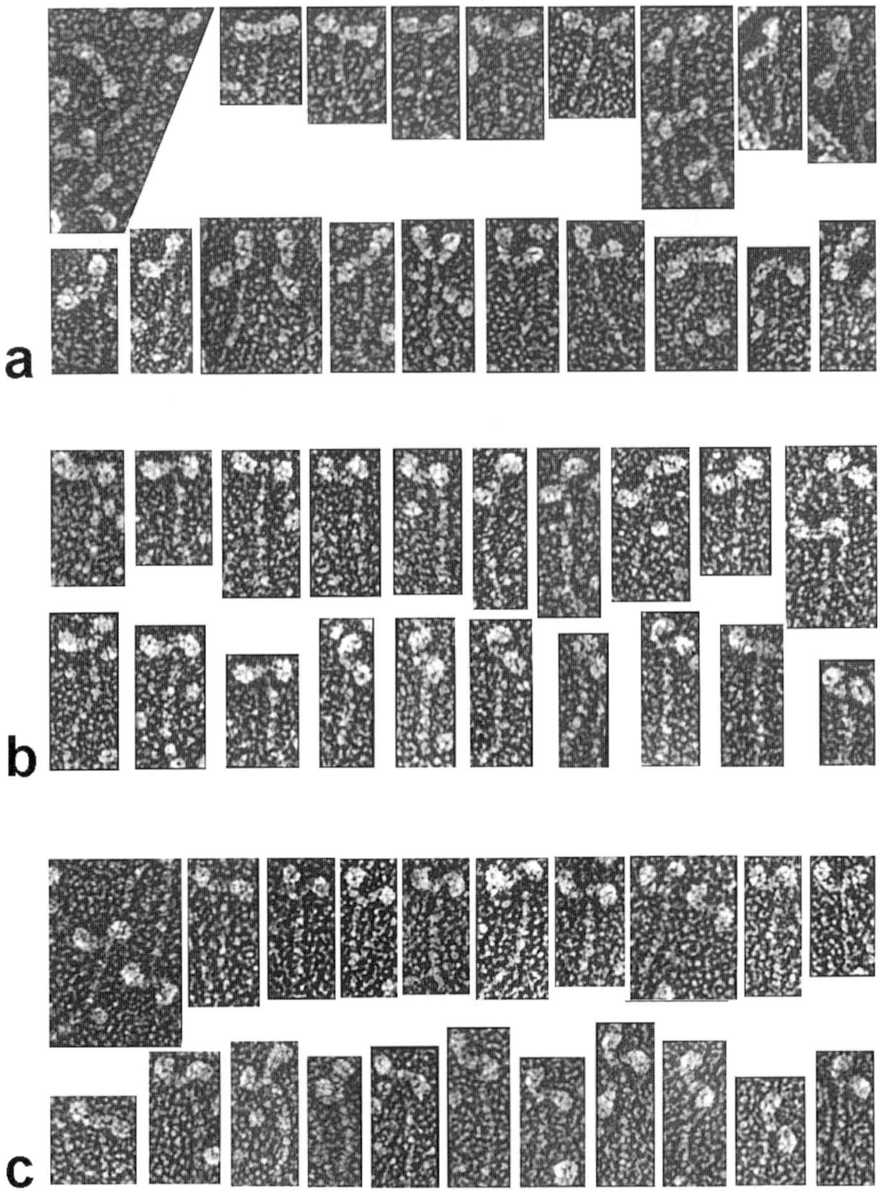

Fig. 11.1 Gallery of electron micrographs indicating the conformational change of HMM heads induced by nucleotides (Katayama 1991, 1993*b*). (a) In the absence of ATP, two pear-shaped heads of individual HMM are elongated and almost straight. (b) In the presence of ATP, they are strongly kinked to the same direction forming a sigmoid shape. (c) By the addition of ADP/inorganic vanadate, HMM head forms an analogous complex to HMM/ADP/Pi and takes the configuration similar to (b). All the images were obtained using quick-freeze deep-etch replication coupled with mica-flake methodology.

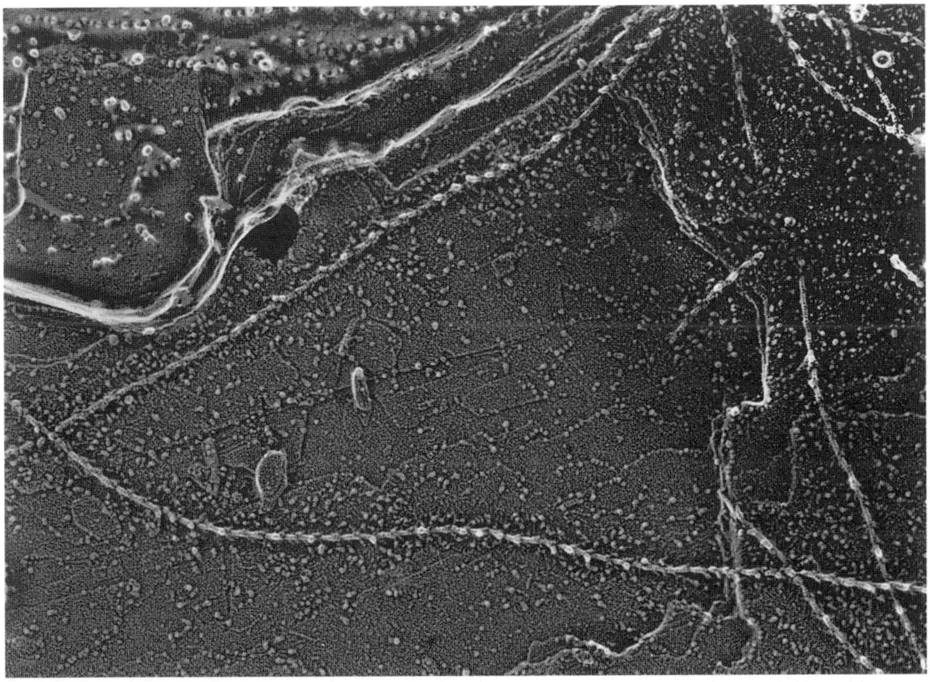

Fig. 11.2 Low magnification view of acto-HMM rigor complex visualized by quick-freeze deep-etch replication coupled with mica-flake methodology. Actin filaments are decorated with HMM to form arrowheaded structures with distinct polarity. Image is reversed to give normal sensation.

literature in which the detailed structural features of cross-bridges under sliding conditions are described, the important points of these observations will be summarized here (the manuscript for the complete publication is now in preparation). Before starting the study, actual spatial resolution of the technique attained here was checked by the image of purified actin as a test material. It was found that the cleft between the subdomains of actin monomer, which measured about 1.5 nm, was clearly discriminated by this methodology. It was also notable that the head moiety of actin-unbound HMM, as visualized by this technique, had a strongly kinked configuration in the presence of ATP, while it was straight and elongated in the absence of ATP (Katayama 1991; Wakabayashi *et al.* 1992; Katayama, 1993*b*).

In this study, acto-HMM rigor complex was used as the starting material for mixing with mica slurry. The mixture was quickly frozen before, or immediately after, addition of ATP. Before sliding started, rigor cross-bridges of HMM were associated with actin mostly through two heads, which were aligned side-by-side forming an approximate 45° tilt-angle with actin filaments. Each head showed

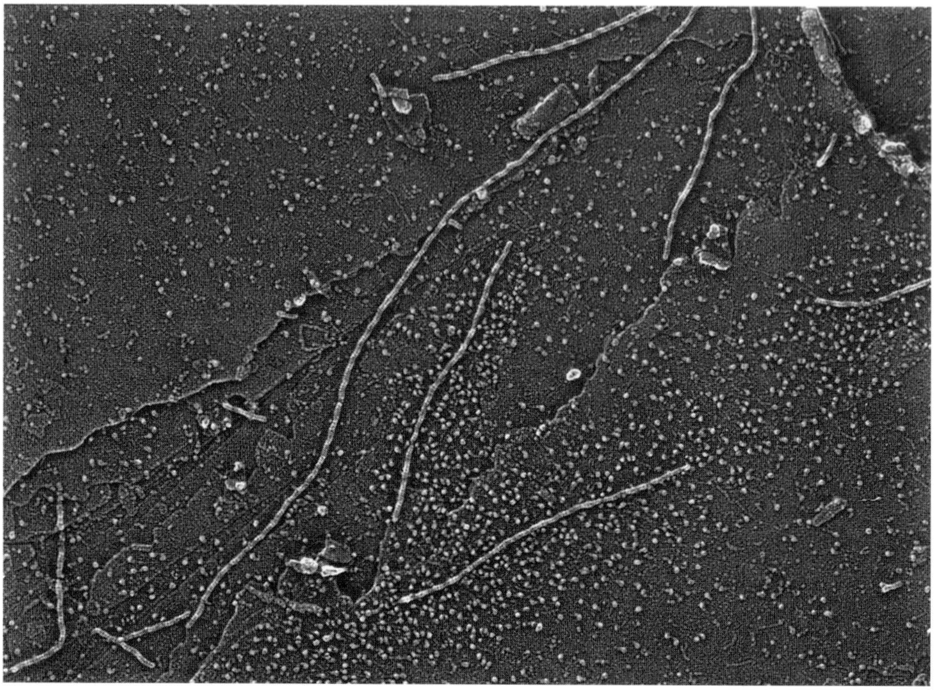

Fig. 11.3 Low magnification view of acto-HMM complex immediately after addition of ATP. Actin filaments are fragmented into short pieces as often seen under *in vitro* sliding conditions. Note the conspicuous structural change from the image shown in Fig. 11.2. Some of the actin filaments have already slid out from HMM track zone, while the others still remain. Observation of HMM molecules supporting such filaments should provide the structural information of cross-bridges in function.

the conventional pear-shaped appearance, which was elongated and almost straight. After addition of ATP, all the HMM cross-bridges supported sliding actin filaments through only one head and exhibited completely different configuration from those in rigor. Their orientation varied from rigor-like to anti-rigor-like tiltangles, though the predominant population of cross-bridges held actin perpendicularly, as if they embraced the filaments in the inner side of the curvature. The other head was often located at the furthest position from actin. Thus, the morphology of actin-attached HMM head in the absence or presence of ATP was very similar to that observed in free HMM or actin-attached S1 under respective conditions. From these observations, it is now apparent that the straight and elongated particles, and short and rounded ones observed in acto-S1 specimens under rigor and activated conditions represent S1s elongated and kinked conformer, respectively.

References

Applegate, D. and Flicker, P.F. (1987) New states of acto-myosin. *J. Biol. Chem.* **262**, 6856–63.

Bridgeman, P.C. and Reese, T.S. (1984) The structure of cytoplasm in directly frozen cultured cells. 1. Filamentous meshworks and the cytoplasmic ground substance. *J. Cell Biol.* **99**, 1655–68.

Craig, R., Greene, L.E. and Eisenberg, E. (1985) Structure of the actin myosin complex in the presence of ATP. *Proc. Natl. Acad. Sci. (USA)* **82**, 3247–51.

Funatsu, T., Kono, E. and Tsukita, S. (1993) Time-resolved electron microscopic analysis of the behavior of myosin heads on actin filaments after photolysis of caged ATP. *J. Cell Biol.* **121**, 1053–64.

Gross, H. (1987) High resolution metal replication of freeze-dried specimens. In *Cryotechniques in biological electron microscopy* (ed. Steinbrecht, R.A. and Zierold, K.), pp. 205–15. Springer-Verlag, Berlin and Heidelberg.

Harada, Y., Noguchi, A., Kishino, A. and Yanagida, T. (1987) Sliding movement of single actin filaments on one-headed myosin filaments. *Nature (London)* **326**, 805–8.

Harada, Y., Sakurada, K., Aoki, T., Thomas, D.D. and Yanagida, T. (1990) Mechanochemical coupling in actomyosin energy transduction studied by *in vitro* movement assay. *J. Mol. Biol.* **216**, 49–68.

Heuser, J.E. (1983) Procedure for freeze-drying molecules adsorbed to mica flakes. *J. Mol. Biol.* **169**, 155–95.

Heuser, J.E. and Cooke, R. (1983) Actin-myosin interactions visualized by the quick-freeze, deep-etch replica technique. *J. Mol. Biol.* **169**, 97–122.

Heuser, J.E. and Salpeter, S.R. (1979) Organization of acetylcholine receptors in quick-frozen, deep-etched, and rotary-replicated "torpedo" postsynaptic membrane. *J. Cell Biol.* **82**, 150–73.

Heuser, J.E., Reese, T.S., Dennis, M.J., Jan, Y., Jan, L. and Evans, L. (1979) Synaptic vesicle exocytosis captured by quick-freezing and correlated with quantal transmitter release. *J. Cell Biol.* **81**, 275–300.

Hirose, K. and Wakabayashi, T. (1993*a*) Structural change of crossbridges of rabbit skeletal muscle during isometric contraction. *J. Muscle Res. Cell Motil.* **14**, 432–45.

Hirose, K., Franzini-Armstrong, C., Goldman, Y.E. and Murray, J.M. (1993*b*) Structural changes in muscle crossbridges accompanying force generation. *J. Cell Biol.* **127**, 763–78.

Hirose, K., Lenart, T.D., Murray, J.M., Franzini-Armstrong, C. and Goldman, Y.E. (1993) Flash and Smash: rapid freezing of muscle fibers activated by photolysis of caged ATP. *Biophys. J.* **65**, 397–408.

Ip, W. and Heuser, J.E. (1983) Direct visualization of the myosin crossbridge helices on relaxed rabbit psoas thick filaments. *J. Mol. Biol.* **171**, 105–9.

Jontes, J.D., Wilson-Kubalek, E.M. and Milligan, R.A. (1995) A 32° tail swing in brush border myosin I on ADP release. *Nature (London)* **378**, 751–3.

Katayama, E. (1989) The effects of various nucleotides on the structure of actin-attached myosin subfragment-1 studied by quick-freeze deep-etch electron microscopy. *J. Biochem.* **106**, 751–70.

Katayama, E. (1991) Conformational change of myosin head induced by nucleotides as studied by quick-freeze deep-etch electron microscopy. *J. Muscle Res. Cell Motil.* **12**, 313.

Katayama, E. (1993*a*) Structure of myosin crossbridges during *in vitro* sliding movement of actin filaments. *J. Muscle Res. Cell Motil.* **14**, 364.

Katayama, E. (1993*b*) Gross structural features of myosin head during sliding movement of actin as studied by quick-freeze deep-etch electron microscopy. *Adv. Exper. Med. Biol.* **332**, 47–55.

Katayama, E., Shiraishi, T., Oosawa, K., Baba, N. and Aizawa, S.-I. (1996) Geometry of the flagellar motor in the cytoplasmic membrane of Salmonella typhimurium as determined by stereo-photogrammetry if quick-freeze deep-etch replica images. *J. Mol. Biol.* **255**, 458–75.

Kishino, A. and Yanagida, T. (1988) Force measurements by micromanipulation of a single actin filament by glass needles. *Nature* (*London*) **334**, 74–6.

Kron, S. and Spudich, J.A. (1986) Fluorescent actin filaments move on myosin fixed to a glass surface. *Proc. Natl. Acad. Sci.* (*USA*) **83**, 6272–6.

Lang, R.D.A. and Bronk, J.R. (1978) A study of rapid mitochondrial structural changes *in vitro* by spray-freeze-etching. *J. Cell Biol.* **77**, 134–47.

Lenart, T.D., Murray, J.M., Franzini-Armstrong, C. and Goldman, Y.E. (1996) Structure and periodicities of cross-bridges in relaxation, in rigor, and during contractions initiated by photolysis of caged Ca^{2+}. *Biophys. J.* **71**, 2289–306.

Ménétret, J.-F. and Craig, R. (1994) Unfixed cryosections of striated muscle to study dynamic molecular events. *Biophys. J.* **67**, 1612–19.

Milligan, R.A. and Flicker, P.F. (1987) Structural relationships of actin, myosin, and tropomyosin revealed by cryo-electron microscopy. *J. Cell Biol.* **105**, 29–39.

Milligan, R.A., Whittaker, M. and Safer, D. (1990) Molecular structure of F-actin and location of surface binding sites. *Nature* (*London*) **348**, 217–21.

Morris, E.P., Katayama, E. and Squire, J.M. (1994) Evaluation of high-resolution shadowing applied to freeze-fractured, deep-etched particles: 3-D helical reconstitution of shadowed actin filaments. *J. Struct. Biol.* **113**, 47–55.

Nicolas, M.-T. and Bassot, J.-M. (1993) Freeze substitution after fast-freeze fixation in preparation for immunocytochemistry. *Microsc. Res. Tech.* **24**, 474–87.

Pollard, T.D., Maupin, P., Sinard, J. and Huxley, H.E. (1990) A stopped-flow/rapid-freezing machine with millisecond time resolution to prepare intermediates in biochemical reactions for electron microscopy. *J. Electron Microsc. Tech.* **16**, 160–6.

Pollard, T.D., Bhandari, D., Maupin, P., Wachsstock, D., Weeds, A. and Zot, H.G. (1993) Direct visualization by electron microscopy of the weakly bound intermediates in the actomyosin adenosine triphosphatase cycle. *Biophys. J.* **67**, 1612–19.

Steinbrecht, R.A. and Zierold, K. (ed.) (1987) *Cryotechniques in biological electron microscopy*. Springer-Verlag, Berlin and Heidelberg.

Stewart, M. and Vigers, G. (1986) Electron microscopy of frozen-hydrated biological material. *Nature* (*London*) **319**, 631–6.

Suzuki, S. and Pollack, G.H. (1986) Bridgelike interconnections between thick filaments in stretched skeletal muscle fibers observed by the freeze-fracture method. *J. Cell Biol.* **102**, 1093–8.

Suzuki, S., Oshimi, Y. and Sugi, H. (1993) Freeze-fracture studies on the cross-bridge angle distribution at various states and the thin filament stiffness in single skinned frog muscle fibers. *J. Electron Microsc.* **42**, 107–16.

Tsukita, S. and Yano, M. (1984) Actomyosin structure in contracting muscle detected by rapid freezing. *Nature* (*London*) **317**, 182–4.

Wakabayashi, K., Tokunaga, M., Kohno, I., Sugimoto, Y., Hamanaka, T., Takezawa, Y. *et al.* (1992) Small-angle synchrotron X-ray scattering reveals distinct shape change of the myosin head during hydrolysis of ATP. *Science (Wash. DC)* **258**, 443–7.

Walker, M., White, H., Belknap, B. and Trinick, J. (1994) Electron cryomicroscopy of acto-myosin-S1 during steady-state ATP hydrolysis. *Biophys. J.* **66**, 1563–72.

Walker, M., Trinick, J. and White, H. (1995) Millisecond time resolution electron cryo-microscopy of the M-ATP transient kinetic state of the acto-myosin ATPase. *Biophys. J.* **68**, 87s–91s.

Whittaker, M., Wilson-Kubalek, E.M., Smith, M., Faust, L., Milligan, R.A. and Sweeney, H.L. (1995) A 35-Å movement of smooth muscle myosin on ADP release. *Nature (London)* **378**, 748–51.

Zingsheim, H.P. (1972) Membrane structure and electron microscopy. The significance of physical problems and techniques (freeze etching). *Biochim. Biophys. Acta* **265**, 339–66.

PART III

Other experiments

12 The use of Ca^{2+} indicators

Masato Konishi and Satoshi Kurihara

12.1 Overview

A change in cytoplasmic free Ca ion (Ca^{2+}) concentration triggers contraction of various types of muscles (Ebashi and Endo 1968). However, the detail of the regulatory mechanism of muscle contraction by Ca^{2+} in living muscles is not fully understood; particularly the mechanism of Ca^{2+} release from sarcoplasmic reticulum (SR) and regulatory mechanisms of cytoplasmic Ca^{2+} concentration ($[Ca^{2+}]_i$ is cytoplasmic free Ca ion concentration) are two of the major interests in muscle physiology. Therefore, the simultaneous measurement of $[Ca^{2+}]_i$ with various cellular functions related to excitation–contraction coupling is essential for the study of muscle physiology, and various types of Ca^{2+} indicators applicable to living muscle cells have been developed. In this chapter, we briefly describe the fundamental properties of optical Ca^{2+} indicators applicable to muscle cells (skeletal, cardiac and smooth muscle cells) and compare their characteristics. Since each Ca^{2+} indicator has various advantages and disadvantages in the measurement of the cytoplasmic Ca^{2+} transients, it is almost impossible to deal with all optical Ca^{2+} indicators within the limited space available. Therefore, we have classified the currently available optical Ca^{2+} indicators into five classes (Table 12.1), and will briefly describe the main characteristics of representative indicators from each class.

The first measurement of cytoplasmic Ca^{2+} transients in skeletal muscle was reported by Jöbsis and O'Connor (1966) who introduced murexide into toad skeletal muscles by intraperitoneal injection. In many subsequent studies, various types of Ca^{2+} indicators have been directly micro-injected into isolated single skeletal muscle fibres.

Shimomura *et al.* (1962) extracted aequorin, the Ca^{2+} sensitive photoprotein, from a jellyfish *Aequorea aequorea* and clarified the fundamental properties of aequorin luminescence. Ashley and Ridgway (1968) injected aequorin into the striated muscles of a barnacle with a large diameter and simultaneously measured the membrane potential, the Ca^{2+} transients and tension. This work clearly demonstrated the central role of the cytoplasmic Ca^{2+}, which links together the membrane excitation and contraction. Thereafter, the properties of aequorin were extensively studied by Blinks *et al.* (1976, 1982) and aequorin was applied to cardiac and smooth muscle cells as well as frog single skeletal muscle fibres.

Table 12.1 List of Ca^{2+} indicators described in the text

Class	Indicator	Optical signals
Photoproteins	aequorin obelin	bioluminescence
Metallochromic dyes	arsenazo III antipyrylazo III	absorbance
Purpurate dyes	murexide tetramethylmurexide PDAA DMPDAA	absorbance
Tetracarboxylate dyes	quin2 fura-2 indo-1 fluo-3 fura red calcium green-1 calcium orange calcium green-5N* calcium orange-5N*	fluorescence and absorbance
Tricarboxylate dyes	furaptra (mag-fura-2) mag-fura-5 mag-indo-1 magnesium orange	fluorescence and absorbance

* Calcium green-5N and calcium orange-5N are low-affinity tetracarboxylates

However, aequorin, with relatively slower kinetics, cannot follow the cytoplasmic Ca^{2+} transients in frog skeletal muscle fibres. In addition, it is difficult with aequorin to obtain measurements of the cytoplasmic Ca^{2+} transients with reasonable signal-to-noise ratio from a small single cardiac and smooth muscle cells under physiological condition.

Metallochromic indicators change absorbance upon binding metal ions, and were first used by Miledi *et al.* (1977) to measure the rapid Ca^{2+} transients in frog single skeletal muscle fibres. The various metallochromic indicators have been applied to skeletal muscle fibres and the cytoplasmic Ca^{2+} transients have been measured. However, different quantitative features of the cytoplasmic Ca^{2+} transients are reported by different metallochromic indicators. This is considered to be due to the fact that some metallochromic indicators (e.g. arsenazo III) track Ca^{2+} with kinetic delays and the intracellularly injected metallochromic dyes heavily bind to intracellular constituents, which alters the properties of the dyes.

These disadvantages of the metallochromic Ca^{2+} indicators were improved by newly synthesized purpurate derivatives, purpurate-3,3′ diacetic acid (PDAA) and 1,1′-dimethylpurpurate-3,3′diacetic acid (DMPDAA). These analogues of murexide have rapid Ca^{2+} reaction kinetics and are less bound to cellular constituents.

Therefore, the absorbance signals of PDAA and DMPDAA are likely to be more reliable and accurate in measuring the fast Ca^{2+} transients in frog single skeletal muscle fibres (Hirota *et al.* 1989). However, the absorbance measurement is, in general, seriously influenced by the movement during muscle contraction, necessitating stretch of muscle fibres to long sarcomere length. Also, these indicators cannot be applied to small cardiac and smooth muscle cells, due to the small Ca^{2+}-related absorbance change.

Tsien and his colleagues have developed high-affinity fluorescent Ca^{2+} indicators, which have a similar structure to EGTA (Tsien 1980; Tsein *et al.* 1982; Grynkiewicz *et al.* 1985). The measurement of fluorescence signals from living cells requires a more complex optical apparatus compared to that for the measurement of luminescence of photoproteins. However, fluorescent Ca^{2+} indicators have various advantages over other indicators. Fluorescent Ca^{2+} indicators with a high affinity for Ca^{2+} provide an estimate of resting $[Ca^{2+}]_i$, which is lower than the detection limit of aequorin and the Ca^{2+} indicators with absorbance measurement. Acetoxy-methyl ester forms of the fluorescent Ca^{2+} indicators, which can cross cell membrane and are converted to the Ca^{2+} sensitive form of the dye by esterase, made the introduction of the dyes into small muscle cells easier. In addition to these advantages, the combined use of fluorescent Ca^{2+} indicators and fluorescence confocal microscope revealed the focal increase in Ca^{2+} concentrations (Ca^{2+} sparks) (Cheng *et al.* 1993). Thus, the fluorescent Ca^{2+} indicators are now widely employed in muscle physiology, although various disadvantages are also indicated (see Section 12.4 for detail).

In the following, we compare the advantages and disadvantages of the currently available optical Ca^{2+} indicators and briefly discuss the following practical points: loading of indicators into cells, Ca^{2+} sensitivity, quantitative estimation (calibration), kinetic delay, influence of intracellular constituents, distribution of indicators, and other points.

12.2 Bioluminescent photoproteins

General

Several photoproteins (aequorin and obelin) have been used to measure $[Ca^{2+}]_i$ but aequorin is the most widely used bioluminescent photoprotein. Aequorin is a single 20 000-dalton macromolecule and all of the components required for luminescence (the apoprotein, chromophore and oxygen) are bound together. Aequorin emits blue light (peak wavelength, 469 nm) upon binding with Ca^{2+}. The luminescent reaction of aequorin is rather specific for Ca^{2+}, but other divalent and trivalent cations can also emit light. The fundamental characteristics of aequorin have been extensively studied (see Blinks *et al.* 1976, 1982 for reference). Recently, recombinant aequorins with different properties have been produced by Shimomura *et al.* (1990) but native aequorin is widely used among users.

Loading of aequorin into cells

Since aequorin, a large molecule, hardly crosses the cell membrane, a glass micropipette with a small diameter (similar to a glass micro-electrode to measure the membrane potential) is used for injection. Pressure injection is commonly employed to introduce aequorin into various muscle cells (Allen and Blinks 1978; Blinks *et al.* 1982; Allen and Kurihara 1982).

For small cells (smooth and cardiac cells), a chemical loading method has been developed and applied (Morgan and Morgan 1982; MacKimmon *et al.* 1987). The idea of the method is that the membrane is permeabilized to aequorin using EGTA, and the membrane is resealed by removing EGTA and the subsequent addition of a high concentration Mg^{2+} after the loading. The chemical loading procedure is tricky, and direct injection into cells is generally more reliable.

Ca^{2+} sensitivity of aequorin

Allen *et al.* (1977) measured the Ca^{2+} concentration–luminescence relationship in a cuvette and showed that the relationship can be fitted into a two-state model. The curve of the relationship, expressed as a double logarithmic plot of Ca^{2+} concentration and the light signal, which is normalized to the maximal light measured at saturating Ca^{2+} concentration, is sigmoid (Fig. 12.1). At a very low concentration range of Ca^{2+} (usually lower than 10^{-7} M, but depending on the Mg^{2+} concentration in the solution), the luminescence of aequorin is insensitive to Ca^{2+} (Ca^{2+}-independent light) and the detectable lower limit of aequorin is close to the resting $[Ca^{2+}]_i$ (Konishi and Kurihara 1987; Blatter and Blinks 1991). Aequorin also cannot respond to higher concentrations of Ca^{2+} than 10^{-4} M, which is beyond the range of $[Ca^{2+}]_i$ under physiological conditions. The slope of the curve is approximately 2.5–3, suggesting that for luminescence one aequorin molecule binds three Ca^{2+}. This relationship between Ca^{2+} concentration and luminescence indicates that the light signal measured from the cells is a non-linear expression of the $[Ca^{2+}]_i$ change and the light signal generally reflects the highest $[Ca^{2+}]$ in the cell. This complicates the quantitative interpretation of the light signals from cells containing aequorin. However, a small change in $[Ca^{2+}]_i$ is more easily detectable due to the non-linearity of the relationship between luminescence of aequorin and Ca^{2+} concentration. This feature can be used for imaging the cytoplasmic Ca^{2+} to observe the inhomogeneity of Ca^{2+} (Konishi and Kurihara 1993). However, fluorescent Ca^{2+} indicators are preferred for imaging a focal change in cytoplasmic Ca^{2+}.

The affinity of aequorin for Ca^{2+} is predicted from the two-state model and aequorin apparently does not work as a strong Ca^{2+} buffer, which binds sufficient amount of Ca^{2+} and influences contraction (Blinks 1978). In practice, it is hard to measure accurately $[Ca^{2+}]_i$ at resting state with aequorin in skeletal muscles, and other high affinity Ca^{2+} inidicators, more sensitive to lower concentrations of Ca^{2+}, should be used for this purpose (Suda and Kurihara 1991). However, the cytoplasmic Ca^{2+} transients can be measured with some limitations.

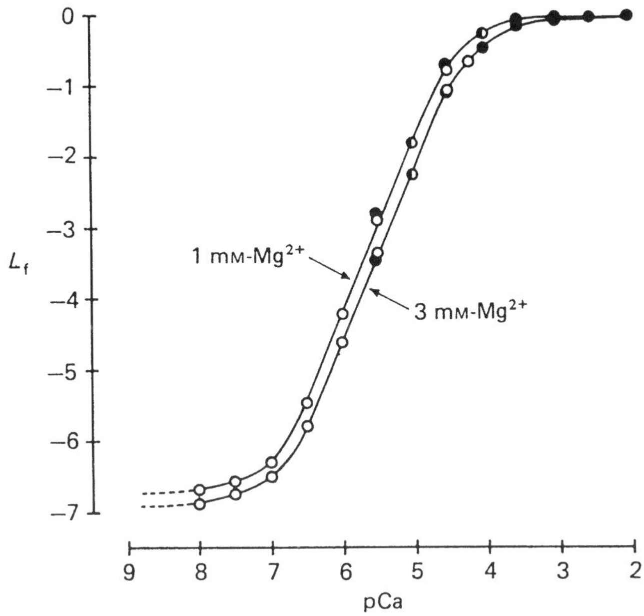

Fig. 12.1 *In vitro* calibration curves of aequorin. Aequorin was pre-equilibrated with Mg^{2+} (1 or 3 mM) and then rapidly injected into solutions of known Ca^{2+} and Mg^{2+} concentrations at 18 °C. Calibration solutions were made up by simple dilution (*closed circles*) in the range of pCa 2.0–5.5 (150 mM KCl; 10 mM MOPS) and by using EGTA in the range of pCa 4.0–8.0 (*open circles*) (ionic strength 0.15 M, containing 10 mM EGTA and 20 mM PIPES). pH of the solutions was 7.0. Aequorin light intensity in each solution (L) was normalized to the maximal light intensity obtained by exposing the aequorin to 10 mM Ca^{2+} without Mg^{2+} pre-equilibration (L_{max}). Fractional luminescence (L_f) defined as $L_f = \log\,(L/L_{max})$, was determined for pCa 2.0–8.0. *Abbreviations*: MOPS, 3-(N-morpholino)propanesulfonic acid; PIPES, peperazine-N,N'-bis(2-ethanesulfonic acid); EGTA, ethylene glycol-bis(β-amino-ethylether)N,N,N',N'-tetraacetic acid. (Reproduced from Konishi and Kurihara, *J. Physiol.* (*Lond.*) **383**, 269–83, 1987 by copyright permission of The Physiological Society.)

Measurement of optical signals and analysis

A photomultiplier with low dark current is used for detection of aequorin luminescence with low intensity from the cells (Blinks 1984). The intensity of the light signals from the cells depends on how much aequorin could be introduced into the cells. The principle of aequorin light detection from the cells is simple, compared with the measurement of absorbance or fluorescence of other indicators. This is one of the advantages of the aequorin method. Another distinct advantage of aequorin over other indicators is that the measurement of luminescence is generally little affected by artifacts due to cell motion. The aequorin light signal can be reliably measured during muscle contraction without overstretching the muscle fibre to eliminate fibre movements. This property, little interference from the movement artifacts, is also helpful for the $[Ca^{2+}]_i$ measurement during mechanical perturbations, such as length changes (Gordon and Ridgway 1990; Kurihara and Komukai 1995).

For the conversion of the aequorin light signals to $[Ca^{2+}]_i$, an *in vitro* calibration curve measured under similar ionic conditions to cytoplasmic environment is used. However, it is difficult to completely mimic the same conditions as in the cytoplasm (see Calibration of the indicator signals). The amount of aequorin should be estimated by discharging all aequorin inside the cells by the application of Triton X-100 to destroy the cells at the end of experiment (Allen and Blinks 1978; Blinks *et al.* 1982). Then, the peak light signal at saturating $[Ca^{2+}]_i$ should be calculated using the rate constant of the decay of Ca^{2+} signal measured when aequorin was rapidly mixed with the solution containing saturating Ca^{2+} in a cuvette under similar conditions to that inside the cells (in the presence of 1 mM Mg^{2+}, the rate constant is about $0.63\,s^{-1}$, at 18 °C, and $1.73\,s^{-1}$, at 30 °C).

If an aequorin molecule binds Ca^{2+}, the chromophore is dissociated and no luminescence is observed upon re-application of Ca^{2+} (consumption). This consumption is a drawback of aequorin but the consumption of aequorin is not serious for even many twitch contractions of frog skeletal muscle fibres (Blinks *et al.* 1978). However, a long lasting increase in cytoplasmic Ca^{2+} with a large amplitude in tetanic contraction and K^+-contracture consumes a large amount of aequorin. Therefore, the consumed aequorin should be back calculated for the conversion of the light signal to $[Ca^{2+}]_i$. The consumption is not serious in cardiac muscle twitch contractions and can be ignored under physiological conditions, except for Ca^{2+} overloaded condition.

Kinetic delay

The speed of aequorin response to Ca^{2+} is not fast enough to follow the Ca^{2+} transients in skeletal muscles. This kinetic delay is a serious disadvantage in measuring the fast Ca^{2+} transients. However, the kinetic delay of aequorin is probably not serious in cardiac muscles.

The kinetics of aequorin for Ca^{2+} is substantially influenced by Mg^{2+} and the temperature of the solution. Cytoplasmic Mg^{2+} concentration ($[Mg^{2+}]_i$) in frog skeletal muscle is reported to be less or around 1 mM (e.g. Konishi *et al.* 1993). In which case, following a step increase in $[Ca^{2+}]$, the time to half-peak of light at 0.5 mM Mg^{2+} and at 20 °C, is about 12 ms (Moore 1986). This is slow for the detection of the Ca^{2+} transients in skeletal muscle fibres but is probably rapid enough for the measurement of the Ca^{2+} transients in mammalian cardiac muscles. Therefore, the slow kinetics of aequorin limits the accurate quantitative measurement of the faster Ca^{2+} transients in skeletal muscle fibres but this is not the case in cardiac and smooth muscles.

Influences of other intracellular ions

As discussed in the previous sections, the reaction of aequorin and Ca^{2+} (the steady-state calibration curve and kinetics) is influenced by Mg^{2+}, which is considered to bind to one of the two states of aequorin (probably binding to the inactive form, which can be converted to the active form by Ca^{2+}). Therefore, if $[Mg^{2+}]_i$ is significantly altered, the light signal of aequorin is seriously influenced and the quantitative estimation of the cytoplasmic Ca^{2+} transients is difficult. However, according to our studies (Konishi *et al.* 1993; Hongo *et al.* 1994), $[Mg^{2+}]_i$ is well-controlled and does not change significantly under physiological conditions, particularly for short periods.

The calibration curve of aequorin is relatively insensitive to pH change in solution. This is advantageous for the measurement of $[Ca^{2+}]_i$ under patho-physiological conditions (acidosis and ischemia which lower cytoplasmic pH). However, a large fall of pH inhibits luminescence.

Aequorin emits light upon binding with other divalent and trivalent cations, which generally do not exist in living cells (Blinks *et al.* 1982). This should be kept in mind when using these ions for experiments.

12.3 Absorbance Ca^{2+} indicators

General

Some indicators undergo changes in absorbance upon binding with Ca^{2+} (we will call these indicators 'absorbance Ca^{2+} indicators' in the following sections). For introduction of the indicator molecules into intact skeletal muscle fibres, the method which has been most widely used is micro-injection, either by pressure or by iontophoresis. For cut fibres, however, the indicator can be also introduced into the fibre by diffusion through the cut ends. Absorbance is usually measured at more than one wavelength from the fibre portion containing the indicator in the resting state and during activity, e.g. resting absorbance at an isosbestic wavelength [$A(\lambda_{iso})$] and the change of absorbance at Ca^{2+}-sensitive wavelength(s) during activity [$\Delta A(\lambda_{Ca})$]. The conversion of the indicator absorbance signals to the change in $[Ca^{2+}]_i$ uses both $A(\lambda_{iso})$ and $\Delta A(\lambda_{Ca})$, and also relies on calibration parameters, such as molar extinction coefficient at λ_{iso}, the change in molar extinction coefficient upon Ca^{2+} binding at λ_{Ca} and the indicator's dissociation constant for Ca^{2+}(K_D). The analysis of absorbance signals also usually requires the subtraction of the fibre intrinsic absorbance and its change during activity. The intrinsic absorbance of the fibre steeply increases at short wavelengths, which limits the use of absorbance indicators at u.v. (or near u.v.) wavelengths (Baylor *et al.* 1986; Irving *et al.* 1987; Konishi *et al.* 1993).

The absorbance Ca^{2+} indicators that have been used to date react with Ca^{2+} with relatively low affinity, and therefore are not suitable for measurement of low levels of $[Ca^{2+}]_i$ in the resting muscles. However, a low affinity for Ca^{2+} appears to be advantageous for tracking rapid change in $[Ca^{2+}]_i$ without kinetic delay (see Low-affinity indicators). One possible advantage of the absorbance measurement over fluorescence measurement is that absorbance is generally less sensitive to the local chemical environment than fluorescence. This is because fluorescence intensity depends on quantum yield, in addition to factors that determine absorbance (e.g. molar extinction coefficient), and because these factors (quantum yield, molar extinction coefficient, etc.) are likely to be altered in the cell interior (see Calibration of the indicator signals). A disadvantage of absorbance measurements is that a relatively large amount of the indicator must be introduced into cells of large size; for reliable measurement of the absorbance signal, it is generally required to have indicator concentration in the order of 0.1–1 mM in $\sim 100\,\mu m$ diameter fibres.

Metallochromic dyes

Metallochromic indicators, such as arsenazo III and antipyrylazo III, have been used widely to monitor the cytoplasmic Ca^{2+} transients during skeletal muscle activity (e.g. Miledi *et al.* 1977; Baylor *et al.* 1982; Palade and Vergara 1982). However, significant difficulties have been noted in attempts to interpret the signals from these indicators in muscle fibres. Firstly, these indicators are subject to the widely recognized problem of reacting with ions other than Ca^{2+}, of which the most important, under intracellular conditions, are probably H^+ and Mg^{2+}. The difficulties are also partly attributed to the complicated stoichiometries of the Ca^{2+} : indicator reaction. In the case of arsenazo III, four different stoichiometric complexes of Ca^{2+} : arsenzo III have been reported; 1 : 1, 1 : 2, 2 : 1 and 2 : 2 (Dorogi and Neumann 1981; Rios and Schneider 1981; Palade and Vergara 1983). *In vitro* measurements of antipyrylazo III absorbance have revealed at least three Ca^{2+} : indicator complexes form, with 1 : 1, 1 : 2 and possibly 2 : 2 stoichiometries, but the 1 : 2 Ca^{2+} : indicator complex appears to be predominantly formed at the concentrations of antipyrylazo III and Ca^{2+} normally encountered in the cytoplasm (Rios and Schneider 1981; Hollingworth *et al.* 1987). The absorbance signal from the 1 : 2 Ca^{2+} : antipyrylazo III complex can be used to monitor the time course of the cytoplasmic Ca^{2+} transients; antipyrylazo III absorbance signal is considered to follow a rapid change in $[Ca^{2+}]_i$ during twitch response with a relatively small (1–2 ms) kinetic delay in frog skeletal muscle fibres (Baylor *et al.* 1985; Maylie *et al.* 1987*a*). On the other hand, the peak amplitude of the cytoplasmic Ca^{2+} transient calibrated from antipyrylazo III is 2–3 μM (Maylie *et al.* 1987*a*; Baylor and Hollingworth 1988), which may be erroneously small. This probably results because $\sim 75\%$ of the cytoplasmic antipyrylazo III molecules appear to be bound to intracellular constituents, and the properties of the bound indicator molecules are likely to be different from those in salt solutions (Baylor *et al.* 1986). The calibration difficulties arising from the binding of indicator to cellular constituents applies not only to metallochromic indicators, but, as discussed below, are general problems shared among all optical indicators.

Purpurate dyes

Purpurate (murexide) derivatives represent another class of absorbance Ca^{2+} indicators. Among them, murexide and tetramethylmurexide have a low affinity for Ca^{2+}(K_D is 3.6 mM and 2.8 mM for murexide and tetramethylmurexide, respectively at 25 °C), and are essentially insensitive to Mg^{2+} (Ohnishi 1978; tetramethylmurexide is also insensitive to pH). Maylie *et al.* (1987*b*) introduced tetramethylmurexide into frog cut muscle fibres by diffusion from the cut ends, and successfully measured the indicator absorbance signals at rest and during muscle activity. From the absorbance spectrum of the indicator, they have found that more than 10% of tetramethylmurexide inside the fibre was complexed with Ca^{2+} at the resting state. This was unexpected from the low affinity of the indicator (2.8 mM) and the low level of $[Ca^{2+}]_i$ in the resting muscle. Furthermore, after action potential stimulation, the indicator's absorbance transiently changed in the direction expected for the increase in $[Ca^{2+}]_i$, which was followed by a maintained change of the reversed direction. These results suggest that tetramethylmurexide is able to enter the sarcoplasmic reticulum of the frog skeletal muscle fibres, as observed in isolated sarcoplasmic reticulum vesicles (Ohnishi 1979; Ogawa *et al.* 1980).

To avoid the complication of being able to cross the sarcoplasmic reticulum membrane observed with tetramethylmurexide (and possibly with murexide), Southwick and Waggoner (1989) synthesized new members of purpurate indicators: purpurate-3,3′-diacetic acid (PDAA) and 1,1′-dimethylpurpurate-3, 3′diacetic acid (DMPDAA), analogues of murexide and tetramethylmurexide, respectively. These indicators have a low affinity for Ca^{2+} (K_D 0.8–1.0 mM range), and PDAA is highly selective for Ca^{2+} over Mg^{2+} and H^+ (Hirota *et al.* 1989). When applied to frog cut muscle fibres, both PDAA and DMPDAA do not appear to enter the sarcoplasmic reticulum, and are thought to track cytoplasmic Ca^{2+} transient without kinetic delay (Fig. 12.2; from Hirota *et al.* 1989). The time course of the Ca^{2+} transients was found to be surprisingly fast; time to peak ~5 ms, half-width 7–9 ms (16 °C). Similar results were obtained in intact frog skeletal muscle fibres injected with PDAA (Konishi and Baylor 1991). Interestingly, these new indicators also reported a large amplitude for the cytoplasmic Ca^{2+} transients, as large as 20 μM, stimulated by an action potential. Another important advantage of these new purpurate indicators is that the percentage of the indicator that appeared to be bound to intracellular constituents is relatively small (~20% in cut fibres, Hirota *et al.* 1989; 24–43% in intact fibres, Konishi and Baylor 1991), in comparison with the large bound fraction (~60–90%) detected with the other indicators. Because an indicator with a smaller bound fraction is likely to yield a more reliable estimate of the amplitude of the cytoplasmic Ca^{2+} transients, the amplitude of Ca^{2+} transients is considered to be more reliable when estimated with PDAA (or DMPDAA) than other indicators. A disadvantage of PDAA and DMPDAA is that the size of the absorbance signals is small, even with millimolar

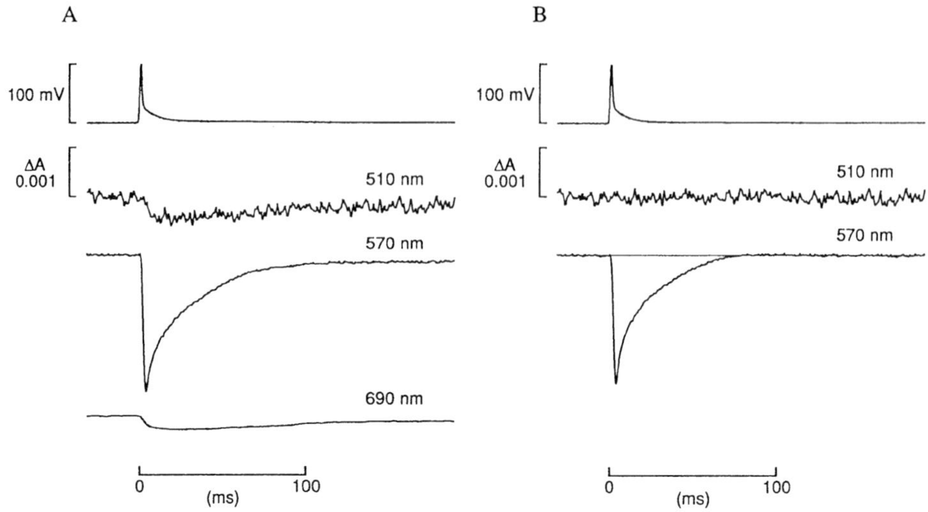

Fig. 12.2 Changes in absorbance, after action potential stimulation, of a frog cut muscle fibre containing 3.27 mM PDAA (16 °C). (A) The *top* trace shows action potential (slightly attenuated by low pass filtering) and the *bottom three* traces show changes in absorbance measured simultaneously at three different wavelengths, as indicated. (B) The *top* trace shows the action potential from A. The *bottom two* traces show the changes in indicator-related absorbance at 510 and 570 nm, obtained from the corresponding traces in A by subtraction of the estimated intrinsic absorbance signal. (Reproduced from Hirota *et al.*, *J. Gen. Physiol.* **94**, 597–631, 1989 by copyright permission of The Rockefeller University Press.)

concentration of the indicators in the cytoplasm of ~ 100 μm diameter fibre. Consequently, the optical signal in the later phase of the Ca^{2+} transient is subject to movement artifacts and fibre intrinsic absorbance changes.

12.4 Fluorescent Ca^{2+} indicators

General

Since Tsien (1980) first synthesized a fluorescent Ca^{2+} indicator with a binding cavity consisting of four carboxyl groups, a number of 'tetracarboxylate indicators' have been developed, including quin2 (Tsien *et al.* 1982), fura-2 (Grynkiewicz *et al.* 1985), indo-1 (Grynkiewicz *et al.* 1985), fluo-3 (Minta *et al.* 1989), fura red, calcium green-1 and calcium orange. Most of these tetracarboxylate indicators react with Ca^{2+} with 1 : 1 stoichiometry, with relatively high affinity (values of K_D for Ca^{2+} of 0.1–0.5 μM) and high selectivity for Ca^{2+} over Mg^{2+} and H^+. Some of these tetracarboxylate indicators exhibit a large shift of either excitation or emission spectrum upon Ca^{2+} binding, and [Ca^{2+}] can be deduced from the ratio of fluorescence intensity at two excitation wavelengths (or two emission wavelengths), rather than absolute fluorescence intensity at one pair of excitation/emission wavelengths. The fluorescence ratio signals are, in principle, independent of the indicator concentration, unless the indicator concentration is sufficiently high to cause 'inner filter effects' by absorption of excitation light. During muscle contraction, fibre movement often causes artifacts in the optical signals by a change in the number of indicator molecules in the optical field. The fluorescence ratio measurements should therefore be useful to reduce the influence of movement artifacts due to muscle contraction in the fluorescence signals.

Another important feature of the tetracarboxylate indicators is that the indicators can be introduced into the cells simply by incubation with the acetoxymethyl (AM) ester form of the indicators; AM ester of the indicator molecules cross the cell membrane, and are trapped inside the cell after hydrolysis by cellular intrinsic esterases. This AM loading method is particularly useful for cells of small size, such as cardiac and smooth muscle cells, which could be easily damaged by injection of the indicator. Even for skeletal muscle fibres, in which the micro-injection can be reliably carried out without substantial cell damage, the AM loading may be a method of choice to introduce the indicator evenly into the entire length of the fibre. With the micro-injection technique, the higher indicator concentration achieved around the site of injection changes as a function of time after injection. A drawback of the AM loading technique is, on the other hand, a possibility that a significant fraction of the indicator molecules is trapped inside intracellular organelles (see Localization of indicators).

High-affinity indicators

High-affinity indicators are advantageous for monitoring the low levels of $[Ca^{2+}]_i$ near the basal level, but are not suitable for tracking large cytoplasmic Ca^{2+} transients during muscle activity. This is because of the:

(1) saturation of indicator with high $[Ca^{2+}]_i$,

(2) slow kinetics due to slow dissociation rate constant (see the next section).

At high intracellular concentrations, these indicators can also act as a Ca^{2+} buffer, which results in the alteration of the cytoplasmic Ca^{2+} transients by the indicator's buffering effect. Because the indicators with high affinity bind more Ca^{2+} than those with low affinity, the buffering effect is more likely to be a problem in the case of high-affinity indicators. In frog skeletal muscle fibres, Baylor and Hollingworth (1988) reported that injection of high concentrations of fura-2 ($> \sim 100\,\mu M$) shortened the time course of, and reduced the peak amplitude of, the cytoplasmic Ca^{2+} transients measured with antipyrylazo III. This stands in contrast to the small buffering effect by furaptra, a low-affinity indicator (see next paragraph), at cytoplasmic concentrations up to $500\,\mu M$ (Konishi *et al.* 1991). The Ca^{2+} buffering effect of the high-affinity indicators can be also utilized as an experimental tool, i.e. as an extrinsic Ca^{2+} buffer whose Ca^{2+} binding can be quantitatively monitored. When the buffering power of the indicator is much larger than that of cell's intrinsic buffers (e.g. with very high indicator concentration in the cytoplasm), the total amount of Ca^{2+} entered in the cytoplasm may be approximated by the amount of Ca^{2+} : indicator complex formed (Baylor and Hollingworth 1988; Jacquemond *et al.* 1991; Hollingworth *et al.* 1992; Jong *et al.* 1993).

Low-affinity indicators

Recently, many low-affinity fluorescent Ca^{2+} indicators have become available, as new members of the tetracarboxylate or tricarboxylate families. These low-affinity indicators share many properties with the high-affinity tetracarboxylate indicators:

(1) They generally retain simple 1 : 1 stoichiometry, and are thought to be relatively insensitive to pH around 7.

(2) AM ester forms are also available.

(3) Fluorescence ratio can be measured from some of these low-affinity indicators.

Furaptra is one of the low-affinity tricarboxylate indicators; although originally designed as a Mg^{2+} indicator (Raju *et al.* 1989), it has been often used in muscle fibres as a Ca^{2+} indicator with rapid kinetics (Konishi *et al.* 1991; Claflin *et al.* 1994; Zhao *et al.* 1996; Hollingworth *et al.* 1996). Although this indicator is commercially available under the name 'mag-fura-2', it has 1.5–2 order higher affinity for Ca^{2+}($K_D \sim 50\,\mu M$ at 16–37 °C) than for Mg^{2+} ($K_D = 5.3\,mM$ and 1.5 mM at 16 °C and 37 °C, respectively) (Raju *et al.* 1989; Konishi *et al.* 1991). As expected from these K_D values, furaptra was found to be a useful indicator for measuring the cytoplasmic [Mg^{2+}] in resting muscle fibres (Konishi *et al.* 1993) and as a Ca^{2+} indicator in stimulated fibres (Konishi *et al.* 1991). In frog muscle fibres, the brief time course of the cytoplasmic Ca^{2+} transient reported by furaptra is very similar to that reported by PDAA. However, the average amplitude of the Ca^{2+} transient calculated from furaptra fluorescence signals based on *in vitro* calibration is only about a half that estimated with PDAA. Because the binding of the indicator to intracellular constituents appears to be somewhat less for PDAA (24–43%) than furaptra (42–51%), the quantitative information on the amplitude of the cytoplasmic Ca^{2+} transients may be more reliably obtained from PDAA. Even though there is a quantitative discrepancy of ~ two-fold in the calibration of the amplitude of Ca^{2+} transients, furaptra seems to be a useful tool to track the cytoplasmic Ca^{2+} transient without a kinetic delay. Furaptra fluorescence signals can be easily detected with much lower cytoplasmic concentrations (30–40 μM) than that required for PDAA absorbance measurements (~ 1 mM) from ~ 100 μm diameter muscle fibres, and can be applied to even single cardiac muscle cells (Konishi and Berlin 1993). Generally, a fluorescence measurement has the advantages of being less susceptible to movement artifacts and interference from intrinsic signals.

From an extensive survey of high- and low-affinity fluorescent indicators, Zhao *et al.* (1996) found that a similar situation has arisen for other commercially available tricarboxylate indicators; mag-indo-1, mag-fura-5, magnesium orange could be also used as Ca^{2+} indicators with little (or slight) detectable kinetic delay. They also found that two low-affinity tetracarboxylate Ca^{2+} indicators, calcium green-5N and calcium orange-5N, can react with Ca^{2+} rapidly (but with some slow component with unknown origin).

The tendency for low-affinity Ca^{2+} indicators to report a faster Ca^{2+}-related signal is expected, if the greater K_D value of the low-affinity indicator is due to a greater dissociation rate constants (K_{-1}) in the face of a similar association rate constants (K_{+1}). The effect of K_{-1} on the time course of the optical signals is displayed in Fig. 12.3; following the cytoplasmic Ca^{2+} transient with brief time course (time to peak 5.3 ms, half-width 9.1 ms) and large peak amplitude (13 μM), Ca^{2+} : indicator binding is computed with varied K_{-1} and fixed K_{+1}. It is clear that the computed waveform of the indicator signal is slowed with smaller values of K_{-1} (and therefore with higher affinity). The broader waveform with smaller K_{-1} (higher affinity) is partly due to the non-linearity of the indicator signal introduced by partial saturation of Ca^{2+} : indicator binding, but is mainly due to the small K_{-1} of the indicator. With the slow dissociation of Ca^{2+} from the indicator, the time to peak of the indicator signal is also prolonged, as shown in Fig. 12.3B. A disadvantage of low-affinity indicators is also seen in Fig. 12.3: with K_D values higher than 1 μM, it becomes progressively more difficult to detect a small change in $[Ca^{2+}]_i$ above basal level. With high affinity, on the other hand, non-zero optical signal is clearly noted even in the resting state before stimulation.

The cytoplasmic Ca^{2+} transients that accompany activation of muscle fibres have been extensively studied in frog skeletal muscle fibres. The Ca^{2+} transients in mammalian skeletal muscle are less well understood but are of increasing interest. Fluorescent Ca^{2+} indicators of tetracarboxylate and tricarboxylate families have been applied to monitor both the steady levels of and transient changes of $[Ca^{2+}]_i$ in mammalian muscle fibres; indo-1 in mouse flexor digitorum brevis muscles (Westerblad and Allen 1993*a*,*b*, 1994); furaptra in cut fibres from rat extensor digitorum longus muscles (Delbono and Stefani 1993); fura-2 in rat extensor digitorum longus muscles (Garcia and Schneider 1993); fura-2 in enzymatically dissociated rat flexor digitorum brevis muscles (Carroll *et al.* 1995); calcium green-1 and fluo-3 in cut fibres from mouse extensor digitorum longus muscles (Shirokova *et al.* 1996); furaptra in mouse extensor digitorum longus muscles (Hollingworth *et al.* 1996). As in frog fibres, furaptra appears to be useful for tracking the brief time course of Ca^{2+} transients in mammalian skeletal muscle fibres, which generally have diameters about half as large as frog fibres. The cytoplasmic Ca^{2+} transients measured in mouse extensor digitorum longus muscle appear to have a very brief time course, even briefer than those in frog fibres (Fig. 12.4); after single action

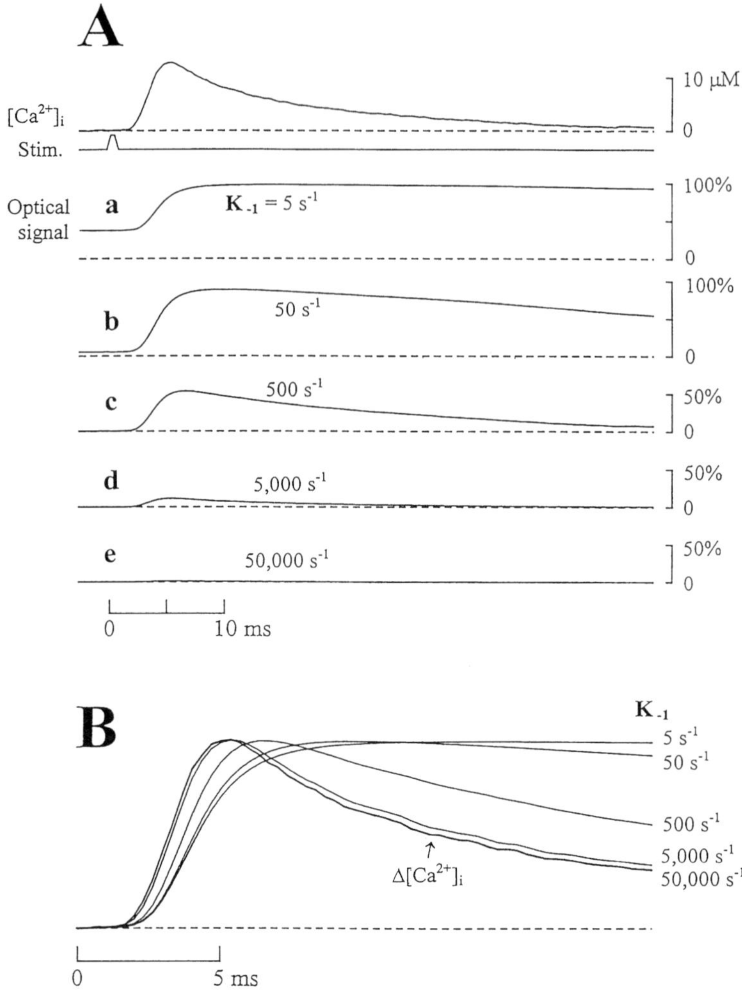

Fig. 12.3 Indicator optical signals during twitch response calculated with various dissociation rate constants (and therefore K_D) for Ca^{2+} : indicator reaction. (A) the *top* trace (labelled $[Ca^{2+}]_i$) shows spatially averaged cytoplasmic Ca^{2+} transient recorded in a frog intact muscle fibre (16.5 °C) with furaptra. Furaptra fluorescence change in response to single stimulus (the *second* trace labelled Stim.) was converted to $[Ca^{2+}]_i$ with two-fold scaling to correct for the calibration error described in the text (also see Konishi *et al.* 1991). Resting $[Ca^{2+}]_i$ before stimulation was set to 60 nM. The *bottom five* traces (a–e) show simulated optical signals with K_{+1} of $5 \times 10^7\ M^{-1} s^{-1}$ and variable K_{-1} indicated near each trace. The K_D values were therefore 100 nM (a), 1 μM (b), 10 μM (c), 100 μM (d) and 1 mM (e). The calculation used a differential equation: $d[CaD]/dt = K_{+1} \cdot [Ca^{2+}] \cdot [D] - K_{-1} \cdot [CaD]$, where $[CaD]$ and $[D]$ denote concentrations of Ca^{2+} : indicator complex and free indicator, respectively. The optical signal (fluorescence or absorbance) was assumed to be proportional to the Ca^{2+}-bound fraction of the indicator between 0 (no Ca^{2+} binding) and 100% (saturated with Ca^{2+}) at any moment. The spatial inhomogeneity of $[Ca^{2+}]_i$ was not taken into account. (B) Comparison of the waveform of changes in $[Ca^{2+}]_i$ ($\Delta[Ca^{2+}]_i$) and optical signals after stimulation at time zero. Each trace shown in part A was scaled to have the same peak height above the basal level before stimulation.

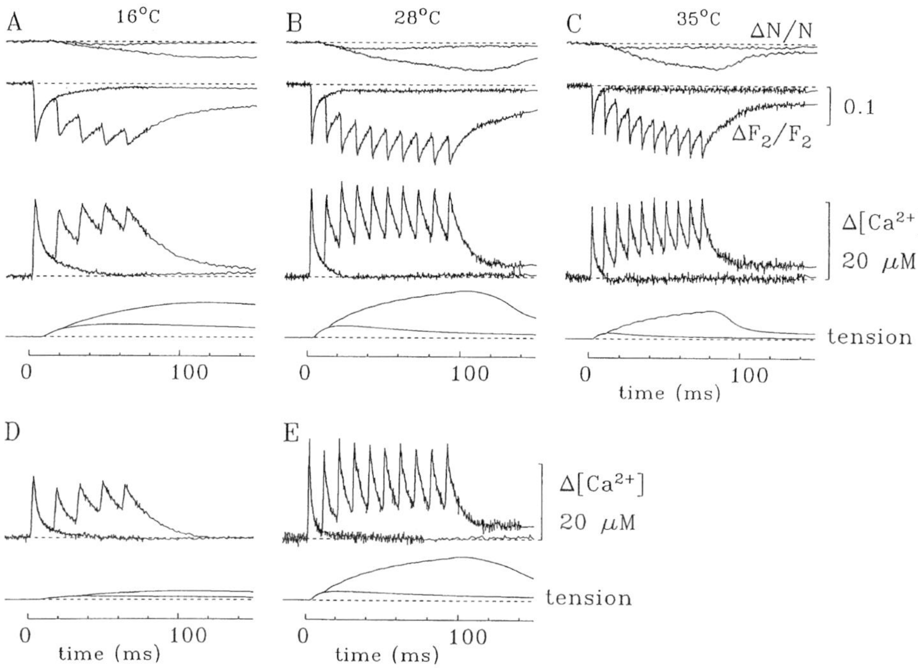

Fig. 12.4 Examples of furaptra Ca^{2+} transients ($\Delta[Ca^{2+}]_i$) and tension in two intact fibres from mouse extensor digitorum longus muscles stimulated by high frequency trains: A, B, and C at 16, 28, and 35 °C, respectively from one fibre; D and E at 16 and 28 °C, respectively from the other fibre. In A, B and C, furaptra fluorescence signals ($\Delta F_2/F_2$) and estimates of movement artifacts in the fluorescence signals ($\Delta N/N$) are also shown. (Reproduced from Hollingworth *et al.*, *J. Gen. Physiol.* **108**, 455–69, 1996 by copyright permission of The Rockefeller University Press.)

potential stimulation, the half-width is 4.6 ms at 16 °C, and 2.0 ms at 28 °C. The results further necessitate the use of low-affinity indicators to faithfully monitor the time course of the cytoplasmic Ca^{2+} transients in skeletal muscle fibres.

Laser scanning confocal microscopy

High sensitivity of fluorescent signals is particularly useful for laser scanning confocal microscopy. The confocal microscopy can detect, in the 'line-scan' mode, localized fluorescence signals with a temporal resolution of a few milliseconds, and with a spatial resolution of submicron along one axis, at the expense of two-directional spatial resolution. With this technique, the elementary events of the Ca^{2+} release process (called 'Ca^{2+} sparks') can be detected in cardiac and skeletal muscle cells (Cheng *et al.* 1993; Tsugorka *et al.* 1995; Klein *et al.* 1996). For this purpose, fluo-3, a tetracarboxylate fluorescent indicator, is widely used, because the fluorescence change of this indicator upon Ca^{2+} binding is very large;

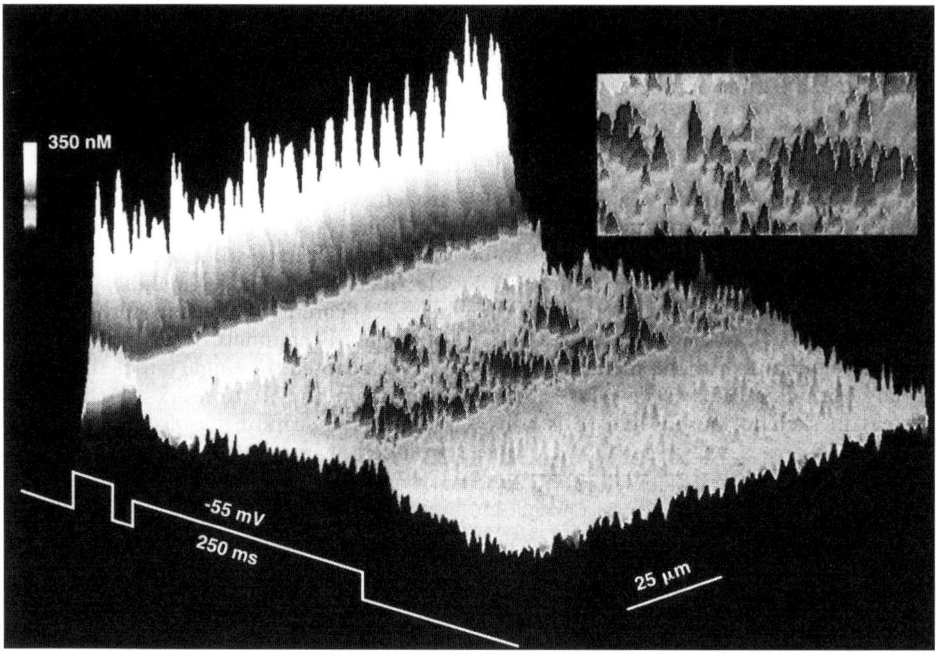

Fig. 12.5 Distribution of Ca^{2+} in a frog cut muscle fibre containing fluo-3. The fibre was voltage clamped at −90 mV, and two pulses to −50 and −55 mV were applied, while line-scan images of fluo-3 fluorescence intensity were obtained as a function of time. Intracellular solution contained 10 mM EGTA, restricting $[Ca^{2+}]$ change to the release site. Intracellular $[Ca^{2+}]$, calculated from fluo-3 fluorescence intensity, is represented as a function of time and position in the scan, derived from a line scan image. The inset expands the plot in a region where discrete events were visible during the second pulse. (Reproduced from Tsugorka *et al.*, *Science* **269**, 1723–6, 1995 by copyright permission of American Association for the Advancement of Science.)

the indicator fluorescence in its Ca^{2+}-bound form is 40–200 times brighter than that in its Ca^{2+}-free form (Minta *et al.* 1989; Harkins *et al.* 1993). Thus, the low fluorescence intensity of Ca^{2+}-free fluo-3 provides a relatively dark background in the resting state. Figure 12.5 shows results from an experiment by Tsugorka *et al.* (1995) on frog cut skeletal muscle fibres under voltage clamp, which reported a local rise of $[Ca^{2+}]_i$ derived from fluo-3 fluorescence measured as a function of time and space with the 'line-scan' mode.

12.5 Problems and limitations

Calibration of the indicator signals

Cytoplasmic $[Ca^{2+}]$ estimated from optical signals of indicators is usually based on the assumption that the indicators behave in the same way in the intracellular environment as in simple salt solutions. It has long been recognized that this assumption may not be valid (e.g. Beeler *et al.* 1980; Konishi *et al.* 1988). Ca^{2+} indicators, including those listed in Table 12.1, are bound to cellular constituents within the cytoplasm; the bound fraction is estimated to exceed 80% for many indicators (see Zhao *et al.* 1996). This binding alters both spectral and binding properties of the indicators, e.g. arsenazo III binds to soluble proteins, which, as a result, changes their absorbance properties (Beeler *et al.* 1980). Blatter and Blinks (1991) have reported that diffusible cytoplasmic constituents of frog fibres influenced aequorin luminescence at very low $[Ca^{2+}]_i$. The binding to soluble proteins changes both fluorescence and Ca^{2+} binding of tetracarboxylate indicators, such as fura-2 (Konishi *et al.* 1988; Uto *et al.* 1991), indo-1 (Baker *et al.* 1994), fluo-3 (Harkins *et al.* 1993) and fura red (Kurebayashi *et al.* 1993). The *in vitro* measurements shown in Fig. 12.6 demonstrate that aldolase, the most abundant soluble protein in muscle fibres, increased K_D of fura-2 for Ca^{2+} in a dose-dependent fashion. Substantial increases in K_D were also found for other tetracarboxylate Ca^{2+} indicators by addition of protein to calibration solutions (Uto *et al.* 1991; Hove-Madsen and Bers 1992; Kurebayashi *et al.* 1993; Harkins *et al.* 1993). It is thus likely that the tetracarboxylate indicators react with Ca^{2+}

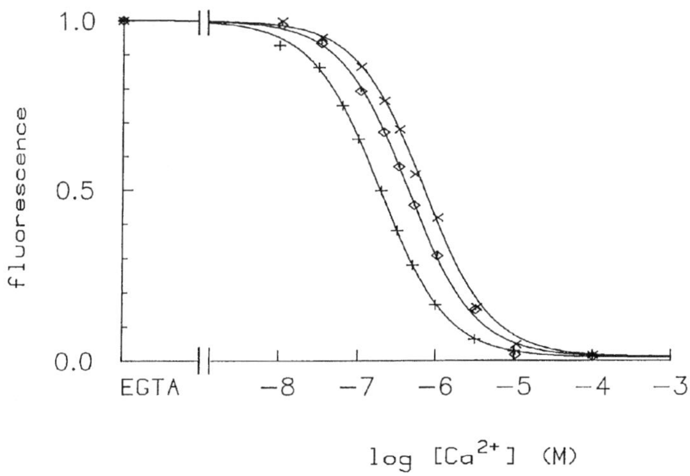

Fig. 12.6 Effect of aldolase on the dissociation constant of fura-2 for Ca^{2+}. *In vitro* fura-2 fluorescence intensity was measured in quartz capillary tubes (internal diameter ~200 μm) as a function of $[Ca^{2+}]$ at 0 mg/ml (+), 27 mg/ml (◇), and 55 mg/ml (×) aldolase. Ionic strength 0.15 M, 1.5 mM free $[Mg^{2+}]$, 10 mM EGTA, 0–10 mM $CaCl_2$, 0.5 mM total [fura-2], pH 7.0 (20 °C). *Solid lines* indicate the least-squares fit of the experimental data by binding curve of 1 : 1 (Ca^{2+} : fura-2) stoichiometry. Fitted dissociation constants were 188 nM (*left-most curve*), 428 nM (*middle curve*), and 687 nM (*right-most curve*). Excitation wavelength 420 ± 15 nm, emission wavelength > 495 nm. (Reproduced from Konishi *et al.*, *Biophys. J.* **54**, 1089–104, 1988 by copyright permission of the Biophysical Society.)

in the cytoplasm with lower affinity than that in simple salt solutions. Furaptra, a tricarboxylate indicator, also appears to have reduced affinity for Ca^{2+} by a factor of about two in the cytoplasm of frog muscle (Konishi *et al.* 1991).

Several methods have been employed to calibrate the indicators' optical signals in the 'intracellular' environment:

(1) Treatment of cells by a Ca^{2+} ionophore (e.g. ionomycin, bromo-A23187) to render the cell membrane permeable to Ca^{2+}, so that $[Ca^{2+}]_i$ can be equilibrated with extracellular $[Ca^{2+}]$ (Li *et al.* 1987; Williams and Fay 1990). Low concentration of β-escin, a saponin ester, may be used to induce more rapid flux of Ca^{2+} across the cell membrane without a gross leak of cellular macromolecules (Konishi and Watanabe 1995).

(2) Injection of muscle fibres with calibration solutions which contain high concentrations of $CaCl_2$, EGTA and their mixture (Westerblad and Allen 1993*a*, 1996). The injection is repeated until the apparent steady-state of optical signals is reached.

(3) Combination of the indicator optical signals measured from intact muscle fibres and *in vitro* solution containing intracellular soluble protein (Harkins *et al.* 1993; Kurebayashi *et al.* 1993).

Unfortunately, none of the above methods have been able to provide conclusive quantitative information on $[Ca^{2+}]_i$. Technical difficulties in calibrating optical signals probably contribute to the large discrepancies between recent estimates of $[Ca^{2+}]_i$ in frog muscle at rest (range between $< 50\,nM$, Blatter and Blinks 1991 and 300 nM, Kurebayashi *et al.* 1993) and in mammalian muscle during tetanus (range between 1 μM, Westerblad and Allen 1993*a,b* and 20 μM, Hollingworth *et al.* 1996; also see Fig. 12.4). Further studies are required to increase the reliability of $[Ca^{2+}]_i$ estimation.

Localization of indicators

For the sake of simple interpretation of the indicator signals, it has been generally assumed that the indicator molecules are homogeneously distributed in the cytoplasm. This may not be the case at least for some (and possibly many) indicators.

It is poorly understood how the injected-aequorin distributes in the cells. We tried to observe the distribution of aequorin in a frog single skeletal muscle fibre using immuno-electronmicroscopy and our preliminary results suggest that aequorin mainly distributes inter-myofibril space rather than the inter-filament space (Yoshioka and Kurihara, unpublished observation). If this is the case, aequorin detects Ca^{2+} near the Ca^{2+} release sites rather than the Ca^{2+} binding sites on troponin-C, which influences the quantitative estimate of $[Ca^{2+}]_i$. However, the detail of the distribution is still not clear.

With scanning confocal microscopy of frog muscle fibres, Tsugorka *et al.* (1995) and Klein *et al.* (1996) reported an oscillating sarcomeric pattern of fluo-3 fluorescence intensity along the length of the fibre. Transverse tubules appeared to be located in the middle of the darker fluorescence bands in the resting muscle, but the band pattern was reversed after depolarization to release Ca^{2+} from sarcoplasmic reticulum, which suggests that $[Ca^{2+}]$ increases more in the triadic regions. This periodicity could result from difference of the indicator binding to specific regions of the sarcomere (the different degree of binding or binding to different proteins), or differences in dye-accessible volume along sarcomeres. The inhomogeneous indicator distribution of this kind would be an additional complication for interpretation of the indicator signals, particularly for early phase of activation when $[Ca^{2+}]_i$ is not uniform within the sarcomeres.

This problem could be more serious, when the indicators are introduced in the cells with the AM loading technique. Membrane permeant AM ester of the indicator can enter intracellular organelles (e.g. sarcoplasmic reticulum, mitochondria) and, as a result of hydrolysis, may be trapped within the organelles. With such 'compartmentalization', a significant fraction of the indicator signals could reflect $[Ca^{2+}]$ in the organelles, thus introducing errors in measurement of $[Ca^{2+}]_i$. In fact, in the case of indo-1 loaded with its AM ester into rat cardiac muscle cells, about one-half of the indicator fluorescence was found to be localized in mitochondria (Miyata *et al.* 1991). Although the compartmentalization is not desirable in the usual circumstances, some well-designed studies have taken advantage of the compartmentalization for monitoring $[Ca^{2+}]$ in the intracellular organelles (Miyata *et al.* 1991; Hirose and Iino 1994). The degree of the compartmentalization appears to depend on both cell type and indicator, as well as on various loading conditions, such as temperature, incubation time, concentration of the AM ester, use of a detergent (e.g. Malgaroli *et al.* 1987); thus this problem needs to be estimated for each individual case. One way to estimate the fraction of the trapped indicator inside the organelles is to treat the indicator loaded cells with either saponin or digitonin, while monitoring the reduction of the indicator fluorescence (e.g. Eisner *et al.* 1989). Since saponin or digitonin, at low concentrations, is known to selectively permeabilize the cell membrane without causing detectable changes in the sarcoplasmic reticulum and mitochondria (Ohtsuki *et al.* 1978; Endo and Iino 1980), the indicator molecules trapped inside the organelles are probably not released by this procedure.

12.6 Future improvement

The binding to intracellular constituents is one general problem that complicates the intracellular use of Ca^{2+} indicators; it causes alterations of optical and Ca^{2+} binding properties of indicators, and probably contributes to the inhomogeneous distribution of the indicators. Because the most accurate quantitative information on $[Ca^{2+}]_i$ should be obtained from indicators that are little bound (or sequestered), it is desirable to have a family of Ca^{2+} indicators which do not interact with cellular constituents. Both high-affinity and low-affinity members of such an indicator family would be very useful to monitor $[Ca^{2+}]_i$ and its change with different dynamic range and kinetics.

Measurement of the Ca^{2+} indicator signals with the fluorescence confocal microscope is undoubtedly a powerful tool for studying excitation–contraction coupling of muscle cells. The peak amplitude of Ca^{2+} sparks derived from fluo-3 fluorescence may be quite different in skeletal muscle and cardiac muscle cells; Tsugorka *et al.* (1995) have reported that the Ca^{2+} sparks in frog skeletal muscle fibres were one-tenth to one-fifth the size of Ca^{2+} sparks in mammalian cardiac muscle cells (Cheng *et al.* 1993), which could imply the fundamental difference in the control of Ca^{2+} release from the sarcoplasmic reticulum in skeletal and cardiac muscle cells. However, Ca^{2+} sparks of somewhat larger amplitude have been also reported in frog skeletal muscle fibres (Klein *et al.* 1996). Because fluo-3 fluorescence exhibits a shift of neither excitation nor emission spectrum, the fluorescence intensity of the sparks at a single pair of excitation/emission wavelengths has to be calibrated in terms of $[Ca^{2+}]_i$ with an additional assumption concerning resting level of $[Ca^{2+}]_i$. For this purpose, it would be therefore helpful to have a 'ratioable' indicator with excitation band in visible wavelengths and large Ca^{2+}-dependent fluorescence changes.

References

Allen, D.G. and Blinks, J.R. (1978) Calcium transients in aequorin-injected frog cardiac muscle. *Nature* **273**, 509–13.

Allen, D.G. and Kurihara, S. (1982) The effects of muscle length on intracellular calcium transients in mammalian cardiac muscle. *J. Physiol.* (*Lond.*) **327**, 79–94.

Allen, D.G., Blinks, J.R. and Prendergast, F.G. (1977) Aequorin luminescence: relation of light emission to calcium concentration—a calcium-independent component. *Science* **195**, 996–8.

Ashley, C.C. and Ridgway, E.B. (1968) Simultaneous recording of membrane potential, calcium transient and tension in single barnacle muscle fibres. *Nature* **219**, 1168–9.

Baker, A.J., Brandes, R., Schreur, J.H.M., Camacho, S.A. and Wiener, M.W. (1994) Protein and acidosis alter calcium-binding and fluorescence spectra of the calcium indicator indo-1. *Biophys. J.* **67**, 1646–54.

Baylor, S.M. and Hollingworth, S. (1988) Fura-2 calcium transients in frog skeletal muscle fibres. *J. Physiol.* (*Lond.*) **405**, 151–92.

Baylor, S.M., Chandler, W.K. and Marshall, M.W. (1982) Use of metallochromic dyes to measure changes in myoplasmic calcium during activity in frog skeletal muscle fibres. *J. Physiol.* (*Lond.*) **331**, 139–77.

Baylor, S.M., Hollingworth, S., Hui, C.S. and Quinta-Ferreira, M.E. (1985) Calcium transients from intact frog skeletal muscle fibres simultaneously injected with Antipyrylazo III and Azo 1. *J. Physiol.* (*Lond.*) **365**, 70P.

Baylor, S.M., Hollingworth, S., Hui, C.S. and Quinta-Ferreira, M.E. (1986) Properties of the metallochromic dyes arsenazo III, antipyrylazo III and azo 1 in frog skeletal muscle fibres at rest. *J. Physiol.* (*Lond.*) **377**, 89–141.

Beeler, T.J., Schibeci, A. and Martonosi, A. (1980) The binding of arsenazo III to cell components. *Biochem. Biophys. Acta* **629**, 317–27.

Blatter, L.A. and Blinks, J.R. (1991) Simultaneous measurement of Ca^{2+} in muscle with Ca electrodes and aequorin. Diffusible cytoplasmic constituent reduces Ca^{2+}-independent luminescence of aequorin. *J. Gen. Physiol.* **98**, 1141–60.

Blinks, J.R. (1978) Measurement of calcium ion concentrations with photoproteins. *Annal. N.Y. Acad. Sci.* (*USA*) **307**, 71–85.

Blinks, J.R. (1984) Methods for monitoring Ca^{2+} concentrations with photoproteins in living cardiac cells. In *Method for studying heart membranes* (ed. N.S. Dhalla), pp. 237–64. CRC Press, Boca Raton.

Blinks, J.R., Prendergast, F.G. and Allen, D.G. (1976) Photoproteins as biological calcium indicators. *Pharmacol. Rev.* **28**, 1–93.

Blinks, J.R., Rüdel, R. and Taylor, S.R. (1978) Calcium transients in isolated amphibian skeletal muscle fibres: detection with aequorin. *J. Physiol.* **277**, 291–323.

Blinks, J.R., Wier, W.G., Hess, P. and Prendergast, F.G. (1982) Measurement of Ca^{2+} concentrations in living cells. *Prog. Biophys. Molec. Biol.* **40**, 1–114.

Carroll, S.L., Klein, M.G. and Schneider, M.F. (1995) Calcium transients in intact rat skeletal muscle fibres in agarose gel. *Am. J. Physiol.* **269**, C28–C34.

Cheng, H., Lederer, W.J. and Cannell, M.B. (1993) Calcium sparks: elementary events underlying excitation–contraction coupling in heart muscle. *Science* **262**, 740–4.

Claflin, D.R., Morgan, D.L., Stephenson, D.G. and Julian, F.J. (1994) The intracellular Ca^{2+} transient and tension in frog skeletal muscle fibres measured with high temporal resolution. *J. Physiol.* (*Lond.*) **475**, 319–25.

Delbono, O. and Stefani, E. (1993) Calcium transients in single mammalian skeletal muscle fibres. *J. Physiol.* (*Lond.*) **463**, 689–707.

Dorogi, P.L. and Newmann, E. (1981) Spectrophotometric determination of reaction stoichiometry and equilibrium constants of metallochromic indicators. II. The Ca^{2+}-arsenazo III complexes. *Biophys. Chem.* **13**, 125–31.

Ebashi, S. and Endo, M. (1968) Calcium ion and muscle contraction. *Prog. Biophys. Molec. Biol.* **18**, 123–83.

Eisner, D.A., Nichols, C.G., O'Neill, S.C., Smith, G.L. and Valdeolmillos, M. (1989) The effects of metabolic inhibition on intracellular calcium and pH in isolated rat ventricular cells. *J. Physiol.* (*Lond.*) **411**, 393–418.

Endo, M. and Iino, M. (1980) Specific perforation of muscle cell membranes with preserved SR functions by saponin treatment. *J. Muscle Res. Cell Motil.* **1**, 89–100.

Garcia, J. and Schneider, M.F. (1993) Calcium transients and calcium release in rat fast-twitch skeletal muscle fibres. *J. Physiol.* (*Lond.*) **463**, 709–28.

Gordon, A.M. and Ridgway, E.B. (1990) Stretch of active muscle during the declining phase of the calcium transient produces biphasic change in calcium binding to the activating sites. *J. Gen. Physiol.* **96**, 1013–5.

Grynkiewicz, G., Poenie, M. and Tsien, R. (1985) A new generation of Ca^{2+} indicators with greatly improved fluorescence properties. *J. Biol. Chem.* **260**, 3440–50.

Harkins, A.B., Kurebayashi, N. and Baylor, S.M. (1993) Resting myoplasmic free calcium in frog skeletal muscle fibres estimated with fluo-3. *Biophys. J.* **65**, 865–81.

Hirose, K. and Iino, M. (1994) Heterogeneity of channel density in inositol-1,4,5-trisphosphate-sensitive Ca^{2+}stores. *Nature* **372**, 791–4.

Hirota, A., Chandler, W.K., Southwick, P.L. and Waggoner, A.S. (1989) Calcium signals recorded from two new purpurate indicators inside frog cut twitch fibres. *J. Gen. Physiol.* **94**, 597–631.

Hollingworth, S., Aldrich, R.W. and Baylor, S.M. (1987) *In vitro* calibration of the equilibrium reactions of the metallochromic indicator dye antipyrylazo III with calcium. *Biophys. J.* **51**, 383–93.

Hollingworth, S., Harkins, A.B., Kurebayashi, N., Konishi, M. and Baylor, S.M. (1992) Excitation–contraction coupling in intact frog skeletal muscle fibres injected with mmolar concentration of fura-2. *Biophys. J.* **63**, 224–34.

Hollingworth, S., Zhao, M. and Baylor, S.M. (1996) The amplitude and time course of the myoplasmic free [Ca^{2+}] transient in fast-twitch fibres of mouse muscle. *J. Gen. Physiol.* **108**, 455–69.

Hongo, K., Konishi, M. and Kurihara, S. (1994) Cytoplasmic free Mg^{2+} in rat ventricular myocytes studied with the fluorescent indicator furaptra. *Jpn. J. Physiol.* **44**, 357–78.

Hove-Madsen, L. and Bers, D.M. (1992) Indo-1 binding to protein in permeabilized ventricular myocytes alters its spectral and Ca binding properties. *Biophys. J.* **63**, 89–97.

Irving, M., Maylie, J., Sizto, N.L. and Chandler, W.K. (1987) Intrinsic optical and passive electrical properties of cut frog twitch fibres. *J. Gen. Physiol.* **89**, 1–40.

Jacquemond, V.L., Csernoch, M.G., Klein, M.G. and Schneider, M.F. (1991) Voltage-gated and calcium-gated calcium release during depolarization of skeletal muscle fibres. *Biophys. J.* **60**, 867–73.

Jöbsis, F.F. and O'Connor, M.J. (1966) Calcium release and reabsorption in the sartorius muscle of the toad. *Biochem. Biophys. Res. Commun.* **25**, 246–52.

Jong, D.-S., Pape, P.C., Chandler, W.K. and Baylor, S.M. (1993) Reduction of calcium inactivation of sarcoplasmic reticulum calcium release by fura-2 in voltage-clamped cut twitch fibres from frog muscle. *J. Gen. Physiol.* **102**, 333–70.

Klein, M.G., Cheng, H., Santana, L.F., Jiang, Y.-H., Lederer, W.J. and Schneider, M.F. (1996) Two mechanisms of quantized calcium release in skeletal muscle. *Nature* **379**, 455–8.

Konishi, M. and Baylor, S.M. (1991) Myoplasmic calcium transients monitored with purpurate indicator dyes injected into intact frog skeletal muscle fibres. *J. Gen. Physiol.* **97**, 245–70.

Konishi, M. and Berlin, J.R. (1993) Ca transients in cardiac myocytes measured with a low affinity fluorescent indicator, furaptra. *Biophys. J.* **64**, 1331–43.

Konishi, M. and Kurihara, S. (1987) Effects of caffeine on intracellular calcium concentration in frog skeletal muscle fibres. *J. Physiol.* (*Lond.*) **383**, 269–83.

Konishi, M. and Kurihara, S. (1993) Radial spread of aequorin Ca^{2+} signal in single frog skeletal muscle fibers. *Mol. Cell. Biochem.* **119**, 59–66.

Konishi, M. and Watanabe, M. (1995) Resting cytoplasmic free Ca^{2+} concentration in frog skeletal muscle measured with fura-2 conjugated to high molecular weight dextran. *J. Gen. Physiol.* **106**, 1125–50.

Konishi, M., Olson, A., Hollingworth, S. and Baylor, S.M. (1988) Myoplasmic binding of fura-2 investigated by steady-state fluorescence and absorbance measurements. *Biophys. J.* **54**, 1089–104.

Konishi, M., Hollingworth, S., Harkins, A.B. and Baylor, S.M. (1991) Myoplasmic calcium transients in intact frog skeletal muscle fibers monitored with the fluorescent indicator furaptra. *J. Gen. Physiol.* **97**, 271–301.

Konishi, M., Suda, N. and Kurihara, S. (1993) Fluorescence signals from the Mg^{2+}/Ca^{2+} indicator furaptra in frog skeletal muscle fibers. *Biophys. J.* **64**, 223–39.

Kurebayashi, N., Harkins, A.B. and Baylor, S.M. (1993) Use of fura red as an intracellular indicator in frog skeletal muscle fibers. *Biophys. J.* **64**, 1934–60.

Kurihara, S. and Komukai, K. (1995) Tension-dependent changes of the intracellular Ca^{2+} transients in ferret ventricular muscles. *J. Physiol.* (*Lond.*) **489**, 617–25.

Li, Q., Altschuld, R.A. and Stokes, B.T. (1987) Quantitation of intracellular free calcium in single adult cardiomyocytes by fura-2 fluorescence microscopy: calibration of fura-2 ratios. *Biochem. Biophys. Res. Commun.* **147**, 120–6.

MacKimmon, R., Gwathmey, J.K. and Morgan, J.P. (1987) Differential effects of reoxygenation on intracellular calcium and isometric tension. *Pflügers Arch.* **409**, 448–52.

Malgaroli, A., Milani, D., Meldolesi, J. and Pozzan, T. (1987) Fura-2 measurement of cytoplasmic free Ca^{2+} in monolayers and suspensions of various types of animal cells. *J. Cell Biol.* **105**, 2145–55.

Maylie, J., Irving, M., Sizto, N.L. and Chandler, W.K. (1987*a*) Calcium signals recorded from cut frog twitch fibers containing antipyrylazo III. *J. Gen. Physiol.* **89**, 83–143.

Maylie, J., Irving, M., Sizto, N.L. and Chandler, W.K. (1987*b*) Calcium signals recorded from cut frog twitch fibres containing tetramethylmurexide. *J. Gen. Physiol.* **89**, 145–76.

Miledi, R., Parker, I. and Schalow, G. (1977) Measurement of calcium transients in frog muscle by the use of arsenazo III. *Proc. R. Soc. Lond.* **B198**, 201–10.

Minta, A., Kao, J.P.Y. and Tsien, R.Y. (1989) Fluorescent indicators for cytosolic calcium based on rhodamine and fluorescein chromophores. *J. Biol. Chem.* **264**, 8171–8.

Miyata, H., Silverman, H.S., Sollott, S.J., Lakatta, E.G., Stern, M.D. and Hansford, R.G. (1991) Measurement of mitochondrial free Ca^{2+} concentration in living single rat cardiac myocytes. *Am. J. Physiol.* **261**, H1123–34.

Moore, E.D.W. (1986) Properties of aequorin relevant to its use in the measurement of intracellular [Ca^{2+}] in skeletal muscle fibres. Ph.D. thesis. University of Minnesota, Rochester, Minnesota, USA.

Morgan, J.P. and Morgan, K.G. (1982) Vascular smooth muscle: the first recorded Ca^{2+} transients. *Pflügers Arch.* **395**, 75–7.

Ogawa, Y., Harafuji, H. and Kurebayashi, N. (1980) Comparison of the characteristics of four metallochromic dyes as potential calcium indicators for biological experiments. *J. Biochem.* **87**, 1293–303.

Ohnishi, S.T. (1978) Characterization of the murexide method: Dual-wavelength spectrophotometry of cations under physiological conditions. *Anal. Chem.* **85**, 165–79.

Ohnishi, S.T. (1979) Interaction of metallochromic indicators with calcium sequestering organelles. *Biochim. Biophys. Acta* **585**, 315–9.

Ohtsuki, I., Manzi, R.M. Palade, G.E. and Jamieson, J.D. (1978) Entry of macromolecular tracers into cells with low concentrations of aldehydes. *Biol. Cell* **31**, 119–26.

Palade, P. and Vergara, J. (1982) Arsenazo III and antipyrylazo III calcium transients in single skeletal muscle fibers. *J. Gen. Physiol.* **79**, 679–707.

Palade, P. and Vergara, J. (1983) Stoichiometries of arsenazo III-Ca complexes. *Biophys. J.* **43**, 355–69.

Raju, B., Murphy, E., Levy, L.A., Hall, R.D. and London, R.E. (1989) A fluorescent indicator for measuring cytosolic free magnesium. *Am. J. Physiol.* **256**, C540–8.

Rios, E. and Schneider, M.F. (1981) Stoichiometry of the reactions of calcium with the metallochromic indicator dyes antipyrylazo III and arsenazo III. *Biophys. J.* **36**, 607–21.

Shimomura, O., Johnson, F.F. and Saiga, Y. (1962) Extraction, purification, and properties of aequorin, a bioluminescent protein from the luminous hydromedusan, *Aequorea. J. Cell. Comp. Physiol.* **59**, 223–39.

Shimomura, O., Inouye, S., Musick, B. and Kishi, Y. (1990) Recombinant aequorin and recombinant semi-synthetic aequorins: cellular Ca^{2+} indicators. *Biochem. J.* **270**, 309–12.

Shirokova, N., Garcia, J., Pizzaro, G. and Rios, E. (1996) Ca^{2+} release from the sarcoplasmic reticulum compared in amphibian and mammalian skeletal muscle. *J. Gen. Physiol.* **107**, 1–18.

Southwick, P.L. and Waggoner, A.S. (1989) Synthesis of purpurate-1,1′-diacetic acid PDAA tripotassium salt: a new calcium indicator for biological applications. *Org. Prep. Proc. Int.* **21**, 493–500.

Suda, N. and Kurihara, S. (1991) Intracellular calcium signals measured with fura-2 and aequorin in frog skeletal muscle fibers. *Jpn. J. Physiol.* **41**, 277–95.

Tsien, R.Y. (1980) New calcium indicators and buffers with high selectivity against magnesium and protons: design, synthesis, and properties of prototype structures. *Biochem.* **19**, 2396–404.

Tsien, R.Y., Pozzan, T. and Rink, T.J. (1982) Calcium homeostasis in intact lymphocytes: cytoplasmic free calcium monitored by a new, intracellularly trapped fluorescent-indicator. *J. Cell. Biol.* **94**, 325–34.

Tsugorka, A., Rios, E. and Blatter, L.A. (1995) Imaging elementary events of calcium release in skeletal muscle cells. *Science* **269**, 1723–6.

Uto, A., Arai, H. and Ogawa, Y. (1991) Reassessment of fura-2 and the ratio method for determination of intracellular Ca^{2+} concentrations. *Cell Calcium* **12**, 29–37.

Westerblad, H. and Allen, D.G. (1993*a*) The influence of intracellular pH on contraction, relaxation and $[Ca^{2+}]_i$ in intact single fibres from mouse muscle. *J. Physiol.* (*Lond.*) **466**, 611–28.

Westerblad, H. and Allen, D.G. (1993*b*) The contribution of $[Ca^{2+}]_i$ to the slowing of relaxation in fatigued single fibres from mouse skeletal muscle. *J. Physiol.* (*Lond.*) **468**, 729–40.

Westerblad, H. and Allen, D.G. (1994) Relaxation, $[Ca^{2+}]_i$ and $[Mg^{2+}]_i$ during prolonged tetanic stimulation of intact single fibres from mouse skeletal muscle. *J. Physiol.* (*Lond.*) **480**, 31–43.

Westerblad, H. and Allen, D.G. (1996) Intracellular calibration of the calcium indicator indo-1 in isolated fibers of *Xenopus* muscle. *Biophys. J.* **71**, 908–17.

Williams, H. and Fay, F.S. (1990) Intracellular calibration of the fluorescent calcium indicator fura-2. *Cell Calcium* **11**, 75–83.

Zhao, M., Hollingworth, S. and Baylor, S.M. (1996) Properties of tri- and tetracarboxylate Ca^{2+} indicators in frog skeletal muscle fibers. *Biophys. J.* **70**, 896–916.

13 Electrophysiological techniques

Jean Mironneau

13.1 Introduction

The voltage-clamp technique has been a powerful tool in the development of membrane biophysics. Applied to large cells, such as squid axons or giant snail neurones, this technique allows reconstruction of an action potential by means of separate Na^+ and K^+ conductances (Hodgkin and Huxley 1952). However, the need to penetrate cells with two micro-electrodes restricts its use only to large cells. This limitation can be partly overcome by the introduction of a voltage-clamp circuit, which only needs one micro-electrode (Brennecke and Lindemann 1974); this has been useful for describing ion currents in small cells (Amédée *et al.* 1987). At the same time, the patch-clamp technique for measuring single currents was developed by Neher and Sakmann (1976). When the improved patch-clamp technique for single-channel and whole-cell recordings was described in detail (Hamill *et al.* 1981) it quickly became the technique for investigating the electrical properties of cell membranes. This method represents a major advance in the ability to monitor cell membrane function. By sealing a small pipette to a clean cell membrane, it is possible to record the ion currents through single channels contained in small patches from intact cell membranes, as well as the whole-cell ion currents. With the latter configuration it is possible to modify the intracellular medium or to introduce specific agents into the cells. Physiological processes including signal transduction and synaptic transmission, as well as pathological conditions that involve ion channel function or dysfunction, can be studied with the patch-clamp technique. Pharmacological agents that act on ion channels are readily examined by the use of this method.

13.2 The patch-clamp technique

The patch-clamp technique can be applied to all types of cells of any size and is particularly suitable for studying small cells, freshly dissociated or in culture, that cannot be investigated with classical micro-electrode techniques. The voltage-clamp method of recording ion channel activity is the most widely used mode of the patch-clamp technique. With this method, a given membrane potential is maintained with a feed-back circuit using a very rapid, low noise differential amplifier. At the same time the membrane current is recorded and is related to the conductance through ion channels in accordance with Ohm's law. The patch-clamp pipette is filled with a conductance solution and acts as an electrode, which maintains the voltage-clamp and also senses the current across the membrane. The quality of the ion current measurement is largely dependent on the resistance between the pipette solution and the bath solution. Therefore, the ability to create high-resistance seals of 10–100 Gohms (gigaseals) is essential for the resolution of single channel ion currents with amplitudes of less than 1 pA. Another important property of the gigaseal is that it is mechanically stable, allowing membrane patches to be excised or broken, thus permitting various recording configurations.

Glasses for patch pipettes are selected on the basis of their electrical, thermal and sealing properties, but cannot be considered as physiologically inert as they have been shown to release components that affect ion channels (Zuazaga and Steinacher 1990; Copello *et al.* 1991). Patch pipettes with a tip inner diameter of 0.5 to 3 μm are generally coated with a hydrophilic material to reduce capacitive currents and are filled with a conducting salt solution whose composition can be controlled. The pipette is connected to the headstage of a low-noise current amplifier. When the pipette is in contact with the cell surface a gentle suction forms a gigaseal, which is signalled by a sudden increase in resistance and virtually no current recorded. Then voltage-clamp experiments can begin by selecting a variety of protocols to study the voltage- and time-dependencies of ion currents.

Different recording configurations can be obtained with patch pipettes (Fig. 13.1) and used to study the different aspects of an experimental problem (Hamill *et al.* 1981; Sakmann and Neher 1984). They can be divided into 'outside-out' and 'inside-out' configurations, depending on whether the outer or inner cell membrane is exposed to the bathing solution. Cell-attached, whole-cell and outside-out excised patch recordings represent outside-out configurations, whereas inside-out excised patch and cell-attached (when the membrane in the bathing solution is perforated) recordings represent inside-out configurations. A consequence of these configurations is that both the external and internal ion environment can be controlled. In contrast, when the native intracellular medium has to be maintained, the perforated-patch recording can be used (Horn and Marty 1988). In this case a pore-forming agent, such as nystatin, is added to the pipette solution and progressively perforates the cell membrane, allowing low-resistance access to the intracellular space but leaving the intracellular medium intact (Horn 1991).

Cell-attached patch configurations

This configuration probably disturbs less the structure and environment of the cell membrane. The membrane potential can be changed without intracellular micro-electrodes, and both transmitter- and voltage-gated channels can be studied in their normal intracellular environment. However, in this configuration, the potential across the patch is dependent on the resting membrane potential of

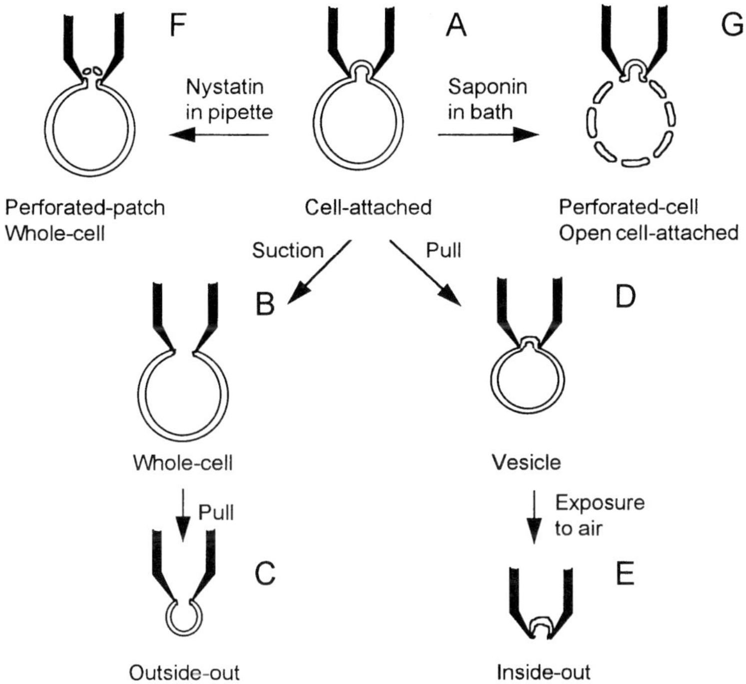

Fig. 13.1 Patch-clamp recording configurations. From the cell-attached patch configuration (A) the whole-cell configuration (B) is obtained by rupturing the cell membrane under the pipette with suction. Outside-out patch configuration (C) is formed by pulling from a whole-cell configuration. By pulling from the cell-attached patch configuration a vesicle is generally formed if Ca^{2+} are present in the bathing solution and inside-out patch configuration is obtained after air exposure (E). Perforated whole-cell configuration is formed by addition of nystatin to the pipette configuration (F); whereas addition of saponin to the bathing solution leads to an open cell-attached configuration (G).

the cell, which cannot be determined. Experimentally, a bath solution with a high-potassium concentration is used to make the membrane potential close to zero, as determined by the Nernst equation. Possible perturbations induced by the gigaseal formation have been previously described (Sakmann and Neher 1984).

Single-channel currents have been studied initially in a normal ionic environment for a variety of channels including acetylcholine-receptor channels, GABA-activated channels, glutamate-activated channels, and the electrically excitable Na^+ and K^+ channels. However, the patch pipette offers the advantage of modifying the external medium and therefore of recording single channel currents, which cannot be seen in normal physiological solution, such as Ca^{2+} and Cl^- channels. Figure 13.2A illustrates single Ca^{2+} channel openings in a vascular myocyte, using the cell-attached configuration under voltage-clamp conditions.

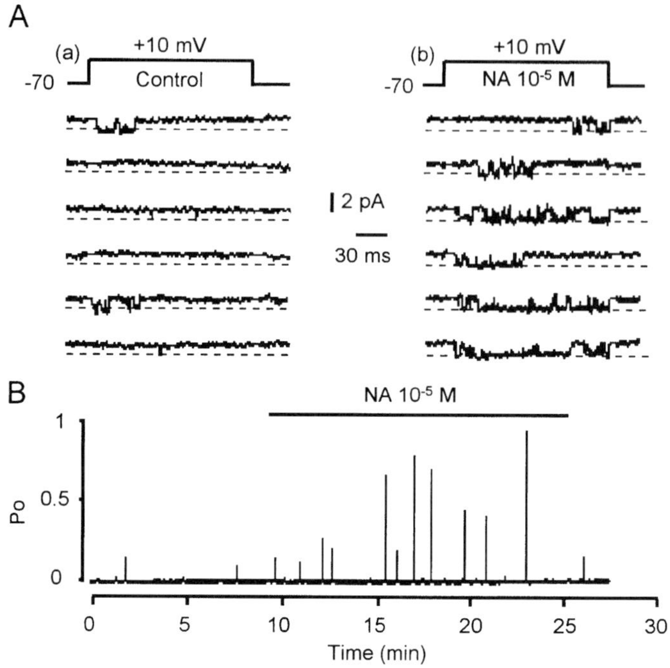

Fig. 13.2 Cell-attached patch configuration. (A) Single L-type Ca^{2+} channel currents recorded in a vascular myocyte held at -70 mV and stepped to a test potential of $+10$ mV in control conditions (a) and during application of 10 μM noradrenaline (b). *Dashed lines* indicate the mean single-channel current level obtained from amplitude histograms. (B) Channel open probability P_o for individual depolarizing steps plotted against time. P_o is determined by dividing the time that the channel spends in the open state during a depolarization by the total time of depolarization. The noradrenaline effect is shown as an increase in P_o and in the frequency of openings. The pipette solution contains 90 mM Ba^{2+}. (From Pacaud *et al.* 1991.)

The bathing solution contains 90 mM Ba^{2+} (Pacaud *et al.* 1991). When an agent that is added to the bathing solution affects single-channel currents under the patch pipette, it is likely that this agent exerts its effects through a second messenger system inside the cell. Conversely, if an agent modifies the activity of an ion channel when added to the pipette solution but has no effect when added to the bathing solution, it is likely that the effect occurs via a direct modulation of the ion channel under the patch pipette. A typical example of the latter case is given by acetylcholine, which is thought to mediate its effects by directly binding to a receptor on a non-specific cation channel at the neuromuscular junction. In vascular myocytes, application of 10 μM noradrenaline in the bathing solution increases the frequency of openings during a depolarizing step and reduces the number of steps without opening (Fig. 13.2A). Therefore, the open probability of the L-type Ca^{2+} channel appears to be increased by noradrenaline (Fig. 13.2B). Because the patch and the bath solution are mechanically isolated, these results indicate that the effect of noradrenaline on Ca^{2+} channels is mediated by an intracellular second messenger. Two transduction pathways activated by noradrenaline binding to α_{1A}- and α_{2A}-adrenoceptors have been identified in vascular myocytes, leading to modulation of Ca^{2+} channels (Leprêtre *et al.* 1994*b*,*c*; Leprêtre and Mironneau 1994).

Inside-out patch configuration

In this configuration, the cytoplasmic membrane faces the bathing solution and the potential difference between the pipette solution and the bathing solution can be set, achieving voltage-clamp conditions. Although some intracellular proteins linked to the plasma membrane may be included in the excised patch, it is generally accepted that the intracellular environment of the cell is lost. The main advantage of this configuration is to manipulate the composition of the bathing solution. The direct effect of second messengers, such as Ca^{2+} ions, phosphorylating kinases and G-proteins on ion channels can be only demonstrated by using the inside-out patch configuration (Marty 1981; Yatani *et al.* 1987).

Whole-cell configuration

This configuration allow recording of the activity of the entire population of ion channels on a cell membrane. It has been powerful to voltage-clamp small cells which are difficult to study with techniques using cell impalement with microelectrodes. Although current recordings obtained in whole-cell configuration resemble those obtained with classical voltage-clamp technique, this method offers several advantages. First, the pipettes are filled with a medium of the ionic strength of the physiological saline solution in order to prevent massive loading of the cell with KCl. Second, the intracellular medium can be manipulated by controlling the composition of the pipette solution. When applied to small cells, this configuration provides the conditions for the dialysis of the cell interior, including the possibility of losing important intracellular components through diffusive depletion.

Manipulation of the intracellular medium is required for recording Ca^{2+} currents, which are usually contaminated by K^{+} currents. In this case, K^{+} currents are removed by replacing intracellular KCl by CsCl. With regard to this method, it should be pointed out that Ca^{2+} currents appear to be up- and down-regulated as a function of the time of dialysis, suggesting the washout of intracellular components.

This configuration is well suited for recording voltage-dependent ion channels and studying the effects of pharmacological agents that interact directly with the external face of the channels. For example, Fig. 13.3 illustrates the blocking effects of elgodipine (a light-resistant dihydropyridine) on the Ca^{2+} channel current in vascular myocytes (Leprêtre *et al.* 1994*a*). The concentration–response curve for the inhibitory effect of elgodipine allows the inhibition constant of this agent

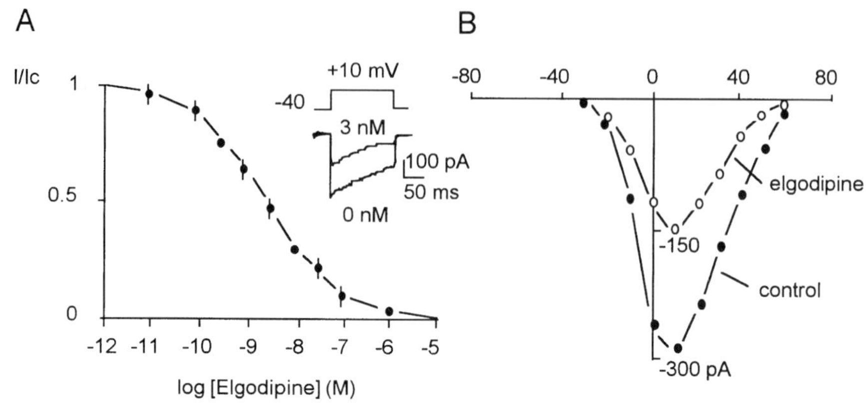

Fig. 13.3 Whole-cell configuration. (A) Inward Ca^{2+} channel currents in single vascular myocytes. Concentration–effect relationship for the inhibitory action of elgodipine. The current elicited from a holding potential of -40 mV to $+10$ mV is expressed as a fraction of control current in the absence of elgodipine (I/Ic) and plotted against the logarithmic value of elgodipine concentration. Each point represents the mean response of three to six cells with SEM shown by vertical lines. *Inset*, Ba^{2+} current traces showing inhibition induced by 3 nM elgodipine. The external Ba^{2+} concentration is 5 mM. (B) Current–voltage relationships obtained in the absence (●) and presence (○) of 3 nM elgodipine. The holding potential is -40 mV. Stimulation frequency, 0.05 Hz. (From Leprêtre *et al.* 1994*a*.)

for the Ca^{2+} channel to be calculated. This result can be compared to the inhibition constant determined from the displacement curve of [^{3}H]isradipine binding by elgodipine. The similarity of the inhibition constants confirms the validity of the electrophysiological analysis to characterize the pharmacological properties of new compounds. In addition, the voltage- and use-dependent properties of Ca^{2+} channel inhibition can be determined.

Because in most preparations Cl^- ions are close to their electrochemical equilibrium at the resting membrane potential, it is difficult to separate Cl^- and K^+ currents. By performing the experiments with pipettes filled with CsCl, the noradrenaline-evoked current in vascular myocytes is maximal at -70 mV, decreases in amplitude with depolarization, is null at 0 mV and reserves at positive membrane potential (Fig. 13.4). The reversal potential of the noradrenaline-induced current is near the equilibrium potential of Cl^- ions for the reference chloride (146 mM : 130 mM) gradient, and is shifted as predicted by the Nernst equation when the extracellular or intracellular chloride concentration is changed (Pacaud *et al.* 1989).

Outside-out patch configuration

In the outside-out patch configuration, the extracellular bathing solution facing the outer cell membrane is easily manipulated, and the potential difference

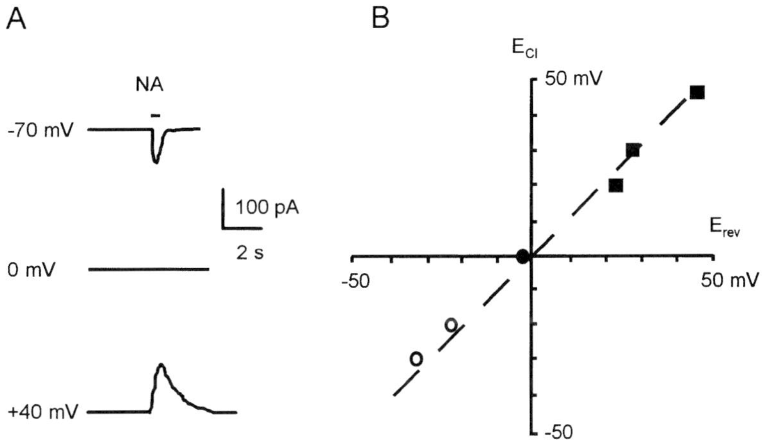

Fig. 13.4 Whole-cell configuration. (A) Membrane current responses to ejections of noradrenaline at different holding potentials (noted near the current traces). The external chloride concentration is 146 mM, and the pipette concentration is 130 mM. The measured reversal potential is close to the calculated equilibrium potential for chloride ions (E_{Cl}). (B) Reversal potential (E_{rev}) of the noradrenaline-induced current as a function of E_{Cl} for reference chloride (146 mM : 130 mM) gradient (●), low external chloride (■) and low pipette chloride concentrations (○). *Dashed lines* predicts the reversal potential for a pure chloride conductance. Each point is the mean of 3–15 cells.

between the pipette solution and the bathing solution can be set, resulting in voltage-clamp conditions. This configuration is well adapted for the single-channel study of neurotransmitters, hormones and agents that bind to ion channels. However, most of the intracellular regulatory components are lost and the physiological processes that regulate ion channel function by way of second messengers cannot be studied.

A variant of the outside-out excised patch configuration is the whole-cell configuration, where the density of channels or the probability of channel opening is low. This is illustrated in Fig. 13.5, where non-selective cation single-channel currents are evoked by an increase in $[Ca^{2+}]_i$ induced by caffeine, noradrenaline or acetylcholine (Loirand *et al.* 1991).

13.3 Applications of the patch-clamp technique

Exocytosis and secretion

Capacitive currents can be accurately measured by the patch-clamp technique (whole-cell configuration) and correlated to cell membrane area, assuming a specific capacitance of $1\,\mu F/cm^2$ (Lindau and Neher 1988). In the process of secretion, exocytosis causes an increase in membrane area when a vesicle fuses

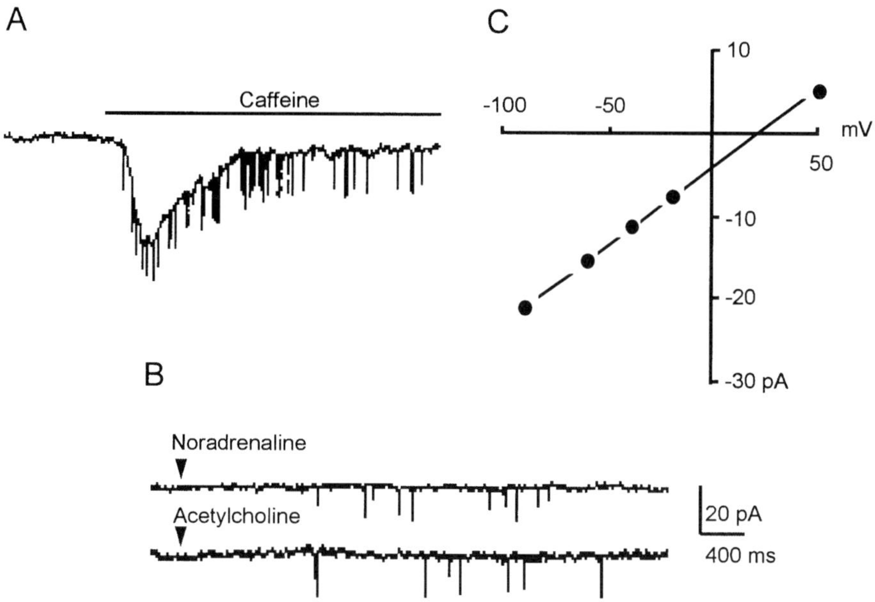

Fig. 13.5 Whole-cell configuration. Single channel currents activated by release of Ca^{2+} induced by caffeine, noradrenaline and acetylcholine in vascular myocytes. (A) Ejection of 10 mM caffeine induces both a large Ca^{2+}-dependent Cl^- current and superimposed single-channel currents. (B) After blocking Cl^- current with 10 mM anthracene-9-carboxylic acid, noradrenaline (10 μM) and acetylcholine (10 μM) activate single-channel currents. (C) Relationship between the amplitude of the caffeine-activated single-channel current and the membrane potential. The mean slope conductance and reversal potential are 190 pS and +12 mV, respectively. (From Loirand *et al.* 1991.)

with the plasma membrane. The increase in membrane area is reflected by the rise in membrane capacitance. During secretion, single exocytosis events can be resolved by recording of cell membrane capacitance (Penner and Neher 1989).

Transduction couplings

Ion channels can be modulated by second messengers acting either directly or indirectly by their effects on G-proteins, intracellular kinases or other regulatory substances. The patch-clamp technique provides a valuable method for studying transduction couplings between membrane receptors and ion channels (Neher 1988). Figure 13.6 shows stimulation of the Ca^{2+} channel current by clonidine (an α_2-adrenoceptor agonist) and phenylephrine (an α_1-adrenoceptor agonist) in vascular myocytes dialysed with a normal pipette solution. When an antibody directed against the α_q/α_{11} subunits of G-protein is dialysed into the cell, the phenylephrine-induced stimulation of the Ca^{2+} channel current is completely

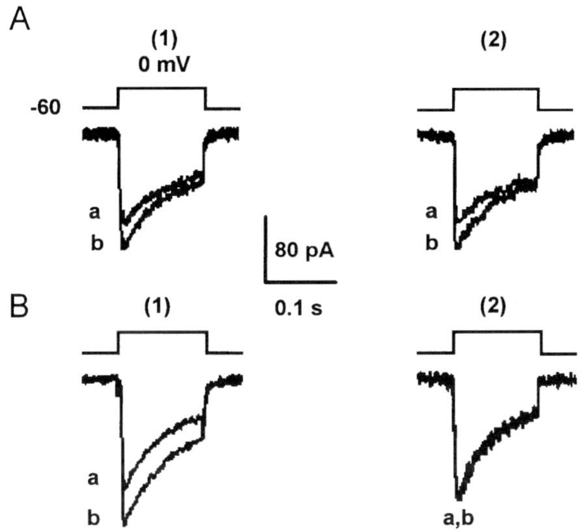

Fig. 13.6 Whole-cell configuration. Inward Ca^{2+} channel currents elicited from a holding potential of -60 mV to 0 mV in control conditions (basic pipette solution, (1)) and in cells dialysed with 12.5 μg/ml anti-$G\alpha_q/\alpha_{11}$ antibody for 3 min (2) before (a) and during the application (b) of 10 μM clonidine (A) or phenylephrine (B). External solution contained 5 mM Ba^{2+}. (From Leprêtre *et al.* 1994*c*.)

removed, whereas the clonidine-induced stimulation of the Ca^{2+} channel current is unchanged, indicating that two different G-proteins are involved in the transduction pathways activated by α_1- and α_2-adrenoceptors (Leprêtre *et al.* 1994*c*; Macrez-Leprêtre *et al.* 1996).

Antisense oligonucleotides have become a powerful technique for studying signal transduction pathways. They can be used for selective and transient knockout of cellular proteins. Nuclear micro-injection of oligonucleotides directed against individual G-protein subunits selectively inhibits the expression of these G-protein subunits and consequently the effects of receptor activation on ion channels (Kleuss *et al.* 1991, 1992, 1993). Using this method it has been shown that in GH_3 cells the M_4 muscarinic receptor couples to the G-protein trimer consisting of $\alpha_{01}\beta_3\gamma_4$, the somatostatin receptor to the trimer $\alpha_{01}\beta_1\gamma_3$, and the galanin receptor to the trimers $\alpha_{01}\beta_2\gamma_2$ and $\alpha_{01}\beta_3\gamma_4$ to inhibit voltage-dependent Ca^{2+} channels.

In addition to the short-term modulation of ion channels, neurotransmitters are believed to trigger long-term adaptive responses by activating specific programs of gene expression. A major signalling mechanism in stimulated neurones involves the membrane depolarization–activated openings of voltage-dependent Ca^{2+} channels. The resulting increase in $[Ca^{2+}]_i$ regulates c-fos expression by affecting both transcriptional initiation and elongation (Thompson *et al.* 1995).

Molecular biology

By combining molecular biology techniques with the patch-clamp technique a better understanding of ion channel structure and function can be advanced (Perez-Reyes and Schneider 1994). The significant amount of amino-acid sequence identity between Na^+ and Ca^{2+} channels suggest that they have similar structures. Each channel has four domains that each contain six membrane spanning segments. K^+ channels have a similar predicted structure, but only one domain; four subunits come together to form the channel. This similarity in structure predicts similarities in function. For example, the segment 4 of K^+, Na^+ and Ca^{2+} channels may play an important role in the voltage sensor. Other regions of similarity include the pore lining domain. Mutations in this region alter ion selectivity and conductance (Yang *et al.* 1993). These studies suggest that the selectivity filter is formed by a spiral staircase of glutamates from both domains (Tang *et al.* 1993).

A very useful strategy in elucidating structure–function relationships has been to make chimeras between the different channel subtypes and to express these chimeras in *Xenopus* oocytes or in a human embryonic kidney line (HEK293). Electrophysiological analysis by patch-clamp methods enables a characterization of the expressed channels. Therefore, it is possible to study the physiological manifestations of alterations at a molecular level in a single cell. For example, in Ca^{2+} channels the loop between domain II and domain III is required for skeletal muscle excitation–contraction coupling, whereas segment 3 and the linker between segment 3 and 4 of domain I contain regions involved in determining activation kinetics. Finally, the segment 6 of domain I is shown to be involved in voltage-dependent inactivation, while the putative E-F hand located after domain IV is implicated in Ca^{2+}-dependent inactivation (Tanabe *et al.* 1991; Nakai *et al.* 1994; Zhang *et al.* 1994; De Leon *et al.* 1995). Dihydropyridine binding regions have also been identified using molecular biology techniques. It has been shown that the dihydropyridine interaction site is located on segment 6 of domain IV (Tang *et al.* 1993).

13.4 Problems and limitations

Limitations of the patch-clamp technique arise first from the use of a single electrode. The unavoidable presence of access resistance, electrode and membrane capacitance distorts the ion currents recorded in response to a given command voltage step. The membrane potential itself is imperfectly clamped to the command potential by combined effects of the same factors. It appears that the patch-clamp technique is the most suitable method for voltage-clamping small cells with small conductance, although the limits between small and large cells cannot be clearly defined (Sala and Sala 1994). Errors associated with a non-isopotential voltage-clamp are experimentally difficult to determine. Large deviations from isopotentiality are disclosed by a discontinuity in the current–voltage relationship for inward currents; however, more discrete voltage deviations are not easily detected, suggesting that ion currents showing both transient and sustained components should be regarded with extreme caution. If pharmacological agents lead to a selective block of one component of current, it can be assumed that the current is composed of multiple components with different sensitivities to the blocking agent (White *et al.* 1995). Furthermore, it should be taken into account that ion channels may be not uniformly distributed on the cell membrane. Visualizing the distribution of ion channels with monoclonal antibodies and ion-sensitive probes may be useful for improving biophysical studies of cell membrane. In addition, liquid junction potentials may persist even after applying the normal zeroing procedure to eleminate them (Barry and Lynch 1991).

The exact physical basis of the gigaseal formed by applying a glass pipette on the cell membrane remains unclear. Distances of a few angströms have been estimated between glass and membrane; molecular interactions (i.e. hydrogen bonds, Van der Waals forces) seem to be the forces involved in gigaseal formation. Frequently, application of a suction to the interior of the pipette is needed to make a gigaseal but causes formation of membrane blebs under the pipette (Milton and Caldwell 1990). Therefore, single channel recordings from these blebs may not reflect normal channel behaviour as cytoskeletal attachements would be altered drastically. This implies that channel modulation requiring interactions with other membrane proteins (e.g. G-proteins) would be lost or altered if not all the necessary components move into the bleb.

Glasses used to construct patch-clamp pipettes release O^{2-}, Na^{+}, Ba^{2+}, Ca^{2+} and Mg^{2+}, all to micromolar concentrations. These elements have been shown to interfere and alter the properties of ion channels (Copello *et al.* 1991). Thus, screening for channel-modifying substances released by the glass may be necessary for adequate interpretation of patch-clamp data.

The whole-cell patch-clamp technique which offers the possibility to modify the ion concentration on both sides of the cell membrane, represents a significant advance in the ability to characterize the ionic dependence of channels and the dependence of ion pumps and exchangers on external and internal ions. However, the concentration of substances may differ notably between the pipette and the cell interior under most conditions, as would be expected if the pipette tip represents the rate-limiting barrier to diffusion (Mathias *et al.* 1990).

13.5 Future prospects

Several optical probes are now available for measuring the ion concentrations in the cytoplasm of single cells. Although these probes may be added to the bathing solution as acetoxymethylesters which are permeant through the cell membrane, addition of the fluorescent dyes to the pipette solution allows a more direct and controlled dialysis into the cells. With the development of laser scanning confocal microscopes it is possible to have a temporal and spatial high-resolution imaging of Ca^{2+} in volumes of a few μm^3. Elementary Ca^{2+} release events arising from openings of ryanodine-sensitive channels, termed Ca^{2+} sparks, can be initiated by Ca^{2+} influx through voltage-dependent Ca^{2+} channels in various cell types (Cannell *et al.* 1995; Tsugorka *et al.* 1995; Nelson *et al.* 1995; Arnaudeau *et al.* 1996). This approach may be developed for other ion channel current and intracellular ion regulation.

In addition, photolabile caged compounds can be diffused into the cells and released by flash illumination. Various second messengers, such as inositol 1,4,5-trisphosphate and Ca^{2+}, are used and their effects on ion channels can be detected by combined patch-clamp and fluorescence measurements.

References

Amédée, T., Mironneau, C. and Mironneau, J. (1987) The calcium channel current of pregnant rat single myometrial cells in short-term primary culture. *J. Physiol.* **392**, 253–72.

Arnaudeau, S., Macrez-Leprêtre, N. and Mironneau, J. (1996) Activation of calcium sparks by angiotensin II in vascular myocytes. *Biochem. Biophys. Res. Comm.* **222**, 809–15.

Barry, P.H. and Lynch, J.W. (1991) Liquid junction potentials and small cell effects in patch-clamp analysis. *J. Membr. Biol.* **121**, 101–18.

Brennecke, R. and Lindemann, B. (1974) Theory of a membrane voltage-clamp with discontinuous feedback through a pulsed current clamp. *Rev. Sci. Instrum.* **45**, 184–8.

Cannell, M.B., Cheng, H. and Lederer, W.J. (1995) The control of calcium release in heart muscle. *Science* **268**, 1045–9.

Copello, J., Simon, B., Segal, Y., Wehner, F., Ramanujam, V.M.S., Alcock, N. *et al.* (1991) Ba^{2+} release from soda glass modifies single maxi K^+ channel activity in patch-clamp experiments. *Biophys. J.* **60**, 931–41.

De Leon, M., Wang, Y., Jones, L., Perez-Reyes, E., Wei, X., Soong, T.W. *et al.* (1995) Essential Ca^{2+}-binding motif for Ca^{2+}-sensitive inactivation of L-type Ca^{2+} channels. *Science* **270**, 1502–6.

Hamill, O.P., Marty, A., Neher, E., Sakmann, B. and Sigworth, F.J. (1981) Improved patch-clamp techniques for high resolution current recording from cells and cell-free membrane patches. *Pflügers Arch.* **391**, 85–100.

Hodgkin, A.L. and Huxley, A.F. (1952) A quantitative description of membrane current and its application to conduction and excitation in nerve. *J. Physiol.* **117**, 500–44.

Horn, R. (1991) Diffusion of nystatin in plasma membrane is inhibited by a glass-membrane seal. *Biophys. J.* **60**, 329–33.

Horn, R. and Marty, A. (1988) Muscarinic activation of ionic currents measured by a new whole-cell recording method. *J. Gen. Physiol.* **92**, 145–59.

Kleuss, C., Hescheler, J., Ewel, C., Rosenthal, W., Schultz, G. and Wittig, B. (1991) Assignment of G-protein subtypes to specific receptors inducing inhibition of calcium currents. *Nature* **353**, 43–8.

Kleuss, C., Scherubl, H., Hescheler, J., Schultz, G. and Wittig, B. (1992) Different β subunits determine G-protein interaction with transmembrane receptors. *Nature* **358**, 424–6.

Kleuss, C., Scherubl, H, Hescheler, J., Schultz, G. and Wittig, B. (1993) Selectivity in signal transduction determined by γ subunits of heterotrimeric G proteins. *Science* **259**, 832–4.

Leprêtre, N. and Mironneau, J. (1994) α_2-adrenoceptors activate dihydropyridine-sensitive calcium channels via G_i-proteins and protein kinase C in rat portal vein myocytes. *Pflügers Arch.* **429**, 253–61.

Leprêtre, N., Arnaudeau, S., Mironneau, J., Rakotoarisoa, L., Mironneau, C. and Galiano, A. (1994*a*) Electrophysiological and radioligand binding studies of elgodipine and derivatives in portal vein myocytes. *J. Pharmacol. Exp. Ther.* **271**, 1209–15.

Leprêtre, N., Mironneau, J., Arnaudeau, S., Tanfin, Z., Harbon, S., Guillon, G. *et al.* (1994*b*) Activation of α_{1A}-adrenoceptors mobilizes calcium from the intracellular stores in myocytes from rat portal vein. *J. Pharmacol. Exp. Ther.* **268**, 167–74.

Leprêtre, N., Mironneau, J. and Morel J.L. (1994*c*) Both α_{1A}-and α_{2A}-adrenoreceptor subtypes stimulate voltage-operated L-type calcium channels in rat portal vein myocytes: evidence for two distinct transduction pathways. *J. Biol. Chem.* **269**, 29 546–52.

Lindau, M. and Neher, E. (1988) Patch-clamp techniques for time-resolved capacitance measurements in single cells. *Pflügers Arch.* **411**, 137–46.

Loirand, G., Pacaud, P., Baron, A., Mironneau, C. and Mironneau, J. (1991) Large conductance calcium-activated non-selective cation channel in smooth muscle cells isolated from rat portal vein. *J. Physiol.* **437**, 461–75.

Macrez-Leprêtre, N., Morel, J.L. and Mironneau, J. (1996) Effects of phospholipase C inhibitors on Ca^{2+} channel stimulation and Ca^{2+} release from intracellular stores evoked by α_{1A}- and α_{2A}-adrenoceptors in rat portal vein myocytes. *Biochem. Biophys. Res. Comm.* **218**, 30–4.

Marty, A. (1981) Ca-dependent K channels with large unitary conductance in chromaffin cell membranes. *Nature* **291**, 497–500.

Mathias, R.T., Cohen, I.S. and Olivia, C. (1990) Limitations of the whole cell patch-clamp technique in the control of intracellular concentrations. *Biophys. J.* **58**, 759–70.

Milton, R.L. and Caldwell, J.H. (1990) How do patch-clamp seals form? A lipid bleb model. *Pflügers Arch.* **416**, 758–60.

Nakai, J., Adams, B.A., Imoto, K. and Beam, K.G. (1994) Critical roles of the S3 segment and S3-S4 linker of repeat-I in activation of L-type calcium channels. *Proc. Nat. Acad. Sci.* (*USA*) **91**, 1014–8.

Nelson, M.T., Cheng, H., Rubart, M., Santana, L.F., Bonev, A.D., Knot, H.J. *et al.* (1995) Relaxation of arterial smooth muscle by calcium sparks. *Science* **270**, 633–7.

Neher, E. (1988) The use of the patch-clamp technique to study second messenger-mediated cellular events. *Neuroscience* **26**, 727–34.

Neher, E. and Sakmann, B. (1976) Single-channel currents recorded from membrane of denervated frog muscle fibres. *Nature* **260**, 799–802.

Pacaud, P., Loirand, G., Mironneau, C. and Mironneau, J. (1989) Noradrenaline activates a calcium-activated chloride conductance and increases the voltage-dependent calcium current in caltured single cells of rat portal vein, *Br. J. Pharmacol.* **97**, 139–46.

Pacaud, P., Loirand, G., Baron, A., Mironneau, C. and Mironneau, J. (1991) Ca^{2+} channel activation and membrane depolarization mediated by Cl^- channels in response to noradrenaline in vascular myocytes. *Br. J. Pharmacol.* **104**, 1000–6.

Penner, R. and Neher, E. (1989) The patch-clamp technique in the study of secretion. *Trends Neurosci.* **12**, 159–63.

Perez-Reyes, E. and Schneider, T. (1994) Calcium channels: structure, function, and classification, *Drug Development Res.* **33**, 295–318.

Sala, F. and Sala, S. (1994) Sources of errors in different single-electrode voltage-clamp techniques: a computer simulation study. *J. Neurosci. Methods* **53**, 189–97.

Sakmann, B. and Neher, E. (1984) Patch-clamp techniques for studying ionic channels in excitable membranes. *Annu. Rev. Physiol.* **46**, 455–72.

Tanabe, T., Adams, B.A., Numa, S. and Beam, K.G. (1991) Repeat I of the dihydropyridine receptor is critical in determining calcium channel activation kinetics *Nature* **352**, 800–3.

Tang, S., Mikala, G., Bahinski, A., Yatani, A., Varadi, G. and Schwartz, A. (1993) Molecular localization of ion selectivity sites within the pore of a human L-type cardiac calcium channel. *J. Biol. Chem.* **268**, 13 026–9.

Thompson, M.A., Ginty, D.D., Bonni, A. and Greenberg, M.E. (1995) L-type voltage-sensitive Ca^{2+} channel activation regulates c-fos transcription at multiple levels. *J. Biol. Chem.* **270**, 4224–35.

Tsugorka, A., Rios, E. and Blatter, L.A. (1995) Imaging elementary events of calcium release in skeletal muscle cells. *Science* **269**, 1723–6.

White, J.A., Sekar, N.S. and Kay, A.R. (1995) Errors in persistent inward currents generated by space-clamp errors: a modeling study. *J. Neurophysiology* **73**, 2369–77.

Yang, J., Ellinor, P.T., Sather, W.A., Zhang, J.F. and Tsien, R.W. (1993) Molecular determinants of Ca^{2+} selectivity and ion permeation in L-type Ca^{2+} channels. *Nature* **366**, 158–61.

Yatani, A., Codina, J., Imoto, Y., Reeves, J.P., Birnbaumer, L. and Brown, A.M. (1987) A G protein directly regulates mammalian cardiac calcium channels. *Science* **238**, 1288–92.

Zuazaga, C. and Steinacker, A. (1990) Patch-clamp recording of ion channels: interfering effects of patch pipette glass. *News Physiol. Sci.* **5**, 155–8.

Zhang, J.F., Ellinor, P.T., Aldrich, R.W. and Tsien, R.W. (1994) Molecular determinants of voltage-dependent inactivation in calcium channels. *Nature* **372**, 97–100.

14 Techniques for muscle energetics

Roger C. Woledge

14.1 What is energetics and how can it be studied?

A list of the functions of muscle might include:

(1) the transduction of energy from chemical to mechanical form (a motor function);

(2) the dissipation of energy from mechanical form to heat (a braking function);

(3) the storage of energy received in mechanical form (a spring function);

(4) the storage of energy received in chemical form (a storehouse function);

(5) the dissipation as heat of energy produced by chemical reactions (a heater function);

(6) the sequestration of potassium ions against osmotic and electrical gradients (a battery function).

Research into how muscle plays these roles and the limitations of its ability in each regard requires that we measure directly or indirectly the energy changes involved; this has been recognized for at least a century (see Hill 1965, pp. 1–18, for the early history of muscle energetics). These experimental activities and the interpretation of the results obtained comprise the science of muscle energetics.

The main forms of energy which are relevant to consider in these contexts are discussed below.

Mechanical energy

The flow of mechanical energy to or from the muscle is found from measurements of force (P, Newtons) and of movement (dL, metres). Work is the integral of force with respect to length over the period of study ($\int P \cdot dL$) results obtained are in Joules. The usual sign convention is that work done by the muscle is positive and work done on the muscle is negative. In some energetic work concerned with locomotion, muscle work has been defined as the force length integral over only those periods of a cyclic process for which power ($P \cdot dL/dt$) is positive. If this convention is used, particular care is needed to avoid confusion with the more usual conventions. Muscles can in principle do work against the pressure surrounding them by changing their volume. However, volume changes during contraction are very small, so this work is negligible at atmospheric pressure. *In vivo* muscle fibres may do work not only by pulling on their tendons as they shorten but also by pushing outwards as their cross-section increases as a consequence of shortening.

Energetic experiments *in vitro* are made with small preparations freed from any constraint on change in cross-section, so this factor also will be negligible *in vitro*; its role *in vivo* as a means of transmission of work from muscles to the environment remains to be studied.

Heat energy

Changes in the heat content of the muscle (dQ) can be measured either by the changes in temperature (adiabatic measurement) that occur or by the corresponding heat flows generated by these temperature changes (isothermal measurements). The former approach will usually be adopted when high time resolution is needed and the latter when slower processes are being studied. Results in Joules are obtained by appropriate calibration of the instruments. For adiabatic measurements this calibration will have to include a determination of the heat capacity (Joules/K) of the system studied. The sign convention, to conform to that used for mechanical energy has to be that a heating up of, or a flow of heat out from, the system of study is taken as positive. This is in spite of the fact that this heat flow indicates a *decrease* in the heat content of the muscle. Thus the sign convention used is not the same as in chemical thermodynamics.

Chemical energy

Measurements are of amounts or relative amounts of metabolites present at the beginning and end of the period of study from which the extents (ζ) of certain specified reactions during the intervening period would be inferred, using independent information about the stoichiometry of the reactions. These results would be expressed in moles of reaction and where the relevant molar enthalpy (ΔH) of the reactions studied are known could also be expressed (in Joules) as enthalpy change (i.e. the sum of heat and work changes). Alternatively, interest might centre on free energy change rather than the enthalpy change, in which case the molar free energy changes (ΔG) rather than the ΔH value would be needed. Most of the reactions of interest in muscle run in both directions at different times in the contraction and recovery cycle and thus the specification of one direction as 'forward' is arbitrary. When the specified reaction runs in the direction taken as forward, ζ would be regarded as a positive quantity; by the normal thermodynamic convention ΔH is negative for an exothermic process, thus the energy yield as heat and work from a particular chemical reaction is given within the sign convention we wish to use by:

$$-1 \cdot \Delta H \cdot \zeta. \tag{14.1}$$

A statement of the 1st Law of Thermodynamics suitable for this discussion is then:

$$dQ + \int P \cdot dL + \sum \Delta H \cdot \zeta = 0, \tag{14.2}$$

where the $\sum$ signifies summation over all the chemical processes occurring. If electrical or osmotic changes were of interest, as with the study of ion transport, then suitable terms would have to be included to cover these forms of energy. More commonly, however, the interest is in the muscle as motor, brake or storehouse for which the above three terms suffice, i.e. it is assumed, probably quite reasonably in practice, that there have been no quantitatively significant energy changes of an osmotic or electrical nature between the time points over which eqn 14.2 is to be applied.

A statement of the 2nd Law of Thermodynamics suitable for use in muscle energetic experiments is:

$$\int P \cdot dL + \sum \Delta G \cdot \zeta < 0. \tag{14.3}$$

Equations 14.2 and 14.3 provide a variety of experimental research strategies of which it is of interest to list the major variations.

Type A experiments

Here the experimenter assumes, implicitly or explicitly, that he knows the major energetic pathways of the system and makes measurements of some forms of energy in order to obtain information of the flow through these assumed pathways. Thus estimates of the amount of energy used in particular contraction(s) or parts of a contraction, could be obtained either through measurement of heat and work ($dQ + \int P \cdot dL$) or through measurement of chemical change ($-1 \cdot \sum \Delta H \cdot \zeta$). The amount of energy used might be related to tension time integral ($\int P \cdot dt$) or to work done ($\int P \cdot dL$) and the ratios economy and efficiency calculated as follows:

$$\text{economy } (Ec) = \int P \cdot dt \bigg/ \left(\int P \cdot dL + dQ \right) \quad \text{or} \quad = \int P \cdot dt \bigg/ \sum \Delta H \cdot \zeta, \tag{14.4}$$

$$\text{efficiency } (Ef_{\mathrm{H}}) = \int P \cdot dL \bigg/ \left(\int P \cdot dL + dQ \right) \quad \text{or} \quad = \int P \cdot dL \bigg/ \sum \Delta H \cdot \zeta. \tag{14.5}$$

These quantities are of use particularly in comparing the properties of different muscles, either within or between species. It could be objected that this approach to energetics makes too many assumptions about the nature of the reactions occurring. There is some justification in this criticism, which applies whether the quantity measured is chemical change or heat + work. For example, to measure change in phosphocreatine level as the sole index of metabolic cost in a situation in which phosphocreatine was significantly resynthesized by oxidative phosphorylation would clearly lead to wrong conclusions. Less obviously perhaps it does not make sense to compare the heat + work output of a tissue in which most of the energy output has been derived from oxidation (70 kJ of enthalpy per mole of ATP used) with one in which the energy source is phosphocreatine splitting (34 kJ of enthalpy per mole of ATP used). Other things being equal, the latter will appear as twice as economical or efficient, because the full energy output associated with the activity, which will not be apparent until the phosphocreatine has been resynthesized, has not been included in the experimental measurements. The onus is on those who make this type of measurement to state what they consider to be the likely chemical sources of the energy they measure, and on those who choose particular chemical reactions to measure, to state the evidence that these reactions are indeed the predominant energy source over the period of observation. However in practice, during short contractions of many vertebrate muscles, it is likely that phosphocreatine splitting is the source of most of the energy output and that variations in economy and efficiency that have been observed are predominantly due to differences in the contractile properties of the myofibrils.

Another question that frequently arises in the discussion of efficiency is how the definitions above relate to the thermodynamic definitions of efficiency:

$$\text{thermodynamic efficiency } (Ef_{\mathrm{T}}) = \int P \cdot dL \Big/ \sum \Delta G \cdot \zeta. \tag{14.6}$$

Where the energy has been provided by a single chemical process the relation is simple:

$$Ef_{\mathrm{T}} = Ef_{\mathrm{H}} \cdot \Delta H / \Delta G. \tag{14.7}$$

Where the measurement of energy has come from chemical measurements Ef_{T} can be calculated directly provided only that ΔG is known. For phosphocreatine splitting in muscle ΔH is -34 and ΔG about -48 kJ/mol and so: $Ef_{\mathrm{T}} = 0.71 \cdot Ef_{\mathrm{H}}$.

This conversion is not commonly applied however. There are two possible reasons for this:

1. To obtain, from observation of heat or chemical change during working contraction, the efficiency for energy conversion by the myofibrils (as opposed to the whole muscle cell) it is necessary to apply another correction: for the ATP used by the calcium pump. This correction is in the other direction to that for the non-unity of the ratio $\Delta H/\Delta G$ so that the sum of both corrections will be rather small, and often within the statistically uncertainty of the measurements.

2. A more interesting reason is that it is not known whether either Ef_{T} or Ef_{H} would remain constant if the ratio $\Delta H/\Delta G$ was changed. For example, changing the buffer present while keeping the pH constant would change ΔH but not ΔG whereas changing the concentration of reactant and/or product, as happens for example in fatigue, changes ΔG but does not change ΔH. While it must be true that a sufficient lowering of ΔG will decrease the work done per ATP split, which can never exceed ΔG, it is not known whether the effect is a linear one or shows a threshold, it would seem worthwhile to find out.

Type B experiments

The more sceptical experimenter might wish to test the completeness of his knowledge of the pathways of energy movement by measuring *all* the forms of energy change involved, which will provide a set of results that will satisfy eqn 14.2 only if all significant pathways have been specified (or if the energy flows through the unspecified pathways sum to approximately zero). A conspicuous failure of this method was that of Hill and Myerhof to detect that part of the energy in a series of muscle contractions that can be produced by phosphocreatine splitting, when they carried out an energy balance study of muscle based on the idea that the only anaerobic energy source was glycolysis. They did in fact note the discrepancy (Myherhof 1930; Hill 1932, 1965, pp. 162–8) but this did not seem to aid their acceptance of the discovery by Embden of the role of phosphagen in muscle which could not be generally appreciated until the use of an inhibitor of glycolysis by Lundsgaard put the issue beyond doubt (for references see Hill 1965, pp. 162–8). Nevertheless the energy balance method has often successfully provided information leading to the discovery of previously unknown pathways of energetic consequence. A good example is the finding that the high heat rate early in a tetanic contraction has a component which is not due to phosphocreatine (PCr) splitting or the reactions that resynthesise PCr (Curtin and Woledge 1979; Homsher *et al.* 1979), now thought to be largely due to the binding of calcium to parvalbumin (Woledge *et al.* 1985, p. 257). However the results of this type of experiment have often been ignored as anomalous, since they did not fit into the current understanding of the subject. An outstanding example is the discovery by Homsher and colleagues (Rall *et al.* 1976; Homsher *et al.* 1981) that live, intact and uninhibited muscles can produce unloaded shortening without ATP splitting using an energy source, perhaps changes in the myofibrillar proteins, that is then recharged by ATP splitting in the subsequent seconds. The relevance of this discovery (that the immediate energy supply for rapid shortening in live muscle is not ATP splitting) has seemingly not been appreciated by those who later measured the distance shortened per ATP in various myosin driven systems other than intact muscle (Yanagida *et al.* 1985; Harada *et al.* 1990; Toyoshima *et al.* 1990; Uyeda *et al.* 1990; Higuchi and Goldman 1991; Yamada *et al.* 1993). Thus the successful experimenter in this field should not only be sceptical but also needs to be a skilful advocate if striking experimental results are to have any impact on the field of muscle physiology.

Definition of the system of study

It is essential in muscle energetic experiments to carefully define the system on which the measurements are being made and what the exchanges of material and energy across its boundaries might be. For example, in Hill's classic study of muscle energetics (Hill 1938) the object of study was a pair of frog sartorius muscles together with a very small amount of Ringer solution and the thermopile of low thermal capacity on which they were mounted. They were surrounded by moist gas and could exchange oxygen, carbon dioxide and water vapour with their surroundings. Heat could flow to and from this system by conduction through the thermopile wires and to a lesser extent through convection and conduction into the surrounding gas. Mechanical energy could enter and leave the system through the wire connection between the muscle and the mechanical recording system located about 20 cm above the thermopile. A potentially troublesome aspect was that the muscles had tendons at either end which were capable of storing energy through mechanical strain. These tendons were not part of the system of study because, although the mechanical energy flowing into the tendons was included in the $\int P \cdot dL$ measurement, the thermal changes in them were not measured, only the central uniform part of the muscle was over the 'active' region of the thermopile, and thus contributed to the recorded temperature changes. Hill was therefore careful to measure and treat separately the properties of the 'series elastic element' which in his experiments was located to a large extent, though not entirely, within these tendons. His experimental conclusions then rested to some extent on the accuracy of this assumption about the nature and location of the series elasticity. The same problem exists with smaller preparations today. The experimenter needs to determine the extent to which the series elasticity resides within the area of the muscle from which heat is to be measured, and to make appropriate corrections for the part that does not. A similar logical problem exists with parallel elasticity. It is common practice to assume that the force exerted by elastic elements in parallel with the muscle is the same whether or not the muscle is active, and on the basis of this assumption to subtract from the force record made during an active movement a control observation made without stimulation. Sometimes control records of heat production are subtracted in a similar way (Homsher *et al.* 1983).

Direct and indirect measurements

When calorimetry was fashionable, measurements of chemical change were sometimes referred to as 'indirect' calorimetry—a term still found in respiratory physiology (Elia and Livesley 1988; Ferrannini 1988; Livesey and Elia 1988). Later, when calorimetry was regarded as old-fashioned, one fully expected to see heat production methods described as indirect biochemistry! The idea that there are two basic methods of studying muscle energetics: one 'direct' and therefore good, and the other 'indirect' and therefore not so good, is ridiculous for the reasons outlined above. Either the experimenter is engaged on a type A experiment, in which case measurements of either chemical change or heat and work can be used, the choice will be a matter of convenience, taking into regard the sensitivity, speed of response of the methods (see Section 14.5) and the resources available to the researcher. *In either case the experimenter is tacitly assuming that the nature of the reactions is known.* Alternatively, the experimenter is engaged on a type B experiment, in which case all types of measurement will be needed to produce a result. Where the result of a type A experiment appears to depend on the choice of method (chemical vs. heat + work), e.g. in the debate about the energetic consequences of rapid shortening between Davies and colleagues (Kushmerick and Davies 1969) and Hill (1938, 1964) this is an indication of the need for an investigation of type B. All our available methods are complementary in the search for understanding of the energetic principles of muscle action; unfortunately there is a regrettable tendency for workers with one technique to ignore evidence obtained with another technique, even when it can support their conclusions.

14.2 Measurement of work output

Methods that might be used for measuring force and length changes of muscle during experiments on their energetics are broadly the same as for mechanical experiments and are reviewed elsewhere in this volume. However the use of the strain gauge type of force transducers would usually be avoided in experiments using thermal measurements because the heat they produce would preclude positioning the transducer close to the muscle fibre. Motors for controlling muscle length inevitably will produce heat and therefore must be positioned some centimetres away from any muscle heat measurements. A muscle moving over the surface of a thermopile will experience some friction and this will limit the use of fast movements such as might be used, for example, in experiments to measure T1 and T2 curves because some frictional heating may occur and it will not be possible to measure accurately the force produced by the muscle (Gilbert and Ford 1988; Linari and Woledge 1995). Heat produced by this friction will heat the thermopile before the bulk of the muscle because the muscle, even a single fibre, is much thicker than the thermopile. Therefore the artefact produced will be a heating followed by a cooling; such processes must be considered when examining the time-course of heat changes during and after rapid movements (Gilbert and Ford 1988; Gilbert and Mathias 1988).

Mechanical energy could be 'stored' in muscle by strained structures located there. In general we shall not be able to distinguish, by energetic experiments alone, between the following two views of this process which are perhaps complementary. The 'mechanical' view would be that stretch causes a change which has a positive ΔH (is endothermic), which is reversible, and for which ΔF is almost equal to ΔH. When the 'stored' mechanical energy of such a system is taken out by a controlled release at any speed there will be no heat change apart from any thermoelastic heat that may occur as a result of the change of force (for discussion of the nature of thermoelastic changes see Woledge *et al.* 1985, p. 237). Study of biological and non-biological tissue has established the range of expectations for these thermoelastic effects. In the 'chemical' view the stretch perturbs the equilibrium between two different states of the cross-bridges in which they produce different forces. This change would also be almost reversible if the stretches and releases were made on a time-scale long compared to the time constant with which the equilibrium is restored after a sudden perturbation. However there is no expectation as to the relative size of the ΔF and ΔH for such a process. Indeed even the sign of ΔH would be uncertain. Probably there is not a real difference between these two views, except that in the latter view muscle is not expected to behave like any other type of elastic material.

14.3 Measurement of heat output

Thermopile methods

The principle of using a thermopile to measure muscle heat production goes back to Helmholz in the mid-19th century, and was developed very considerably by A.V. Hill, whose book (Hill 1965) gives a wealth of information, both historical and practical, about the techniques of measuring initial and recovery heat production of whole isolated amphibian muscles. An updated and more compact account of the techniques (though a less characterful one) is given in Woledge *et al.* (1985, Chapter 4). The following is a brief description of the methods currently in use for experiments on smaller muscle preparations than Hill used, which has been made possible through the development of metal-film thermopiles (Mulieri *et al.* 1977). For further details consult the following papers describing current practice in the main laboratories now using the technique and exemplary of the results that can be obtained: Curtin *et al.* 1986, 1988; Elzinga *et al.* 1987; Curtin and Woledge 1991, 1993; Barclay 1994; Barclay *et al.* 1993, 1995; Lannergren *et al.* 1993. In addition to this work with skeletal muscle fibres and fibre bundles, the techniques are also used to study cardiac muscle; for recent examples see Mast and Elzinga (1989) and Alpert *et al.* (1992).

The heat production of a single intact muscle fibre or of a bundle of intact fibres can be studied by mounting them in contact with a plastic film on the other side of which has been evaporated a series of metal film thermocouples arranged in series so that the hot junctions are directly under the muscle fibres and the cold junctions lie on a frame holding the film. The film may be mounted vertically or horizontally, but the latter is preferred where the muscle will shorten. The muscle fibre(s) have only a trace of Ringer solution around them, and are kept from drying by enclosure in a small volume of moist air. On contraction the heat produced is shared rapidly (within 10 ms) with the plastic and metal films and the small volume of solution present and heat loss occurs principally via conduction of heat through the metal films to the frame. The time constant of heat loss (τ) depends on the thermal mass of the preparation and is typically in the region of 1 to 10 s. As an example the output of the thermopile for a 0.5 s tetanic contraction of a frog single fibre will be in the region of 1 μV, which is well above the Johnson noise for the thermopile (typically 2 nV/$\sqrt{\text{Hz}}$ or 20 nV for a bandwidth of 100 Hz). As with any recording of small voltages, a low noise amplifier with an input impedance well matched to the impedance of the thermopile is needed and attention must be paid to screening out extraneous voltage fluctuations. As the muscle is stimulated by applications of voltages of at least a million times the thermopile signal, some stimulus artefact is common; however, since the stimulus is brief and its time of occurrence is known, it is often possible to eliminate these artefacts by 'masking' the signal for a few milliseconds at the time of each stimulus.

Calibration of thermopiles is of special importance because the design of experiments usually requires adding together work and heat which are measured in quite separate ways, and possibly comparing their sum to the product, $\sum \Delta H \cdot \zeta$. The problem is seen at its most acute in experiments with stretches of active muscle when the work done is negative and the heat produced positive; the sum of the two may well be very small compared to the absolute value of either quantity and is then exquisitely sensitive to any calibration error. Calibration methods using the Peltier effect have become routine since their introduction to the thermopile field by Kretzschmar and Wilkie (1972, 1975). (Similar methods had previously been proposed for heat flow calorimeters; Calvet and Prat 1963.) A series of metal bars of known heat capacity are needed which can be mounted in place of the fibres. A small trace of glycerol is usually added to ensure good thermal contact (allowance is duly made for this extra heat capacity). Each bar is heated by the Peltier effect, i.e. by passing a known current (I) through the thermopile, for a sufficient time that a steady temperature will have been reached, about $4 \cdot \tau$, (where τ is the time constant for heat loss) and the thermopile output during cooling after this heating is observed. The initial rate of voltage drop (dV/dt) immediately after turning off the current will be related to the heat capacity of the bar (C), the thermopile sensitivity (S), and the temperature (T) by:

$$dV/dt = I \cdot T \cdot S^2/(C + C_T), \tag{14.8}$$

where C_T is the heat capacity of the thermopile itself. The purpose of using several values of C is so that C_T can be found, usually from the intercept of a plot of $1/(dV/dt)$ against C. Once the thermopile sensitivity (S) is known, the heat capacity of each preparation, together with the thermopile and Ringer with which the heat is shared, is found by applying the same principle.

An interesting variant on the principle of Peltier calibration is provided by mounting on the thermopile an electrical heater of resistance R. Heat can then be applied to the thermopile by passing current (I_J) through this heater, and removed from the thermopile by Peltier cooling with a current I_P. If the currents are adjusted so that there is no net heating or cooling the sensitivity can be found from the equation:

$$S = I_J^2 \cdot R/I_P \cdot T. \tag{14.9}$$

This method does not require heat capacity standards.

As the above brief description of modern thermopile methods should make clear, the methods used today do not differ *in principle* from those developed by Hill in the 1930s, although greater sensitivity has been achieved by the development of apparatus of lower heat capacity. Central to this has been the use of evaporated or sputtered thermopiles deposited onto thin plastic substrates, which are small enough to allow the measurement of heat production in single muscle fibres; this was pioneered by Mulieri and colleagues (Mulieri *et al.* 1977). As these devices work so well for measurement of the heat produced during muscle contractions there has been little incentive to develop other methods such as myothermic radiometry or other possible methods which all probably have the capability of providing measurements of the temperature changes during muscle contraction.

These thermopiles have also been used for the measurement of heat during the much longer periods, from several minutes, upto 2 h over which muscle 'recovery' occurs, i.e. the time for which metabolic rate remains measurably above the precontraction or basal level. For this purpose the slower heat-conduction calorimeters might be equally or better suited. In interpreting the results of these observations the fact that the muscle can exchange water vapour with its surroundings when on a Hill thermopile must be considered among the other processes which can produce/absorb heat. Such an exchange can be started, for example, by an increase in the osmolarity of the muscle, which will occur when it contracts because of the splitting of phosphocreatine, water will then condense from the vapour phase onto the muscle causing heat production. This process will continue until the muscle has again reached the same osmolarity as its surroundings, either because the metabolic changes have been reversed and/or sufficient water has condensed to reduce the osmolarity to that of the surroundings. The latter process (concentration of the myoplasm by evaporation) will produce an amount of heat which is very large compared with the former (metabolic recovery). Fortunately the evaporation process is quite slow because:

(1) of the limited rate at which water can diffuse to the surface of the muscle;

(2) because the consequent temperature increase raises the vapour pressure of the muscle to almost match that of the vapour phase.

A fuller account of this phenomenon, including a description of its exploitation to measure osmotic changes, is given by Hill (1931). It might be attractive to further exploit it to construct a calorimetric system which could simultaneously measure the osmotic changes and the metabolic heat production. This could be done by providing, in close proximity to the usual muscle and thermopile, a second such system with the muscle replaced by a equivalent mass of inert solution of equal osmolarity. The sum of the outputs of the two systems would give the metabolic changes uncontaminated by the osmotic effects and the dummy system output would indicate the extent of the osmotic change alone.

Calorimeter methods

Heat-flow calorimetry

In a heat-flow calorimeter, the heat flowing between a heat sink and a relatively small enclosed box is measured by thermoelectric means, usually using semiconductor thermopiles which are commercially available. Such systems can be used either to study reactions in solution or to study the heat production of tissues enclosed within them. It is possible to have some flow through the experimental chamber without disrupting the operation of the heat flow measuring system and so a muscle can be supplied, for instance, with oxygen by this means. These systems have a relatively large heat capacity and therefore the speed of their response is typically 1000 times less than the thermopile systems described above. The principles of design and operation of such systems are described well by Calvet and Prat (1963) and by Wadso (1974). These systems are immune from any osmotic errors such as those described above, because they can be entirely filled with solution and thus no condensation can occur within them, and thus this system may be a good method to use for studying slow recovery heat. This type of calorimetry has been used, for example, by Chinet and his colleagues (Chinet *et al.* 1977, 1992, 1993; Chinet 1993; Decrouy *et al.* 1993; Dulloo *et al.* 1994; Even *et al.* 1994) to study the resting metabolic rate of mouse skeletal muscles, and to demonstrate the contribution to this process from calcium pumping, from Na/K pumping and the influence of substrate availability. Also they have compared normal and mdx (dystrophic) mouse muscles from these points of view. Fagher *et al.* (1994) have applied calorimetric methods similar to those described by Wadso (1974) to the study of the metabolic rate of small unstimulated muscle preparations obtained by needle biopsy from human subjects. Also Lonnbro and Hellstrand (1991) used such systems to study chemically skinned *Taenia coli* smooth muscle and they used an energy balance principle to show that the heat production was that expected from the hypothesis that the net reaction occurring was phosphocreatine synthesis.

Flow calorimetry

In this method the temperature of a flowing liquid is measured before and after it passes over the active tissue, whose metabolic rate is to be measured. The method is described in detail, and its limitations discussed by Daut and Elzinga (1988, 1989). The predominant route of removal of the metabolic heat in these experiments is through the flowing liquid, in contrast to the heat-flow calorimeter, which may also have superfusion solution flowing through it, but in which the heat is removed predominantly by conduction through the walls of the chamber. The measurement that is made is of the difference in temperature of the flowing liquid before and after it has passed over the preparation. For example, Dietrich and Elzinga (1993) measured this temperature difference by 16 pairs of thermocouples embedded in the walls of the inlet and outlet tubes. The systems are not 100% efficient; about 20% of the heat is lost through routes other than the flowing liquid. However, accurate calibration is possible by replacing the preparation by a suitable small electrical heater. These systems are very suitable for measuring rates of energy production by unstimulated preparations, because a base-line can be established by withdrawing the preparation along the tube, so that it no longer lies between the two sets of thermocouples. Thermopile methods are less suited to this purpose because of the difficulty of providing an adequate base-line level. This flow calorimetry has proved very successful for the study of the heat produced by small cardiac muscle preparations, with a moderate time resolution (Daut and Elzinga 1988, 1989; Dietrich *et al.* 1990; Dietrich and Elzinga 1993; Schramm *et al.* 1994).

Heat production *in vivo*

The heat produced by human muscle *in situ* during voluntary contractions can be measured by inserting temperature measuring probes, either thermocouples or thermistors, into the muscle. Edwards *et al.* (1974, 1975), who give references to earlier work, applied this technique to human quadriceps, using a single probe to study electrically stimulated contractions. As these authors point out, the circulation is arrested by internal pressure in many muscles when the contraction force reaches about 20% of the maximum voluntary force, so that heat is no longer removed from the muscles by the blood. During isometric contractions the temperature record after an initial transient deflection, perhaps associated with relative movement of the probe and the muscle as force is developed, showed an almost linear rise for the next 10 to 20 s probably because the heat production occurred at a steady rate and heat loss was negligible on this time-scale. These authors show that the heat produced can be accounted for by the phosphocreatine splitting and net glycolysis that has occurred, measured from biopsy samples before and after the contractions. They also showed that during prolonged stimulation (96 s) the rate of temperature rise fell more than force did, suggesting a slowing of the metabolic processes. Saugen and Vollestad (1995) used an adaptation of the technique to follow the changes that occurred in the rate of heat production in voluntary contractions of the quadriceps muscle during a fatiguing series of contractions (see also Vollestad *et al.* 1990; Vollestad and Verburg 1990). They used a series of thermistors (rather than a single probe) inserted at intervals to sample this large mass of muscle adequately. The striking result of this experiment is that the rate of heat production is actually increased by fatigue in this voluntary fatiguing protocol in contrast to the results with electrical stimulation.

The heat production by human leg muscle *in situ* can also be obtained by observing the blood flow and the arterio-venous temperature difference, from which heat outflow from the leg can be calculated by the Fick principle (B. Saltin, personal communication). These observations require the insertion of arterial and venous catheters into human volunteers and are thus somewhat invasive. However, such catheterization is regarded as routinely acceptable in a number of Scandinavian laboratories (Anderson and Saltin 1985). If the skin temperature is maintained close to the muscle temperature little heat will leave the leg through the skin. In the steady state, which will take some minutes of uniform exercise to achieve, all the metabolic heat will be removed through the circulation. Since, for example, arterio-venous oxygen and lactate can also be measured in this type of experiment, the possibility of performing energy balance (Type B) experiments exists.

14.4 Measurements of chemical changes

Magnetic resonance spectroscopy

Since the pioneering work in the 1970s (Hoult *et al.* 1974; Dawson *et al.* 1977, 1978) the use of magnetic resonance spectroscopy (MRS also known as NMR) to measure the levels (usually the relative levels) of metabolites in muscle has become extremely popular in spite of (or perhaps, in some cases, because of) the relatively high cost of the necessary equipment. The huge attraction of the method has been that it is non-invasive and non-destructive and that the worries about changes that occur in muscle extracts between sampling and analysis can be wholly avoided. Also, compared to biopsy methods, the MRS measurements sample a large volume of muscle and therefore uncertainties, inherent in the use of biopsy samples, about how representative the sample is of the whole muscle are greatly reduced. For these reasons the method has been particularly appropriate for studies of human muscle. A recent account of the principles and the application of the MRS methods, particularly its application to the study of human tissues, is provided by Gadian (1995).

MRS signals can be obtained either from muscles *in vivo*, which requires a large bore magnet into which a whole animal, a human limb or a whole human can be inserted, or from perfused or superfused muscle(s) in a chamber inserted into a smaller bore magnet. With the latter arrangement, higher fields can be used which can compensate for the lower sensitivity associated with use of a smaller sample. Wiseman *et al.* (1993) point out that an important determinant of the sensitivity is the extent to which the coil volume is filled with sample, and have designed apparatus which will give good sensitivity for single, quite small muscles, such as mouse soleus (Phillips *et al.* 1993*a*,*b*).

The MRS method in principle gives an almost instantaneous measurement of the metabolite levels, and thus potentially has a very high time-resolution, but it is relatively insensitive and averaging is usually required to measure levels of metabolites with reasonable precision, for which purpose time-resolution may have to be sacrificed. Three types of averaging have been used, sometimes in combination:

1. The use of many muscles within a superfusion chamber at one time, so that the signal obtained represents the average of the behaviour of these muscles (Dawson *et al.* 1977, 1978; Yamada and Tanokura 1983; Kawano *et al.* 1988). For example, Phillips *et al.* (1993) used 16 rat soleus muscles simultaneously.
2. The combination of many individual observations (FIDs) to obtain each spectrum so that the result is an average over some period of time, typically 100 or more FIDs obtained at 2- to 4-s intervals might be averaged giving a time resolution of about 5 min.
3. The averaging of a number of cycles of contraction and recovery occurring sequentially. By this means, Kawano *et al.* (1988) were able to obtain a time

resolution of 16 s in the study of bullfrog sartorii by averaging within each 16 s interval four FIDs for each of 128 cycles of contraction and recovery. Phillips *et al.* (1993) used only one FID per time interval in a similar experimental plan and obtained a time resolution of 4 s in their study of rat soleus.

As the muscle studies can either be *in vivo* or an isolated perfused or superfused preparation, which can be kept in a stable condition for a long time, many MRS observations can be made on each preparation, but the frequency with which they can be made is limited by the phenomenon of saturation which, simply stated, is that the sensitivity is diminished by the process of making a measurement and then recovers exponentially. Although the experimenter could wait a sufficient time for full recovery of sensitivity before making the next measurement, more efficient use of the technique is made if the measurements are repeated more frequently, when recovery of sensitivity is partial. It then becomes necessary to correct the results for the loss of sensitivity. This correction requires a knowledge of time constant for the recovery of sensitivity. The value is usually different for each metabolite measured, and may differ also between different types of preparations. Once these values have been determined, then the MRS results are readily converted to *relative* amounts, or levels, of metabolites. As definition of the exact boundaries of the sensitive volume from which the measurements were obtained is not easy, calibration of the signals in terms of the absolute amounts or levels of metabolites is usually carried out by chemically analysing the tissue used for some component (such as ATP) which is readily extracted and analysed and thought to be not present to a significant extent in bound form in the tissue. Sometimes however this step is omitted and a literature value is used as a standard for the ATP or other metabolite level. It is important when quoting absolute values from MRS publications to read the paper carefully to find out if this has been done, otherwise an inappropriate source of the primary data for concentration will have been quoted. Cady (1992) reviews the methods available for obtaining absolute concentrations of metabolites from MRS.

MRS could not be used to discover that the basic processes occurring in most muscles during contraction is of phosphocreatine splitting, and that this is reversed in the subsequent recovery period, since this was well known at the time the technique was first applied to muscle. MRS has however been used to demonstrate that these statements are true also for a wide variety of tissues previously untested, and it is noteworthy that no new pathways of energetic importance have been revealed during this work. The great potential of the MRS method is that it is insensitive to bound components within the muscle, unlike most analytic techniques, for which the preparatory preparation of an extract will usually liberate all bound components (however, see Seraydarian *et al.* 1962). Thus the apparent failure of expected stoichiometry to explain the results obtained by MRS might indicate such a process of binding or unbinding of metabolites. Just such effects have been reported by Yamada and Tanokura (1983) and further studied by

Kawano *et al.* (1988). These exciting results could not have been obtained by other methods currently available.

The position in the MRS spectrum of the inorganic phosphate (Pi) peak, relative to the PCr peak, is influenced by the pH in the cells; MRS can therefore be used to obtain a measure of the intracellular pH of muscle and observe how it changes during contraction and recovery. pH changes will result from PCr splitting, which will be measured in such experiments, and also from any accumulation of lactic acid in the cell. After allowance for the former, information about the latter can be obtained, on the assumption that no other significant processes occur which can change intracellular pH (Kemp *et al.* 1992, 1993*a,b*). If lactate does not leave the muscle cells during the experiment, which can be checked by appropriate analysis of the per- or superfusate, and if the buffer capacity (that is the rate with which pH changes with added acid) is known, an estimate of the lactate produced could be obtained by this means. This buffer capacity of the live muscle must not be confused with the much larger buffer capacity determined in muscle *extracts* in which considerable, perhaps total, splitting of phosphocreatine has occurred (Adams *et al.* 1990), which greatly raises the inorganic phosphate level and thus also the buffer capacity.

Just three independent quantities are usually measured from each ^{31}PMRS spectrum: the *relative* levels of three metabolites (ATP, PCr and Pi) and the pH. All of these change in contraction and recovery. Many further calculations can be made from these results if certain assumptions are made (e.g. Kemp and Radda 1994). For example, if it assumed that the creatine kinase reaction is in equilibrium and that its equilibrium constant is known, then given a value for the free creatine level, it is possible to calculate the free energy for the splitting of phosphocreatine, which on these assumptions is equal to the free energy of ATP splitting. The level of ADP, which is much too low to be directly measured by MRS, can also be calculated. Because of the likely importance of ADP levels in the control of metabolism, a great deal of stimulating information can be obtained in this way, but the assumptions made in these calculations do need clear statement. It could be very misleading, for example, to refer to the *measurement* by MRS of, for example, ADP level, when what has actually been measured is the ratio of PCr to Pi and a theoretical link between this ratio and the ADP level has been assumed.

Several laboratories have worked out protocols for assessing on human subjects the capacity for ATP splitting, anaerobic recovery and aerobic recovery from periods of muscular exercise (Arnold *et al.* 1984; Boska 1991; Conley *et al.* 1992; Blei *et al.* 1993*a,b*; Kemp *et al.* 1994) and have used the results of NMR studies to discuss how the recovery processes might be controlled (Kemp *et al.* 1993*a*; Kushmerick *et al.* 1992, 1993). Turner *et al.* (1993) have compared the metabolic costs of working compared to isometric contractions using MRS.

A valuable feature of MRS measurements is that they may give information about compartmentation of metabolites. For example, if Pi were to exist in muscle in two separate pools at different pH values, a splitting of the Pi peak might be apparent in the MRS spectrum or a broadening of the Pi peak relative to the width of the PCr peak. Where the preparation contains fibres of different types, the compartments might turn out to be these distinct fibre types, and subcellular compartmentation might also be detected. Reports of such splitting or broadening of the Pi peak have been made by Kushmerick *et al.* (1983) using cat biceps, Achten *et al.* (1990) and Parck *et al.* (1987) both using human muscle. Vanderborne *et al.* (1993) have reported similar observations using a technique in which the spectra are obtained from several defined areas within the triceps surae muscle group. There is good evidence from MRS experiments that different mammalian fibre types have characteristically different metabolite concentrations (Kushmerick *et al.* 1983, 1993; Phillips *et al.* 1993*a*).

MRS measurements of non-phosphorous containing metabolites are possible through the signals obtained from ^{1}H, which can be used to measure lactate concentration (Pan *et al.* 1991). However the measurement is difficult in the presence of fat and has therefore been of more use in the study of brain metabolism than of muscle metabolism (see Gadian 1995, Chapter 3, for references and review). ^{13}C nuclei are present in very low natural abundance but can nevertheless be used to provide MRS signals from which measurements of muscle glycogen have been made (Price *et al.* 1991).

Analysis of muscle extracts

The major metabolic pathways involved in the supply of energy for muscle contraction were all established by analysis of extracts of muscles rapidly frozen at various times during contractions with and without pre-treatment with various metabolic inhibitors. The history of these discoveries can be found in the following reviews: Carlson (1963), Wilkie (1966), Mommaerts (1969), Woledge (1971), Curtin and Woledge (1978), and Kushmerick (1983).

Two types of muscle extracts have been commonly used in studies of chemical energetics:

1. Perchloric acid extracts in which rapidly frozen muscles are ground, thawed in cold perchloric acid to denature the proteins, and then quickly neutralized with potassium carbonate before significant breakdown of phosphocreatine (which is acid labile) has occurred. The supernatant above the resulting precipitate of protein and potassium perchlorate is suitable for analysis.
2. Alcohol extracts in which the frozen muscles are incubated at $-30\,^{\circ}C$ for some days in a solution containing ethanol or methanol as an antifreeze and EDTA as calcium chelator. After removal of the muscle residue the extract is then freeze-dried, which destroys residual enzymatic activity. Unlike perchloric acid extracts, this technique does not extract all of the nucleotides bound to the contractile proteins and thus may give a more accurate indication of the amounts in the cytoplasm.

More recently, perchloric extracts have been made of muscle fibres which have been freeze-dried after rapid freezing (Nagesser *et al.* 1992, 1993). This has allowed the possibility of obtaining information about how metabolite levels and rates of various reactions differ among fibres.

Linked assay measurements of ATP-splitting in skinned fibres

Steinen, Potma and their colleagues in Amsterdam (Steinen *et al.* 1996) have developed and improved the methods, used earlier by Pybus and Tregear (1975) and by Glyn and Sleep (1985) by which the production of ADP by skinned muscle fibres, which comes of course from the splitting of ATP, can be followed by enzymatically linking the ADP production to the oxidation of NADH to NAD^+. NADH breakdown can then be followed by u.v. absorbance changes. The reactions are catalysed by pyruvate kinase and lactate dehydrogenase and a crucial part of the experimental arrangements is that the activities of the enzymes present should be sufficient that the ATP splitting by the muscle is the rate limiting step in the removal of NADH. The skinned fibre segment is mounted in a small chamber, which is continuously stirred, and a beam of u.v. light passes through the chamber close to, but not through, the fibre. Calibration is made simply by injecting known amounts of ADP into the system, while the preparation is still *in situ*. The chamber used by Steinen *et al.* (1990) had a volume of 30 μl and could be used, for example, to measure the ATP splitting, of about 0.1 nmol, by segments of human muscle fibres obtained by needle biopsy, during isometric contractions of 10 to 30 s duration at temperatures between 10 and 20 °C (Steinen *et al.* 1996). The energetics of length changes have also been studied with this equipment (Potma *et al.* 1994*a*) but repeated length changes were necessary to obtain sufficient sensitivity. Since skinned fibres were being used, the fibres could not be relaxed between successive shortenings. The result was that the Fenn effect was only demonstrable in fibres at high levels of activation, but not at low levels of activation. The Fenn effect (Fenn 1923, 1924) is an increase in the ATPase rate caused by shortening, above that during isometric contraction. Since this is the largest part of the energy use during working contraction the result was very striking. Similar equipment has been used to study how rates of ATP splitting vary between different fibre types from rat muscles (Bottinelli *et al.* 1994).

Steinen, Zaremba and Elzinga (1995) used this technique of studying ATP-ase activity in skinned fibres to investigate the proportion of the activity that was due to the SR Ca-ATPase. Treatment of the skinned fibres with Triton X-100 was used to abolish this component of the activity, which was found to be 25% of the total ATPase and fast twitch fibres, and 40% in slow twitch fibres. These values agree with determinations made in living muscle and muscle fibres, in which the myofibrillar ATPase was varied by changing filament overlap (Burchfield and Rall 1985; Lou *et al.* 1997).

The same group have gone on to construct a miniature version of their apparatus with a chamber volume of only 4 μl (Potma and Steinen 1997), which has about 10-fold better sensitivity and thus can measure changes during as little as 1-s of contraction. By this means the extra ATP split, as a result of a period of shortening, has been clearly demonstrated. The records show the extra signal detected after the shortening is over. The delay will be due, at least in part, to the time required for the diffusion from the core of the fibre(s) to the part of the chamber where u.v. absorption is measured. The possibility exists, however, that part of the delay is due to the processes in the cross-bridges between the performance of work and the release of ADP, perhaps analogous to the processes shown by Homsher and colleagues (1981) by energy balance experiments to follow rapid shortening and precede the consequent ATP splitting. This experimental story has thus reached an extremely interesting stage.

Ferenczi *et al.* (1995) have made a preliminary report concerning a novel method with great potential for revealing new information about muscle energetics. Their measurements are of phosphate release, which is monitored by its binding to a genetically engineered phosphate binding protein that has a fluorescent label which reports phosphate binding. The binding process is quick enough to give information on a millisecond time-scale and the sensitivity is sufficient to detect release of phosphate by less than 10% of the cross-bridges. The protein was introduced into permeabilized rabbit muscle fibres along with caged-ATP; contraction was induced by release of ATP from the cage. This resulted in a very rapid initial release of phosphate, which then slowed during the few hundred milliseconds for which ATP splitting could be monitored. This initial rate of phosphate release is at about ten times the rate found in the steady state in comparable experiments using ADP release (Potma *et al.* 1994*a*,*b*). The amount of ATP split in this early phase seems to be more than one ATP per myosin site. The significance of these results for muscle energetics remains to be discussed when fuller accounts of the work are published. (*Added in proof*: now published, see He *et al.* 1997.)

Caged-ATP methods

The relationship between muscle shortening and ATP used has been investigated not only by the experimenter fixing the shortening distance and then measuring the ATP split (Homsher *et al.* 1981) but also by experiments in which the experimenter provides a known amount of ATP and then measures the distance shortened (Yamada *et al.* 1993*a*,*c*; Higuchi and Goldman 1995). These experiments are not exactly energetic experiments because the interest has largely been to determine the distance shortened per ATP split, a quantity that is not thermodynamically constrained, rather than the work done per ATP split which subject to eqn 14.3 above. These experiments are possible because of the availability of caged-ATP (McCray *et al.* 1980), which can be diffused into a skinned muscle fibre in rigor.

A flash of u.v. light will then liberate ATP within the myoplasm. The amount of ATP liberated can be accurately determined and the speed of the subsequent shortening controlled.

Measurement of oxygen consumption and lactate efflux *in vitro*

Oxygen electrodes, when maintained at constant voltage, pass a current that is proportional to the dissolved oxygen concentration. By placing such an electrode in a chamber with a muscle the oxygen consumption by the muscle can be followed. The electrode itself uses oxygen, which creates oxygen gradients within the system, as does the usage of oxygen by the muscle tissue. To obtain useful measurements, these gradients need to be minimized by good stirring of the chamber, by using very small electrodes, which use less oxygen, and by choosing the materials used for the chamber and the stirrer with great care. Precise thermostatting of the entire apparatus is also essential. An example of the use of the technique to study whole muscle oxygen consumption is the work of Heglund and Cavanga (1987). Elzinga *et al.* (1984), Van der Laarse and Elzinga (1987) and Elzinga and Van der Laarse (1988) have made equipment suitable for smaller preparations. They have adopted the system of having a closed circulation around channels in a glass chamber, which in the later version of the apparatus had a volume of 136 μl and was used to measure the oxygen consumption by isolated intact *Xenopus* muscle fibres. Solution passes over the muscle fibre and then passes over the oxygen electrode. The response time of the system, determined by injecting small volumes of solution equilibrated at either very high or very low oxygen tension, was found to be about 4 s. The equipment was used to measure the maximum rates of oxygen consumption of fibres of different types in response to stimulation and also in response to the uncoupling agent CCCP, which gave similar maximum rates. It has long been recognized that in superfused isolated whole muscles the rate of oxygen usage is limited by diffusion (Hill 1965, p. 204). This is not the case in isolated single fibres since substantial changes can be made in the oxygen tension in the superfusate without change in the maximum rate of oxygen use.

Lactate formed during net glycolysis accumulates in muscle cells and is released both by passive diffusion and, much more rapidly, through a facilitated transfer through the membrane (for reviews see Mainwood *et al.* 1987 and Poole and Hallestrap 1993). An estimate of the extent of glycolysis can be made through measurement of the lactate released into the solution. Nagesser *et al.* (1994) applied this technique to the study of the energetics of isolated *Xenopus* muscle fibres during intermittent stimulation under aerobic conditions. The rate of lactate found corresponded to rates of glycolytic ATP synthesis, which were up to 30% of the rate of ATP synthesis that could be achieved by oxidative phosphorylation. It was shown that the rate of release was unchanged by stirring the solution close to the fibre and, as expected, greatly reduced by inhibition of the lactate transported by cinnamate. The authors point out, from a comparison of their results with earlier work using whole isolated muscles, that the latter have probably underestimated the rates at which lactate can be released from muscle fibres, because of the accumulation of lactate in the extracellular space in the absence of circulatory perfusion.

Measurement of oxygen consumption and lactate production *in vivo*

The methods for the study of the chemical processes of metabolism at the whole-animal level *in vivo* are beyond the scope of this article, but the following reference might be found useful for interested readers. Lammara and Whipp (1995) review modern methods for the measurement of oxygen and carbon dioxide exchange through external respiration. This method can be used to assess the metabolic

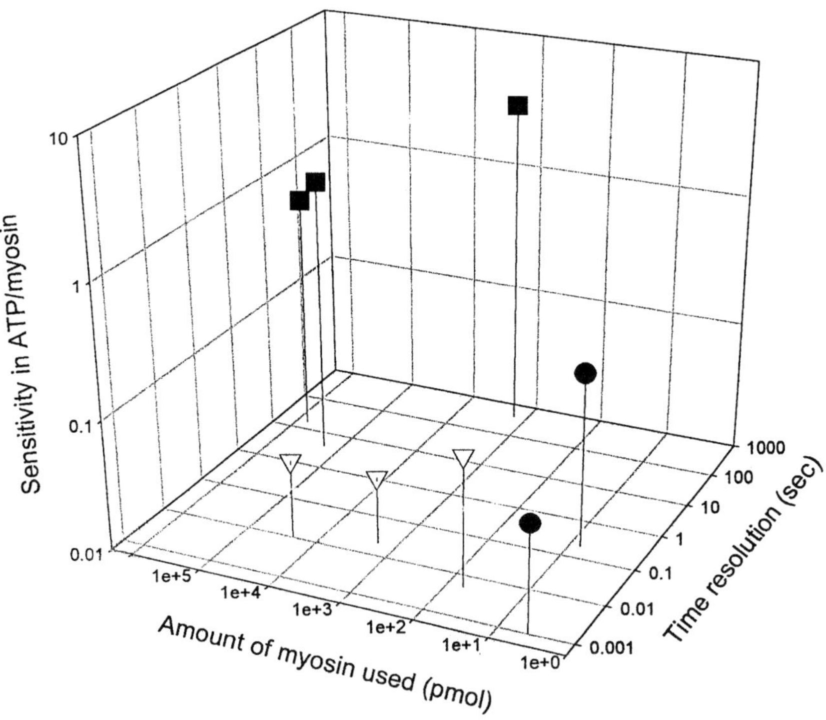

Fig. 14.1 A comparison of the sensitivity of different methods that have been used for the study of muscle energetics: (*closed squares*) MRS studies (from *left to right*) Kawano *et al.* (1988), Phillips *et al.* (1993*c*) and Phillips *et al.* (1993*b*); (*open triangle*) heat production measurements (from *left to right*) Hill (1948), Curtin and Woledge (1993) and Linari and Woledge (1995); (*closed circles*) linked assay studies (from *left to right*) Ferenczi *et al.* (1995) and Potma and Steinen (1996). Note that the MRS studies are averages, whereas the other studies are single observations. The difference between the sensitivities of the methods are therefore underestimated in this diagram.

demand of exercise in the steady state when it can be assumed that the overall reaction is oxidation of various substrates within the muscle. Something can be inferred about the nature of the substrate from the value of the respiratory exchange ratio (also known as RQ). It is well recognized that *in vivo* the muscles are less constrained as to substrate and product than they may be *in vitro*. Also the patterns of stimulation and shortening that occur during natural activities, such as locomotion, are usually much more complex than those studied in experiments with isolated muscles. Among these complexities are likely to be strategies that optimize the use of muscles either as producers of power or as efficient producers of work (Curtin and Woledge 1996). It cannot be assumed, therefore, that the factors governing the energetics of muscle contraction are the same *in vivo* as *in vitro*. It would be desirable to make a systematic attempt to understand the influence of these extra factors and to bring together the study of muscle contraction *in vivo* and *in vitro*.

14.5 Comparison of sensitivity and speed of methods currently available

To give some idea of the relative sensitivity of the different methods available for studying muscle energetics some selected results, chosen as exemplary of the state of the art, have been plotted in Fig. 14.1. On the ordinate is shown the energy per myosin molecule measured and on the abscissa the amount of muscle needed, expressed as myosin molecules. The graph is given a third dimension by showing the time-resolution used. All the axes are logarithmic. Of course compromise between these dimensions is usually possible through averaging or other means, and the MRS results shown are in fact averages whereas the other results refer to single measurements.

References

Achten, E., Van Cauteren, M., Willem, R., Luypaert, R., Malaisse, W.J., Van Bosch, G. *et al.* (1990) ^{31}P NMR spectroscopy and the metabolic properties of different muscle fibers. *Journal of Applied Physiology* **68**, 644–9.

Adams, G.R., Foley, J.M. and Meyer, R.A. (1990) Muscle buffer capacity estimated from pH changes during rest-to-work transitions. *Journal of Applied Physiology* **69**, 968–72.

Alpert, N.R., Hasenfuss, G., Mulieri, L.A., Blanchard, E.M., Leavitt, B.J. and Ittleman, F. (1992) The reorganization of the human and rabbit heart in response to haemodynamic overload. *European Heart Journal* **13** (Suppl. D), 9–16.

Andersen, P. and Saltin, B. (1985) Maximal perfusion of skeletal muscle in man. *Journal of Physiology* **366**, 233–49.

Arnold, D.L., Matthews, P.M. and Radda, G.K. (1984) Metabolic recovery after exercise and the assessment of mitochondrial function *in vivo* in human skeletal muscle by means of ^{31}P NMR. *Magnetic Resonance in Medicine* **1**, 307–15.

Barclay, C.J. (1994) Efficiency of fast- and slow-twitch muscles of the mouse performing cyclic contractions. *Journal of Experimental Biology* **193**, 65–78.

Barclay, C.J., Constable, J.K. and Gibbs, C.L. (1993) Energetics of fast- and slow-twitch muscles of the mouse. *Journal of Physiology* **472**, 61–80.

Barclay, C.J., Arnold, P.D. and Gibbs, C.L.S.O. (1995) Fatigue and heat production in repeated contractions of mouse skeletal muscle. *Journal of Physiology* **488**, 741–52.

Blei, M.L., Conley, E.E. and Kushmerick, M.J. (1993*a*) Separate measures of ATP utilization and recovery in human skeletal muscle. *Journal of Physiology* **465**, 203–22. (See also erratum (1994) *Journal of Physiology* **475**, 548P.)

Blei, M.L., Conley, K.E., Odderson, I.B., Esselman, P.C. and Kushmerick, M.J. (1993*b*) Individual variation in contractile cost and recovery in a human skeletal muscle. *Proceedings of the National Academy of Sciences* (*USA*) **90**, 7396–400.

Boska, M. (1991) Estimating the ATP cost of force production in the human gastrocnemius/soleus muscle group using ^{31}P MRS and ^{1}H MRI. *NMR in Biomedicine* **4**, 173–81.

Bottinelli, R., Canepari, M., Reggiani, C. and Stienen, G.J.M. (1994) Myofibrillar ATPase activity during isometric contraction and isomyosin composition in rat single skinned muscle fibres. *Journal of Physiology* **481**, 663–75.

Burchfield, D.M. and Rall, J.A. (1985) Energetics of activation in frog skeletal muscle at sarcomere lengths beyond myofilament overlap. *Biophysical Journal* **48**, 1049–51.

Cady, E.B. (1992) Determination of absolute concentrations of metabolites from NMR spectra. In *NMR basic principles and progress* (ed. P. Diehl *et al.*), Vol. 26, pp. 249–81. Springer-Verlag, Berlin.

Calvet, E. and Prat, H. (1963) *Recent progress in microcalorimetry.* Pergamon Press, Oxford.

Carlson, F.D. (1963) The mechanochemistry of muscular contraction, a critical revaluation of *in vivo* studies. *Progress in Biophysics* **13**, 262–314.

Chinet, A. (1993) Ca(2+)-dependent heat production by rat skeletal muscle in hypertonic media depends on Na(+)-Cl-co-transport stimulation. *Journal of Physiology* **461**, 689–703.

Chinet, A., Clausen, T. and Girardier, L. (1977) Microcalorimetric determination of energy expenditure due to active sodium-potassium transport in the soleus muscle and brown adipose tissue of the rat. *Journal of Physiology* **265**, 43–61.

Chinet, A., Decrouy, A. and Even, P.C. (1992) Ca(2+)-dependent heat production under basal and near-basal conditions in the mouse soleus muscle. *Journal of Physiology* **455**, 663–78.

Chinet, A., Ventura-Clapier, R. and Vassort, G. (1993) Energetics of ionic contracture in rat-heart papillary muscles. *Journal of Molecular and Cellular Cardiology* **25**, 145–57.

Conley, K.E., Blei, M.L., Odderson, I.R., Esselman, P.C., Cress, M.E. and Kushmerick, M.J. (1992) ATP synthesis rates and oxidative capacity: variation in normal human forearm muscle. *Proceedings of the Society for Magnetic Resonance in Medicine, 11th Annual Meeting* 774.

Curtin, N.A. and Woledge, R.C. (1978) Energy changes and muscular contraction. *Physiologocical Reviews* **58**, 690–761.

Curtin, N.A. and Woledge, R.C. (1979) Chemical change and energy production during contraction in frog muscle: how are their time courses related? *Journal of Physiology* **288**, 353–66.

Curtin, N.A. and Woledge, R.C. (1991) Efficiency of energy conversion during shortening of muscle fibres from the dogfish, *Scyliorhinus canicula Journal of Experimental Biology* **158**, 343–53.

Curtin, N.A. and Woledge, R.C. (1993) Efficiency of energy conversion during sinusoidal movement of white muscle fibres from the dogfish *Scyliorhinus canicula. Journal of Experimental Biology* **183**, 137–47.

Curtin, N.A. and Woledge, R.C. (1996) Power at the expense of efficiency in contraction of white muscle fibres from dogfish *Scyliorhinus canicula. Journal of Experimental Biology* **199**, 593–601.

Curtin, N.A., Howarth, J.V., Rall, J.A., Wilson, M.G.A. and Woledge, R.C. (1986) Absolute values of myothermic measurements on single muscle fibres from frog. *Journal of Muscle Research and Cell Motility* **7**, 327–32.

Curtin, N.A., Kometani, K. and Woledge, R.C. (1988) Effect of intracellular pH on force and heat production in isometric contraction of frog muscle fibres. *Journal of Physiology* **396**, 93–104.

Daut, J. and Elzinga, G. (1988) Heat production of quiescent ventricular trabeculi isolated from guinea pig hearts. *Journal of Physiology* **398**, 259–75.

Daut, J. and Elzinga, G. (1989) Substrate dependence of energy metabolism in isolayted guinea pig cardiac muscle: a microcalorimetric study. *Journal of Physiology* **413**, 379–97.

Dawson, M.J., Gadian, D.G. and Wilkie, D.R. (1977) Contraction and recovery of living muscles studied by ^{31}P nuclear magnetic resonance. *Journal of Physiology* **267**, 703–35.

Dawson, M.J., Gadian, D.G. and Wilkie, D.R. (1978) Muscular fatigue investigated by phosphorous nuclear magnetic resonance. *Nature* **274**, 861–6.

Decrouy, A., Even, P.C. and Chinet, A. (1993) Decreased rates of Ca(2+)-dependent heat production in slow- and fast-twitch muscles from the dystrophic (mdx) mouse. *Experientia* **49**, 843–9.

Dietrich, D.L.L. and Elzinga, G. (1993) Heat produced by rabbit papillary muscle during anoxia and reoxygenation. *Circulation Research* **73**, 1177–87.

Dietrich, D.L.L., Mast, F. and Elzinga, G. (1990) Energy demand, supply, and utilization in hypoxia, and force recovery after reoxygenation in rabbit heart muscle. *Circulation Research* **67**, 1089–96.

Dulloo, A.G., Decrouy, A. and Chinet, A. (1994) Suppression of Ca(2+)-dependent heat production in mouse skeletal muscle by high fish oil consumption. *Metabolism* **43**, 931–4.

Edwards, R.H.T., Hill, D.K. and McDonnell, M. (1974) A thermistor probe for myothermal measurements in man. *Journal of Applied Physiology* **36**, 511–3.

Edwards, R.H.T., Hill, D.K. and Jones, D.A. (1975) Heat production and chemical changes during isometric contractions of the human quadriceps muscle. *Journal of Physiology* **251**, 303–15.

Elia, M. and Livesey, G. (1988) Theory and validity of indirect calorimetry during net lipid synthesis *American Journal of Clinical Nutrition* **47**, 591–607.

Elzinga, G. and Van Der Laarse, W.J. (1988) Oxygen consumption of single muscle fibres of *Rana temporaria* and *Xenopus laevis* at 20 °C. *Journal of Physiology* **399**, 405–18.

Elzinga, G., Langewouters, G.J., Westerhof, N. and Wiechmann, A.H.C.A. (1984) Oxygen uptake of frog skeletal muscle fibres following tetanic contractions at 18 °C. *Journal of Physiology* **346**, 365–77.

Elzinga, G., Lannergren, J. and Stienen, G.J.M. (1987) Stable maintenance heat rate and contractile properties of different single muscle fibres from *Xenopus laevis* at 20 °C. *Journal of Physiology* **393**, 399–412.

Even, P.C., Decrouy, A. and Chinet, A. (1994) Defective regulation of energy metabolism in mdx-mouse skeletal muscles. *Biochemical Journal* **304**, 649–54.

Fagher, B., Thysell, H. and Monti, M. (1994) Effect of erythropoietin on muscle metabolic rate, as measured by direct microcalorimetry, and ATP in hemodialysis patients. *Nephron* **67**, 167–71.

Fenn, W.O. (1923) A quantitative comparison between the energy liberated and the work performed by the isolated sartorius muscle of the frog. *Journal of Physiology* **58**, 175–203.

Fenn, W.O. (1924) The relation between the work performed and the energy liberated in muscular contraction. *Journal of Physiology* **58**, 373–95.

Ferenczi, M.E., Zhen-He, H., Chillingworth, R.K., Brune, M., Corrie, E.T., Trentham, D.R. *et al.* (1995) A new method for the time-resolved measurement of phosphate release in permeabilized muscle fibres. *Biophysical Journal* **68**, 191s–3s.

Ferrannini, E. (1988) The theoretical bases of indirect calorimetry: a review. *Metabolism* **37**, 287–301.

Gadian, D.G. (1995) *NMR and its application to living systems*, 2nd edn. Oxford University Press.

Gilbert, S.H. and Ford, L.E. (1988) Heat changes during transient tension responses to small releases in active frog muscle. *Biophysical Journal* **54**, 611–7.

Gilbert, S.H. and Mathias, R.T. (1988) Analysis of diffusion delay in a layered medium. Application to heat measurements from muscle. *Biophysical Journal* **54**, 603–10.

Glyn, H. and Sleep, J. (1985) Dependence of adenosine triphospatase activity of rabbit psoas muscle fibres and myofibrils on substrate concentration. *Journal of Physiology* **365**, 259–76.

Harada, Y., Sakurada, K., Acki, T., Thomas, D.D. and Yanagida, T.J. (1990) Mechanochemical coupling in actomyosin energy transduction studied by *in-vitro* movement assay. *Journal of Molecular Biology* **216**, 49–68.

He, Z.-H., Chillingworth, R.K., Brune, M., Corrie, J.E.T., Trentham, D.R., Webb M.R. and Ferenczi, M.A. (1997) ATPase kinetics on activation of rabbit and frog permeabilized isometric muscle fibres: a real time phosphate assay. *Journal of Physiology* **501**, 125–48.

Heglund, N.C. and Cavagna, G.A. (1987) Mechanical work, oxygen consumption, and efficiency in isolated frog and rat muscle. *American Journal Physiology* **253**, C22–9.

Higuchi, H. and Goldman, Y.E. (1991) Sliding distance between actin and myosin filaments per ATP molecule hydrolysed in skinned muscle fibres. *Nature* **352**, 352–4.

Higuchi, H. and Goldman, Y.E. (1995) Sliding distance per ATP molecule hydrolyzed by myosin heads during isotonic shortening of skinned muscle fibers. *Biophysical Journal* **69**, 1491–507.

Hill, A.V. (1931) *Adventures in biophysics*. University of Pennsylvania Press.

Hill, A.V. (1932) The revolution in muscle physiology. *Physiological Reviews* **12**, 56–7.

Hill, A.V. (1938) The heat of shortening and the dynamic constants of muscle. *Proceedings of the Royal Society* **126**, 136–95.

Hill, A.V. (1948) The onset of contraction. *Proceedings of the Royal Society* **136**, 242–154.

Hill, A.V. (1964) The effect of load on the heat of shortening of muscle. *Proceedings of the Royal Society* **159**, 297–318.

Hill, A.V. (1965) *Trails and trials in physiology.* Edward Arnold, London.

Homsher, E., Kean, C.J., Wallner, A. and Garibian-Sarian, V. (1979) The time course of energy balance in an isometric tetanus. *Journal of General Physiology* **73**, 553–67.

Homsher, E., Irving, M. and Wallner, A. (1981) High-energy phosphate metabolism and energy liberation associated with rapid shortening in frog skeletal muscle. *Journal of Physiology* **321**, 423–36.

Homsher, E., Irving, M. and Lebacq, J. (1983) The variation in shortening heat with sacromere length in frog muscle. *Journal of Physiology* **345**, 107–21.

Hoult, D.I., Busby, S.J.W., Gadian, D.G., Radda, G.K., Richards, R.E. and Seely, P.J. (1974) Observation of tissue metabolites using ^{31}P nuclear magnetic resonance. *Nature* **252**, 285–7.

Kawano, Y., Tanokura, M. and Yamada, K. (1988) Phosphorus nuclear magnetic resonance studies on the effects of duration of contraction in bull frog skeletal muscles. *Journal of Physiology* **407**, 243–61.

Kemp, G.J. and Radda, G.K. (1994) Quantitative interpretation of bioenergetic data from ^{31}P and ^{1}H magnetic resonance spectroscopic studies of skeletal muscle: an analytical review. *Magnetic Resonance Quarterly* **10**, 43–63.

Kemp, G.J., Thompson, C.H. and Radda, G.K. (1992) Proton efflux from skeletal muscle during recovery from exercise: evidence for increase in hypertension. *Clinical Science* **82**, 489–91.

Kemp, G.J., Taylor, D.J. and Radda, G.K. (1993*a*) Control of phosphocreatine resynthesis during recovery from exercise in human skeletal muscle. *NMR in Biomedicine* **6**, 66–72.

Kemp, G.J., Taylor, D.J., Styles, P. and Radda, G.K. (1993*b*) The production, buffering and efflux of protons in human skeletal muscle during exercise and recovery. *NMR in Biomedicine* **6**, 73–83.

Kemp, G.J., Thompson, C.H., Barnes, P.R.J. and Radda, G.K. (1994) Comparisons of ATP turnover in human muscle during ischemic and aerobic exercise using ^{31}P magnetic resonance spectroscopy. *Magnetic Resonance in Medicine* **31**, 248–58.

Kretzschmar, K.M. and Wilkie, D.R. (1972) A new method for absolute heat measurement utilizing the Peltier effect. *Journal of Physiology* **224**, 18–21.

Kretzschmar, K.M. and Wilkie, D.R. (1975) The use of the Peltier effect for simple and accurate calibration of thermoelectric devices. *Proceedings of the Royal Society* **190**, 315–21.

Kushmerick, M.J. (1983) Energetics of muscle contraction. *Handbook of Physiology, Skeletal Muscle* **10**, 189–236.

Kushmerick, M.J. and Davies, R.E. (1969) The chemical energetics of muscle contraction. II. The chemistry, efficiency and power of maximally working sartorius muscles. *Proceedings of the Royal Society* **174**, 315–53.

Kushmerick, M.J., Meyer, R.A. and Brown, T.R. (1983) Phosphorus NMR spectroscopy of cat biceps and soleus muscles. *Advances in Experimental Medicine and Biology* **159** 303–25.

Kushmerick, M.J., Meyer, R.A. and Brown, T.R. (1992) Regulation of oxygen consumption in fast- and slow-twitch muscle. *American Journal of Physiology* **263**, C598–606.

Kushmerick, M.J., Moerland, T.S. and Wiseman, R.W. (1993) Two classes of mammalian skeletal muscle fibers distinguished by metabolite content. *Advances in Experimental and Medical Biology* **332**, 749–61.

Lamarra, N. and Whipp, B. (1995) Measurement of pulmonary gas exchange. In *Physical assessment of human fitness* (ed. Maud, P.J. and Foster, C.). Human Kinetics Champaign IL. USA.

Lannergren, J., Elzinga, G. and Stienen, G.J. (1993) Force relaxation, labile heat and parvalbumin content of skeletal muscle fibres of *Xenopus laevis*. *Journal of Physiology* **463**, 123–40.

Linari, M. and Woledge, R.C. (1995) Comparison of energy output during ramp and staircase shortening in frog muscle fibres. *Journal of Physiology* **487**, 699–710.

Livesey, G. and Elia, M. (1988) Estimation of energy expenditure, net carbohydrate utilization, and net fat oxidation and synthesis by indirect calorimetry: evaluation of errors with special reference to the detailed composition of fuels. *American Journal of Clinical Nutrition* **47**, 608–28.

Lonnbro, P. and Hellstrand, P. (1991) Heat production in chemically skinned smooth muscle of guinea-pig *Taenia coli. Journal of Physiology* **440**, 385–402.

Lou, F., Curtin, N.A. and Woledge, R.C. (1997) The energetic cost of activation of white muscle fibres from the dogfish *Scyliorhinus canicula. Journal of Experimental Biology* **200**, 495–501.

Mainwood, G.W., Renaud, J.M. and Mason, M.J. (1987) The pH dependence of the contractile response of fatigued skeletal muscle. *Canadian Journal of Physiology and Pharmacology* **65**, 648–58.

Mast, F. and Elzinga, G. (1989) Metabolic recovery of acidotic rabbit cardiac muscle: effects of low pH and oxygen shortage. *Circulation Research* **65**, 909–16.

McCray, J.A., Herbette, L., Kihara, T. and Trentham, D.R. (1980) A new approach to time-resolved studies on ATP-requiring biological systems: laser flash photolysis of caged ATP. *Proceedings of the National Academy of Sciences* (*USA*) **77**, 7237–41.

Meyerhof, O. (1930) *Die chemischen Vorgenge in Muskel.* Springer-Verlag, Berlin.

Mommaerts, W.F.H.M. (1969) Energetics of muscular contraction. *Physiological Reviews* **49**, 427–508.

Mulieri, L.A., Luhr, G., Trefry, J. and Alpert, N.R. (1977) Metal film thermopiles for use with rabbit right ventricular papillary muscles. *American Journal of Physiology* **233**, C146–56.

Nagesser, A.S., Van Der Laarse, W.J. and Elzinga, G. (1992) Metabolic changes with fatigue in different types of single muscle fibres of Xenopus laevis. *Journal of Physiology* **448**, 511–23.

Nagesser, A.S., Van Der Laarse, W.J. and Elzinga, G. (1993) ATP formation and ATP hydrolysis during fatiguing, intermittent stimulation of different types of single muscle fibres from *Xenopus laevis. Journal of Muscle Research and Cell Motility* **14**, 608–18.

Nagesser, A.S., Van Der Laarse, W.J. and Elzinga, G. (1994) Lactate efflux from fatigued fast-twitch muscle fibres of *Xenopus laevis* under various extracellular conditions. *Journal of Physiology* **481**, 139–47.

Pan, J.W., Hamm, J.R., Hetherington, H.P., Rothman, D.L. and Shulman, R.G. (1991) Correlation of lactate and pH in human skeletal muscle after exercise by ^{1}H NMR. *Magnetic Resonance in Medicine* **20**, 57–65.

Park, J.H., Brown, R.L., Park, C.R., McCully, K., Cohn, M., Haselgrove, J. *et al.* (1987) Functional pools of oxidative and glycolytic fibers in human muscle observed by ^{31}P magnetic resonance spectroscopy during exercise. *Proceedings of the National Academy of Sciences* (*USA*) **84**, 8976–80.

Phillips, S.K., Wiseman, R.W., Woledge, R.C. and Kushmerick, M.J. (1993*a*) The effects of metabolic fuel on force production and resting inorganic phosphate levels in isolated mouse skeletal muscle. *Journal of Physiology* **462**, 135–46.

Phillips, S.K., Wiseman, R.W., Woledge, R.C. and Kushmerick, M.J. (1993*b*) Neither changes in phosphorous metabolite levels nor myosin isoforms can explain the weakness in aged mouse muscle. *Journal of Physiology* **463**, 129–34.

Phillips, S.K., Takei, M. and Yamada, K. (1993*c*) The time course of phosphate metabolites and intracellular pH using ^{31}P NMR compared to recovery heat in rat soleus muscle. *Journal of Physiology* **460**, 693–704.

Poole, R.C. and Halestrap, A.P. (1993) Transport of lactate and other monocarboxylates across mammalian plasma membranes. *American Journal of Physiology* **264**, C761–28.

Potma, E.J. and Stienen, G.J.M. (1997) Increase in ATP consumption during shortening in skinned fibres from rabbit psoas muscle: effects of inorganic phosphate. *Journal of Physiology* **496**, 1–12.

Potma, E.J., Stienen, G.J.M., Barends, J.P.F. and Elzinga, G. (1994*a*) Myofibrillar ATPase activity and mechanical performance of skinned fibres from rabbit psoas muscle. *Journal of Physiology* **474**, 303–17.

Potma, E.J., van Graas, I.A. and Stienen, G.J.M. (1994*b*) Effects of pH on myofibrillar ATPase activity in fast and slow skeletal muscle fibers of the rabbit. *Biophysical Journal* **67**, 2404–10.

Price, T.B., Rothman, D.L., Avison, M.J., Buonamico, P. and Shulman, R.G. (1991) ^{13}C-NMR measurements of muscle glycogen during low-intensity exercise. *Journal of Applied Physiology* **70**, 1836–40.

Pybus, J. and Tregear, R.T. (1975) The relationship of adenosine triphosphatase activity to tension and power output of insect flight muscle. *Journal of Physiology* **247**, 71–89.

Rall, J.A., Homsher, E., Wallner, A. and Mommaerts, W.F.H.M. (1976) A temporal dissociation of energy liberation and high energy phosphate splitting during shortening in frog skeletal muscle. *Journal of General Physiology* **68**, 13–27.

Saugen, E. and Vollestad, N.K. (1995) Metabolic heat production during fatigue from voluntary repetitive isometric contractions in humans. *Journal of Applied Physiology* **76**, 2043–9.

Schramm, M., Klieber, H.G. and Daut, J. (1994) The energy expenditure of actomyosin-ATPase, Ca(2+)-ATPase and Na+, K(+)-ATPase in guinea-pig cardiac ventricular muscle. *Journal of Physiology* **481**, 647–62.

Seraydarian, K., Mommaerts, W.F.H.M., Wallner, A. and Guillory, T.J. (1962) The amount and compartmentalization of adenosinediphosphate in muscle. *Biochimica Biophysica Acta* **65**, 443–60.

Stienen, G.J., Roosemalen, M.C.M., Wilson, M.G.A. and Elzinga, G. (1990) Depression of force by phosphate in skinned skeletal muscles of the frog. *American Journal of Physiology* **259**, C349–57.

Stienen, G.J., Zaremba, R. and Elzinga, G. (1995) ATP utilization for calcium uptake and force production in skinned muscle fibres of *Xenopus laevis. Journal of Physiology* **482**, 109–22.

Stienen, G.J.M., Kiers, J.L., Bottinelli, R. and Reggiani, C. (1996) Myofibrillar ATPase activity in skinned human skeletal muscle fibres: fibre type and temperature dependence. *Journal of Physiology* **493**, 299–307.

Toyoshima, Y.Y., Kron, S.J. and Spudich, J.A. (1990) The myosin step size: measurement of the unit displacement per ATP hydrolyzed in an *in vitro* assay. *Proceedings of the National Academy of Sciences* (*USA*) **87**, 7130–4.

Turner, D.L., Jones, D.A., Mcintyre, D.B. and Newham, D.J. (1993) ATP turnover measured by ^{31}P-magnetic resonance spectroscopy in human adductor pollicis muscle during isometric and shortening contractions. *Journal of Physiology* **459**, 154P.

Uyeda, T.Q.P., Kron, S.J. and Spudich, J.A. (1990) Myosin step size. Estimation from slow sliding movement of actin over low densities of heavy meromyosin. *Journal of Molecular Biology* **214**, 699–710.

Vandenborne, K., Walter, G., Leigh, J.S. and Goelman, G. (1993) pH heterogeneity during exercise in localized spectra from single human muscles. *American Journal of Physiology* **265**, C1332–9.

Van Der Laarse, W.J. and Elzinga, G. (1987) Oxygen consumption of single muscle fibres from *Xenopus laevis. Journal of Muscle Research and Cell Motility* **8**, 92P.

Vollestad, N.K., Wesche, J. and Sejersted, O.M. (1990) Gradual increase in leg oxygen uptake during repeated submaximal contractions in humans. *Journal of Applied Physiology* **68**, 1150–6.

Vollestad, N.K. and Verburg, E. (1996) Muscular function, metabolism and electrolyte shifts during prolonged repetitive exercise in humans. *Acta Physiologica Scandinavica* **156**, 271–8.

Wadso, I. (1974) A microcalorimeter for biological analysis. *Science Tools* **21**, 18–21.

Wilkie, D.R. (1996) Muscle. *Annual Reviews of Physiology* **28**, 17–34.

Wiseman, R.W., Moerland, T.S. and Kushmerick, M.J. (1993) Biological applications for small solenoids: NMR spectroscopy of microliter volumes at high fields. *NMR in Biomedicine* **6**, 153–6.

Woledge, R.C. (1971) Heat production and chemical change in muscle. *Progress in Biophysics and Molecular Biology* **22**, 37–74.

Woledge, R.C., Curtin, N.A. and Homsher, E. (1985) Energetic aspects of muscle contractions. *Monographs of the Physiological Society* **41**, Academic Press.

Yamada, K. and Tanokura, M. (1983) Post-contractile phosphocreatine splitting in muscle as revealed by time-resolved 31P nuclear magnetic resonance. *Japanese Journal of Physiology* **33**, 909–19.

Yamada, T., Abe, O., Kobayashi, T. and Sugi, H. (1993*a*) Distance of myofilament sliding per ATP molecule in skeletal muscle fibers studied using laser flash photolysis of caged ATP. *Advances in Experimental Medicine and Biology* **332**, 505–11.

Yamada, T., Kikuchi, K. and Sugi, H. (1993*b*) 31P nuclear magnetic resonance studies on the glycogenolysis regulation in resting and contracting frog skeletal muscle. *Journal of Physiology* **460**, 273–86.

Yamada, T., Abe, O., Kobayashi, T. and Sugi, H. (1993*c*) Myofilament sliding per ATP molecule in rabbit muscle fibres studied using laser flash photolysis of caged ATP. *Journal of Physiology* **466**, 229–43.

Yanagida, T., Arata, T. and Oosawa, F. (1985) Sliding distance of actin filament induced by a myosin cross-bridge during one ATP hydrolysis cycle. *Nature* **316**, 366–9.

Index